Cognitive Technologies

Managing Editors: D. M. Gabbay   J. Siekmann

Editorial Board: A. Bundy   J. G. Carbonell
M. Pinkal   H. Uszkoreit   M. Veloso   W. Wahlster
M. J. Wooldridge

Ben Goertzel
Cassio Pennachin (Eds.)

Artificial General Intelligence

With 42 Figures and 16 Tables

 Springer

Editors:

Ben Goertzel
Cassio Pennachin
AGIRI – Artificial General Intelligence Research Institute
1405 Bernerd Place
Rockville, MD 20851
USA
ben@agiri.org
cassio@agiri.org

Managing Editors:

Prof. Dov M. Gabbay
Augustus De Morgan Professor of Logic
Department of Computer Science, King's College London
Strand, London WC2R 2LS, UK

Prof. Dr. Jörg Siekmann
Forschungsbereich Deduktions- und Multiagentensysteme, DFKI
Stuhlsatzenweg 3, Geb. 43, 66123 Saarbrücken, Germany

ACM Computing Classification (1998): F.1, F.4, H.5, I.2, I.6

ISSN 1611-2482
ISBN-13 978-3-642-06267-4   e-ISBN-13 978-3-540-68677-4

Springer is a part of Springer Science+Business Media
springer.com

© Springer-Verlag Berlin Heidelberg 2007
Softcover reprint of the hardcover 1st edition 2007

Cover Design: KünkelLopka, Heidelberg

Printed on acid-free paper   45/3100/YL   5 4 3 2 1 0

Preface

"Only a small community has concentrated on general intelligence. No one has tried to make a thinking machine ...
The bottom line is that we really haven't progressed too far toward a truly intelligent machine. We have collections of dumb specialists in small domains; the true majesty of general intelligence still awaits our attack. ...
We have got to get back to the deepest questions of AI and general intelligence..."

– Marvin Minsky
as interviewed in *Hal's Legacy*, edited by David Stork, 2000.

Our goal in creating this edited volume has been to fill an apparent gap in the scientific literature, by providing a coherent presentation of a body of contemporary research that, in spite of its integral importance, has hitherto kept a very low profile within the scientific and intellectual community. This body of work has not been given a name before; in this book we christen it "Artificial General Intelligence" (AGI). What distinguishes AGI work from run-of-the-mill "artificial intelligence" research is that it is explicitly focused on engineering general intelligence in the short term. We have been active researchers in the AGI field for many years, and it has been a pleasure to gather together papers from our colleagues working on related ideas from their own perspectives. In the Introduction we give a conceptual overview of the AGI field, and also summarize and interrelate the key ideas of the papers in the subsequent chapters.

Of course, "general intelligence" does not mean exactly the same thing to all researchers. In fact it is not a fully well-defined term, and one of the issues raised in the papers contained here is how to define general intelligence in a way that provides maximally useful guidance to practical AI work. But, nevertheless, there is a clear qualitative meaning to the term. What is meant by AGI is, loosely speaking, AI systems that possess a reasonable degree of self-understanding and autonomous self-control, and have the ability to solve a variety of complex problems in a variety of contexts, and to learn to solve new problems that they didnt know about at the time of their creation. A marked distinction exists between practical AGI work and, on the other hand:

- Pragmatic but specialized "narrow AI" research which is aimed at creating programs carrying out specific tasks like playing chess, diagnosing diseases, driving cars and so forth (most contemporary AI work falls into this category.)
- Purely theoretical AI research, which is aimed at clarifying issues regarding the nature of intelligence and cognition, but doesnt involve technical details regarding actually realizing artificially intelligent software.

Some of the papers presented here come close to the latter (purely theoretical) category, but we have selected them because the theoretical notions they contain seem likely to lead to such technical details in the medium-term future, and/or resonate very closely with the technical details of AGI designs proposed by other authors.

The audience we intend to reach includes the AI community, and also the broader community of scientists and students in related fields such as philosophy, neuroscience, linguistics, psychology, biology, sociology, anthropology and engineering. Significantly more so than narrow AI, AGI is interdisciplinary in nature, and a full appreciation of the general intelligence problem and its various potential solutions requires one to take a wide variety of different perspectives.

Not all significant AGI researchers are represented in these pages, but we have sought to bring together a multiplicity of perspectives, including many that disagree with our own. Bringing a diverse body of AGI research together in a single volume reveals the common themes among various researchers work, and makes clear what the big open questions are in this vital and critical area of research. It is our hope that this book will interest more researchers and students in pursuing AGI research themselves, thus aiding in the progress of science.

In the three years that this book has been in the making, we have noticed a significant increase in interest in AGI-related research within the academic AI community, including a number of small conference workshops with titles related to "Human-Level Intelligence." We consider this challenge to the overwhelming dominance of narrow-AI an extremely positive move; however, we submit that "Artificial General Intelligence" is a more sensible way to conceptualize the problem than "Human-Level Intelligence." The AGI systems and approaches described in these pages are not necessarily oriented towards emulating the human brain; and given the heterogeneity of the human mind/brain and its highly various levels of competence at various sorts of tasks, it seems very difficult to define "Human-Level Intelligence" in any way that is generally

applicable to AI systems that are fundamentally non-human-like in conception. On the other hand, the work of Hutter and Schmidhuber reported here
provides a reasonable, abstract mathematical characterization of general intelligence which, while not in itself providing a practical approach to AGI design
and engineering, at least provides a conceptually meaningful formalization of
the ultimate goal of AGI work.

The grand goal of AGI remains mostly unrealized, and how long it will
be until this situation is remedied remains uncertain. Among scientists who
believe in the fundamental possibility of strong AI, the most optimistic serious estimates we have heard are in the range of 5-10 years, and the most
pessimistic are in the range of centuries. While none of the articles contained
here purports to present a complete solution to the AGI problem, we believe
that they collectively embody meaningful conceptual progress, and indicate
clearly that the direct pursuit of AGI is an endeavor worthy of significant
research attention.

Contents

The Novamente Artificial Intelligence Engine
Ben Goertzel, Cassio Pennachin

Essentials of General Intelligence:
The Direct Path to Artificial General Intelligence
Peter Voss

Artificial Brains
Hugo de Garis

The New AI: General & Sound & Relevant for Physics
Jürgen Schmidhuber

Gödel Machines: Fully Self-Referential Optimal Universal Self-improvers
Jürgen Schmidhuber

Universal Algorithmic Intelligence: A Mathematical Top→Down Approach
Marcus Hutter

Levels of Organization in General Intelligence
Eliezer Yudkowsky

Contemporary Approaches to Artificial General Intelligence

Cassio Pennachin and Ben Goertzel

AGIRI – Artificial General Intelligence Research Institute
1405 Bernerd Place, Rockville, MD 20851, USA
cassio@agiri.org, ben@agiri.org – http://www.agiri.org

1 A Brief History of AGI

The vast bulk of the AI field today is concerned with what might be called "narrow AI" – creating programs that demonstrate intelligence in one or another specialized area, such as chess-playing, medical diagnosis, automobile-driving, algebraic calculation or mathematical theorem-proving. Some of these narrow AI programs are extremely successful at what they do. The AI projects discussed in this book, however, are quite different: they are explicitly aimed at artificial *general* intelligence, at the construction of a software program that can solve a variety of complex problems in a variety of different domains, and that controls itself autonomously, with its own thoughts, worries, feelings, strengths, weaknesses and predispositions.

Artificial General Intelligence (AGI) was the original focus of the AI field, but due to the demonstrated difficulty of the problem, not many AI researchers are directly concerned with it anymore. Work on AGI has gotten a bit of a bad reputation, as if creating digital general intelligence were analogous to building a perpetual motion machine. Yet, while the latter is strongly implied to be impossible by well-established physical laws, AGI appears by all known science to be quite possible. Like nanotechnology, it is "merely an engineering problem", though certainly a very difficult one.

The presupposition of much of the contemporary work on "narrow AI" is that solving narrowly defined subproblems, in isolation, contributes significantly toward solving the overall problem of creating real AI. While this is of course true to a certain extent, both cognitive theory and practical experience suggest that it is not so true as is commonly believed. In many cases, the best approach to implementing an aspect of mind in isolation is very different from the best way to implement this same aspect of mind in the framework of an integrated AGI-oriented software system.

The chapters of this book present a series of approaches to AGI. None of these approaches has been terribly successful yet, in AGI terms, although several of them have demonstrated practical value in various specialized domains (narrow-AI style). Most of the projects described are at an early stage of engineering development, and some are still in the design phase. Our aim is not to present AGI as a mature field of computer science – that would be

impossible, for it is not. Our goal is rather to depict some of the more exciting ideas driving the AGI field today, as it emerges from infancy into early
childhood.

In this introduction, we will briefly overview the AGI approaches taken
in the following chapters, and we will also discuss some other historical and
contemporary AI approaches not extensively discussed in the remainder of
the book.

1.1 Some Historical AGI-Related Projects

Generally speaking, most approaches to AI may be divided into broad categories such as:

- symbolic;
- symbolic and probability- or uncertainty-focused;
- neural net-based;
- evolutionary;
- artificial life;
- program search based;
- embedded;
- integrative.

This breakdown works for AGI-related efforts as well as for purely narrow-
AI-oriented efforts. Here we will use it to structure a brief overview of the AGI
field. Clearly, there have been many more AGI-related projects than we will
mention here. Our aim is not to give a comprehensive survey, but rather to
present what we believe to be some of the most important ideas and themes in
the AGI field overall, so as to place the papers in this volume in their proper
context.

The majority of ambitious AGI-oriented projects undertaken to date have
been in the symbolic-AI paradigm. One famous such project was the General
Problem Solver [42], which used heuristic search to solve problems. GPS did
succeed in solving some simple problems like the Towers of Hanoi and *crypto-
arithmetic*,[1] but these are not really general problems – there is no learning
involved. GPS worked by taking a general goal – like solving a puzzle – and
breaking it down into subgoals. It then attempted to solve the subgoals, breaking them down further into even smaller pieces if necessary, until the subgoals
were small enough to be addressed directly by simple heuristics. While this
basic algorithm is probably necessary in planning and goal satisfaction for
a mind, the rigidity adopted by GPS limits the kinds of problems one can
successfully cope with.

---

[1]Crypto-arithmentic problems are puzzles like DONALD + GERALD = ROBERT. To
solve such a problem, assign a number to each letter so that the equation comes out
correctly.

Probably the most famous and largest symbolic AI effort in existence today is Doug Lenat's CYC project.[2] This began in the mid-80's as an attempt to create true AI by encoding all common sense knowledge in first-order predicate logic. The encoding effort turned out to require a large effort, and soon Cyc deviated from a pure AGI direction. So far they have produced a useful knowledge database and an interesting, highly complex and specialized inference engine, but they do not have a systematic R&D program aimed at creating autonomous, creative interactive intelligence. They believe that the largest subtask required for creating AGI is the creation of a knowledge base containing all human common-sense knowledge, in explicit logical form (they use a variant of predicate logic called CycL). They have a large group of highly-trained knowledge encoders typing in knowledge, using CycL syntax.

We believe that the Cyc knowledge base may potentially be useful eventually to a mature AGI system. But we feel that the kind of reasoning, and the kind of knowledge embodied in Cyc, just scratches the surface of the dynamic knowledge required to form an intelligent mind. There is some awareness of this within Cycorp as well, and a project called CognitiveCyc has recently been initiated, with the specific aim of pushing Cyc in an AGI direction (Stephen Reed, personal communication).

Also in the vein of "traditional AI", Alan Newell's well-known SOAR project[3] is another effort that once appeared to be grasping at the goal of human-level AGI, but now seems to have retreated into a role of an interesting system for experimenting with limited-domain cognitive science theories. Newell tried to build "Unified Theories of Cognition", based on ideas that have now become fairly standard: logic-style knowledge representation, mental activity as problem-solving carried out by an assemblage of heuristics, etc. The system was by no means a total failure, but it was not constructed to have a real autonomy or self-understanding. Rather, it's a disembodied problem-solving tool, continually being improved by a small but still-growing community of SOAR enthusiasts in various American universities.

The ACT-R framework [3], though different from SOAR, is similar in that it's an ambitious attempt to model human psychology in its various aspects, focused largely on cognition. ACT-R uses probabilistic ideas and is generally closer in spirit to modern AGI approaches than SOAR is. But still, similarly to SOAR, many have argued that it does not contain adequate mechanisms for large-scale creative cognition, though it is an excellent tool for the modeling of human performance on relatively narrow and simple tasks.

Judea Pearl's work on Bayesian networks [43] introduces principles from probability theory to handle uncertainty in an AI scenario. Bayesian networks are graphical models that embody knowledge about probabilities and dependencies between events in the world. Inference on Bayesian networks is possible using probabilistic methods. Bayesian nets have been used with

---

[2]See www.cyc.com and [38].
[3]See http://ai.eecs.umich.edu/soar/ and [37].

success in many narrow domains, but, in order to work well, they need a reasonably accurate model of the probabilities and dependencies of the events being modeled. However, when one has to *learn* either the structure or the probabilities in order to build a good Bayesian net, the problem becomes very difficult [29].

Pei Wang's NARS system, described in this volume, is a very different sort of attempt to create an uncertainty-based, symbolic AI system. Rather than using probability theory, Wang uses his own form of uncertain logic – an approach that has been tried before, with fuzzy logic, certainty theory (see, for example, [50]) and so forth, but has never before been tried with such explicit AGI ambitions.

Another significant historical attempt to "put all the pieces together" and create true artificial general intelligence was the Japanese 5th Generation Computer System project. But this project was doomed by its pure engineering approach, by its lack of an underlying theory of mind. Few people mention this project these days. In our view, much of the AI research community appears to have learned the wrong lessons from the 5th generation AI experience – they have taken the lesson to be that integrative AGI is bad, rather than that integrative AGI should be approached from a sound conceptual basis.

The neural net approach has not spawned quite so many frontal assaults on the AGI problem, but there have been some efforts along these lines. Werbos has worked on the application of recurrent networks to a number of problems [55, 56]. Stephen Grossberg's work [25] has led to a host of special neural network models carrying out specialized functions modeled on particular brain regions. Piecing all these networks together could eventually lead to a brain-like AGI system. This approach is loosely related to Hugo de Garis's work, discussed in this volume, which seeks to use evolutionary programming to "evolve" specialized neural circuits, and then piece the circuits together into a whole mind. Peter Voss's a2i2 architecture also fits loosely into this category – his algorithms are related to prior work on "neural gasses" [41], and involve the cooperative use of a variety of different neural net learning algorithms. Less biologically oriented than Grossberg or even de Garis, Voss's neural system net does not try to closely model biological neural networks, but rather to emulate the sort of thing they do on a fairly high level.

The evolutionary programming approach to AI has not spawned any ambitious AGI projects, but it has formed a part of several AGI-oriented systems, including our own Novamente system, de Garis's CAM-Brain machine mentioned above, and John Holland's classifier systems [30]. Classifier systems are a kind of hybridization of evolutionary algorithms and probabilistic-symbolic AI; they are AGI-oriented in the sense that they are specifically oriented toward integrating memory, perception, and cognition to allow an AI system to act in the world. Typically they have suffered from severe performance problems, but Eric Baum's recent variations on the classifier system theme seem to have partially resolved these issues [5]. Baum's Hayek systems were tested on a simple "three peg blocks world" problem where any disk may be placed on any other; thus the required number of moves grows only linearly with the number of disks, not exponentially. The chapter authors were able to replicate their results only for n up to 5 [36].

The artificial life approach to AGI has remained basically a dream and a vision, up till this point. Artificial life simulations have succeeded, to a point, in getting interesting mini-organisms to evolve and interact, but no one has come close to creating an Alife agent with significant general intelligence. Steve Grand made some limited progress in this direction with his work on the Creatures game, and his current R&D efforts are trying to go even further [24]. Tom Ray's Network Tierra project also had this sort of ambition, but seems to have stalled at the stage of the automated evolution of simple multicellular artificial lifeforms.

Program search based AGI is a newer entry into the game. It had its origins in Solomonoff, Chaitin and Kolmogorov's seminal work on algorithmic information theory in the 1960s, but it did not become a serious approach to practical AI until quite recently, with work such as Schmidhuber's OOPS system described in this volume, and Kaiser's dag-based program search algorithms. This approach is different from the others in that it begins with a formal theory of general intelligence, defines impractical algorithms that are provably known to achieve general intelligence (see Hutter's chapter on AIXI in this volume for details), and then seeks to approximate these impractical algorithms with related algorithms that are more practical but less universally able.

Finally, the integrative approach to AGI involves taking elements of some or all of the above approaches and creating a combined, synergistic system. This makes sense if you believe that the different AI approaches each capture some aspect of the mind uniquely well. But the integration can be done in many different ways. It is not workable to simply create a modular system with modules embodying different AI paradigms: the different approaches are too different in too many ways. Instead one must create a unified knowledge representation and dynamics framework, and figure out how to manifest the core ideas of the various AI paradigms within the universal framework. This is roughly the approach taken in the Novamente project, but what has been found in that project is that to truly integrate ideas from different AI paradigms, most of the ideas need to be in a sense "reinvented" along the way.

Of course, no such categorization is going to be complete. Some of the papers in this book do not fit well into any of the above categories: for instance, Yudkowsky's approach, which is integrative in a sense, but does not involve integrating prior AI algorithms; and Hoyes's approach, which is founded on the notion of 3D simulation. What these two approaches have in common is that they both begin with a maverick cognitive science theory, a bold new explanation of human intelligence. They then draw implications and designs for AGI from the respective cognitive science theory.

None of these approaches has yet proved itself successful – this book is a discussion of promising approaches to AGI, not successfully demonstrated ones. It is probable that in 10 years a different categorization of AGI approaches will seem more natural, based on what we have learned in the interim. Perhaps one of the approaches described here will have proven successful, perhaps more than one; perhaps AGI will still be a hypothetical achievement, or perhaps it will have been achieved by methods totally unrelated to those described here. Our own belief, as AGI researchers, is that an integrative approach such as the one embodied in our Novamente AI Engine has an excellent chance of making it to the AGI finish line. But as the history of AI shows, researchers' intuitions about the prospects of their AI projects are highly chancy. Given the diverse and inter-contradictory nature of the different AGI approaches presented in these pages, it stands to reason that a good percentage of the authors have got to be significantly wrong on significant points! We invite the reader to study the AGI approaches presented here, and others cited but not thoroughly discussed here, and draw their own conclusions. Above all, we wish to leave the reader with the impression that AGI is a vibrant area of research, abounding with exciting new ideas and projects – and that, in fact, it is AGI rather than narrow AI that is properly the primary focus of artificial intelligence research.

2 What Is Intelligence?

What do we mean by *general intelligence*? The dictionary defines intelligence with phrases such as "The capacity to acquire and apply knowledge", and "The faculty of thought and reason." General intelligence implies an ability to acquire and apply knowledge, and to reason and think, in a variety of domains, not just in a single area like, say, chess or game-playing or languages or mathematics or rugby. Pinning down general intelligence beyond this is a subtle though not unrewarding pursuit. The disciplines of psychology, AI and control engineering have taken differing but complementary approaches, all of which are relevant to the AGI approaches described in this volume.

2.1 The Psychology of Intelligence

The classic psychological measure of intelligence is the "g-factor" [7], although this is quite controversial, and many psychologists doubt that any available IQ test really measures human intelligence in a general way. Gardner's [15] theory of multiple intelligences argues that human intelligence largely breaks down into a number of specialized-intelligence components (including linguistic, logical-mathematical, musical, bodily-kinesthetic, spatial, interpersonal, intra-personal, naturalist and existential).

Taking a broad view, it is clear that, in fact, human intelligence is not all that general. A huge amount of our intelligence is focused on situations that have occurred in our evolutionary experience: social interaction, vision processing, motion control, and so forth. There is a large research literature in support of this fact. For instance, most humans perform poorly at making probabilistic estimates in the abstract, but when the same estimation tasks are presented in the context of familiar social situations, human accuracy becomes much greater. Our intelligence is general "in principle", but in order to solve many sorts of problems, we need to resort to cumbersome and slow methods such as mathematics and computer programming. Whereas we are vastly more efficient at solving problems that make use of our in-built specialized neural circuitry for processing vision, sound, language, social interaction data, and so forth. Gardner's point is that different people have particularly effective specialized circuitry for different specializations. In principle, a human with poor social intelligence but strong logical-mathematical intelligence could solve a difficult problem regarding social interactions, but might have to do so in a very slow and cumbersome over-intellectual way, whereas an individual with strong innate social intelligence would solve the problem quickly and intuitively.

Taking a somewhat different approach, psychologist Robert Sternberg [53] distinguishes three aspects of intelligence: componential, contextual and experiential. Componential intelligence refers to the specific skills people have that make them intelligent; experiential refers to the ability of the mind to learn and adapt through experience; contextual refers to the ability of the mind to understand and operate within particular contexts, and select and modify contexts.

Applying these ideas to AI, we come to the conclusion that, to roughly emulate the nature of human general intelligence, an artificial general intelligence system should have:

- the ability to solve general problems in a non-domain-restricted way, in the same sense that a human can;
- most probably, the ability to solve problems in particular domains and particular contexts with particular efficiency;
- the ability to use its more generalized and more specialized intelligence capabilities together, in a unified way;
- the ability to learn from its environment, other intelligent systems, and teachers;
- the ability to become better at solving novel types of problems as it gains experience with them.

These points are based to some degree on human intelligence, and it may be that they are a little too anthropomorphic. One may envision an AGI system that is so good at the "purely general" aspect of intelligence that it doesn't need the specialized intelligence components. The practical possibility of this type of AGI system is an open question. Our guess is that the multiple-specializations nature of human intelligence will be shared by any AGI system operating with similarly limited resources, but as with much else regarding AGI, only time will tell.

One important aspect of intelligence is that it can only be achieved by a system that is capable of learning, especially autonomous and incremental learning. The system should be able to interact with its environment and other entities in the environment (which can include teachers and trainers, human or not), and learn from these interactions. It should also be able to build upon its previous experiences, and the skills they have taught it, to learn more complex actions and therefore achieve more complex goals.

The vast majority of work in the AI field so far has pertained to highly specialized intelligence capabilities, much more specialized than Gardner's multiple intelligence types – e.g. there are AI programs good at chess, or theorem verification in particular sorts of logic, but none good at logical-mathematical reasoning in general. There has been some research on completely general non-domain-oriented AGI algorithms, e.g. Hutter's AIXI model described in this volume, but so far these ideas have not led to practical algorithms (Schmidhuber's OOPS system, described in this volume, being a promising possibility in this regard).

2.2 The Turing Test

Next, no discussion of the definition of intelligence in an AI context would be complete without mention of the well-known Turing Test. Put loosely, the Turing test asks an AI program to *simulate a human in a text-based conversational interchange*. The most important point about the Turing test, we believe, is that it is a *sufficient* but not *necessary* criterion for artificial general intelligence. Some AI theorists don't even consider the Turing test as a sufficient test for general intelligence – a famous example is the Chinese Room argument [49].

Alan Turing, when he formulated his test, was confronted with people who believed AI was impossible, and he wanted to prove the existence of an intelligence test for computer programs. He wanted to make the point that intelligence is defined by behavior rather than by mystical qualities, so that if a program could act like a human it should be considered as intelligent as a human. This was a bold conceptual leap for the 1950's. Clearly, however, general intelligence does not necessarily require the accurate simulation of human intelligence. It seems unreasonable to expect a computer program without a human-like body to be able to emulate a human, especially in conversations regarding body-focused topics like sex, aging, or the experience of having the flu. Certainly, humans would fail a "reverse Turing test" of emulating computer programs – humans can't even emulate pocket calculators without unreasonably long response delays.

2.3 A Control Theory Approach to Defining Intelligence

The psychological approach to intelligence, briefly discussed above, attempts to do justice to the diverse and multifaceted nature of the notion of intelligence. As one might expect, engineers have a much simpler and much more practical definition of intelligence.

The branch of engineering called control theory deals with ways to cause complex machines to yield desired behaviors. Adaptive control theory deals with the design of machines which respond to external and internal stimuli and, on this basis, modify their behavior appropriately. And the theory of intelligent control simply takes this one step further. To quote a textbook of automata theory [2]:

> *[An] automaton is said to behave "intelligently" if, on the basis of its "training" data which is provided within some context together with information regarding the desired action, it takes the correct action on other data within the same context not seen during training.*

This is the sense in which contemporary artificial intelligence programs are intelligent. They can generalize within their limited context; they can follow the one script which they are programmed to follow. Of course, this is not really general intelligence, not in the psychological sense, and not in the sense in which we mean it in this book.

On the other hand, in their treatise on robotics, [57] presented a more general definition:

> *Intelligence is the ability to behave appropriately under unpredictable conditions.*

Despite its vagueness, this criterion does serve to point out the problem with ascribing intelligence to chess programs and the like: compared to our environment, at least, the environment within which they are capable of behaving appropriately is very predictable indeed, in that it consists only of certain (simple or complex) patterns of arrangement of a very small number of specifically structured entities. The *unpredictable conditions* clause suggests the experiential and contextual aspects of Sternberg's psychological analysis of intelligence.

Of course, the concept of appropriateness is intrinsically subjective. And unpredictability is relative as well – to a creature accustomed to living in interstellar space and inside stars and planets as well as on the surfaces of planets, or to a creature capable of living in 10 dimensions, our environment might seem just as predictable as the universe of chess seems to us. In order to make this folklore definition precise, one must first of all confront the vagueness inherent in the terms "appropriate" and "unpredictable".

In some of our own past work [17], we have worked with a variant of the Winkless and Browning definition,

> *Intelligence is the ability to achieve complex goals in complex environments.*

In a way, like the Winkless and Browning definition, this is a subjective rather than objective view of intelligence, because it relies on the subjective identification of what is and is not a complex goal or a complex environment. Behaving "appropriately", as Winkless and Browning describe, is a matter of achieving organismic goals, such as getting food, water, sex, survival, status, etc. Doing so under unpredictable conditions is one thing that makes the achievement of these goals complex.

Marcus Hutter, in his chapter in this volume, gives a rigorous definition of intelligence in terms of algorithmic information theory and sequential decision theory. Conceptually, his definition is closely related to the "achieve complex goals" definition, and it's possible the two could be equated if one defined *achieve, complex* and *goals* appropriately.

Note that none of these approaches to defining intelligence specify any particular properties of the *internals* of intelligent systems. This is, we believe, the correct approach: "intelligence" is about what, not how. However, it is possible that what implies how, in the sense that there may be certain structures and processes that are necessary aspects of any sufficiently intelligent system. Contemporary psychological and AI science are nowhere near the point where such a hypothesis can be verified or refuted.

2.4 Efficient Intelligence

Pei Wang, a contributor to this volume, has proposed his own definition of intelligence, which posits, basically, that "Intelligence is the ability to work and adapt to the environment with insufficient knowledge and resources." More concretely, he believes that an intelligent system is one that works under the *Assumption of Insufficient Knowledge and Resources* (AIKR), meaning that the system must be, at the same time,

A finite system The system's computing power, as well as its working and storage space, is limited.

A real-time system The tasks that the system has to process, including the assimilation of new knowledge and the making of decisions, can arrive at any time, and all have deadlines attached with them.

An ampliative system The system not only can retrieve available knowledge and derive sound conclusions from it, but also can make refutable hypotheses and guesses based on it when no certain conclusion can be drawn.

An open system No restriction is imposed on the relationship between old knowledge and new knowledge, as long as they are representable in the system's interface language.

A self-organized system The system can accommodate itself to new knowledge, and adjust its memory structure and mechanism to improve its time and space efficiency, under the assumption that future situations will be similar to past situations.

Wang's definition[4] is not purely behavioral: it makes judgments regarding the internals of the AI system whose intelligence is being assessed. However, the biggest difference between this and the above definitions is its emphasis on the limitation of the system's computing power. For instance, Marcus Hutter's AIXI algorithm, described in this volume, assumes infinite computing power (though his related AIXItl algorithm works with finite computing power). According to Wang's definition, AIXI is therefore unintelligent. Yet, AIXI can solve any problem at least as effectively as any finite-computing-power-based AI system, so it seems in a way unintuitive to call it "unintelligent".

We believe that what Wang's definition hints at is a new concept, that we call *efficient intelligence*, defined as:

> *Efficient intelligence is the ability to achieve intelligence using severely limited resources.*

Suppose we had a computer IQ test called the CIQ. Then, we might say that an AGI program with a CIQ of 500 running on 5000 machines has more intelligence, but less efficient-intelligence, than a machine with a CIQ of 100 that runs on just one machine.

According to the "achieving complex goals in complex environments" criterion, AIXI and AIXItl are the most intelligent programs described in this book, but not the ones with the highest efficient intelligence. According to Wang's definition of intelligence, AIXI and AIXItl are not intelligent at all, they only emulate intelligence through simple, inordinately wasteful program-search mechanisms.

As editors, we have not sought to impose a common understanding of the nature of intelligence on all the chapter authors. We have merely requested that authors be clear regarding the concept of intelligence under which they have structured their work. At this early stage in the AGI game, the notion of intelligence most appropriate for AGI work is still being discovered, along with the exploration of AGI theories, designs and programs themselves.

3 The Abstract Theory of General Intelligence

One approach to creating AGI is to formalize the problem mathematically, and then seek a solution using the tools of abstract mathematics. One may begin by formalizing the notion of intelligence. Having defined intelligence, one may then formalize the notion of computation in one of several generally-accepted ways, and ask the rigorous question: How may one create intelligent computer programs? Several researchers have taken this approach in recent years, and while it has not provided a panacea for AGI, it has yielded some

---

[4]In more recent work, Wang has modified the details of this definition, but the theory remains the same.

very interesting results, some of the most important ones are described in Hutter's and Schmidhuber's chapters in this book.

From a mathematical point of view, as it turns out, it doesn't always matter so much exactly how you define intelligence. For many purposes, any definition of intelligence that has the general form *"Intelligence is the maximization of a certain quantity, by a system interacting with a dynamic environment"* can be handled in roughly the same way. It doesn't always matter exactly what the quantity being maximized is (whether it's "complexity of goals achieved", for instance, or something else).

Let's use the term "behavior-based maximization criterion" to characterize the class of definitions of intelligence indicated in the previous paragraphs. Suppose one has some particular behavior-based maximization criterion in mind – then Marcus Hutter's work on the AIXI system, described in his chapter here, gives a software program that will be able to achieve intelligence according to the given criterion. Now, there's a catch: this program may require infinite memory and an infinitely fast processor to do what it does. But he also gives a variant of AIXI which avoids this catch, by restricting attention to programs of bounded length l and bounded time t. Loosely speaking, the AIXItl variant will provably be as intelligent as any other computer program of length up to l, satisfying the maximization criterion, within a constant multiplicative factor and a constant additive factor.

Hutter's work draws on a long tradition of research in statistical learning theory and algorithmic information theory, mostly notably Solomonoff's early work on induction [51, 52] and Levin's [39, 40] work on computational measure theory. At the present time, this work is more exciting theoretically than pragmatically. The "constant factor" in his theorem may be very large, so that, in practice, AIXItl is not really going to be a good way to create an AGI software program. In essence, what AIXItl is doing is searching the space of all programs of length L, evaluating each one, and finally choosing the best one and running it. The "constant factors" involved deal with the overhead of trying every other possible program before hitting on the best one!

A simple AI system behaving somewhat similar to AIXItl could be built by creating a program with three parts:

- the data store;
- the main program;
- the meta-program.

The operation of the meta-program would be, loosely, as follows:

- At time t, place within the data store a record containing the complete internal state of the system, and the complete sensory input of the system.

- Search the space of all programs P of size $|P| < l$ to find the one that, based on the data in the data store, has the highest expected value for the given maximization criterion.[5]
- Install P as the main program.

Conceptually, the main value of this approach for AGI is that it solidly establishes the following contention:

> *If you accept any definition of intelligence of the general form "maximization of a certain function of system behavior,"*
> *then the problem of creating AGI is basically a problem of dealing with the issues of space and time efficiency.*

As with any mathematics-based conclusion, the conclusion only follows if one accepts the definitions. If someone's conception of intelligence fundamentally can't be cast into the form of a behavior-based maximization criterion, then these ideas aren't relevant for AGI as that person conceives it. However, we believe that the behavior-based maximization criterion approach to defining intelligence is a good one, and hence we believe that Hutter's work is highly significant.

The limitations of these results are twofold. Firstly, they pertain only to AGI in the "massive computational resources" case, and most AGI theorists feel that this case is not terribly relevant to current practical AGI research (though, Schmidhuber's OOPS work represents a serious attempt to bridge this gap). Secondly, their applicability to the physical universe, even in principle, relies on the Church-Turing Thesis. The editors and contributors of this volume are Church-Turing believers, as are nearly all computer scientists and AI researchers, but there are well-known exceptions such as Roger Penrose. If Penrose and his ilk are correct, then the work of Hutter and his colleagues is not necessarily informative about the nature of AGI in the physical universe.

For instance, consider Penrose's contention that non-Turing quantum gravity computing (as allowed by an as-yet unknown incomputable theory of quantum gravity) is necessary for true general intelligence [44]. This idea is not refuted by Hutter's results, because it's possible that:

- AGI is in principle possible on ordinary Turing hardware;
- AGI is only pragmatically possible, given the space and time constraints imposed on computers by the physical universe, given quantum gravity powered computer hardware.

The authors very strongly doubt this is the case, and Penrose has not given any convincing evidence for such a proposition, but our point is merely that in spite of recent advances in AGI theory such as Hutter's work, we have

---

[5]There are some important details here; for instance, computing the "expected value" using probability theory requires assumption of an appropriate prior distribution, such as Solomonoff's universal prior.

no way of ruling such a possibility out mathematically. At points such as this, uncertainties about the fundamental nature of mind and universe rule out the possibility of a truly definitive theory of AGI.

From the perspective of computation theory, most of the chapters in this book deal with ways of achieving *reasonable degrees of intelligence given reasonable amounts of space and time resources*. Obviously, this is what the human mind/brain does. The amount of intelligence it achieves is clearly limited by the amount of space in the brain and the speed of processing of neural wetware.

We do not yet know whether the sort of mathematics used in Hutter's work can be made useful for defining practical AGI systems that operate within our current physical universe – or, better yet, on current or near-future computer hardware. However, research in this direction is proceeding vigorously. One exciting project in this area is Schmidhuber's OOPS system [48], which is a bit like AIXItl, but has the capability of operating with realistic efficiency in some practical situations. As Schmidhuber discusses in his first chapter in this book, OOPS has been applied to some classic AI problems such as the Towers of Hanoi problem, with highly successful results.

The basic idea of OOPS is to run all possible programs, but interleaved rather than one after the other. In terms of the "meta-program" architecture described above, here one has a meta-program that doesn't run each possible program one after the other, but rather lines all the possible programs up in order, assigns each one a probability, and then at each time step chooses a single program as the "current program", with a probability proportional to its estimated value at achieving the system goal, and then executes one step of the current program. Another important point is that OOPS freezes solutions to previous tasks, and may reuse them later.

As opposed to AIXItl, this strategy allows, in the average case, brief and effective programs to rise to the top of the heap relatively quickly. The result, in at least some practical problem-solving contexts, is impressive. Of course, there are many ways to solve the Towers of Hanoi problem. Scaling up from toy examples to real AGI on the human scale or beyond is a huge task for OOPS as for other approaches showing limited narrow-AI success. But having made the leap from abstract algorithmic information theory to limited narrow-AI success is no small achievement.

Schmidhuber's more recent Gödel Machine, which is fully self-referential, is in principle capable of proving and subsequently exploiting performance improvements to its own code. The ability to modify its own code allows the Gödel Machine to be more effective. Gödel Machines are also more flexible in terms of the utility function they aim to maximize while searching.

Lukasz Kaiser's chapter follows up similar themes to Hutter's and Schmidhuber's work. Using a slightly different computational model, Kaiser also takes up the algorithmic-information-theory motif, and describes a program search problem which is solved through the combination of program construction and the proof search – the program search algorithm itself, represented as a directed acyclic graph, is continuously improved.

4 Toward a Pragmatic Logic

One of the primary themes in the history of AI is formal logic. However, there are strong reasons to believe that classical formal logic is not suitable to play a central role in an AGI system. It has no natural way to deal with uncertainty, or with the fact that different propositions may be based on different amounts of evidence. It leads to well-known and frustrating logical paradoxes. And it doesn't seem to come along with any natural "control strategy" for navigating the combinatorial explosion of possible valid inferences.

Some modern AI researchers have reacted to these shortcomings by rejecting the logical paradigm altogether; others by creating modified logical frameworks, possessing more of the flexibility and fluidity required of components of an AGI architecture.

One of the key issues dividing AI researchers is the degree to which logical reasoning is fundamental to their artificial minds. Some AI systems are built on the assumption that basically every aspect of mental process should be thought about as a kind of logical reasoning. Cyc is an example of this, as is the NARS system reviewed in this volume. Other systems are built on the premise that logic is irrelevant to the task of mind-engineering, that it is merely a coarse, high-level description of the results of mental processes that proceed according to non-logical dynamics. Rodney Brooks' work on subsumption robotics fits into this category, as do Peter Voss's and Hugo de Garis's neural net AGI designs presented here. And there are AI approaches, such as Novamente, that assign logic an important but non-exclusive role in cognition – Novamente has roughly two dozen cognitive processes, of which about one-fourth are logical in nature.

One fact muddying the waters somewhat is the nebulous nature of "logic" itself. *Logic* means different things to different people. Even within the domain of formal, mathematical logic, there are many different kinds of logic, including forms like fuzzy logic that encompass varieties of reasoning not traditionally considered "logical". In our own work we have found it useful to adopt a very general conception of logic, which holds that logic:

- has to do with forming and combining estimations of the (possibly probabilistic, fuzzy, etc.) truth values of various sorts of relationships based on various sorts of evidence;
- is based on incremental processing, in which pieces of evidence are combined step by step to form conclusions, so that at each stage it is easy to see which pieces of evidence were used to give which conclusion

This conception differentiates logic from mental processing in general, but it includes many sorts of reasoning besides typical, crisp, mathematical logic.

The most common form of logic is predicate logic, as used in Cyc, in which the basic entity under consideration is the *predicate*, a function that maps argument variables into Boolean truth values. The argument variables are quantified universally or existentially. An alternate form of logic is term logic, which predates predicate logic, dating back at least to Aristotle and his notion of the syllogism. In term logic, the basic element is a subject-predicate statement, denotable as $A \to B$, where $\to$ denotes a notion of inheritance or specialization. Logical inferences take the form of *syllogistic rules*, which give patterns for combining statements with matching terms, such as the deduction rule

$$(A \to B \wedge B \to C) \Rightarrow A \to C.$$

The NARS system described in this volume is based centrally on term logic, and the Novamente system makes use of a slightly different variety of term logic. Both predicate and term logic typically use variables to handle complex expressions, but there are also variants of logic, based on combinatory logic, that avoid variables altogether, relying instead on abstract structures called "higher-order functions" [10].

There are many different ways of handling uncertainty in logic. Conventional predicate logic treats statements about uncertainty as predicates just like any others, but there are many varieties of logic that incorporate uncertainty at a more fundamental level. Fuzzy logic [59, 60] attaches fuzzy truth values to logical statements; probabilistic logic [43] attaches probabilities; NARS attaches degrees of uncertainty, etc. The subtle point of such systems is the transformation of uncertain truth values under logical operators like AND, OR and NOT, and under existential and universal quantification.

And, however one manages uncertainty, there are also multiple varieties of speculative reasoning. Inductive [4], abductive [32] and analogical reasoning [31] are commonly discussed. Nonmonotonic logic [8] handles some types of nontraditional reasoning in a complex and controversial way. In ordinary, monotonic logic, the truth of a proposition does not change when new information (axioms) is added to the system. In nonmonotonic logic, on the other hand, the truth of a proposition may change when new information (axioms) is added to or old information is deleted from the system. NARS and Novamente both use logic in an uncertain and nonmonotonic way.

Finally, there are special varieties of logic designed to handle special types of reasoning. There are temporal logics designed to handle reasoning about time, spatial logics for reasoning about space, and special logics for handling various kinds of linguistic phenomena. None of the approaches described in this book makes use of such special logics, but it would be possible to create an AGI approach with such a focus. Cyc comes closest to this notion, as its reasoning engine involves a number of specialized reasoning engines oriented toward particular types of inference such as spatial, temporal, and so forth.

When one gets into the details, the distinction between logical and non-logical AI systems can come to seem quite fuzzy. Ultimately, an uncertain logic rule is not that different from the rule governing the passage of *activation* through a node in a neural network. Logic can be cast in terms of semantic networks, as is done in Novamente; and in that case uncertain logic formulas are arithmetic formulas that take in numbers associated with certain nodes and links in a graph, and output numbers associated with certain other nodes and links in the graph. Perhaps a more important distinction than logical vs. non-logical is whether a system gains its knowledge experientially or via being given *expert rule* type propositions. Often logic-based AI systems are fed with knowledge by human programmers, who input knowledge in the form of textually-expressed logic formulas. However, this is not a necessary consequence of the use of logic. It is quite possible to have a logic-based AI system that forms its own logical propositions by experience. On the other hand, there is no existing example of a non-logical AI system that gains its knowledge from explicit human knowledge encoding. NARS and Novamente are both (to differing degrees) logic-based AI systems, but their designs devote a lot of attention to the processes by which logical propositions are formed based on experience, which differentiates them from many traditional logic-based AI systems, and in a way brings them closer to neural nets and other traditional non-logical AI systems.

5 Emulating the Human Brain

One almost sure way to create artificial general intelligence would be to exactly copy the human brain, down to the atomic level, in a digital simulation. Admittedly, this would require brain scanners and computer hardware far exceeding what is currently available. But if one charts the improvement curves of brain scanners and computer hardware, one finds that it may well be plausible to take this approach sometime around 2030-2050. This argument has been made in rich detail by Ray Kurzweil in [34, 35]; and we find it a reasonably convincing one. Of course, projecting the future growth curves of technologies is a very risky business. But there's very little doubt that creating AGI in this way is physically possible.

In this sense, creating AGI is "just an engineering problem." We know that general intelligence is possible, in the sense that humans – particular configurations of atoms – display it. We just need to analyze these atom configurations in detail and replicate them in the computer. AGI emerges as a special case of nanotechnology and in silico physics.

Perhaps a book on the same topic as this one, written in 2025 or so, will contain detailed scientific papers pursuing the detailed-brain-simulation approach to AGI. At present, however, it is not much more than a futuristic speculation. We don't understand enough about the brain to make detailed simulations of brain function. Our brain scanning methods are improving rapidly but at present they don't provide the combination of temporal and spatial acuity required to really map thoughts, concepts, percepts and actions as they occur in human brains/minds.

It's still possible, however, to use what we know about the human brain to structure AGI designs. This can be done in many different ways. Most simply, one can take a neural net based approach, trying to model the behavior of nerve cells in the brain and the emergence of intelligence therefrom. Or one can proceed at a higher level, looking at the general ways that information processing is carried out in the brain, and seeking to emulate these in software.

Stephen Grossberg [25, 28] has done extensive research on the modeling of complex neural structures. He has spent a great deal of time and effort in creating cognitively-plausible neural structures capable of spatial perception, shape detection, motion processing, speech processing, perceptual grouping, and other tasks. These complex brain mechanisms were then used in the modeling of learning, attention allocation and psychological phenomena like schizophrenia and hallucinations.

From the experiences modeling different aspects of the brain and the human neural system in general, Grossberg has moved on to the linking between those neural structures and the mind [26, 27, 28]. He has identified two key computational properties of the structures: *complementary computing* and *laminar computing*.

Complementary computing is the property that allows different processing streams in the brain to compute complementary properties. This leads to a hierarchical resolution of uncertainty, which is mostly evident in models of the visual cortex. The complementary streams in the neural structure interact, in parallel, resulting in more complete information processing. In the visual cortex, an example of complementary computing is the interaction between the *what* cortical stream, which learns to recognize what events and objects occur, and the *where* cortical stream, which learns to spacially locate those events and objects.

Laminar computing refers to the organization of the cerebral cortex (and other complex neural structures) in layers, with interactions going bottom-up, top-down, and sideways. While the existence of these layers has been known for almost a century, the contribution of this organization for control of behavior was explained only recently. [28] has recently shed some light on the subject, showing through simulations that laminar computing contributes to learning, development and attention control.

While Grossberg's research has not yet described complete minds, only neural models of different parts of a mind, it is quite conceivable that one could use his disjoint models as building blocks for a complete AGI design. His recent successes explaining, to a high degree of detail, how mental processes can emerge from his neural models is definitely encouraging.

Steve Grand's Creatures [24] are social agents, but they have an elaborate internal architecture, based on a complex neural network which is divided into several lobes. The original design by Grand had explicit AGI goals, with attention paid to allow for symbol grounding, generalization, and limited language processing. Grand's creatures had specialized lobes to handle verbal input, and to manage the creature's internal state (which was implemented as a simplified biochemistry, and kept track of feelings such as pain, hunger and others). Other lobes were dedicated to adaptation, goal-oriented decision making, and learning of new concepts.

Representing the neural net approach in this book, we have Peter Voss's paper on the a2i2 architecture. a2i2 is in the vein of other modern work on reinforcement learning, but it is unique in its holistic architecture focused squarely on AGI. Voss uses several different reinforcement and other learning techniques, all acting on a common network of artificial neurons and synapses. The details are original, but are somewhat inspired by prior neural net AI approaches, particularly the "neural gas" approach [41], as well as objectivist epistemology and cognitive psychology. Voss's theory of mind abstracts what would make brains intelligent, and uses these insights to build artificial brains.

Voss's approach is incremental, involving a gradual progression through the "natural" stages in the complexity of intelligence, as observed in children and primates – and, to some extent, recapitulating evolution. Conceptually, his team is adding ever more advanced levels of cognition to its core design, somewhat resembling both Piagetian stages of development, as well as the evolution of primates, a level at which Voss considers there is enough complexity on the neuro-cognitive systems to provide AGI with useful metaphors and examples.

His team seeks to build ever more complex virtual primates, eventually reaching the complexity and intelligence level of humans. But this metaphor shouldn't be taken too literally. The perceptual and action organs of their initial proto-virtual-ape are not the organs of a physical ape, but rather visual and acoustic representations of the Windows environment, and the ability to undertake simple actions within Windows, as well as various *probes* for interaction with the real world through vision, sound, etc.

There are echoes of Rodney Brooks's subsumption robotics work, the well-known Cog project at MIT [1], in the a2i2 approach. Brooks is doing something a lot more similar to actually building a virtual cockroach, with a focus on the robot body and the pragmatic control of it. Voss's approach to AI could easily be nested inside robot bodies like the ones constructed by Brooks's team; but Voss doesn't believe the particular physical embodiment is the key, he believes that the essence of experience-based reinforcement learning can be manifested in a system whose inputs and outputs are "virtual."

6 Emulating the Human Mind

Emulating the atomic structure of the brain in a computer is one way to let the brain guide AGI; creating virtual neurons, synapses and activations is another. Proceeding one step further up the ladder of abstraction, one has approaches that seek to emulate the overall architecture of the human brain, but not the details by which this architecture is implemented. Then one has approaches that seek to emulate the human mind, as studied by cognitive psychologists, ignoring the human mind's implementation in the human brain altogether.

Traditional logic-based AI clearly falls into the "emulate the human mind, not the human brain" camp. We actually have no representatives of this approach in the present book; and so far as we know, the only current research that could fairly be described as lying in the intersection of traditional logic-based AI and AGI is the Cyc project, briefly mentioned above.

But traditional logic-based AI is far from the only way to focus on the human mind. We have several contributions in this book that are heavily based on cognitive psychology and its ideas about how the mind works. These contributions pay greater than zero attention to neuroscience, but they are clearly more mind-focused than brain-focused.

Wang's NARS architecture, mentioned above, is the closest thing to a formal logic based system presented in this book. While it is not based specifically on any one cognitive science theory, NARS is clearly closely motivated by cognitive science ideas; and at many points in his discussion, Wang cites cognitive psychology research supporting his ideas.

Next, Hoyes's paper on 3D vision as the key to AGI is closely inspired by the human mind and brain, although it does not involve neural nets or other micro-level brain-simulative entities. Hoyes is not proposing to copy the precise wiring of the human visual system in silico and use it as the core of an AGI system, but he is proposing that we should copy what he sees as the basic architecture of the human mind. In a daring and speculative approach, he views the ability to deal with changing 3D scenes as the essential capability of the human mind, and views other human mental capabilities largely as offshoots of this. If this theory of the human mind is correct, then one way to achieve AGI is to do as Hoyes suggests and create a robust capability for 3D simulation, and build the rest of a digital mind centered around this capability.

Of course, even if this speculative analysis of the human mind is correct, it doesn't intrinsically follow that 3D simulation centric approach is the only approach to AGI. One could have a mind centered around another sense, or a mind that was more cognitively rather than perceptually centered. But Hoyes' idea is that we already have one example of a thinking machine – the human brain – and it makes sense to use as much of it as we can in designing our new digital intelligences.

Eliezer Yudkowsky, in his chapter, describes the conceptual foundations of his AGI approach, which he calls "deliberative general intelligence" (DGI). While DGI-based AGI is still at the conceptual-design phase, a great deal of analysis has gone into the design, so that DGI essentially amounts to an original and detailed cognitive-science theory, crafted with AGI design in mind. The DGI theory was created against the backdrop of Yudkowsky's futurist thinking, regarding the notions of:

- a *Seed AI*, an AGI system that progressively modifies and improves its own codebase, thus projecting itself gradually through exponentially increasing levels of intelligence; [58]
- a *Friendly AI*, an AGI system that respects positive ethics such as the preservation of human life and happiness, through the course of its progressive self-improvements.

However, the DGI theory also may stand alone, independently of these motivating concepts.

The essence of DGI is a functional decomposition of general intelligence into a complex supersystem of interdependent internally specialized processes. Five successive levels of functional organization are posited:

Code The source code underlying an AI system, which Yudkowsky views as roughly equivalent to neurons and neural circuitry in the human brain.

Sensory modalities In humans: sight, sound, touch, taste, smell. These generally involve clearly defined stages of information-processing and feature-extraction. An AGI may emulate human senses or may have different sorts of modalities.

Concepts *Categories* or *symbols* abstracted from a system's experiences. The process of abstraction is proposed to involve the recognition and then reification of a similarity within a group of experiences. Once reified, the common quality can then be used to determine whether new mental imagery satisfies the quality, and the quality can be imposed on a mental image, altering it.

Thoughts Conceived of as being built from structures of concepts. By imposing concepts in targeted series, the mind builds up complex mental images within the workspace provided by one or more sensory modalities. The archetypal example of a thought, according to Yudkowsky, is a human *sentence* – an arrangement of concepts, invoked by their symbolic tags, with internal structure and targeting information that can be reconstructed from a linear series of words using the constraints of syntax, constructing a complex mental image that can be used in reasoning. Thoughts (and their corresponding mental imagery) are viewed as disposable one-time structures, built from reusable concepts, that implement a non-recurrent mind in a non-recurrent world.

Deliberation Implemented by sequences of thoughts. This is the *internal narrative* of the conscious mind – which Yudkowsky views as the core of intelligence both human and digital. It is taken to include explanation, prediction, planning, design, discovery, and the other activities used to solve knowledge problems in the pursuit of real-world goals.

Yudkowsky also includes an interesting discussion of probable differences between humans and AI's. The conclusion of this discussion is that, eventually, AGI's will have many significant advantages over biological intelligences. The lack of motivational peculiarities and cognitive biases derived from an evolutionary heritage will make *artificial psychology* quite different from, and presumably far less conflicted than, human psychology. And the ability to fully observe their own state, and modify their own underlying structures and dynamics, will give AGI's an ability for self-improvement vastly exceeding that possessed by humans. These conclusions by and large pertain not only to AGI designs created according to the DGI theory, but also to many other AGI designs as well. However, according to Yudkowsky, AGI designs based too closely on the human brain (such as neural net based designs) may not be able to exploit the unique advantages available to digital intelligences.

Finally, the authors' Novamente AI project has had an interesting relationship with the human mind/brain, over its years of development. The Webmind AI project, Novamente's predecessor, was more heavily human brain/mind based in its conception. As Webmind progressed, and then as Novamente was created based on the lessons learned in working on Webmind, we found that it was more and more often sensible to depart from human-brain/mind-ish approaches to various issues, in favor of approaches that provided greater efficiency on available computer hardware. There is still a significant cognitive psychology and neuroscience influence on the design, but not as much as there was at the start of the project.

One may sum up the diverse relationships between AGI approaches and the human brain/mind by distinguishing between:

- approaches that draw their primary structures and dynamics from an attempt to model biological brains;
- approaches like DGI and Novamente that are explicitly guided by the human brain as well as the human mind;
- approaches like NARS that are inspired by the human mind much more than the human brain;
- approaches like OOPS that have drawn very little on known science about human intelligence in any regard.

7 Creating Intelligence by Creating Life

If simulating the brain molecule by molecule is not ambitious enough for you, there is another possible approach to AGI that is even more ambitious, and even more intensely consumptive of computational resources: simulation of the sort of evolutionary processes that gave rise to the human brain in the first place.

Now, we don't have access to the primordial soup from which life presumably emerged on Earth. So, even if we had an adequately powerful supercomputer, we wouldn't have the option to simulate the origin of life on Earth molecule by molecule. But we can try to emulate the type of process by which life emerged – cells from organic molecules, multicellular organisms from unicellular ones, and so forth.

This variety of research falls into the domain of artificial life rather than AI proper. Alife is a flourishing discipline on its own, highly active since the early 1990's. We will briefly review some of the best known projects in the area. While most of this research still focuses on the creation and evolution of either very unnatural or quite simplistic creatures, there are several projects that have managed to give rise to fascinating levels of complexity.

Tierra, by Thomas Ray [45] was one of the earlier proposals toward an artificial evolutionary process that generates life. Tierra was successful in giving rise to unicellular organisms (actually, programs encoded in a 32-instruction machine language). In the original Tierra, there was no externally defined fitness function – the fitness emerged as a consequence of each creature's ability to replicate itself and adapt to the presence of other creatures.

Eventually, Tierra would converge to a stable state, as a consequence of the creature's optimization of their replication code. Ray then decided to explore the emergence of multicellular creatures, using the analogy of parallel processes in the digital environment. Enter Network Tierra [46], which was a distributed system providing a simulated landscape for the creatures, allowing migration and exploitation of different environments. Multicellular creatures emerged, and a limited degree of cell differentiation was observed in some experiments [47]. Unfortunately, the evolvability of the system wasn't high enough to allow greater complexity to emerge.

The Avida platform, developed at Caltech, is currently the most used ALife platform, and work on the evolution of complex digital creatures continues.

Walter Fontana's AlChemy [14, 13] project focuses on addressing a different, but equally important and challenging issue – defining *a theory of biological organization* which allows for self-maintaining organisms, i.e., organisms which possess a metabolic system capable of sustaining their persistence. Fontana created an artificial chemistry based on two key abstractions: *constructiveness* (the interaction between components can generate new components. In chemistry, when two molecules collide, new molecules may arise as a consequence.) and the existence of *equivalence classes* (the property that the same final result can be obtained by different reaction chains). Fontana's artificial chemistry uses lambda calculus as a minimal system presenting those key features.

From this chemistry, Fontana develops his theory of biological organization, which is a theory of self-maintaining systems. His computer simulations have shown that networks of interacting lambda-expressions arise which are self-maintaining and robust, being able to repair themselves when components are removed. Fontana called these networks organizations, and he was able to generate organizations capable of self-duplication and maintenance, as well as the emergence of self-maintaining metaorganizations composed of single organizations.

8 The Social Nature of Intelligence

All the AI approaches discussed so far essentially view the mind as something associated with a single organism, a single computational system. Social psychologists, however, have long recognized that this is just an approximation. In reality the mind is social – it exists, not in isolated individuals, but in individuals embedded in social and cultural systems.

One approach to incorporating the social aspect of mind is to create individual AGI systems and let them interact with each other. For example, this is an important part of the Novamente AI project, which involves a special language for Novamente AI systems to use to interact with each other. Another approach, however, is to consider sociality at a more fundamental level, and to create systems from the get-go that are at least as social as they are intelligent.

One example of this sort of approach is Steve Grand's neural-net architecture as embodied in the Creatures game [24]. His neural net based creatures are intended to grow more intelligent by interacting with each other – struggling with each other, learning to outsmart each other, and so forth.

John Holland's classifier systems [30] are another example of a multi-agent system in which competition and cooperation are both present. In a classifier system, a number of rules co-exist in the system at any given moment. The system interacts with an external environment, and must react appropriately to the stimuli received from the environment. When the system performs the appropriate actions for a given perception, it is rewarded. While the individuals in Holland's system are quite primitive, recent work by Eric Baum [5] has used a similar metaphor with more complex individuals, and promising results on some large problems.

In order to decide how to answer to the perceived stimuli, the system will perform multiple rounds of competition, during which the rules bid to be activated. The winning rule will then either perform an internal action, or an external one. Internal actions change the system's internal state and affect the next round of bidding, as each rule's right to bid (and, in some variations, the amount it bids) depends on how well it matches the system's current state.

Eventually, a rule will be activated that will perform an external action, which may trigger reward from the environment. The reward is then shared by all the rules that have been active since the stimuli were perceived. The credit assignment algorithm used by Holland is called *bucket brigade*. Rules that receive rewards can bid higher in the next rounds, and are also allowed to reproduce, which results in the creation of new rules.

Another important example of social intelligence is presented in the research inspired by social insects. *Swarm Intelligence* [6] is the term that generically describes such systems. Swarm Intelligence systems are a new class of biologically inspired tools.

These systems are self-organized, relying on direct and indirect communication between agents to lead to emergent behavior. Positive feedback is given by this communication (which can take the form of a dance indicating the direction of food in bee colonies, or pheromone trails in ant societies), which biases the future behavior of the agents in the system. These systems are naturally stochastic, relying on multiple interactions and on a random, exploratory component. They often display highly adaptive behavior to a dynamic environment, having thus been applied to dynamic network routing [9]. Given the simplicity of the individual agents, Swarm Intelligence showcases the value of cooperative emergent behavior in an impressive way.

Ant Colony Optimization [11] is the most popular form of Swarm Intelligence. ACO was initially designed as a heuristic for NP-hard problems [12], but has since been used in a variety of settings. The original version of ACO was developed to solve the famous Traveling Salesman problem. In this scenario, the environment is the graph describing the cities and their connections, and the individual agents, called *ants*, travel in the graph.

Each ant will do a tour of the cities in the graph, iteratively. At each city it will choose the next city to visit, based on a *transition rule*. This rule considers the amount of pheromone in the links connecting the current city and each of the possibilities, as well as a small random component. When the ant completes its tour, it updates the pheromone trail in the links it has used, laying an amount of pheromone proportional to the quality of the tour it has completed. The new trail will then influence the choices of the ants in the next iteration of the algorithm.

Finally, an important contribution from Artificial Life research is the *Animat* approach. Animats are biologically-inspired simulated or real robots, which exhibit adaptive behavior. In several cases [33] animats have been evolved to display reasonably complex artificial nervous systems capable of learning and adaptation. Proponents of the Animat approach argue that AGI is only reachable by embodied autonomous agents which interact on their own with their environments, and possibly other agents. This approach places an emphasis on the developmental, morphological and environmental aspects of the process of AI creating.

Vladimir Red'ko's self-organizing agent-system approach also fits partially into this general category, having some strong similarities to Animat projects. He defines a large population of simple agents guided by simple neural networks. His chapter describes two models for these agents. In all cases, the agents live in a simulated environment in which they can move around, looking for resources, and they can mate – mating uses the typical genetic operators of uniform crossover and mutation, which leads to the evolution of the agent population.

In the simpler case, agents just move around and eat virtual food, accumulating resources to mate. The second model in Red'ko's work simulates more complex agents. These agents communicate with each other, and modify their behavior based on their experience. None of the agents individually are all that clever, but the population of agents as a whole can demonstrate some interesting collective behaviors, even in the initial, relatively simplistic implementation. The agents communicate their knowledge about resources in different points of the environment, thus leading to the emergence of adaptive behavior.

9 Integrative Approaches

We have discussed a number of different approaches to AGI, each of which has – at least based on a cursory analysis – strengths and weaknesses compared to the others. This gives rise to the idea of integrating several of the approaches together, into a single AGI system that embodies several different approaches.

Integrating different ideas and approaches regarding something as complex and subtle as AGI is not a task to be taken lightly. It's quite possible to integrate two good ideas and obtain a bad idea, or to integrate two good software systems and get a bad software system. To successfully integrate different approaches to AGI requires deep reflection on all the approaches involved, and unification on the level of conceptual foundations as well as pragmatic implementation.

Several of the AGI approaches described in this book are integrative to a certain extent. Voss's a2i2 system integrates a number of different neural-net-oriented learning algorithms on a common, flexible neural-net-like data structure. Many of the algorithms he integrated have been used before, but only in an isolated way, not integrated together in an effort to make a "whole mind." Wang's NARS-based AI design is less strongly integrative, but it still may be considered as such. It posits the NARS logic as the essential core of AI, but leaves room for integrating more specialized AI modules to deal with perception and action. Yudkowsky's DGI framework is integrative in a similar sense: it posits a particular overall architecture, but leaves some room for insights from other AI paradigms to be used in filling in roles within this architecture.

By far the most intensely integrative AGI approach described in the book, however, is our own Novamente AI approach.

The Novamente AI Engine, the work of the editors of this volume and their colleagues, is in part an original system and in part an integration of ideas from prior work on narrow AI and AGI. The Novamente design incorporates aspects of many previous AI paradigms such as genetic programming, neural networks, agent systems, evolutionary programming, reinforcement learning, and probabilistic reasoning. However, it is unique in its overall architecture, which confronts the problem of creating a holistic digital mind in a direct and ambitious way.

The fundamental principles underlying the Novamente design derive from a novel complex-systems-based theory of mind called the *psynet model*, which was developed in a series of cross-disciplinary research treatises published during 1993-2001 [17, 16, 18, 19, 20]. The psynet model lays out a series of properties that must be fulfilled by any software system if it is going to be an autonomous, self-organizing, self-evolving system, with its own understanding of the world, and the ability to relate to humans on a *mind-to-mind* rather than a *software-program-to-mind* level. The Novamente project is based on many of the same ideas that underlay the Webmind AI Engine project carried out at Webmind Inc. during 1997-2001 [23]; and it also draws to some extent on ideas from Pei Wang's Non-axiomatic Reasoning System (NARS) [54].

At the moment, a complete Novamente design has been laid out in detail [21], but implementation is only about 25% complete (and of course many modifications will be made to the design during the course of further implementation). It is a C++ software system, currently customized for Linux clusters, with a few externally-facing components written in Java. The overall mathematical and conceptual design of the system is described in a paper [22] and a forthcoming book [21]. The existing codebase implements roughly a quarter of the overall design. The current, partially-complete codebase is being used by the startup firm Biomind LLC, to analyze genetics and proteomics data in the context of information integrated from numerous biological databases. Once the system is fully engineered, the project will begin a phase of interactively teaching the Novamente system how to respond to user queries, and how to usefully analyze and organize data. The end result of this teaching process will be an autonomous AGI system, oriented toward assisting humans in collectively solving pragmatic problems.

10 The Outlook for AGI

The AGI subfield is still in its infancy, but it is certainly encouraging to observe the growing attention that it has received in the past few years. Both the number of people and research groups working on systems designed to achieve general intelligence and the interest from outsiders have been growing.

Traditional, narrow AI does play a key role here, as it provides useful examples, inspiration and results for AGI. Several such examples have been mentioned in the previous sections in connection with one or another AGI approach. Innovative ideas like the application of complexity and algorithmic information theory to the mathematical theorization of intelligence and AI provide valuable ground for AGI researchers. Interesting ideas in logic, neural networks and evolutionary computing provide both tools for AGI approaches and inspiration for the design of key components, as will be seen in several chapters of this book.

The ever-welcome increase in computational power and the emergence of technologies like Grid computing also contribute to a positive outlook for AGI. While it is possible that, in the not too distant future, regular desktop machines (or whatever form the most popular computing devices take 10 or 20 years from now) will be able to run AGI software comfortably, today's AGI prototypes are extremely resource intensive, and the growing availability of world-wide computing farms would greatly benefit AGI research. The popularization of Linux, Linux-based clusters that extract considerable horsepower from stock hardware, and, finally, Grid computing, are seen as great advances, for one can never have enough CPU cycles.

We hope that the precedent set by these pioneers in AGI research will inspire young AI researchers to stray a bit off the beaten track and venture into the more daring, adventurous and riskier path of seeking the creation of truly general artificial intelligence. Traditional, narrow AI is very valuable, but, if nothing else, we hope that this volume will help create the awareness that AGI research is a very present and viable option. The complementary and related fields are mature enough, the computing power is becoming increasingly easier and cheaper to obtain, and AGI itself is ready for popularization. We could always use yet another design for an artificial general intelligence in this challenging, amazing, and yet friendly race toward the awakening of the world's first real artificial intelligence.

Acknowledgments

Thanks are due to all the authors for their well-written collaborations and patience during a long manuscript preparation process. Also, we are indebted to Shane Legg for his careful reviews and insightful suggestions.

References

1. Bryan Adams, Cynthia Breazeal, Rodney Brooks, and Brian Scassellati. Humanoid Robots: A New Kind of Tool. *IEEE Intelligent Systems*, 15(4):25–31, 2000.
2. Igor Aleksander and F. Keith Hanna. *Automata Theory: An Engineering Approach*. Edward Arnold, 1976.
3. J. R. Anderson, M. Matessa, and C. Lebiere. ACT-R: A Theory of Higher-Level Cognition and its Relation to Visual Attention. *Human Computer Interaction*, 12(4):439–462, 1997.
4. D. Angluin and C. H. Smith. Inductive Inference, Theory and Methods. *Computing Surveys*, 15(3):237–269, 1983.
5. Eric Baum and Igor Durdanovic. An Evolutionary Post Production System. 2002.
6. Eric Bonabeau, Marco Dorigo, and Guy Theraulaz. *Swarm Intelligence: From Natural to Artificial Systems*. Oxford University Press, 1999.
7. Christopher Brand. *The G-Factor: General Intelligence and its Implications*. John Wiley and Sons, 1996.
8. Gerhard Brewka, Jürgen Dix, and Kurt Konolige. *Nonmonotonic Reasoning: An Overview*. CSLI Press, 1995.
9. G. Di Caro and M. Dorigo. AntNet: A Mobile Agents Approach to Adaptive Routing. Technical Report Tech. Rep. IRIDIA/97-12, Université Libre de Bruxelles, 1997.
10. Haskell Curry and Robert Feys. *Combinatory Logic*. North-Holland, 1958.

11. M. Dorigo and L. M. Gambardella. Ant Colonies for the Traveling Salesman Problem. *BioSystems*, 43:73–81, 1997.
12. M. Dorigo and L. M. Gambardella. Ant Colony Systems: A Cooperative Learning Approach to the Traveling Salesman Problem. *IEEE Trans. Evol. Comp.*, 1:53–66, 1997.
13. W. Fontana and L. W. Buss. The Arrival of the Fittest: Toward a Theory of Biological Organization. *Bull. Math. Biol.*, 56:1–64, 1994.
14. W. Fontana and L. W. Buss. What would be conserved if 'the tape were played twice'? *Proc. Natl. Acad. Sci. USA*, 91:757–761, 1994.
15. Howard Gardner. *Intelligence Reframed: Multiple Intelligences for the 21st Century*. Basic Books, 2000.
16. Ben Goertzel. *The Evolving Mind*. Gordon and Breach, 1993.
17. Ben Goertzel. *The Structure of Intelligence*. Springer-Verlag, 1993.
18. Ben Goertzel. *Chaotic Logic*. Plenum Press, 1994.
19. Ben Goertzel. *From Complexity to Creativity*. Plenum Press, 1997.
20. Ben Goertzel. *Creating Internet Intelligence*. Plenum Press, 2001.
21. Ben Goertzel. *Novamente: Design for an Artificial General Intelligence*. 2005. In preparation.
22. Ben Goertzel, Cassio Pennachin, Andre Senna, Thiago Maia, and Guilherme Lamacie. Novamente: An Integrative Approach for Artificial General Intelligence. In *IJCAI-03 (International Joint Conference on Artificial Intelligence) Workshop on Agents and Cognitive Modeling*, 2003.
23. Ben Goertzel, Ken Silverman, Cate Hartley, Stephan Bugaj, and Mike Ross. The Baby Webmind Project. In *Proceedings of AISB 00*, 2000.
24. Steve Grand. *Creation: Life and How to Make It*. Harvard University Press, 2001.
25. Stephen Grossberg. *Neural Networks and Natural Intelligence*. MIT Press, 1992.
26. Stephen Grossberg. Linking Mind to Brain: The Mathematics of Biological Inference. *Notices of the American Mathematical Society*, 47:1361–1372, 2000.
27. Stephen Grossberg. The Complementary Brain: Unifying Brain Dynamics and Modularity. *Trends in Cognitive Science*, 4:233–246, 2000.
28. Stephen Grossberg. How does the Cerebral Cortex Work? Development, Learning, Attention and 3D Vision by Laminar Circuits of Visual Cortex. Technical Report CAS/CNS TR-2003-005, Boston University, 2003.
29. D. Heckerman, D. Geiger, and M. Chickering. Learning Bayesian Networks: the Combination of Knowledge and Statistical Data. Technical Report Tech. Rep. MSR-TR-94-09, Microsoft Research, 1994.
30. John Holland. A Mathematical Framework for Studying Learning in Classifier Systems. *Physica D*, 2, n. 1-3, 1986.
31. Bipin Indurkhya. *Metaphor and Cognition: An Interactionist Approach*. Kluwer Academic, 1992.
32. John Josephson and Susan Josephson. *Abductive Inference: Computation, Philosophy, Technology*. Cambridge University Press, 1994.
33. J. Kodjabachian and J. A. Meyer. Evolution and Development of Control Architectures in Animats. *Robotics and Autonomous Systems*, 16:2, 1996.
34. Ray Kurzweil. *The Age of Spiritual Machines*. Penguin Press, 2000.
35. Ray Kurzweil. *The Singularity is Near*. Viking Adult, 2005.
36. I. Kwee, M. Hutter, and J. Schmidhuber. Market-based reinforcement learning in partially observable worlds. *Proceedings of the International Conference on Artificial Neural Networks (ICANN-2001)*, (IDSIA-10-01, cs.AI/0105025), 2001.

37. J. E. Laird, A. Newell, and P. S. Rosenbloom. SOAR: An Architecture for General Intelligence. *Artificial Intelligence*, 33(1):1–6, 1987.
38. D. B. Lenat. Cyc: A Large-Scale Investment in Knowledge Infrastructure. *Communications of the ACM*, 38, no. 11, November 1995.
39. L. A. Levin. Laws of information conservation (non-growth) and aspects of the foundation of probability theory. *Problems of Information Transmission*, 10:206–210, 1974.
40. L. A. Levin. On a concrete method of assigning complexity measures. *DAN SSSR: Soviet Mathematics Doklady*, 17(2):727–731, 1977.
41. T. M. Martinetz and K. J. Schulten. *A "neural-gas" network learns topologies*, pages 397–402. North-Holland, 1991.
42. A. Newell and H. A. Simon. *GPS, a Program that Simulates Human Thought*, pages 109–124. 1961.
43. Judea Pearl. *Probabilistic Reasoning in Intelligent Systems: Networks of Plausible Inference*. Morgan-Kaufmann, 1988.
44. Roger Penrose. *Shadows of the Mind*. Oxford University Press, 1997.
45. Thomas S. Ray. An Approach to the Synthesis of Life. In *Artificial Life II: Santa Fe Studies in the Sciences of Complexity*, pages 371–408. Addison-Wesley, 1991.
46. Thomas S. Ray. A Proposal to Create a Network-Wide Biodiversity Reserve for Digital Organisms. Technical Report ATR Technical Report TR-H-133, ATR, 1995.
47. Thomas S. Ray and Joseph Hart. Evolution of Differentiated Multi-threaded Digital Organisms. In *Artificial Life VI Proceedings*. MIT Press, 1998.
48. Juergen Schmidhuber. Bias-Optimal Incremental Problem Solving. In *Advances in Neural Information Processing Systems - NIPS 15*. MIT Press, 2002.
49. John R. Searle. Minds, Brains, and Programs. *Behavioral and Brain Sciences*, 3:417–457, 1980.
50. E. H. Shortliffe and B. G. Buchanan. *A model of inexact reasoning in medicine*, pages 233–262. Addison-Wesley, 1984.
51. Ray Solomonoff. A Formal Theory of Inductive Inference, Part I. *Information and Control*, 7:2:1–22, 1964.
52. Ray Solomonoff. A Formal Theory of Inductive Inference, Part II. *Information and Control*, 7:2:224–254, 1964.
53. Robert Sternberg. *What Is Intelligence? Contemporary Viewpoints on its Nature and Definition*. Ablex Publishing, 1989.
54. Pei Wang. *Non-Axiomatic Reasoning System: Exploring the Essence of Intelligence*. PhD thesis, University of Indiana, 1995.
55. Paul Werbos. Advanced forecasting methods for global crisis warning and models of intelligence. *General Systems Yearbook*, 22:25–38, 1977.
56. Paul Werbos. Generalization of backpropagation with application to a recurrent gas market model. *Neural Networks*, 1, 1988.
57. Nels Winkless and Iben Browning. *Robots on Your Doorstep*. Robotics Press, 1978.
58. Eliezer Yudkowsky. General Intelligence and Seed AI. 2002.
59. Lotfi A. Zadeh. *Fuzzy Sets and Applications: Selected Papers by L. A. Zadeh*. John Wiley and Sons, 1987.
60. Lotfi A. Zadeh and Janusz Kacprzyk, editors. *Fuzzy Logic for the Management of Uncertainty*. John Wiley and Sons, 1992.

The Logic of Intelligence

Pei Wang

Department of Computer and Information Sciences, Temple University
Philadelphia, PA 19122, USA
pei.wang@temple.edu - http://www.cis.temple.edu/~pwang/

Summary. Is there an "essence of intelligence" that distinguishes intelligent systems from non-intelligent systems? If there is, then what is it? This chapter suggests an answer to these questions by introducing the ideas behind the NARS (Non-axiomatic Reasoning System) project. NARS is based on the opinion that the essence of intelligence is the ability to adapt with insufficient knowledge and resources. According to this belief, the author has designed a novel formal logic, and implemented it in a computer system. Such a "logic of intelligence" provides a unified explanation for many cognitive functions of the human mind, and is also concrete enough to guide the actual building of a general purpose "thinking machine".

1 Intelligence and Logic

1.1 To Define Intelligence

The debate on the essence of intelligence has been going on for decades, but there is still little sign of consensus (this book itself is evidence of this).

In "mainstream AI", the following are some representative opinions:

> "AI is concerned with methods of achieving goals in situations in which the information available has a certain complex character. The methods that have to be used are related to the problem presented by the situation and are similar whether the problem solver is human, a Martian, or a computer program." [19]

> Intelligence usually means "the ability to solve hard problems". [22]

> "By 'general intelligent action' we wish to indicate the same scope of intelligence as we see in human action: that in any real situation behavior appropriate to the ends of the system and adaptive to the demands of the environment can occur, within some limits of speed and complexity." [23]

Maybe it is too early to define intelligence. It is obvious that, after decades of study, we still do not know very much about it. There are more questions than answers. Any definition based on current knowledge is doomed to be revised by future work. We all know that a well-founded definition is usually the result, rather than the starting point, of scientific research. However, there are still reasons for us to be concerned about the definition of intelligence at the current time. Though clarifying the meaning of a concept always helps communication, this problem is especially important for AI. As a community, AI researchers need to justify their field as a scientific discipline. Without a (relatively) clear definition of intelligence, it is hard to say why AI is different from, for instance, computer science or psychology. Is there really something novel and special, or just fancy labels on old stuff? More vitally, every researcher in the field needs to justify his/her research plan according to such a definition. Anyone who wants to work on artificial intelligence is facing a two-phase task: to choose a working definition of intelligence, then to produce it in a computer.

A *working definition* is a definition concrete enough that you can directly work with it. By accepting a working definition of intelligence, it does not mean that you really believe that it fully captures the concept "intelligence", but that you will take it as a goal for your current research project.

Therefore, the lack of a consensus on what intelligence is does not prevent each researcher from picking up (consciously or not) a working definition of intelligence. Actually, unless you keep one (or more than one) definition, you cannot claim that you are working on artificial intelligence.

By accepting a working definition of intelligence, the most important commitments a researcher makes are on the acceptable assumptions and desired results, which bind all the concrete work that follows. The defects in the definition can hardly be compensated by the research, and improper definitions will make the research more difficult than necessary, or lead the study away from the original goal.

Before studying concrete working definitions of intelligence, we need to set up a general standard for what makes a definition better than others.

Carnap met the same problem when he tried to clarify the concept "probability". The task "consists in transforming a given more or less inexact concept into an exact one or, rather, in replacing the first by the second", where the first may belong to everyday language or to a previous stage in the scientific language, and the second must be given by explicit rules for its use [4].

According to Carnap, the second concept, or the *working definition* as it is called in this chapter, must fulfill the following requirements [4]:

1. It is *similar* to the concept to be defined, as the latter's vagueness permits.
2. It is defined in an *exact* form.
3. It is *fruitful* in the study.
4. It is *simple*, as the other requirements permit.

It seems that these requirements are also reasonable and suitable for our current purpose. Now let us see what they mean concretely to the working definition of intelligence:

Similarity (to standard usage). Though "intelligence" has no exact meaning in everyday language, it does have some common usages with which the working definition should agree. For instance, normal human beings are intelligent, but most animals and machines (including ordinary computer systems) are either not intelligent at all or much less intelligent than human beings.

Exactness (or well-definedness). Given the working definition, whether (or how much) a system is intelligent should be clearly decidable. For this reason, intelligence cannot be defined in terms of other ill-defined concepts, such as *mind, thinking, cognition, intentionality, rationality, wisdom, consciousness*, and so on, though these concepts do have close relationships with intelligence.

Fruitfulness (and instructiveness). The working definition should provide concrete guidelines for the research based on it – for instance, what assumptions can be accepted, what phenomena can be ignored, what properties are desired, and so on. Most importantly, the working definition of intelligence should contribute to the solving of fundamental problems in AI.

Simplicity. Although intelligence is surely a complex mechanism, the working definition should be simple. From a theoretical point of view, a simple definition makes it possible to explore a theory in detail; from a practical point of view, a simple definition is easy to use.

For our current purpose, there is no "right" or "wrong" working definition for intelligence, but there are "better" and "not-so-good" ones. When comparing proposed definitions, the four requirements may conflict with each other. For example, one definition is more fruitful, while another is simpler. In such a situation, some weighting and trade-off become necessary. However, there is no evidence showing that in general the requirements cannot be satisfied at the same time.

1.2 A Working Definition of Intelligence

Following the preparation of the previous section, we propose here a working definition of intelligence:

Intelligence is the capacity of a system to adapt to its environment while operating with insufficient knowledge and resources.

The *environment* of a system may be the physical world, or other information processing systems (human or computer). In either case, the interactions can be described by the *experiences* (or *stimuli*) and *responses* of the system, which are streams of input and output information, respectively. For the system, perceivable patterns of input and producible patterns of output constitute its *interface language*.

To *adapt* means that the system learns from its experiences. It adjusts its internal structure to approach its goals, as if future situations will be similar to past situations. Not all systems adapt to their environment. For instance, a traditional computing system gets all of its knowledge during its design phase. After that, its experience does not contribute to its behaviors. To acquire new knowledge, such a system would have to be redesigned.

Insufficient knowledge and resources means that the system works under the following restrictions:

Finite. The system has a constant information-processing capacity.
Real-time. All tasks have time requirements attached.
Open. No constraints are put on the knowledge and tasks that the system
    can accept, as long as they are representable in the interface language.

The two main components in the working definition, *adaptation* and *insufficient knowledge and resources*, are related to each other. An adaptive system must have some insufficiency in its knowledge and resources, for otherwise it would never need to change at all. On the other hand, without adaptation, a system may have insufficient knowledge and resources, but make no attempt to improve its capacities.

Not all systems take their own insufficiency of knowledge and resources into full consideration. Non-adaptive systems, for instance, simply ignore new knowledge in their interactions with their environment. As for artificial adaptive systems, most of them are not finite, real-time, and open, in the following senses:

1. Though all actual systems are finite, many theoretical models (for example, the Turing Machine) neglect the fact that the requirements for processor time and/or memory space may go beyond the supply capacity of the system.
2. Most current AI systems do not consider time constraints at run time. Most real-time systems can handle time constraints only if they are essentially deadlines [35].
3. Various constraints are imposed on what a system can experience. For example, only questions that can be answered by retrieval and deduction from current knowledge are acceptable, new knowledge cannot conflict with previous knowledge, and so on.

Many computer systems are designed under the assumption that their knowledge and resources, though *limited* or *bounded*, are still *sufficient* to fulfill the tasks that they will be called upon to handle. When facing a situation where this assumption fails, such a system simply panics or crashes, and asks for external intervention by a human user.

For a system to work under the assumption of insufficient knowledge and resources, it should have mechanisms to handle the following types of situation, among others:

- a new processor is required when all existent processors are occupied;
- extra memory is required when all available memory is already full;

- a task comes up when the system is busy with something else;
- a task comes up with a time requirement, so exhaustive search is not an option;
- new knowledge conflicts with previous knowledge;
- a question is presented for which no sure answer can be deduced from available knowledge.

For traditional computing systems, these types of situations usually require human intervention or else simply cause the system to refuse to accept the task or knowledge involved. However, for a system designed under the assumption of insufficient knowledge and resources, these are *normal situations*, and should be managed smoothly by the system itself. According to the above definition, intelligence is a "highly developed form of mental adaptation" [26].

When defining intelligence, many authors ignore the complementary question: what is unintelligent? If everything is intelligent, then this concept is empty. Even if we agree that intelligence, like almost all properties, is a matter of degree, we still need criteria to indicate what makes a system more intelligent than another. Furthermore, for AI to be an (independent) discipline, we require the concept "intelligence" to be different from other established concepts, because otherwise we are only talking about some well-known stuff with a new name, which is not enough to establish a *new branch of science*. For example, if every computer system is intelligent, it is better to stay within the theory of computation. Intuitively, "intelligent system" does not mean a faster and bigger computer. On the other hand, an unintelligent system is not necessarily incapable or gives only wrong results. Actually, most ordinary computer systems and many animals can do something that human beings cannot. However, these abilities do not earn the title "intelligent" for them. What is missing in these capable-but-unintelligent systems? According to the working definition of intelligence introduced previously, an *unintelligent* system is one that does not adapt to its environment. Especially, in artificial systems, an *unintelligent* system is one that is designed under the assumption that it only works on problems for which the system has sufficient knowledge and resources. An intelligent system is not always "better" than an unintelligent system for practical purposes. Actually, it is the contrary: when a problem can be solved by both of them, the unintelligent system is usually better, because it guarantees a correct solution. As Hofstadter said, for tasks like adding two numbers, a "reliable but mindless" system is better than an "intelligent but fallible" system [13].

1.3 Comparison With Other Definitions

Since it is impossible to compare the above definition to each of the existing working definitions of intelligence one by one, we will group them into several categories.

Generally speaking, research in artificial intelligence has two major motivations. As a field of science, we want to learn how the human mind, and "mind" in general, works; and as a branch of technology, we want to apply computers to domains where only the human mind works well currently. Intuitively, both goals can be achieved if we can build computer systems that are "similar to the human mind". But in what sense they are "similar"? To different people, the desired similarity may involve *structure, behavior, capacity, function,* or *principle.* In the following, we discuss typical opinions in each of the five categories, to see where these working definitions of intelligence will lead AI.

To Simulate the Human Brain

Intelligence is produced by the human brain, so maybe AI should attempt to simulate a brain in a computer system as faithfully as possible. Such an opinion is put in its extreme form by neuroscientists Reeke and Edelman, who argue that "the ultimate goals of AI and neuroscience are quite similar" [28].

Though it sounds reasonable to identify AI with *brain model,* few AI researchers take such an approach in a very strict sense. Even the "neural network" movement is "not focused on *neural modeling* (i.e., the modeling of neurons), but rather ... focused on *neurally inspired* modeling of cognitive processes" [30]. Why? One obvious reason is the daunting *complexity* of this approach. Current technology is still not powerful enough to simulate a huge neural network, not to mention the fact that there are still many mysteries about the brain. Moreover, even if we were able to build a brain model at the neuron level to any desired accuracy, it could not be called a success of AI, though it would be a success of neuroscience.

AI is more closely related to the concept "model of mind" – that is, a *high-level* description of brain activity in which biological concepts do not appear [32]. A high-level description is preferred, not because a low-level description is impossible, but because it is usually simpler and more general. A distinctive characteristic of AI is the attempt to "get a mind without a brain" – that is, to describe mind in a medium-independent way. This is true for all models: in building a model, we concentrate on certain properties of an object or process and ignore irrelevant aspects; in so doing, we gain insights that are hard to discern in the object or process itself. For this reason, an accurate duplication is not a model, and a model including unnecessary details is not a good model. If we agree that "brain" and "mind" are different concepts, then *a good model of brain is not a good model of mind,* though the former is useful for its own sake, and helpful for the building of the latter.

To Duplicate Human Behaviors

Given that we always judge the intelligence of other people by their behavior, it is natural to use "reproducing the behavior produced by the human mind as accurately as possible" as the aim of AI. Such a working definition of intelligence asks researchers to use the Turing Test [36] as a *sufficient and necessary* condition for having intelligence, and to take psychological evidence seriously.

Due to the nature of the Turing Test and the resource limitations of a concrete computer system, it is out of question for the system to have pre-stored in its memory all possible questions and proper answers in advance, and then to give a convincing imitation of a human being by searching such a list. The only realistic way to imitate human performance in a conversation is to produce the answers in real time. To do this, it needs not only cognitive faculties, but also much prior "human experience" [9]. Therefore, it must have a body that feels human, it must have all human motivations (including biological ones), and it must be treated by people as a human being – so it must simply be an "artificial human", rather than a computer system with artificial intelligence.

As French points out, by using behavior as evidence, the Turing Test is a criterion solely for *human* intelligence, not for intelligence in general [9]. Such an approach can lead to good psychological models, which are valuable for many reasons, but it suffers from "human chauvinism" [13] – we would have to say, according to the definition, that the science-fiction alien creature E. T. was not intelligent, because it would definitely fail the Turing Test.

Though "reproducing human (verbal) behavior" may still be a sufficient condition for being intelligent (as suggested by Turing), such a goal is difficult, if not impossible, to achieve. More importantly, it is not a necessary condition for being intelligent, if we want "intelligence" to be a more general concept than "human intelligence".

To Solve Hard Problems

In everyday language, "intelligent" is usually applied to people who can solve hard problems. According to this type of definition, intelligence is the *capacity* to solve hard problems, and *how* the problems are solved is not very important.

What problems are "hard"? In the early days of AI, many researchers worked on intellectual activities like game playing and theorem proving. Nowadays, expert-system builders aim at "real-world problems" that crop up in various domains. The presumption behind this approach is: "Obviously, experts are intelligent, so if a computer system can solve problems that only experts can solve, the computer system must be intelligent, too". This is why many people take the success of the chess-playing computer Deep Blue as a success of AI.

This movement has drawn in many researchers, produced many practically useful systems, attracted significant funding, and thus has made important contributions to the development of the AI enterprise. Usually, the systems are developed by analyzing domain knowledge and expert strategy, then building them into a computer system. However, though often profitable, these systems do not provide much insight into how the mind works. No wonder people ask, after learning how such a system works, "Where's the AI?" [31] – these systems look just like ordinary computer application systems, and still suffer from great rigidity and brittleness (something AI wants to avoid).

If intelligence is defined as "the capacity to solve hard problems", then the next question is: "Hard for whom?" If we say "hard for human beings", then most existing computer software is already intelligent – no human can manage a database as well as a database management system, or substitute a word in a file as fast as an editing program. If we say "hard for computers," then AI becomes "whatever hasn't been done yet," which has been dubbed "Tesler's Theorem" [13]. The view that AI is a "perpetually extending frontier" makes it attractive and exciting, which it deserves, but tells us little about how it differs from other research areas in computer science – is it fair to say that the problems there are easy? If AI researchers cannot identify other commonalities of the problems they attack besides mere difficulty, they will be unlikely to make any progress in understanding and replicating intelligence.

To Carry out Cognitive Functions

According to this view, intelligence is characterized by a set of cognitive functions, such as reasoning, perception, memory, problem solving, language use, and so on. Researchers who subscribe to this view usually concentrate on just one of these functions, relying on the idea that research on all the functions will eventually be able to be combined, in the future, to yield a complete picture of intelligence. A "cognitive function" is often defined in a general and abstract manner. This approach has produced, and will continue to produce, tools in the form of software packages and even specialized hardware, each of which can carry out a function that is similar to certain mental skills of human beings, and therefore can be used in various domains for practical purposes. However, this kind of success does not justify claiming that it is the proper way to study AI. To define intelligence as a "toolbox of functions" has serious weaknesses.

When specified in isolation, an implemented function is often quite different from its "natural form" in the human mind. For example, to study analogy without perception leads to distorted cognitive models [5]. Even if we can produce the desired tools, this does not mean that we can easily combine them, because different tools may be developed under different assumptions, which prevents the tools from being combined.

The basic problem with the "toolbox" approach is: without a "big picture" in mind, the study of a cognitive function in an isolated, abstracted, and often distorted form simply does not contribute to our understanding of intelligence.

A common counterargument runs something like this: "Intelligence is very complex, so we have to start from a single function to make the study tractable." For many systems with weak internal connections, this is often a good choice, but for a system like the mind, whose complexity comes directly from its tangled internal interactions, the situation may be just the opposite. When the so-called "functions" are actually phenomena produced by a complex-but-unified mechanism, reproducing all of them together (by duplicating the mechanism) is simpler than reproducing only one of them.

To Develop New Principles

According to this type of opinions, what distinguishes intelligent systems and unintelligent systems are their *postulations*, applicable *environments*, and basic *principles* of information processing.

The working definition of intelligence introduced earlier belongs to this category. As a system adapting to its environment with insufficient knowledge and resources, an intelligent system should have many cognitive *functions*, but they are better thought of as emergent phenomena than as well-defined tools used by the system. By learning from its experience, the system potentially can acquire the *capacity* to solve hard problems – actually, *hard problems* are those for which a solver (human or computer) has insufficient knowledge and resources – but it has no such built-in capacity, and thus, without proper training, no capacity is guaranteed, and acquired capacities can even be lost. Because the human mind also follows the above principles, we would hope that such a system would behave similarly to human beings, but the similarity would exist at a more abstract level than that of concrete *behaviors*. Due to the fundamental difference between human experience/hardware and computer experience/hardware, the system is not expected to accurately reproduce masses of psychological data or to pass a Turing Test. Finally, although the internal *structure* of the system has some properties in common with a description of the human mind at the subsymbolic level, it is not an attempt to simulate a biological neural network.

In summary, the *structure* approach contributes to neuroscience by building brain models, the *behavior* approach contributes to psychology by providing explanations of human behavior, the *capacity* approach contributes to application domains by solving practical problems, and the *function* approach contributes to computer science by producing new software and hardware for various computing tasks. Though all of these are valuable for various reasons, and helpful in the quest after AI, these approaches do not, in my opinion, concentrate on the *essence* of intelligence.

To be sure, what has been proposed in my definition of intelligence is not entirely new to the AI community. Few would dispute the proposition that adaptation, or learning, is essential for intelligence. Moreover, "insufficient knowledge and resources" is the focus of many subfields of AI, such as heuristic search, reasoning under uncertainty, real-time planning, and machine learning. Given this situation, what is *new* in this approach? It is the following set of principles:

1. an explicit and unambiguous definition of intelligence as "adaptation under insufficient knowledge and resources";
2. a further definition of the phrase "with insufficient knowledge and resources" as *finite*, *real-time*, and *open*;

3. the design of all formal and computational aspects of the project keeping the two previous definitions foremost in mind.

1.4 Logic and Reasoning Systems

To make our discussion more concrete and fruitful, let us apply the above working definition of intelligence to a special type of information processing system – reasoning system.

A reasoning system usually has the following components:

1. *a formal language* for knowledge representation, as well as communication between the system and its environment;
2. *a semantics* that determines the meanings of the words and the truth values of the sentences in the language;
3. *a set of inference rules* that match questions with knowledge, infer conclusions from promises, and so on;
4. *a memory* that systematically stores both questions and knowledge, and provides a working place for inferences;
5. *a control mechanism* that is responsible for choosing premises and inference rules for each step of inference.

The first three components are usually referred to as a *logic*, or the *logical part* of the reasoning system, and the last two as the *control part* of the system.

According to the previous definition, being a reasoning system is neither necessary nor sufficient for being intelligent. However, an *intelligent reasoning system* does provide a suitable framework for the study of intelligence, for the following reasons:

- It is a general-purpose system. Working in such a framework keeps us from being bothered by domain-specific properties, and also prevents us from cheating by using domain-specific tricks.
- Compared with cognitive activities like low-level perception and motor control, reasoning is at a more abstract level, and is one of the cognitive skills that collectively make human beings so qualitatively different from other animals.
- The framework of reasoning system is highly flexible and expendable. We will see that we can carry out many other cognitive activities in it when the concept of "reasoning" is properly extended.
- Most research on reasoning systems is carried out within a paradigm based on assumptions directly opposed to the one presented above. By "fighting in the backyard of the rival", we can see more clearly what kinds of effects the new ideas have.

Before showing how an intelligent reasoning system is designed, let us first see its opposite – that is, a reasoning system designed under the assumption that its knowledge and resources are *sufficient* to answer the questions asked by its environment (so no adaptation is needed). By definition, such a system has the following properties:

1. No new knowledge is necessary. All the system needs to know to answer the questions is already there at the very beginning, expressed by a set of axioms.
2. The axioms are *true*, and will remain true, in the sense that they correspond to the actual situation of the environment.
3. The system answers questions by applying a set of formal rules to the axioms. The rules are sound and complete (with respect to the valid questions), therefore they guarantee correct answers for all questions.
4. The memory of the system is so big that all axioms and intermediate results can always be contained within it.
5. There is an algorithm that can carry out any required inference in finite time, and it runs so fast that it can satisfy all time requirements that may be attached to the questions.

This is the type of system dreamed of by Leibniz, Boole, Hilbert, and many others. It is usually referred to as a "decidable axiomatic system" or a "formal system". The attempt to build such systems has dominated the study of logic for a century, and has strongly influenced the research of artificial intelligence. Many researchers believe that such a system can serve as a model of human thinking.

However, if intelligence is defined as "to adapt under insufficient knowledge and resources", what we want is the *contrary*, in some sense, to an axiomatic system, though it is still *formalized* or *symbolized* in a technical sense. Therefore *Non-axiomatic Reasoning System* (NARS) is chosen as the name for the intelligent reasoning system to be introduced in the following sections.

Between "pure-axiomatic" systems and "non-axiomatic" ones, there are also "semi-axiomatic" systems. They are designed under the assumption that knowledge and resources are insufficient in some, but not all, aspects. Consequently, adaptation is necessary. Most current reasoning systems developed for AI fall into this category. According to our working definition of intelligence, pure-axiomatic systems are not intelligent at all, non-axiomatic systems are intelligent, and semi-axiomatic systems are intelligent in certain aspects.

Pure-axiomatic systems are very useful in mathematics, where the aim is to idealize knowledge and questions to such an extent that the revision of knowledge and the deadlines of questions can be ignored. In such situations, questions can be answered so accurately and reliably that the procedure can be reproduced by a Turing Machine. We need intelligence only when no such pure-axiomatic method can be used, due to the insufficiency of knowledge and resources. For the same reason, the performance of a non-axiomatic system is not necessarily better than that of a semi-axiomatic system, but it can work in environments where the latter cannot be used.

Under the above definitions, intelligence is still (as we hope) a matter of degree. Not all systems in the "non-axiomatic" and "semi-axiomatic" categories are equally intelligent. Some systems may be more intelligent than some other systems due to a higher resources efficiency, using knowledge in more ways, communicating with the environment in a richer language, adapting more rapidly and thoroughly, and so on.

"Non-axiomatic" does not mean "everything changes". In NARS, nothing is fixed as far as the *content* of knowledge is concerned, but as we will see in the following sections, how the changes happen is fixed, according to the inference rules and control strategy of the system, which remain constant when the system is running. This fact does not make NARS "semi-axiomatic", because the fixed part is not in the "object language" level, but in the "meta-language" level. In a sense, we can say that the "meta-level" of NARS is not non-axiomatic, but pure-axiomatic. For a reasoning system, a fixed inference rule is not the same as an axiom.

Obviously, we can allow the "meta-level" of NARS to be non-axiomatic, too, and therefore give the system more flexibility in its adaptation. However, that approach is not adopted in NARS at the current stage, for the following reasons:

- "Complete self-modifying" is an illusion. As Hofstadter put it, "below every tangled hierarchy lies an inviolate level" [13]. If we allow NARS to modify its meta-level knowledge, i.e., its inference rules and control strategy, we need to give it (fixed) meta-meta-level knowledge to specify how the modification happens. As flexible as the human mind is, it cannot modify its own "law of thought".
- Though high-level self-modifying will give the system more flexibility, it does not necessarily make the system more intelligent. Self-modifying at the meta-level is often dangerous, and it should be used only when the same effect cannot be produced in the object-level. To assume "the more radical the changes can be, the more intelligent the system will be" is unfounded. It is easy to allow a system to modify its own source code, but hard to do it right.
- In the future, we will explore the possibility of meta-level learning in NARS, but will not attempt to do so until the object-level learning is mature. To try everything at the same time is just not a good engineering approach, and this does not make NARS less non-axiomatic, according to the above definition.

Many arguments proposed previously against logical AI [2, 20], symbolic AI [7], or AI as a whole [32, 25], are actually against a more specific target: pure-axiomatic systems. These arguments are powerful in revealing that many aspects of intelligence cannot be produced by a pure-axiomatic system (though these authors do not use this term), but some of them are misleading by using such a system as the prototype of AI research. By working on a reasoning system, with its formal language and inference rules, we do not necessarily bind ourselves with the commitments accepted by the traditional "logical AI" paradigms. As we will see in the following, NARS shares more philosophical opinions with the *subsymbolic*, or *connectionist* movement [15, 17, 30, 34].

What is the relationship of artificial intelligence and computer science? What is the position of AI in the whole science enterprise? Traditionally, AI is referred to as a branch of computer science. According to our previous definitions, AI can be implemented with the tools provided by computer science, but from a *theoretical* point of view, they make opposite assumptions: computer science focuses on pure-axiomatic systems, while AI focuses on non-axiomatic systems. The fundamental assumptions of computer science can be found in mathematical logic (especially first-order predicate logic) and computability theory (especially Turing Machine). These theories take the sufficiency of knowledge and resources as implicit postulates, therefore adaptation, plausible inference, and tentative solutions of problems are neither necessary nor possible.

Similar assumptions are often accepted by AI researchers with the following justification: "We know that the human mind usually works with insufficient knowledge and resources, but if you want to set up a *formal* model and then a *computer* system, you must somehow *idealize* the situation." It is true that every formal model is an idealization, and so is NARS. The problem is what to omit and what to preserve in the idealization. In the current implementation of NARS, many factors that should influence reasoning are ignored, but the insufficiency of knowledge and resources is strictly assumed throughout. Why? Because it is a *definitive* feature of intelligence, so if it were lost through the "idealization", the resulting study would be about something else.

2 The Components of NARS

Non-axiomatic Reasoning System (NARS) is designed to be an intelligent reasoning system, according to the working definition of intelligence introduced previously.

In the following, let us see how the major components of NARS (its formal language, semantics, inference rules, memory, and control mechanism) are determined, or strongly suggested, by the definition, and how they differ from the components of an axiomatic system. Because this chapter is concentrated in the philosophical and methodological foundation of the NARS project, formal descriptions and detailed discussions for the components are left to other papers [39, 40, 42].

2.1 Experience-Grounded Semantics

Axiomatic reasoning systems (and most semi-axiomatic systems) use "model-theoretic semantics". Informally speaking, a model is a description of a domain, with relations among objects specified. For a reasoning system working on the domain, an "interpretation" maps the terms in the system to the objects in the model, and the predicates in the systems to the relations in the model. For a given term, its *meaning* is its image in the model under the interpretation. For a given proposition, its *truth value* depends on whether it corresponds to a fact in the model. With such a semantics, the reasoning system gets a constant "reference", the model, according to which truth and meaning within the system is determined. Though model-theoretic semantics comes in different forms, and has variations, this "big picture" remains unchanged.

This kind of semantics is not suitable for NARS. As an adaptive system with insufficient knowledge and resources, the system cannot judge the truthfulness of its knowledge against a static, consistent, and complete model. Instead, truth and meaning have to be grounded on the system's *experience* [40]. Though a section of experience is also a description of the system's environment, it is fundamentally different from a model, since experience changes over time, is never complete, and is often inconsistent. Furthermore, experience is directly accessible to the system, while model is often "in the eye of an observer".

According to an experience-grounded semantics, truth value becomes a function of the amount of available evidence, therefore inevitably becomes changeable and subjective, though not arbitrary. In such a system, no knowledge is "true" in the sense that it is guaranteed to be confirmed by future experience. Instead, the truth value of a statement indicates the degree to which the statement is confirmed by past experience. The system will use such knowledge to predict the future, because it is exactly what "adaptive", and therefore "intelligent", means. In this way, "truth" has quite different (though closely related) meanings in non-axiomatic systems and axiomatic systems.

Similarly, the *meaning* of a term, that is, what makes the term different from other terms to the system, is determined by its relationships to other terms, according to the system's experience, rather than by an interpretation that maps it into an object in a model.

With insufficient resources, the truth value of each statement and the meaning of each term in NARS is usually grounded on part of the experience. As a result, even without new experience, the inference activity of the system will change the truth values and meanings, by taking previously available-but-ignored experience into consideration. On the contrary, according to model-theoretic semantics, the internal activities of a system have no effects on truth value and meaning of the language it uses.

"Without an interpretation, a system has no access to the semantics of a formal language it uses" is the central argument in Searle's "Chinese room" thought experiment against strong AI [32]. His argument is valid for model-theoretic semantics, but not for experience-grounded semantics. For an intelligent reasoning system, the latter is more appropriate.

2.2 Inheritance Statement

As discussed above, "adaptation with insufficient knowledge and resources" demands an experience-grounded semantics, which in turn requires a formal knowledge representation language in which *evidence* can be naturally defined and measured.

For a non-axiomatic reasoning system, it is obvious that a binary truth value is not enough. With past experience as the only guidance, the system not only needs to know whether there is counter example (negative evidence), but also needs to know its amount, with respect to the amount of positive evidence. To have a domain-independent method to compare competing answers, a numerical truth value, or a *measurement of uncertainty*, becomes necessary for NARS, which quantitatively records the relationship between a statement and available evidence. Furthermore, "positive evidence" and "irrelevant stuff" need to be distinguished too.

Intuitively speaking, the simplest case to define evidence is for a general statement about many cases, while some of them are confirmed by past experience (positive evidence), and some others are disconfirmed by past experience (negative evidence). Unfortunately, the most popular formal language for knowledge representation, first-order predicate calculus, cannot be easily used in this way. In this language, a "general statement", such as "Ravens are black", is represented as a "universal proposition", such as "$(\forall x)(Raven(x) \rightarrow Black(x))$". In the original form of first-order predicate calculus, there is no such a notion as "evidence", and the proposition is either true or false, depending on whether there is such an object x in the domain that makes $Raven(x)$ true and $Black(x)$ false. It is natural to define constants that make the proposition true as its positive evidence, and the constants that make it false its negative evidence. However, such a naive solution has serious problems [40, 44]:

- Only the existence of negative evidence contributes to the truth value of the universal proposition, while whether there is "positive evidence" does not matter. This is the origin Popper's refutation theory [27].
- Every constant is either a piece of positive evidence or a piece of negative evidence, and nothing is irrelevant. This is related to Hempel's conformation paradox [11].

Though evidence is hard to define in predicate calculus, it is easy to do in a properly designed *categorical logic*. Categorical logics, or term logics, is another family of formal logic, exemplified by Aristotle's Syllogism [1]. The major formal features that distinguish it from predicate logic are the use of subject–predicate statements and syllogistic inference rules. Let us start with the first feature.

NARS uses a categorical language that is based on an *inheritance* relation, "$\rightarrow$". The relation, in its ideal form, is a reflexive and transitive binary relation defined on *terms*, where a term can be thought as the name of a concept. For example, "*raven* → *bird*" is an inheritance statement with "*raven*" as *subject term* and "*bird*" as *predicate term*. Intuitively, it says that the subject is a *specialization* of the predicate, and the predicate is a *generalization* of the subject. The statement roughly corresponds to the English sentence "Raven is a kind of bird". Based on the inheritance relation, the *extension* and *intension* of a term are defined as the set of its specializations and generalizations, respectively. That is, for a given term T, its extension T^E is the set $\{x \mid x \to T\}$, and its intension T^I is the set $\{x \mid T \to x\}$. Given the reflexivity and transitivity of the inheritance relation, it can be proved that for any terms S and P, "$S \to P$" is true if and only if S^E is included in P^E, and P^I is included in S^I. In other words, "There is an inheritance relation from S to P" is equivalent to "P inherits the extension of S, and S inherits the intension of P".

When considering "imperfect" inheritance statements, the above theorem naturally gives us the definition of (positive and negative) evidence. For a given statement "$S \to P$", if a term M in both S^E and P^E, or in both P^I and S^I, then it is a piece of positive evidence for the statement, because as far as M is concerned, the proposed inheritance is true; if M in S^E but not in P^E, or in P^I but not in S^I, then it is a piece of negative evidence for the statement, because as far as M is concerned, the proposed inheritance is false; if M is neither in S^E nor in P^I, it is not evidence for the statement, and whether it is also in P^E or S^I does not matter. Let us use w^+, w^-, and w for the amount of positive, negative, and total evidence, respectively, then we have $w^+ = |S^E \cap P^E| + |P^I \cap S^I|$, $w^- = |S^E - P^E| + |P^I - S^I|$, $w = w^+ + w^- = |S^E| + |P^I|$. Finally, we define the truth value of a statement to be a pair of numbers $<f, c>$. Here f is called the *frequency* of the statement, and $f = w^+/w$. The second component c is called the *confidence* of the statement, and $c = w/(w+k)$, where k is a system parameter with 1 as the default value. For a more detailed discussion, see [43].

Now we have the technical basics of the experience-grounded semantics: If the experience of the system is a set of inheritance statements defined above, then for any term T, we can determine its *meaning*, which is its extension and intension (according to the experience), and for any inheritance statement "$S \to P$", we can determine its positive evidence and negative evidence (by comparing the meaning of the two terms), then calculate its truth value according to the above definition.

Of course, the actual experience of NARS is not a set of binary inheritance statements, nor does the system determine the truth value of a statement in the above way. The actual experience of NARS is a stream of statements, with their truth values represented by the $<f, c>$ pairs. Within the system, new statements are derived by the inference rules, with truth-value functions calculating the truth values of the conclusions from those of the premises. The purpose of the above definitions is to *define* the truth value in an idealized situation, and to provide a foundation for the truth value functions (to be discussed in the following).

2.3 Categorical Language

Based on the inheritance relation introduced above, NARS uses a powerful "categorical language", obtained by extending the above core language in various directions:

Derived inheritance relations: Beside the *inheritance* relation defined previously, NARS also includes several of its variants. For example,
- the *similarity* relation $\leftrightarrow$ is symmetric inheritance;
- the *instance* relation $\circ\!\!\rightarrow$ is an inheritance relation where the subject term is treated as an atomic instance of the predicate term;
- the *property* relation $\rightarrow\!\!\circ$ is an inheritance relation where the predicate term is treated as a primitive property of the subject term.

Compound terms: In inheritance statements, the (subject and predicate) terms not only can be simple names (as in the above examples), but also can be compound terms formed by other terms with logical operator. For example, if A and B are terms, we have
- their *extensional intersection* $(A \cap B)$ is a compound term, defined by $(A \cap B)^E = (A^E \cap B^E)$ and $(A \cap B)^I = (A^I \cup B^I)$.
- their *intensional intersection* $(A \cup B)$ is a compound term, defined by $(A \cup B)^E = (A^E \cup B^E)$ and $(A \cup B)^I = (A^I \cap B^I)$;

With compound terms, the expressive power of the language is greatly extended.

Ordinary relation: In NARS, only the inheritance relation and its variants are defined as logic constants that are directly recognized by the inference rules. All other relations are converted into inheritance relations with compound terms. For example, an arbitrary relation R among three terms A, B, and C is usually written as $R(A, B, C)$, which can be equivalently rewritten as one of the following inheritance statements (i.e., they have the same meaning and truth value):
- "$(A, B, C) \rightarrow R$", where the subject term is a compound (A, B, C), an ordered tuple. This statement says "The relation among A, B, C (in that order) is an instance of the relation R."
- "$A \rightarrow R(*, B, C)$", where the predicate term is a compound $R(*, B, C)$ with a "wild-card", $*$. This statement says "A is such an x that satisfies $R(x, B, C)$."
- "$B \rightarrow R(A, *, C)$". Similarly, "B is such an x that satisfies $R(A, x, C)$."
- "$C \rightarrow R(A, B, *)$". Again, "C is such an x that satisfies $R(A, B, x)$."

Higher-order term: In NARS, a statement can be used as a term, which is called a "higher-order" term. For example, "Bird is a kind of animal" is represented by statement "*bird* $\rightarrow$ *animal*", and "Tom knows that bird is a kind of animal" is represented by statement "(*bird* $\rightarrow$ *animal*)$\circ\!\!\rightarrow$ *know*(*Tom*, *)", where the subject term is a statement. Compound higher-order terms are also defined: if A and B are higher-order terms, so do their negations ($\neg A$ and $\neg B$), disjunction ($A \vee B$), and conjunction ($A \wedge B$).

Higher-order relation: Higher-order relations are those whose subject term and predicate term are both higher-order terms. In NARS, there are two defined as logic constants:

- *implication*, "$\Rightarrow$", which is intuitively correspond to "if–then";
- *equivalence*, "$\Leftrightarrow$", which is intuitively correspond to "if and only if".

Non-declarative sentences: Beside the various types of statements introduced above, which represent the system's declarative knowledge, the formal language of NARS uses similar formats to represent non-declarative sentences:

- a *question* is either a statement whose truth value needs to be evaluated ("yes/no" questions), or a statement containing variables to be instantiated ("what" questions);
- a *goal* is a statement whose truthfulness needs to be established by the system through the execution of relevant operations.

For each type of statements, its truth value is defined similarly to how we define the truth value of an inheritance statement.

With the above structures, the expressive power of the language is roughly the same as a typical natural language (such as English or Chinese). There is no one-to-one mapping between sentences in this language and those in first-order predicate calculus, though approximate mapping is possible for many sentences. While first-order predicate calculus may still be better to represent mathematical knowledge, this new language will be better to represent empirical knowledge.

2.4 Syllogistic Inference Rules

Due to insufficient knowledge, the system needs to do non-deductive inference, such as induction, abduction, and analogy, to extend past experience to novel situations. In this context, deduction becomes fallible, too, in the sense that its conclusion may be revised by new knowledge, even if the premises remain unchallenged. According to the experience-grounded semantics, the definition of *validity* of inference rules is changed. Instead of generating infallible conclusions, a valid rule should generate conclusions whose truth values are evaluated against (and only against) the evidence provided by the premises.

As mentioned previously, a main feature that distinguish term logics from predicate/propositional logics is the use of *syllogistic* inference rules, each of which takes a pair of premises that share a common term. For inference among inheritance statements, there are three possible combinations if the two premises share exactly one term:

deduction	induction	abduction
$M \to P <f_1, c_1>$	$M \to P <f_1, c_1>$	$P \to M <f_1, c_1>$
$S \to M <f_2, c_2>$	$M \to S <f_2, c_2>$	$S \to M <f_2, c_2>$
$S \to P <f, c>$	$S \to P <f, c>$	$S \to P <f, c>$

Each inference rule has its own truth-value function to calculate the truth value of the conclusion according to those of the premises. In NARS, these functions are designed in the following way:

1. Treat all relevant variables as binary variables taking 0 or 1 values, and determine what values the conclusion should have for each combination of premises, according to the semantics.
2. Represent the truth values of the conclusion obtained above as Boolean functions of those of the premises.
3. Extend the Boolean operators into real number functions defined on $[0, 1]$ in the following way:

$$not(x) = 1 - x$$
$$and(x_1, ..., x_n) = x_1 * ... * x_n$$
$$or(x_1, ..., x_n) = 1 - (1 - x_1) * ... * (1 - x_n)$$

4. Use the extended operators, plus the relationship between truth value and amount of evidence, to rewrite the above functions.

The result is the following:

deduction   **induction**   **abduction**

$f = f_1 f_2$ $f = f_1$ $f = f_2$
$c = c_1 c_2 f_1 f_2$ $c = f_2 c_1 c_2/(f_2 c_1 c_2 + 1)$ $c = f_1 c_1 c_2/(f_1 c_1 c_2 + 1)$

When the two premises have the same statement, but comes from different sections of the experience, the revision rule is applied to merge the two into a summarized conclusion:

revision

$$S \rightarrow P <f_1, c_1>$$
$$S \rightarrow P <f_2, c_2>$$
$$\overline{\qquad\qquad\qquad\qquad}$$
$$S \rightarrow P <f, c>$$

Since in revision the evidence for the conclusion is the sum of the evidence in the premises, the truth-value function is

$$f = \frac{f_1 c_1/(1-c_1) + f_2 c_2/(1-c_2)}{c_1/(1-c_1) + c_2/(1-c_2)}$$

$$c = \frac{c_1/(1-c_1) + c_2/(1-c_2)}{c_1/(1-c_1) + c_2/(1-c_2) + 1}.$$

Beside the above four basic inference rules, in NARS there are inference rules for the variations of inheritance, as well as for the formation and transformation of the various compound terms. The truth-value functions for those rules are similarly determined.

Beside the above *forward* inference rules by which new knowledge is derived existing knowledge, NARS also has *backward* inference rules, by which a piece of knowledge is applied to a question or a goal. If the knowledge happens to provide an answer for the question or an operation to realize the goal, it is accepted as a tentative solution, otherwise a derived question or goal may be generated, whose solution, combined with the knowledge, will provide a solution to the original question or goal. Defined in this way, for each forward rule, there is a matching backward rule. Or, conceptually, we can see them as two ways to use the same rule.

2.5 Controlled Concurrency in Dynamic Memory

As an open system working in real time, NARS accepts new tasks all the time. A new task may be a piece of knowledge to be digested, a question to be answered, or a goal to be achieved. A new task may come from a human user or from another computer system.

Since in NARS no knowledge is absolutely true, the system will try to use as much knowledge as possible to process a task, so as to provide a better (more confident) solution. On the other hand, due to insufficient resources, the system cannot use all relevant knowledge for each task. Since new tasks come from time to time, and the system generates derived tasks constantly, at any moment the system typically has a large amount of tasks to process. For this situation, it is too rigid to set up a static standard for a satisfying solution [35], because no matter how careful the standard is determined, sometimes it will be too high, and sometimes too low, given the ever changing resources demand of the existing tasks. What NARS does is to try to find the best solution given the current knowledge and resources restriction [40] — similar to what an "anytime algorithm" does [6].

A "Bag" is a data structure specially designed in NARS for resource allocation. A bag can contain certain type of items with a constant capacity, and maintains a priority distribution among the items. There are three major operations defined on bag:

- Put an item into the bag, and if the bag is already full, remove an item with the lowest priority.
- Take an item out of the bag by key (i.e., its unique identifier).
- Take an item out of the bag by priority, that is, the probability for an item to be selected is proportional to its priority value.

Each of the operations takes a constant time to finish, independent of the number of items in the bag.

NARS organizes knowledge and tasks into *concepts*. In the system, a term T has a corresponding concept C_T, which contains all the knowledge and tasks in which T is the subject term or predicate term. For example, knowledge "*bird* → *animal* $<1, 0.9>$" is stored within the concept C_{bird} and the concept

C_{animal}. In this way, the memory of NARS can be seen roughly as a *bag* of concepts, and each concept is named by a (simple or compound) term, and contains a *bag* of knowledge and a *bag* of tasks, all of them are directly about the term.

NARS runs by repeatedly carrying out the following working cycle:

1. Take a concept from the memory by priority.
2. Take a task from the task bag of the concept by priority.
3. Take a piece of knowledge from the knowledge bag of the concept by priority.
4. According to the combination of the task and the knowledge, call the applicable inference rules on them to derive new tasks and new knowledge — in a term logic, every inference step happens within a concept.
5. Adjust the priority of the involved task, knowledge, and concept, according to how they behave in this inference step, then put them back into the corresponding bags.
6. Put the new (input or derived) tasks and knowledge into the corresponding bags. If certain new knowledge provides the best solution so far for a user-assigned task, report a solution.

The priority value of each item reflects the amount of resources the system plans to spend on it in the near future. It has two factors:

Long-term factor. The system gives higher priority to more *important* items, evaluated according to past experience. Initially, the user can assign priority values to the input tasks to indicate their relative importance, which will in turn determine the priority value of the concepts and knowledge generated from it. After each inference step, the involved items have their priority values adjusted. For example, if a piece of knowledge provides a best-so-far solution for a task, then the priority value of the knowledge is increased (so that it will be used more often in the future), and the priority value of the task is decreased (so that less time will be used on it in the future).

Short-term factor. The system gives higher priority to more *relevant* items, evaluated according to current context. When a new task is added into the system, the directly related concepts are *activated*, i.e., their priority values are increased. On the other hand, the priority values decay over time, so that if a concept has not been relevant for a while, it becomes less active.

In this way, NARS processes many tasks in parallel, but with different speeds. This "controlled concurrency" control mechanism is similar to Hofstadter's "parallel terraced scan" strategy [14]. Also, how a task is processed depends on the available knowledge and the priority distribution among concepts, tasks, and knowledge. Since these factors change constantly, the solution a task gets is context-dependent.

3 The Properties of NARS

As a project aimed at general-purpose artificial intelligence, NARS addresses many issues in AI and cognitive science. Though it is similar to many other approaches here or there, the project as a whole is unique in its theoretical foundation and major technical components. Designed as above, NARS shows many properties that make it more similar to human reasoning than other AI systems are.

3.1 Reasonable Solutions

With insufficient knowledge and resources, NARS cannot guarantee that all the solutions it generates for tasks are *correct* in the sense that they will not be challenged by the system's future experience. Nor can it guarantee that the solutions are *optimum* given all the knowledge the system has at the moment. However, the solutions are *reasonable* in the sense that they are the best summaries of the past experience, given the current resources supply. This is similar to Good's "Type II rationality" [10].

NARS often makes "reasonable mistakes" that are caused by the insufficiency of knowledge and resources. They are reasonable and inevitable given the working condition of the system, and they are not caused by the errors in the design or function of the system.

A conventional algorithm provides a single solution to each problem, then stops working on the problem. On the contrary, NARS may provide no, one, or more than one solution to a task — it reports every solution that is the best it finds, then looks for a better one (if resources are still available). Of course, eventually the system will end its processing of the task, but the reason is neither that a satisfying solution has been found, nor that a deadline is reached, but that the task has lost in the resources competition.

Like trial-and-error procedures [18], NARS can "change its mind". Because truth values are determined according to experience, a later solution is judged as "better" than a previous one, because it is based on more evidence, though it is not necessarily "closer to the objective fact".

When a solution is found, usually there is no way to decide whether it is the last the system can get. In NARS, there is no "final solution" that cannot be updated by new knowledge and/or further consideration, because all solutions are based on partial experience of the system. This *self-revisable* feature makes NARS a more general model than the various *non-monotonic logics*, in which only binary statements are processed, and only the conclusions derived from default rules can be updated, but the default rules themselves are not effected by the experience of the system [29].

3.2 Unified Uncertainty Processing

As described previously, in NARS there are various types of uncertainty, in concepts, statements, inference rules, and inference processes. NARS has a unified uncertainty measurement and calculation sub-system.

What makes this approach different from other proposed theories on uncertainty is the experience-grounded semantics. According to it, all uncertainty comes from the insufficiency of knowledge and resources. As a result, the evaluation of uncertainty is changeable and context-dependent.

From our previous definition of truth value, it is easy to recognize its relationship with probability theory. Under a certain interpretation, the frequency measurement is similar to probability, and the confidence measurement is related to the size of sample space. If this is the case, why not directly use probability theory to handle uncertainty?

Let us see a concrete case. The deduction rule takes "$M \rightarrow P <f_1, c_1>$" and "$S \rightarrow M <f_2, c_2>$" as premises, and derives "$S \rightarrow P <f, c>$" as conclusion. A direct way to apply probability theory would be treating each term as a set, then turning the rule into one that calculates conditional probability $Pr(P|S)$ from $Pr(P|M)$ and $Pr(M|S)$ plus additional assumptions about the probabilistic distribution function $Pr()$. Similarly, the sample size of the conclusion would be estimated, which gives the confidence value.

Such an approach cannot be applied in NARS for several reasons:

- For an inheritance relation, evidence is defined both extensionally and intensionally, so the frequency of "$S \rightarrow P$" cannot be treated as $Pr(P|S)$, since the latter is purely extensional.
- Each statement has its own evidence space, defined by the extension of its subject and the intension of its predicate.
- Since pieces of knowledge in input may come from different sources, they may contain inconsistency.
- When new knowledge comes, usually the system cannot afford the time to update all of the previous beliefs accordingly.

Therefore, though each statement can be treated as a probabilistic judgment, different statements correspond to different evidence space, and their truth values are evaluated against different bodies of evidence. As a result, they correspond to different probability distributions. For example, if we treat frequency as probability, the deduction rule should calculate $Pr_3(S \rightarrow P)$ from $Pr_1(M \rightarrow P)$ and $Pr_2(S \rightarrow M)$. In standard probability theory, there is few result that can be applied to this kind of cross-distribution calculation.

NARS solves this problem by going beyond probability theory, though still sharing certain intuition and result with it [43].

In NARS, the amount of evidence is defined in such a way that it can be used to indicate *randomness* (see [37] for a comparison with Bayesian network [24]), *fuzziness* (see [41] for a comparison with fuzzy logic [45]), and *ignorance* (see [38] for a comparison with Dempster-Shafer theory [33]). Though different types of uncertainty have different origins, they usually co-exist, and are tangled with one another in practical situations. Since NARS makes no restrictions on what can happen in its experience, and needs to make justifiable decisions when the available knowledge is insufficient, such a unified measurement of uncertainty is necessary.

There may be belief conflicts in NARS, in the sense that the same statement is assigned different truth values when derived from different parts of the experience. With insufficient resources, NARS cannot find and eliminate all implicit conflicts within its knowledge base. What it can do is, when a conflict is found, to generate a summarized conclusion whose truth value reflects the combined evidence. These conflicts are normal, rather than exceptional. Actually, their existence is a major driving force of learning, and only by their solutions some types of inference, like induction and abduction, can have their results accumulated [39]. In first-order predicate logic, a pair of conflicting propositions implies all propositions. This does not happen in a term logic like NARS, because in term logics the conclusions and premises must have shared terms, and statements with the same truth value cannot substitute one another in a derivation (as does in predicate logic). As a result, NARS tolerates implicitly conflicting beliefs, and resolves explicit conflicts by evidence combination.

The concepts in NARS are uncertain because the meaning of a concept is not determined by an interpretation that links it to an external object, but by its relations with other concepts. The relations are in turn determined by the system's experience and its processing of the experience. When a concept is involved in the processing of a task, usually only part of the knowledge associated with the concept is used. Consequently, concepts become "fluid" [16]:

1. No concept has a clear-cut boundary. Whether a concept is an instance of another concept is a matter of degree. Therefore, all the concepts in NARS are "fuzzy".
2. The membership evaluations are revisable. The priority distribution among the relations from a concept to other concepts also changes from time to time. Therefore, what a concept actually means to the system is variable.
3. However, the meaning of a concept is not arbitrary or random, but relatively stable, bounded by the system's experience.

3.3 NARS as a Parallel and Distributed Network

Though all the previous descriptions present NARS as a reasoning system with formal language and rules, in fact the system can also be described as a network. We can see each term as a *node*, and each statement as a *link* between two nodes, and the corresponding truth value as the *strength* of the link. Priorities are defined among nodes and links. In each inference step, two adjacent links generate new links, and different types of inference correspond to different combinations of the links [22, 39]. To answer a question means to determine the strength of a link, given its beginning and ending node, or to locate a node with the strongest link from or to a given node. Because by applying rules, the topological structure of the network, the strength of the links, and the priority distribution are all changed, what the system does is much more than searching a static network for the desired link or node.

Under such an interpretation, NARS shows some similarity to the other network-based AI approaches, such as the connectionist models.

Many processes coexist at the same time in NARS. The system not only processes input tasks in parallel, but also does so for the derived subtasks. The fact that the system can be implemented in a single-processor machine does not change the situation, because what matters here is not that the processes run exactly at the same time on several pieces of hardware (though it is possible for NARS to be implemented in a multiple-processor system), but that they are not run in a one-by-one way, that is, one process begins after another ends.

Such a parallel processing model is adopted by NARS, because given the insufficiency of knowledge and resources, as well as the dynamic nature of the memory structure and resources competition, it is impossible for the system to process tasks one after another.

Knowledge in NARS is represented *distributedly* in the sense that there is no one-to-one correspondence between the input/output in the experience/response and the knowledge in the memory [12]. When a piece of new knowledge is provided to the system, it is not simply inserted into the memory. Spontaneous inferences will happen, which generate derived conclusions. Moreover, the new knowledge may be revised when it is in conflict with previous knowledge. As a result, the coming of new knowledge may cause non-local effects in memory.

On the other hand, the answer of a question can be generated by non-local knowledge. For example, in answering the question "Is dove a kind of bird?", a piece of knowledge *"dove → bird"* (with its truth value) stored in concepts *dove* and *bird* provides a ready-made answer, but the work does not stop. Subtasks are generated (with lower priority) and sent to related concepts. Because there may be implicit conflicts within the knowledge base, the previous "local" answer may be revised by knowledge stored somewhere else.

Therefore, the digestion of new knowledge and the generation of answers are both non-local events in memory, though the concepts corresponding to terms that appear directly in the input knowledge/question usually have larger contributions. How "global" such an event can be is determined both by the available knowledge and the resources allocated to the task.

In NARS, information is not only stored distributively and with duplications, but also processed through multiple pathways. With insufficient knowledge and resources, when a question is asked or a piece of knowledge is told, it is usually impossible to decide whether it will cause redundancy or what is the best method to process it, so multiple copies and pathways become inevitable. Redundancy can help the system recover from partial damage, and also make the system's behaviors depend on statistical events. For example, if the same question is repeatedly asked, it will get more processor time.

Unlike many symbolic AI systems, NARS is not "brittle" [17] — that is, being easily "killed" by improper inputs. NARS is open and domain-independent, so any knowledge and question, as long as they can be expressed in the system's interface language, can be accepted by the system. The conflict between new knowledge and previous knowledge will not cause the "implication paradox" (i.e., from an inconsistence, any propositions can be derived). All mistakes in input knowledge can be revised by future experience to various extents. The questions beyond the system's current capacity will no longer cause a "combinatorial explosion", but will be abandoned gradually by the system, after some futile efforts. In this way, the system may fail to answer a certain question, but such a failure will not cause a paralysis.

According to the working manner of NARS, each concept as a processing unit only takes care of its own business, that is, only does inferences where the concept is directly involved. As a result, the answering of a question is usually the cooperation of several concepts. Like in connectionist models [30], there is no "global plan" or "central process" that is responsible for each question. The cooperation is carried out by message-passing among concepts. The generating of a specific solution is the *emergent result* of lots of local events, not only caused by the events in its derivation path, but also by the activity of other tasks that adjust the memory structure and compete for the resources. For this reason, each event in NARS is influenced by all the events that happen before it.

What directly follows from the above properties is that the solution to a specific task is context-sensitive. It not only depends on the task itself and the knowledge the system has, but also depends on how the knowledge is organized and how the resources are allocated at the moment. The *context* under which the system is given a task, that is, what happens before and after the task in the system's experience, strongly influences what solution the task receives. Therefore, if the system is given the same task twice, the solutions may be (though not necessarily) different, even though there is no new knowledge provided to the system in the interval. Here "context" means the current working environment in which a task is processed. Such contexts are dynamic and continuous, and they are not predetermined situations indexed by labels like "bank" and "hotel".

3.4 Resources Competition

The system does not treat all processes as equal. It distributes its resources among the processes, and only allows each of them to progress at certain speed and to certain "depth" in the knowledge base, according to how much resources are available to the system. Also due to insufficient knowledge, the resource distribution is maintained dynamically (adjusted while the processes are running), rather than statically (scheduled before the processes begin to run), because the distribution depends on how they work.

As a result, the processes compete with one another for resources. To speed up one process means to slow down the others. The priority value of a task reflects its (relative) priority in the competition, but does not determine its (absolute) actual resources consumption, which also depends on the priority values of the other coexisting tasks.

With insufficient processing time, it is inefficient for all the knowledge and questions to be equally treated. In NARS, some of them (with higher priority values) get more *attention*, that is, are more active or accessible, while some others are temporarily forgotten. With insufficient memory space, some knowledge and questions will be permanently forgotten — eliminated from the memory. Like in human memory [21], in NARS forgetting is not a deliberate action, but a side-effect caused by resource competition.

In traditional computing systems, the amount of time spent on a task is determined by the system designer, and the user provides tasks at run time without time requirements. On the other hand, many real-time systems allow users to attach a deadline to a task, and the time spent on the task is determined by the deadline [35]. A variation of this approach is that the task is provided with no deadline, but the user can interrupt the process at any time to get a best-so-far answer [3].

NARS uses a more flexible method to decide how much time to spend on a task, and both the system and the user influence the decision. The user can attaches an initial priority value to a task, but the actual allocation also depends on the current situation of the system, as well as on how well the task processing goes. As a result, the same task, with the same initial priority, will get more processing when the system is "idle" than when the system is "busy".

3.5 Flexible Behaviors

In NARS, how an answer is generated is heavily dependent on what knowledge is available and how it is organized. Facing a task, the system does not choose a method first, then collect knowledge accordingly, but lets it interact with available knowledge. In each inference step, what method is used to process a task is determined by the type of knowledge that happens to be picked up at that moment.

As a result, the processing path for a task is determined dynamically at run time, by the current memory structure and resource distribution of the system, not by a predetermined problem-oriented algorithm. In principle, the behavior of NARS is unpredictable from an input task alone, though still predictable from the system's initial state and complete experience.

For practical purposes, the behavior of NARS is not accurately predictable to a human observer. To exactly predict the system's solution to a specific task, the observer must know all the details of the system's initial state, and closely follow the system's experience until the solution is actually produced. When the system is complex enough (compared with the information processing capacity of the predictor), nobody can actually do this. However, it does not mean that the system works in a random manner. Its behaviors are still determined by its initial state and experience, so approximate predictions are possible.

If NARS is implemented in a von Neumann computer, can it go beyond the scope of computer science? Yes, it is possible because a computer system is a hierarchy with many levels [13]. Some critics implicitly assume that because a certain level of a computer system can be captured by first-order predicate logic and Turing machine, these theories also bind all the performances the system can have [7, 25]. This is not the case. When a system A is implemented by a system B, the former does not necessarily inherit all properties of the latter. For example, we cannot say that a computer cannot process decimal numbers (because they are implemented by binary numbers), cannot process symbols (because they are coded by digits), or cannot use functional or logical programming language (because they are eventually translated into procedural machine language).

Obviously, with its fluid concepts, revisable knowledge, and fallible inference rules, NARS breaks the regulations of classic logics. However, as a virtual machine, NARS can be based on another virtual machine which is a pure-axiomatic system, as shown by its implementation practice, and this fact does not make the system "axiomatic". If we take the system's complete *experience* and *response* as input and output, then NARS is still a Turing Machine that definitely maps inputs to outputs in finite steps. What happens here has been pointed out by Hofstadter as "something can be computational at one level, but not at another level" [15], and by Kugel as "cognitive processes that, although they involve more than computing, can still be modeled on the machines we call 'computers' " [18]. On the contrary, traditional computer systems are Turing Machines either globally (from experience to response) or locally (from question to answer).

3.6 Autonomy and Creativity

The global behavior NARS is determined by the "resultant of forces" of its internal tasks. Initially, the system is driven only by input tasks. The system then derives subtasks recursively by applying inference rules to the tasks and available knowledge.

However, it is not guaranteed that the achievement of the derived tasks will turn out to be really helpful or even related to the original tasks, because the knowledge, on which the derivation is based, is revisable. On the other hand, it is impossible for the system to always determine correctly which tasks are more closely related to the original tasks. As a result, the system's behavior will to a certain extent depend on "its own tasks", which are actually more or less independent of the original processes, even though historically derived from them. This is the *functional autonomy* phenomena [22]. In the extreme form, the derived tasks may become so strong that they even prevent the input tasks from being fulfilled. In this way, the derived tasks are *alienated*.

The alienation and unpredictability sometimes result in the system to be "out of control", but at the same time, they lead to *creative and original* behaviors, because the system is pursuing goals that are not directly assigned by its environment or its innateness, with methods that are not directly deduced from given knowledge.

By *creativity*, it does not mean that all the results of such behaviors are of benefit to the system, or excellent according to some outside standards. Nor does it mean that these behaviors come from nowhere, or from a "free will" of some sort. On the contrary, it means that the behaviors are novel to the system, and cannot be attributed either to the designer (who determines the system's initial state and skills) or to a tutor (who determines part of the system's experience) alone. Designers and tutors only make the creative behaviors possible. What turns the possibility into reality is the system's experience, and for a system that lives in a complex environment, its experience is not completely determined by any other systems (human or computer). For this reason, these behaviors, with their results, are better to be attributed to the system *itself*, than to anyone else [13].

Traditional computer systems always repeat the following "life cycle":

- waiting for tasks
- accepting a task
- working on the task according to an algorithm
- reporting a solution for the task
- waiting for tasks
- · · ·

On the contrary, NARS has a "life-time of its own" [8]. When the system is experienced enough, there will be many tasks for the system to process. On the other hand, new input can come at any time. Consequently, the system's history is no longer like the previous loop. The system usually works on its "own" tasks, but at the same time, it is always ready to respond to new tasks provided by the environment. Each piece of input usually attracts the system's attention for a while, and also causes some long-term effects. The system never reaches a "final state" and stops there, though it can be reset by a human user to its initial state. In this way, each task-processing activity is part of the system's life-time experience, and is influenced by the other activities. In comparison with NARS, traditional computer systems take each problem-solving activity as a separate life cycle with a predetermined end.

4 Conclusions

The key difference between NARS and the mainstream AI projects is not in the technical details, but in the philosophical and methodological position. The NARS project does not aim at a certain practical problem or cognitive function, but attempts to build a general-purpose intelligent system by identifying the "essence of intelligence", i.e., the underlying information processing principle, then designing the components of the system accordingly.

As described above, in the NARS project, it is assumed that "intelligence" means "adaptation with insufficient knowledge and resources", and then a reasoning system is chosen as the framework to apply this assumption. When designing the system, we found that all relevant traditional theories (including first-order predicate logic, model theory, probability theory, computability theory, computational complexity theory, ...) are inconsistent with the above assumption, so all major components need to be redesigned. These components, though technically simple, are fundamentally different from the traditional components in nature.

Built in this way, NARS provides a unified model for many phenomena observed in human cognition. It achieves this not by explicitly fitting psychological data, but by reproducing them from a simple and unified foundation. In this way, we see that these phenomena share a common functional explanation, and all intelligent systems, either natural or artificial, will show these phenomena as long as they are adaptive systems working with insufficient knowledge and resources.

The NARS project started in 1983 at Peking University. Several working prototypes have been built, in an incremental manner (that is, each with more inference rules and a more complicated control mechanism). Currently first-order inference has been finished, and higher-order inference is under development. Though the whole project is still far from completion, past experience has shown the feasibility of this approach. For up-to-date information about the project and the latest publications and demonstrations, please visit http://www.cogsci.indiana.edu/farg/peiwang/papers.html.

References

1. Aristotle (1989) *Prior Analytics*. Hackett Publishing Company, Indianapolis, Indiana. Translated by R Smith.
2. Birnbaum L (1991) Rigor mortis: a response to Nilsson's "Logic and artificial intelligence". *Artificial Intelligence*, 47:57–77.
3. Boddy M, Dean T (1994) Deliberation scheduling for problem solving in time-constrained environments. *Artificial Intelligence*, 67:245–285.
4. Carnap R (1950) *Logical Foundations of Probability*. The University of Chicago Press, Chicago.
5. Chalmers D, French R, Hofstadter D (1992) High-level perception, representation, and analogy: a critique of artificial intelligence methodology. *Journal of Experimental and Theoretical Artificial Intelligence*, 4:185–211.

6. Dean T, Boddy M (1988) An analysis of time-dependent planning. In *Proceedings of AAAI-88*, pages 49–54.
7. Dreyfus H (1992) *What Computers Still Can't Do*. The MIT Press, Cambridge, MA.
8. Elgot-Drapkin J, Miller M, Perlis D (1991) *Memory, reason, and time: the step-logic approach*. In: Cummins R, Pollock J (eds) Philosophy and AI, MIT Press, Cambridge, MA.
9. French R (1990) Subcognition and the limits of the Turing test. *Mind*, 99:53–65.
10. Good I (1983) *Good Thinking: The Foundations of Probability and Its Applications*. University of Minnesota Press, Minneapolis.
11. Hempel C (1943) A purely syntactical definition of confirmation. *Journal of Symbolic Logic*, 8:122–143.
12. Hinton G, McClelland J, Rumelhart D (1986) *Distributed representation*. In: Rumelhart D, McClelland J (eds) Parallel Distributed Processing: Exploration in the Microstructure of cognition, Vol. 1, Foundations, MIT Press, Cambridge, MA
13. Hofstadter D (1979). *Gödel, Escher, Bach: an Eternal Golden Braid*. Basic Books, New York.
14. Hofstadter D (1984) The copycat project: An experiment in nondeterminism and creative analogies. AI memo, MIT Artificial Intelligence Laboratory.
15. Hofstadter D (1985). Waking up from the Boolean dream, or, subcognition as computation. In: Hofstadter D *Metamagical Themas: Questing for the Essence of Mind and Pattern*, Basic Books, New York.
16. Hofstadter D, Mitchell M (1994) The Copycat project: a model of mental fluidity and analogy-making. In: Holyoak K, Barnden J (eds) *Advances in Connectionist and Neural Computation Theory, Volume 2: Analogical Connections*, Ablex, Norwood, NJ
17. Holland J (1986) Escaping brittleness: the possibilities of general purpose learning algorithms applied to parallel rule-based systems. In Michalski R, Carbonell J, Mitchell T (eds) *Machine Learning: an artificial intelligence approach*, volume II, Morgan Kaufmann, Los Altos, CA.
18. Kugel P (1986) Thinking may be more than computing. *Cognition*, 22:137–198.
19. McCarthy J (1988) Mathematical logic in artificial intelligence. *Dædalus*, 117(1):297–311.
20. McDermott D (1987) A critique of pure reason. *Computational Intelligence*, 3:151–160.
21. Medin D, Ross B (1992) *Cognitive Psychology*. Harcourt Brace Jovanovich, Fort Worth, TX.
22. Minsky M (1985). *The Society of Mind*. Simon and Schuster, New York.
23. Newell A, Simon H (1976) Computer science as empirical inquiry: symbols and search. The Tenth Turing Lecture. First published in *Communications of the Association for Computing Machinery* 19.
24. Pearl J (1988) *Probabilistic Reasoning in Intelligent Systems*. Morgan Kaufmann, San Mateo, CA.
25. Penrose R (1994) *Shadows of the Mind*. Oxford University Press, Oxford.
26. Piaget J (1960) *The Psychology of Intelligence*. Littlefield, Adams & Co., Paterson, NJ.
27. Popper K (1959) *The Logic of Scientific Discovery*. Basic Books, New York.
28. Reeke G, Edelman G (1988) Real brains and artificial intelligence. *Dædalus*, 117(1):143–173.

29. Reiter R (1987) Nonmonotonic reasoning. *Annual Review of Computer Science*, 2:147–186.
30. Rumelhart D, McClelland J (1986) PDP models and general issues in cognitive science. In: Rumelhart D, McClelland J (eds) *Parallel Distributed Processing: Explorations in the Microstructure of Cognition, Vol. 1, Foundations*, MIT Press, Cambridge, MA.
31. Schank R (1991) Where is the AI. *AI Magazine*, 12(4):38–49.
32. Searle J (1980) Minds, brains, and programs. *The Behavioral and Brain Sciences*, 3:417–424.
33. Shafer G (1976) *A Mathematical Theory of Evidence*. Princeton University Press, Princeton, NJ.
34. Smolensky P (1988). On the proper treatment of connectionism. *Behavioral and Brain Sciences*, 11:1–74.
35. Strosnider J, Paul C (1994) A structured view of real-time problem solving. *AI Magazine*, 15(2):45–66.
36. Turing A (1950) Computing Machinery and Intelligence. *Mind*, LIX(236):433–460.
37. Wang P (1993) Belief revision in probability theory. In *Proceedings of the Ninth Conference on Uncertainty in Artificial Intelligence*, Morgan Kaufmann, San Mateo, CA.
38. Wang P (1994) A defect in Dempster-Shafer theory. In *Proceedings of the Tenth Conference on Uncertainty in Artificial Intelligence*, Morgan Kaufmann, San Mateo, CA.
39. Wang P (1994) From inheritance relation to nonaxiomatic logic. *International Journal of Approximate Reasoning*, 11(4):281–319.
40. Wang P (1995) *Non-Axiomatic Reasoning System: Exploring the Essence of Intelligence*. PhD thesis, Indiana University, Bloomington, IN.
41. Wang P (1996) The interpretation of fuzziness. *IEEE Transactions on Systems, Man, and Cybernetics*, 26(4):321–326.
42. Wang P (2001) Abduction in non-axiomatic logic. In *Working Notes of the IJCAI workshop on Abductive Reasoning*, pages 56–63, Seattle, WA.
43. Wang P (2001) Confidence as higher-order uncertainty. In *Proceedings of the Second International Symposium on Imprecise Probabilities and Their Applications*, pages 352–361, Ithaca, NY.
44. Wang P (2001) Wason's cards: what is wrong? In *Proceedings of the Third International Conference on Cognitive Science*, pages 371–375, Beijing.
45. Zadeh L (1965) Fuzzy sets. *Information and Control*, 8:338–353.

The Novamente Artificial Intelligence Engine

Ben Goertzel and Cassio Pennachin

AGIRI – Artificial General Intelligence Research Institute
1405 Bernerd Place, Rockville, MD 20851, USA
ben@agiri.org, cassio@agiri.org - http://www.agiri.org

Summary. The *Novamente AI Engine*, a novel AI software system, is briefly reviewed. Novamente is an integrative artificial general intelligence design, which integrates aspects of many prior AI projects and paradigms, including symbolic, probabilistic, evolutionary programming and reinforcement learning approaches; but its overall architecture is unique, drawing on system-theoretic ideas regarding complex mental dynamics and associated emergent patterns. The chapter reviews both the conceptual models of mind and intelligence which inspired the system design, and the concrete architecture of Novamente as a software system.

1 Introduction

We present in this chapter the *Novamente AI Engine*, an integrative design for an AGI. Novamente is based on over a decade of research (see [27, 26, 28, 29, 30]) and has been developed, on conceptual and software design levels, to a significant extent. Through a decade and a half of research, we have created a theoretical foundation for the design of AI systems displaying adaptive, autonomous artificial intelligence, and we are in the midst of developing a highly original, unprecedented software system atop this foundation.

Novamente incorporates aspects of many previous AI paradigms such as agent systems, evolutionary programming, reinforcement learning, automated theorem-proving, and probabilistic reasoning. However, it is unique in its overall architecture, which confronts the problem of creating a holistic digital mind in a direct way that has not been done before. Novamente combines a common, integrative-AI friendly representation of knowledge, with a number of different cognitive processes, which cooperate while acting on that knowledge. This particular combination results in a complex and unique software system: an autonomous, self-adaptive, experientially learning system, in which the cooperation between the cognitive processes enables the emergence of general intelligence. In short, Novamente is a kind of "digital mind."

One way that Novamente differs from many other approaches to AGI is that it is being developed primarily in a commercial, rather than academic, context. While this presents some challenges in terms of prioritizing development of different aspects of the system, we feel it has been a highly valuable approach, for it has meant that, at each stage of the system's development, it has been tested on challenging real-world applications. Through our work on commercial applications of early, partial versions of the Novamente system, we have become acutely aware of the urgent need for Artificial General Intelligence, in various industries. Much is said about the information age, knowledge discovery, and the need for tools that are smart enough to allow human experts to cope with the unwieldy amounts of information in today's business and scientific worlds. We believe that the real answer for these analytical demands lies in AGI, as the current narrow techniques are unable to properly integrate heterogeneous knowledge, derive intelligent inferences from that knowledge and, most important, spontaneously generate new knowledge about the world.

At the time of writing, the Novamente system is completely designed and partially implemented. It can be applied to complex problems in specific domains like bioinformatics and knowledge discovery right now, and will yield ever greater functionality as more of the design is implemented and tested. Of course, the design is continually changing in its details, in accordance with the lessons inevitably learned in the course of implementation. However, these detail-level improvements occur within the overall framework of the Novamente design, which has – so far – proved quite powerful and robust.

1.1 The Novamente AGI System

Given the pressing need for AGI from a practical perspective, there has been surprisingly little recent R&D oriented specifically toward the AGI problem [64]. The AI discipline began with AGI dreams, but for quite some time has been dominated by various forms of narrow AI, including logical-inference-based AI, neural networks, evolutionary programming, expert systems, robotics, computer vision, and so forth. Many of these narrow-AI software systems are excellent at what they do, but they have in common a focus on one particular aspect of mental function, rather than the integration of numerous aspects of mental function to form a coherent, holistic, autonomous, situated cognitive system. Artificial General Intelligence requires a different sort of focus. Table 2 briefly compares key properties of AGI and narrow AI systems.

The authors and their colleagues have been working together for several years on the problem of creating an adequate design for a true AGI system, intended especially to lay the groundwork for AGI. We worked together during 1998-2001 on a proto-AGI system called Webmind [32], which was applied in the knowledge management and financial prediction domains; and since 2001 we have been collaborating on Novamente.

The Novamente design incorporates aspects of many previous AI paradigms such as evolutionary programming, symbolic logic, agent systems, and probabilistic reasoning. However, it is extremely innovative in its overall architecture, which confronts the problem of "creating a whole mind" in a direct way that has not been done before. The fundamental principles underlying the system design derive from a novel complex-systems-based theory of mind called the "psynet model," which was developed by the author in a series of cross-disciplinary research treatises published during 1993-2001 [27, 28, 29, 30]. What the psynet model has led us to is not a conventional AI program, nor a conventional multi-agent-system framework. Rather, we are talking about an autonomous, self-organizing, self-evolving AGI system, with its own understanding of the world, and the ability to relate to humans on a "mind-to-mind" rather than a "software-program-to-mind" level.

The Novamente design is a large one, but the currently deployed implementation already incorporates many significant aspects. Due to the depth of detail in the design, and the abundant pertinent prototyping experience the Novamente engineering team had during the period 1997-2004, the time required to complete the implementation will be less than one might expect given the magnitude of the task: we estimate 1-2 years. The engineering phase will be followed by a phase of interactively teaching the Novamente system how to respond to user queries, and how to usefully analyze and organize data. The end result of this teaching process will be an autonomous AGI system, oriented toward assisting humans in collectively solving pragmatic problems.

This chapter reviews the Novamente AGI design and some of the issues involved in its implementation, teaching and testing. Along the way we will also briefly touch on some practical-application issues, and discuss the ways in which even early versions of Novamente will provide an innovative, strikingly effective solution to the problem of helping human analysts comprehend, organize and analyze data in multiple, complex domains.

1.2 Novamente for Knowledge Management and Data Analysis

The Novamente AGI framework in itself is highly general, and may be applied in a variety of application contexts. For example, one could imagine Novamente being used as the cognitive engine of an advanced robotic system; in fact, a preliminary design for the hybridization of Novamente with James R. Albus's "Reference Model Architecture" for robotics [2] has been developed. Initially, however, our plan is to implement and deploy Novamente in the context of knowledge management and data analysis. We believe that Novamente has some important benefits for these application areas, summarized in Table 1. The current Novamente version is being used for management and analysis of bioinformatic information, specifically genomic and proteomic databases and experimental datasets; and for text understanding in the national security domain. Over the next few years, while continuing our current application work, we envision a significantly broader initiative to apply the system to the management and analysis of information in multiple domains.

The deployment of Novamente for knowledge management and analysis involves attention to many different issues, most falling into the general categories of data sources and human-computer interaction. The optimal way of handling such issues is domain-dependent. For the bioinformatics applications, we have taken an approach guided by the particular needs of bioscientists

Features of the Novamente Approach	Benefits for Knowledge Management, Querying and Analytics
Mixed natural/formal language conversational querying	Flexible, agile, information-rich user interactions. System learns from each user interaction
Integrative knowledge representation	Compact, manipulable representation of all common forms of data, enables integrative analysis across data items regardless of source or type
Powerful integrative cognition toolkit, including probabilistic inference and evolutionary concept creation	Recognizes subtle patterns in diverse data. Combines known patterns to form new ones. Interprets semantically rich user queries
Probabilistic inference based, nonlinear-dynamical attention-focusing	System shifts its focus of cognition based on user queries, and also based on changing trends in the world itself
DINI Distributed Architecture	Enables implementation of massive self-organizing Atom network on a network of commodity PC's

Table 1: Features and benefits of the Novamente approach

analyzing datasets generated via high-throughput genomics and proteomics equipment.

In terms of data sources, once one commits to take a knowledge integration approach, the trickiest issue that remains is the treatment of natural language data ("unstructured text"). Novamente may be used in two complementary ways:

- "Information retrieval" oriented, wherein a text is taken as a series of characters or a series of words, and analyzed statistically;
- Natural Language Processing (NLP) oriented, wherein an attempt is made to parse the sentences in the texts and extract their meanings into semantic-relationship form.

The information retrieval approach is appropriate when one has a large volume of text, and limited processing time for handling it. The NLP approach is more sophisticated and more computationally expensive.

The common weak point of existing NLP algorithms and frameworks is the integration of semantic and pragmatic understanding into syntactic language analysis. The Novamente design overcomes this problem by carrying out syntactic analysis via logical unification, a process that automatically incorporates available semantic and pragmatic knowledge into its behavior. At time of writing, we have not yet integrated the NLP component of Novamente with logical unification, but our experience with a similar implementation in the Webmind system gives us reasonable confidence here.

Regarding interaction with human users, the Novamente design supports a variety of different modalities, including conventional search-engine and NLP queries, Web forms queries, dynamic visualization and automated report generation. The most innovative design we have conceived for human-Novamente interaction, however, involves interactive conversation using a combination of natural language and formal language. Table 3 gives a speculative example of what we envision a dialogue in this vein might look like. The reality of mixed human/formal language dialogue will be defined through practical experimentation with the Novamente system in each different context.

System Characteristic	Narrow AI	AGI
Ability to Adapt Flexibly over Time	NO	YES
Ability to Integrate Dramatically Diverse Data Sources	NO	YES
Ability to Deal with Unforeseen Situations	Sometimes	YES
Ability to Dialogue with Humans	Minimal	YES
Ability to Perceive Subtle Data Patterns	Sometimes	YES
Simple Algorithms	Sometimes	NO
Small Demand on Computing Resources	Sometimes	NO
Requires Massive Human Knowledge Encoding Effort	Sometimes	NO

Table 2: AGI *versus* narrow AI

User-interface-wise, AGI systems may live on the back end of conventional software applications – but they may also be used to support radical new forms of human-computer interaction. As an example of this, we propose to develop an interface for mixed human/formal language conversation. A hypothetical example of this sort of dialogue is given in Table 3. This sort of dialogue does not require humans to learn to think like computers, and nor does it require computers to try to precisely emulate humans. Rather, it involves humans and computers each entering into one another's modes of cognition and description, and teaching and learning from each other in the process.

2 Enabling Software Technologies

The deep concepts underlying AGI are mathematical and theoretical, not dependent on particular software or hardware technologies. However, the viability of engineering, testing and deploying AGI systems, and applying them in a data analysis/querying/management context, is less of an abstract matter, and depends critically on the presence of appropriate enabling technologies.

```
User: ''I want information on Chinese-speaking residents of African
nations.''
NM: Mandarin, Cantonese, or both?
User: Mandarin
NM: See the file ChineseAfrica.txt
User: Thanks
NM: Would you like information on African residents of Chinese
nations?
User: Not now.  What I want is information on Chinese-speaking
residents of African nations who send their children to schools in
South America.
NM:   Could you rephrase please?
User:
> Find Y:
> Inheritance(X, ''African nation'')
> resident(Y, X)
> speaks(Y, Chinese)
> child_of(Z, Y)
> attends(Z, W)
> Inheritance(W, ''school'')
> part_of(X: location(W, X), ''South America'')
NM: By
> child_of(Z, Y)
do you mean that Z is a child of Y?
User: Yes
NM:   Thanks.  See the file ChineseAfricaChildren.txt
```

Table 3: Hypothetical example of a mixed human/formal language conversation

Two categories of enabling technologies – *distributed software architecture* and *database integration* – are particularly critical to practical AGI deployments. These are the subjects of the present section.

2.1 A Distributed Software Architecture for Integrative AI

Marvin Minsky, one of the fathers of AI, is on record conjecturing that a human-level artificial general intelligence could probably be made to run on a 1 Megahertz PC, if one merely knew the right design. Our AGI optimism, while strong, is slightly different in nature. We believe that computing technology is just now reaching the point where advanced AGI software becomes possible. Having the correct design is still the most important thing; but the right design without an adequate hardware platform and operating system will never leave the world of theory.

In the 1980's, specialized supercomputing hardware was one of the themes of the day. Cray's vector processing machines revolutionized computational physics and related areas, and Thinking Machines Corp.'s MIMD parallel Connection Machine [38] architecture appeared poised to do the same thing for artificial intelligence. What happened, however, was that the Connection Machine was unable to keep pace with the incredibly rapid development of conventional von Neumann hardware, and technology for networking traditional machines together. The last Connection Machine created before Thinking Machine Corp.'s dissolution, the CM-5, was less radical than its predecessors, being based on traditional processors coupled in an unusually tight way. And similarly, today's most powerful supercomputers, IBM's [11], are actually distributed computers underneath – they're specially-constructed networks of relatively conventional processors rather than unique processors.

Given a blank slate, it's clear that one could design a vastly more AGI-appropriate hardware platform than the von Neumann architecture. Conceptually speaking, we believe the Connection Machine was on the right track. However, modern networking technology and distributed software architecture have brought the von Neumann architecture a long way from its roots, and we believe that it is possible to use contemporary technology to create distributed AI platforms of significant power and elegance.

Fig. 1 depicts the DINI (Distributed INtegrative Intelligence) architecture, a generic distributed-processing-based framework for AGI-based data analysis/querying/management, designed by the authors as a platform for large-scale Novamente deployment. The mathematical structures and dynamics of Novamente could be implemented in many ways besides DINI; and DINI could be used as a platform for many software systems different from Novamente. But, Novamente and DINI are a natural fit.

The key components of DINI, as shown in Fig. 1, are:

- "Analytic clusters" of machines – each cluster carrying out cognitive analysis of data, and creating new data accordingly
- Massive-scale data haven integrating multiple DBs and providing a unified searchable interface
- "Fisher" process, extracting appropriate data from the data bank into the Analytic Clusters
- "Miner" processes, extracting information from external databases into the data bank
- Web spiders continually gathering new information
- "Mediator" process merging results from multiple analytic clusters into the data bank
- Interfaces for knowledge entry by human beings
- Interfaces for simple and advanced querying
- J2EE middleware for inter-process communication, scalability, transaction control, load balancing, overall adaptive system control

The subtlest processes here are the Fisher and the Mediator.

The Fisher may respond to specific queries for information submitted by the analytic clusters. But it also needs to be able to act autonomously – to

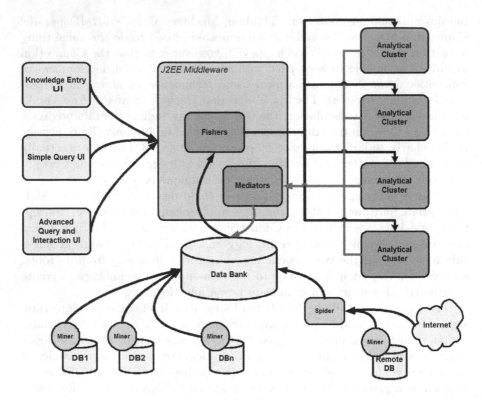

Fig. 1: The DINI architecture

use heuristics to guess what data may be interesting to the analytic clusters, based on similarity to the highest-priority data in the analytic clusters.

The Mediator exists due to the fact that diverse analytic clusters, acting on the same data and thinking about the same problems, may produce contradictory or complementary conclusions. Reconciliation of these conclusions into a single view in the centralized DB is required. When reconciliation is implausible, multiple views are stored in the centralized DB. Reconciliation is carried via a logical process of "belief revision," using formulas derived from Novamente's first-order inference component.

2.2 Database Integration and Knowledge Integration

A large role is played in the DINI architecture by the "data bank" component. Much of the information in a DINI data bank will be created by AGI processes themselves. However, there will also, generally speaking, be a large amount of data from other sources. There is a massive number of databases out there,

created by various organizations in academia, industry and government[1] –
these are useful to an AGI in carrying out practical knowledge management,
querying and analysis functions, and also in building up its intelligence and
its understanding of the world.

However, the information in databases is rarely in a format that can be fed
directly into an AGI that is still at the learning phase. Ultimately, a mature
AGI should be able to digest a database raw, figuring out the semantics of
the schema structure on its own. At the present stage, however, databases
require significant preprocessing in order to be useful for AGI systems. This is
a variant of the "database integration" problem: how to take the information
in multiple databases and make it available in a unified way.

Through surveying the approaches to database integration taken in different domains, we have come to distinguish four different general strategies:

Federation create a common GUI for separate DB's

Amalgamation create formal mappings between the schema of different
    DB's

Schema translation create a new RDB combining information from multi-
    ple DB's

Knowledge integration create a translator mapping DB contents into a
    "universal formal knowledge representation."

Applying AGI systems to database information requires the most robust
approach: knowledge integration. In this approach, knowledge is extracted
from databases into a schema-independent formal language. An example of
this is Cycorp's approach to knowledge integration, which involves the conversion of knowledge into their CycL language [18]. However, for technical
reasons we feel that the CycL approach is not sufficiently flexible to support
non-formal-logic-centric AI approaches.

One practical, and extremely flexible, form that knowledge integration
may take involves the XML language. We have created a special XML DTD
for Novamente, which consists of a set of tags corresponding to Novamente's
internal knowledge representation. To integrate a database into Novamente,
the primary step required is to write code that exports the relational data tables involved into XML structured by the Novamente DTD. However, for best
results, a significant "amalgamation" process must be carried out beforehand,
to be sure that different overlapping databases are exported into Novamente
structures in a fully semantically compatible way. The same software framework could be used to support AI approaches different from Novamente; one
would merely have to create appropriate XML transformation schemata to
translate a Novamente DTD into a DTD appropriate for the other AI system.

---

[1] Of course, the robotics and DB oriented approaches are not contradictory; they
could both be pursued simultaneously. Here however we are focusing on the DB
option, which is our focus at present and in the near future.

3 What Is Artificial General Intelligence?

To understand why and how we pursue the holy grail of AGI, it's necessary to understand what AGI is, and how it's different from what the bulk of researchers in the AI field have come to refer to as "intelligence." If narrow AI did not exist, we wouldn't need the term "general intelligence" at all – we'd simply use the term "intelligence." When we speak of human intelligence, after all, we implicitly mean general intelligence. The notion of IQ arose in psychology as an attempt to capture a "general intelligence" factor or *g-factor* [14], abstracting away from ability in specific disciplines. Narrow AI, however, has subtly modified the meaning of "intelligence" in a computing context, to mean, basically, the ability to carry out *any particular task* that is typically considered to require significant intelligence in humans (chess, medical diagnosis, calculus, ...). For this reason we have introduced the explicit notion of Artificial General Intelligence, to refer to something roughly analogous to what the g-factor is supposed to measure in humans.

When one distinguishes narrow intelligence from general intelligence, the history of the AI field takes on a striking pattern. AI began in the mid-twentieth century with dreams of artificial general intelligence – of creating programs with the ability to generalize their knowledge across different domains, to reflect on themselves and others, to create fundamental innovations and insights. But by the early 1970's, AGI had not come to anything near fruition, and researchers and commentators became frustrated. AGI faded into the background, except for a handful of research projects. In time AGI acquired a markedly bad reputation, and any talk of AGI came to be treated with extreme skepticism.

Today, however, things are a good bit different than in the early 1970s when AGI lost its lustre. Modern computer networks are incomparably more powerful than the best supercomputers of the early 1970s, and software infrastructure has also advanced considerably. The supporting technologies for AGI are in place now, to a much greater extent than at the time of the early failures of the AGI dream. And tremendously more is now known about the mathematics of cognition, partly due to work on narrow AI, but also due to revolutionary advances in neuroscience and cognitive psychology. We believe the time is ripe to overcome the accumulated skepticism about AGI and make a serious thrust in the AGI direction. The implication is clear: the same advances in computer technology that have given us the current information glut enable the AGI technology that will allow us to manage the glut effectively, and thus turn it into an advantage rather than a frustration.

We find it very meaningful to compare AGI to the currently popular field of nanotechnology. Like nanotechnology, we believe, AGI is "merely an engineering problem," though certainly a very difficult one. Brain science and theoretical computer science clearly suggest that AGI is possible if one arrives at the right design[2]. The Novamente project is not the only existing effort to use the "right design" to create a true AGI, but it is one of a handful of such efforts, and we believe it is more advanced than any other.

Because of the confusing history of AI, before launching into the details of the Novamente AGI design, we feel it is worthwhile to spend a few paragraphs clarifying our notion of general intelligence. The reader is asked to bear in mind that "intelligence" is an informal human language concept rather than a rigorously defined scientific concept; its meaning is complex, ambiguous and multifaceted. In order to create useful AGI applications, however, we require a practical working definition of the AGI goal – not a comprehensive understanding of all the dimensions of the natural language concept of intelligence.

3.1 What Is General Intelligence?

One well-known characterization of artificial general intelligence is Alan Turing's famous "Turing Test" – "write a computer program that can simulate a human in a text-based conversational interchange" [67]. This test serves to make the theoretical point that intelligence is defined by behavior rather than by mystical qualities, so that if a program could act like a human, it should be considered as intelligent as a human[3]. However, Turing's test is not useful as a guide for practical AGI development. Our goal is not to create a simulated human, but rather to create a nonhuman digital intelligent system – one that will complement human intelligence by carrying out data analysis and management tasks far beyond the capability of the human mind; and one that will cooperate with humans in a way that brings out the best aspects of both the human and the digital flavors of general intelligence.

Similarly, one might think that human IQ tests – designed to assess human general intelligence – could be of some value for assessing the general intelligence of software programs. But on closer inspection this turns out to be a dubious proposition as. Human IQ tests work fairly well within a single culture, and much worse across cultures [54] – how much worse will they work across different types of AGI programs, which may well be as different as different species of animals?

In [27], a simple working definition of intelligence was given, building on various ideas from psychology and engineering. The mathematical formalization of the definition requires more notation and machinery than we can introduce here, but verbally, the gist is as follows:

General Intelligence is the ability to achieve complex goals in complex environments.

---

[2]Though a small minority of scientists disagree with this, suggesting that there is somethign noncomputational going on in the brain. See [36, 57]

[3]Although Searle's Chinese Room argument attempts to refute this claim, see [59]

The Novamente AI Engine work has also been motivated by a closely related vision of intelligence provided by Pei Wang in his PhD thesis and related works ([69], also this volume.) Wang's definition posits that general intelligence is

> *"[T]he ability for an information processing system to adapt to its environment with insufficient knowledge and resources."*

The Wang and Goertzel definitions are complementary. In practice, an AGI system must be able to achieve complex goals in complex environments with insufficient knowledge and resources. AI researcher Shane Legg has suggested[4] that this notion of intelligence should be labeled "cybernance" to avoid entanglement with the ambiguities of the informal language notion of "intelligence."

A primary aspect of the "complex goals in complex environments" definition is the plurality of the words "goals" and "environments." A single complex goal is not enough, and a single narrow environment is not enough. A chess-playing program is not a general intelligence, nor is a datamining engine that does nothing but seek for patterns in consumer information databases, and nor is a program that can extremely cleverly manipulate the multiple facets of a researcher-constructed microworld (unless the microworld is vastly more rich and diverse one than any yet constructed). A general intelligence must be able to carry out a variety of different tasks in a variety of different contexts, generalizing knowledge from one context to another, and building up a context and task independent pragmatic understanding of itself and the world.

One may also probe one level deeper than these definitions, delving into the subtlety of the relationship between generalized and specialized intelligence. Drawing on ideas from the formal theory of complexity (see [29]; for related, more rigorously developed ideas, see [42]), one may define a system as *fully generally intelligent* for complexity N if it can achieve any goal of complexity N in any environment of complexity N. And this is where things get interesting, because it's clear that full general intelligence is only one aspect of human general intelligence.

The way the human brain seems to work is:

- some of its architecture is oriented towards achieving full general intelligence for small N (i.e. humans can solve any reasonably simple; problem)
- some of its architecture is oriented towards increasing problem-solving ability for goals and environments with complexity N so large that the human brain's full general intelligence for complexity N is effectively zero.

For example, human visual cognition is specialized to deal with environments of great complexity, and the human brain is certainly not able to deal equally well with all phenomena of comparable complexity. The human brain is

---

[4]Personal communication

specialized for visual cognition, even though it brings its "general intelligence capability" to bear on the problem in many ways. The same phenomenon exists in many other areas, from human social cognition [15] to mathematical problem-solving (humans are not good at proving randomly generated mathematical theorems).

Any real-world-useful general intelligence will, like the human brain, display a mix of "full general intelligence" methods focused on boosting full general intelligence for small N, and "general intelligence leveraging specialized intelligence methods" (GILSIM) that are different from narrow-AI methods in that they specifically leverage a combination of specialized heuristics and small-N full-general-intelligence methods.

As it turns out, the hard part of the practical general intelligence problem is not the small-N full-general-intelligence part, but rather the GILSIM part. Achieving "small-N general intelligence" is a mathematics problem, solvable via algorithms such as genetic programming [49], reinforcement learning [66], or Schmidhuber's OOPS algorithm [58]. Novamente uses a combination of several approaches here, as will be briefly discussed below.

On the other hand, contemporary mathematics has less to offer when it comes to the task of *building a system capable of supporting multiple specialized intelligences that combine task-appropriate heuristics with limited-complexity full general intelligence.* And this is the central challenge of AGI design as we see it. It is the challenge the Novamente design addresses.

3.2 The Integrative Approach to AGI

At least three basic approaches to AGI are possible:

1. close emulation of the human brain in software;
2. conception of a novel AGI architecture, highly distinct from the brain and also from narrow AI programs;
3. an integrative approach, synthesizing narrow AI algorithms and structures in a unique overall framework, perhaps guided to some extent by understanding of the human brain.

The Novamente approach lies falls on the continuum between approach 2 and approach 3. Roughly 2/3 of the Novamente design is based on existing narrow AI approaches, and the rest was conceived de novo with AGI in mind.

Novamente emphatically does *not* fall into Category 1: it is not a human-brain emulation. While the human brain was a conceptual inspiration for Novamente, particularly in the early design phase, the Novamente design makes a concerted effort to do things in ways that are efficient for software running on networks of von Neumann machines, and this is often profoundly different from the ways that are efficient on neural wetware. Further along this chapter, Table 9 reviews some of the parallels between human brain structures and processes and Novamente structures and processes.

The integrative approach is based on the idea that many narrow AI approaches embody good ideas about how some particular aspect of intelligence may be implemented computationally. For instance, logic-based AI contains many insights as to the nature of logical reasoning. Formal neural networks embody many insights about memory, perception, classification, and reinforcement learning of procedures. Evolutionary programming is an excellent technique for procedure learning, and for the creation of complex new concepts. Clustering algorithms are good ways of creating speculative new categories in a poorly-understood domain. Et cetera. The observation that narrow AI approaches often model particular aspects of intelligence well leads to the idea of synthesizing several narrow AI approaches to form an AGI architecture.

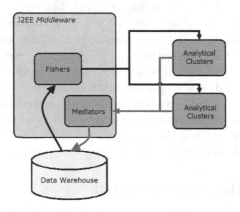

Fig. 2: Loose integration

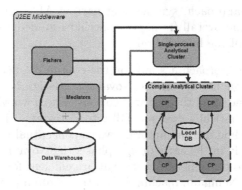

Fig. 3: Tight integration

This kind of synthesis could be conducted in two ways:

Loose integration, in which different narrow AI techniques reside in separate software processes or software modules, and exchange the results of their analysis with each other;

Tight integration, in which multiple narrow AI processes interact in real-time on the same evolving integrative data store, and dynamically affect one another's parameters and control schemata.

The manifestation of these two types of integration in a DINI context is shown in Figures 2 and 3. The "loose integration" approach manifests itself in DINI as an architecture in which separate analytical clusters, embodying separate narrow AI techniques, interact via the central data warehouse. The "tight integration" approach manifests itself in terms of a complex analytical cluster containing its own local DB, involving multiple narrow AI algorithms inextricably interlinked.

Tight integration is more difficult to design, implement, test and tune, but provides the opportunity for greater intelligence via emergent, cooperative effects. Novamente is based on tight integration, and we believe that this is the only approach that is viable for genuine AGI. Novamente essentially consists of a framework for tightly integrating various AI algorithms in the context of a highly flexible common knowledge representation, and a specific assemblage of AI algorithms created or tweaked for tight integration in an integrative AGI context.

3.3 Experiential Interactive Learning and Adaptive Self-modification

We have been discussing AGI as a matter of complex software systems embodying complex mathematical AI algorithms. This is an important perspective, but it must be remembered that AGI is not simply another form of engineering – it is also a deeply philosophical and conceptual pursuit. Novamente was not designed based on engineering and mathematical considerations alone. Rather, it owes its ultimate origin to an abstract, complex-systems-theoretic psychological/philosophical theory of mind – the "psynet model," which was presented by the first author in five research monographs published between 1993 and 2001 [27, 26, 28, 29, 30].

Based on the premise that a mind is the set of patterns in a brain, the psynet model describes a specific set of high-level structures and dynamics for mind-patterns, and proposes that these are essential to any sort of mind, human or digital. These are not structures that can be programmed into a system; rather they are structures that must emerge through the situated evolution of a system – through experiential interactive learning. Novamente's specific structures and dynamics tie in closely with the more general ones posited by the psynet model.

The psynet model also contains a theory of the relation between learning and mind that is different from the most common perspectives expressed in the AI literature. Namely, it posits that:

> *Software and mathematics alone, no matter how advanced, cannot create an AGI.*

What we *do* believe software and mathematics can do, however, is to set up a framework within which artificial general intelligence can emerge through interaction with humans in the context of a rich stream of real-world data. That is:

> *Intelligence most naturally emerges through situated and social experience.*

It is clear that human intelligence does not emerge solely through human neural wetware. A human infant is not so intelligent, and an infant raised without proper socialization will never achieve full human intelligence [22]. Human brains learn to think through being taught, and through diverse social interactions. We suggest the situation will be somewhat similar with AGI's. The basic AGI algorithms in Novamente are not quite adequate for practical general intelligence, because they give only the "raw materials" of thought. What is missing in a Novamente "out of the box" are context-specific control mechanisms for the diverse cognitive mechanisms. The system has the capability to learn these, but just as critically, it has the capability to *learn how to learn these*, through social interaction. A Novamente "out of the box" will be much smarter than narrow AI systems, but not nearly as robustly intelligent as a Novamente that has refined its ability to learn context-specific control mechanisms through meaningful interactions with other minds. For instance, once it's been interacting in the world for a while, it will gain a sense of how to reason about conversations, how to reason about network intrusion data, how to reason about bioinformatics data – by learning context-dependent inference control schemata for each case, according to a schema learning process tuned through experiential interaction.

These considerations lead us straight to the concepts of *autonomy, experiential interactive learning*, and *goal-oriented self-modification* – concepts that lie right at the heart of the notion of Artificial General Intelligence. In order for a software system to demonstrate AGI, we believe, it must demonstrate:

- a coherent autonomy as an independent, self-perceiving, self-controlling system;
- the ability to modify and improve itself based on its own observations and analyses of its own performance;
- the ability to richly interact with, and learn from, other minds (such as human minds).

These general points evoke some very concrete issues, to do with the difference between conventional data analysis and knowledge management systems, and AGI systems applied to data analysis, management and querying.

A tightly-coupled, integrative AI software system *may* be supplied with specific, purpose-oriented control schemata and in this way used as a datamining and/or query processing engine. This is the approach taken, for example, in the current applications of the Novamente engine in the bioinformatics domain. But this kind of deployment of the Novamente software does not permit it to develop anywhere near its maximum level of general intelligence.

For truly significant AGI to emerge, a software system must be deployed somewhat differently. It must be supplied with general goals, and then allowed to learn its own control schemata via execution of its procedure learning dynamics in the context of interaction with a richly structured environment, and in the context of extensive meaningful interactions with other minds. This path is more difficult than the "hard-wired control schemata" route, but it is necessary for the achievement of genuine AGI.

The Novamente system, once fully engineered and tuned, will gain its intelligence through processing practically-relevant data, answering humans' questions about this data, and providing humans with reports summarizing patterns it has observed. In addition to EIL through interactive data analysis/management, we have created a special "EIL user interface" called *Shape-World*, which involves interacting with Novamente in the context of a simple drawing panel on which the human teacher and Novamente may draw shapes and talk about what they're doing and what they see. We have also designed an environment called *EDEN (EDucational Environment for Novamente)*, a virtual-reality world in which Novamente will control simulated agents that interact with human-controlled agents in a simulated environment.

This process of "experiential interactive learning" has been one of the primary considerations in Novamente design and development. It will continually modify not only its knowledge base, but its control schemata based on what it's learned from its environment and the humans it interacts with.

The ultimate limits of this process of self-improvement are hard to foresee – if indeed there are any. It is worth remembering that source code itself is a formal object, which may easily be represented in the knowledge-representation-schema of an AGI system such as Novamente. Inferences about source code and its potential variations and improvements would appear to lie within the domain of computationally-achievable probabilistic reasoning. There seems no basic reason why an AGI system could not study its own source code and figure out how to make itself smarter. And there is an appealing exponential logic to this process: the smarter it gets, the better it will be at improving itself. Of course the realization of this kind of ultimate self-adaptation lies some distance in the future. There may be significant obstacles, unforeseeable at the current point. But, on the conceptual level at least, these ideas are a natural outgrowth of the processes of goal-directed self-improvement that we will be deploying in Novamente in the near term, as part of the AGI tuning and teaching process. The Novamente system has been designed with a clear focus on fulfilling short-term data analysis, management and querying needs, but also with an eye towards the full grandeur of the long-term AGI vision.

4 The Psynet Model of Mind

In this section we will delve a little more deeply into the psynet model of mind, the conceptual and philosophical foundation for the Novamente system.

For starters, we must clarify our use of the term "mind." Our view is that "mind," like "intelligence," is a human language concept, with a rich abundance of overlapping meanings. The psynet model does not aim to fully capture the human-language notion of "mind." Rather, it aims to capture a useful subset of that notion, with a view toward guiding AGI engineering and the analysis of human cognition.

The psynet model is based on what Ray Kurzweil calls a "patternist" philosophy [50]. It rests on the assumption that a mind is neither a physical system, nor completely separate from the physical – rather, a mind is something associated with the set of *patterns* in a physical system. In the case of an intelligent computational system, the mind of the system is not in the source code, but rather in the patterns observable in the dynamic trace that the system creates over time in RAM and in the registers of computer processors.

The concept of pattern used here is a rigorous one, which may be grounded mathematically in terms of algorithmic information theory [29, 16]. In essence, a pattern in an entity is considered as an abstract computer program that is smaller than the entity, and can rapidly compute the entity. For instance, a pattern in a picture of the Mandelbrot set, might be a program that could compute the picture from a formula. Saying "mind is pattern" is thus tantamount to positioning mind in the mathematical domain of abstract, nonphysical computer programs. As cautioned above, we are not asserting this as a complete explanation of all aspects of the concept of "mind" – but merely as a pragmatic definition that allows us to draw inferences about the minds of AGI systems in a useful way.

The "mind is pattern" approach to AI theory is not in itself original; similar ideas can be found in the thinking of contemporary philosophers such as Gregory Bateson [9], Douglas Hofstadter [39] and Daniel Dennett [20]. The psynet model, however, takes the next step and asks how the set of patterns comprising a mind is structured, and how it evolves over time. It seeks to understand mind in terms of pattern dynamics, and the emergent structures arising from pattern dynamics.

According to the psynet model, the patterns constituting a mind function as semi-autonomous "actors," which interact with each other in a variety of ways. Mental functions like perception, action, reasoning and procedure learning are described in terms of interactions between mind-actors (which are patterns in some underlying physical substrate, e.g., a brain or a computer program). And hypotheses are made regarding the large-scale structure and dynamics of the network of mind-patterns.

Consistent with the "complex goals in complex environments" character-ization of intelligence, an intelligent system, at a given interval of time, is assumed to have a certain *goal system* (which may be expressed explicitly and/or implicitly in the system's mind[5]). This goal system may alter over time, either through "goal drift" or through the system's concerted activity (some goals may explicitly encourage their own modification). It is important that an intelligent system has both general and specific goals in its goal sys-tem. Furthermore, one particular general goal is posited as critical to the goal system of any intelligent system: *the creation and recognition of new patterns.* With this goal in its goal system, an intelligence will seek to perceive and creation new structures in itself, as it goes about the business of achieving its other goals; and this self-perception/creation will enhance its intelligence in the long term.

The pattern dynamics of a cognitive system is understood to be governed by two main "forces": spontaneous self-organization and goal-oriented behav-ior.

More specifically, several primary dynamical principles are posited, includ-ing:

Association, in which patterns, when given attention, spread some of this attention to other patterns that they have previously been associated with in some way.

Differential attention allocation, in which patterns that have been valu-able for goal achievement are given more attention, and are encouraged to participate in giving rise to new patterns.

Pattern creation, in which patterns that have been valuable for goal-achievement are mutated to yield new patterns, and are combined with each other to yield new patterns.

Relationship reification,] in which habitual patterns in the system that are found valuable for goal-achievement, are explicitly reinforced and made more habitual.

For example, it is proposed that, for a system to display significant intel-ligence, the network of patterns observable in the system must give rise to several large-scale emergent structures:

Hierarchical network, in which patterns are habitually in relations of con-trol over other patterns that represent more specialized aspects of them-selves

Heterarchical network, in which the system retains a memory of which patterns have previously been associated with each other in any way

---

[5]Parenthetically, it is important that a goal set be defined over an interval of time rather than a single point of time; otherwise the definition of "implicit goal sets" is more difficult.

Dual network, in which hierarchical and heterarchical structures are combined, the dynamics of the two structures working together harmoniously
"Self" structure, in which a portion of the network of patterns forms into an approximate (fractal) image of the overall network of patterns.

The psynet model is a very general construct. It does not tell you how to build an AGI system in the engineering sense; it only tells you, in general terms, "what an AGI system should be like." Novamente is the third AGI-oriented software system created with the psynet model in mind, and it is very different from the previous two efforts. The differences between these systems may be summarized as follows:

1994: Antimagicians, which was an experimental psynet-inspired program in the pure self-organizing-systems vein [29, 68, 46], with very few built-in structures and an intention for the structures and dynamics of mind to emerge via experience. The anticipated emergence was not observed, and it was decided to take a more engineering-oriented approach in which more initial structures and dynamics are implanted as a "seed" for intelligent self-organization.

1996-2001: The Webmind AI Engine, "Webmind," developed at Webmind Inc., was a large-scale Java software system that derived its software design from the psynet model in a very direct way. Portions of Webmind were successfully applied in the domains of financial prediction and information retrieval; and a great amount of useful prototyping was done. But it was found that directly mapping the psynet model's constructs into object-oriented software structures leads to serious problems with computational efficiency.

Since 2001: Novamente, which represents an entirely different approach, embodying a highly flexible, computationally efficient AGI framework, which could be used to implement a variety of different AI systems. This framework includes three main aspects: the DINI architecture, the philosophy of tightly-coupled integrative AI, and the Novamente "Mind OS" architecture to be described below. Novamente also embodies a particular choice of software objects within this framework, whose selection is heavily *shaped* by the ideas in the psynet model of mind.

The relation between the psynet model of mind and Novamente is somewhat like the relationship between evolutionary theory and contemporary evolutionary programming algorithms. Evolutionary theory provides the conceptual underpinnings for evolutionary programming, and the first evolutionary programming algorithm, the traditional bit string GA [33], arose as a fairly direct attempts to emulate biological evolution by natural selection [41]. But contemporary evolutionary programming approaches such as the Bayesian Optimization Algorithm [56] and Genetic Programming [49] achieve superior pragmatic functionality by deviating fairly far from the biological model, and are only more indirectly mappable back into their conceptual inspiration. Similarly, Novamente represents the basic concepts involved in the psynet model, but in an indirect form that owes equally much to issues of pragmatic functionality in a contemporary computing context.

5 The Novamente AGI Design

The Novamente AI Engine ("Novamente") is a large-scale, object-oriented, multithreaded software system, intended to operate within the DINI framework. It is a C++ software system, with a few externally-facing components written in Java. Currently, development is primarily on the Linux operating system, but porting to other varieties of Unix or to Windows would not be problematic[6]. In DINI terms, a Novamente system is a collection of analytical clusters, most of them tightly-integrated, some of them more simple and specialized. It embodies a tightly-coupled integrative approach to AGI, in which a number of narrow AI approaches are combined with several innovative structural and dynamical ideas, in the context of a common "universal knowledge representation." The structures and processes chosen for Novamente are intended to allow the system to realize the abstract dynamics and emergent structures described in the psynet model of mind.

In this section we will paint the Novamente design in broad strokes, illustrating each aspect discussed in the context of data analysis, querying or management. Later on we will delve into a few of the more important AI processes in the system in slightly more detail. The AGIRI website contains a periodically updated page which gives yet more depth to the portrayal, systematically enumerating some of the key structures and dynamics of the system.

Below we briefly describe the major aspects of Novamente design:

Nodes. Nodes may symbolize entities in the external world, they may embody simple executable processes, they may symbolize abstract concepts, or they may serve as components in relationship-webs signifying complex concepts or procedures.

Links. Links may be n-ary, and may point to nodes or links; they embody various types of relationships between concepts, percepts or actions. The network of links is a web of relationships.

MindAgents. A MindAgent is a software object embodying a dynamical process such as importance updating, concept creation, or first-order logical inference. It acts directly on individual Atoms, but is intended to induce and guide system-wide dynamical patterns.

---

[6]In fact the system has been tested on FreeBSD, and a partial Windows port exists.

Mind OS. The Mind OS, living within the DINI framework, enables diverse MindAgents to act efficiently on large populations of Nodes and Links distributed across multiple machines.

Maps. A map represents declarative or procedural knowledge, as a pattern distributed across many Nodes and Links.

Units. A Unit is a collection of Nodes, Links and MindAgents, living on a cluster of machines, collectively devoted to carrying out a particular function such as: vision processing, language generation, highly-focused concentration, ...

5.1 An Integrative Knowledge Representation

Knowledge representation is one of the huge, classic AI problems. Of course, it is intimately bound up with the problem of cognitive algorithms – different cognitive algorithms have different requirements for knowledge representation, and different knowledge representations suggest different cognitive algorithms. Novamente's knowledge representation arose out of a search for the simplest, most conveniently manipulable knowledge representation that was easily compatible with all the different AI processes in the Novamente system. Like the Novamente system itself, Novamente's knowledge representation is a synthesis of ideas from existing narrow AI paradigms – with a significant number of original elements added in as well, to fill roles not addressed by existing ideas (including some roles, like system-wide attention allocation, that intrinsically could not be filled by narrow AI approaches).

Knowledge is represented in Novamente on two levels:

Atoms, software objects that come in two species: Nodes or Links.

Maps, sets of Atoms that tend to be activated together, or tend to be activated according to a certain pattern (e.g. an oscillation, or a strange attractor.)

Generally speaking the same types of knowledge are represented on the Atom level and on the map level. Atom level representation is more precise and more reliable, but map level representation is more amenable to certain types of learning, and certain types of real-time behavior.

Figure 5 gives a graphical example of a map – the map for the concept of "New York" as it might occur in a Novamente system. This map is a fuzzy node set containing the ConceptNode corresponding to the *New_York* concept, and also a host of other related nodes.

On the Atom level, the essential mathematical structure of Novamente's knowledge representation is that of a *hypergraph* (a graph whose edges can span $k > 2$ nodes [13]). We call this hypergraph an *Atomspace*, meaning that it is a hypergraph with the special properties that:

- the nodes and links are weighted with complex weight structures (TruthValue and AttentionValue objects);

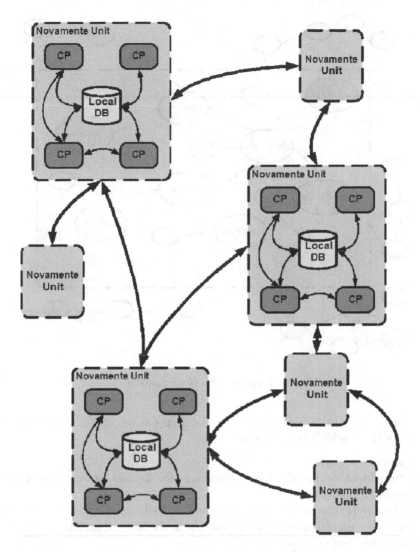

Fig. 4: A Novamente instance as a distributed system (each Novamente Unit is a DINI Analytical Cluster)

- the nodes and links are labeled with different "type" labels;
- some of the nodes can contain data objects (characters, numbers, color values, etc);
- some of the nodes can contain small hypergraphs internally.

Conceptually, the two weight structures associated with Novamente Atoms involved represent the two primary schools of AI research – logic (TruthValue) and neural networks (AttentionValue).

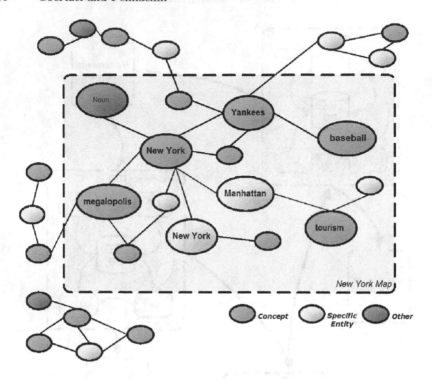

Fig. 5: Example of a Novamente map

The TruthValue indicates, roughly, the degree to which an Atom correctly describes the world. The object contains:

- a probability value;
- a "weight of evidence" value indicating the amount of evidence used to derive the probability;
- optionally further information such as a probability distribution function;
- optionally special information about the probability of an Atom in a given perception/action stream.

The AttentionValue is a bundle of information telling how much attention of various kinds an Atom should get and is getting. This includes:

- Long-Term-Importance (LTI), an estimate of the value of keeping the Atom in memory instead of paging it to disk;
- Recent Utility, a measure of how much value has been obtained from processing the Atom recently;
- Importance, a measure of how much CPU time the Atom should get, which is based on activation, LTI, and recent utility.

This special Atomspace hypergraph is used in many different ways. For instance:

1. all nodes and links are intended to be interpreted logically, using probabilistic logic;
2. some nodes and links can be seen to reflect processes of causation, and are
   used for "assignment of credit" which is a key part of adaptive attention
   allocation;
3. some nodes and links can be interpreted as executable programs.

Enabling all these interpretations simultaneously requires some care.

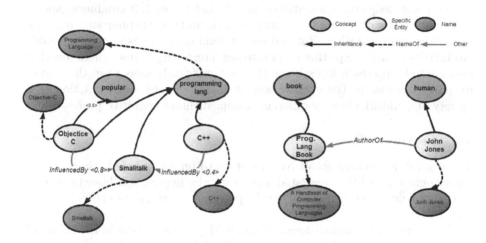

Fig. 6: Predicate expressions represented as nodes and links

What about the map level of knowledge representation? Because maps are
implicitly rather than explicitly encoded in the system, there is less that can
be said about them in a compact way. But one key point is that the network
of maps in the system is also conceivable as a hypergraph – one whose nodes
are fuzzy sets of Atoms. Map-level links are defined in the natural way: the
map-level link of a certain type T, between map A and map B, is defined
as the bundle of links of type T going between Atoms in A and Atoms in B
that are simultaneously active. Map-level links are defined implicitly by Atom-
level links. They represent a more general, diffuse kind of knowledge, which
interacts with Atom-level knowledge via a complex set of feedback effects.

In the language of the psynet model, maps are patterns, the "mind-stuff"
corresponding to the "brain-stuff" that is the Novamente software code and its
dynamic image in RAM. Atoms (nodes and links) exist on an interesting intermediate level that we call "concretely-implemented mind." That is, Atoms are
not mind-stuff, but they are parts of brain-stuff that that are "mind-indexers,"
in the sense that many Atoms are associated with specific patterns in the system (specific instances of mind-stuff), and the rest are directly included as
components in many patterns in the system.

The relation between Novamente structures and human brain structures is interesting but indirect, and will be reviewed in Section 5.8 below. In brief, there is no Novamente correlate of neurons and synapses – Novamente does not emulate the brain on such a low level. However, there is a rough intuitive mapping between Novamente nodes and what neuroscientist Gerald Edelman calls "neuronal groups"[23] – tightly connected clusters of 10,000-50,000 neurons. Novamente links are like bundles of synapses joining neuronal groups. And Novamente maps are something like Edelman's "neural maps."

Viewed against the background of contemporary narrow AI theory, the Novamente knowledge representation is not all that unusual. It combines aspects of semantic network, attractor neural network, and genetic programming style knowledge representation. But it does not combine these aspects in a "multi-modular" way that keeps them separate but interacting: it fuses them together into a novel representation scheme that is significantly more than the sum of its parts, because of the specific way it allows the cooperative action of a variety of qualitatively very different, complementary cognitive processes.

5.2 The Mind OS

The central design concept of Novamente is to implement multiple cognitive algorithms in a tightly-integrated way, using the hypergraph knowledge representation described just above, in the practical context of the DINI software architecture.

The crux of Novamente design from an AI perspective lies in the choice of cognitive algorithms and their manner of tight integration. Before we get there, however, there is one missing link to be filled in – the computational mechanics of actually managing a collection of tightly integrated AI processes. This is handled by a software component that we call the Mind OS, "Novamente core," or simply "the core."

As the "OS" moniker suggests, the Mind OS carries out many of the functions of an operating system. In fact it may be considered as a generic C++ server-side framework for multi-agent systems, optimized for complex and intensive tasks involving massive agent cooperation. While it is customized for Novamente AI, like DINI it is broadly extensible and could be used for many other purposes as well.

The Mind OS is itself a distributed processing framework, designed to live within the larger distributed processing framework of the DINI architecture[7]. It is designed to operate across a cluster of tightly-coupled machines, in such a way that a node living on one machine in the cluster may have links relating it to nodes living on other machines in the cluster. In DINI, the Mind OS is intended to live inside a complex analytic cluster. A complex Novamente configuration will involve multiple functionally-specialized analytic clusters, each one running the Mind OS.

---

[7]The current version of the Mind OS is restricted to a single SMP machine, but has been constructed with easy extension to distributed processing in mind.

On each machine in a Mind OS-controlled cluster, there is a table of Atoms (an AtomTable object, which comes with a collection of specialized indices for rapid lookup), and then a circular queue of objects called MindAgents. The MindAgents are cycled through, and when one gets its turn to act, it acts for a brief period and then cedes the CPU to the next MindAgent in the queue. Most of the MindAgents embody cognitive processes, but some embody "system-level" processes, like periodically caching the AtomTable to disk, polling for external input (such as input from a UI), or gathering system statistics. On an SMP machine, the Mind OS may allocate different MindAgents to the different processors concurrently, obeying a fixed table of exclusion relationships in doing so.

The distinction between MindAgents and psynet-model "mind actors" may be confusing here. This distinction reflects the subtlety of the system design, according to which the abstract mathematical structure of the system is different from the *implementation structure*. The software agents (MindAgents) are not the "mind actors" of the psynet model; rather, they are dynamic objects designed to elicit the emergence of the mind actors of the psynet model (the emergent maps). This separation between implementation agents and emergent agents is a compromise necessary to achieve acceptable computational efficiency.

Currently, communication with a Mind OS can be done either through a customized Unix shell called `nmshell`, which is appropriate for interactive communication, submission of control commands and debugging); through XML, using the Novamente DTD; or through a Java/J2EE middleware layer.

A third communication medium, via a Novamente-specific functional-logical programming language called *Sasha*, has been designed but not implemented. There is also a Novamente knowledge encoding language called NQL (Novamente Query Language), a modification of the KNOW language used with the Webmind system; but this interacts with the core indirectly via `nmshell` or XML.

In sum, the Novamente core is a C++ multi-agent system "OS" that supports:

- Multi-threading
- Flexible plugging and scheduling of heterogeneous agents
- Distributed knowledge with local proxies and caches
- Transaction control
- Communication with external software agents through XML and scripts
- Task and query processing through ticketing system
- Adaptive parameter control
- Dynamic, adaptive load balancing

In designing and implementing the core, great care has been taken to ensure computational time and space efficiency. We learned a lot in this regard from the flaws of the Webmind AI Engine, a distributed AI architecture designed in the late 1990s by a team overlapping with the current Novamente

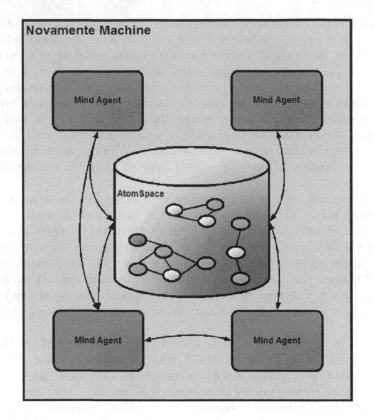

Fig. 7: Conceptual architecture of the Novamente "Mind OS" layer

team. The Webmind architecture was based on a fairly direct mapping of the psynet model into object-oriented software structures. It was a massive multi-agent system [70], using a hypergraph knowledge representation in which each node was implemented as an autonomous software agent. These node-agents carried out many of the same AI processes embodied in Novamente. However, the massive multi-agent system architecture proved difficult to tune and optimize. A moderately-sized distributed Webmind instance had millions of autonomous software agents in it (nodes, mainly); whereas by contrast, a moderately sized distributed Novamente instance will have merely *hundreds* (MindAgents).

Novamente is still a multi-agent system, but with a different architecture – and this architectural change makes a huge difference in the sorts of efficiency optimizations one can implement, resulting in an improvement of three orders of magnitude in speed and two orders of magnitude in memory use. We are extremely pleased that the Novamente Mind OS, in spite of its complexity, is efficient and robust enough to be used at the core of Biomind LLC's Hproduct line. At this stage, after many years of experimenting with this sort of software system, we are at the relatively happy point where the "Mind OS" level practical problems are all solved, and we can focus on the really hard part, fine-tuning the tight dynamical integration of the cognitive MindAgents.

5.3 Atom Types

Now we turn to a review of the specific types of nodes and links utilized in Novamente. As with the choice of MindAgents, this assemblage of node and link types has been chosen partly on pragmatic grounds, and partly on theoretical grounds. We have chosen data structures and dynamics based mainly on the following criteria:

- demonstrated power in narrow AI applications;
- mutual coherence as an integrative AGI framework;
- propensity for embodying the dynamics and structures posited by the Psynet Model of Mind.

Novamente contains a couple dozen node and link types, and a nearly-complete list is given in the AGIRI website. However, there is a layer of conceptual abstraction between the concept of "nodes and links" and the specific node and link types. We call this layer "node and link varieties" – each variety denotes a *conceptual* function rather than a mathematical or implementational category; and each variety may contain many different specific types. Tables 4 and 5 describe the node and link varieties currently used in Novamente.

Node Variety	Description
Perceptual Nodes	These correspond to particular perceived items, like WordInstanceNode, CharacterInstanceNode, NumberInstanceNode, PixelInstanceNode
Procedure Nodes	These contain small programs called "schema," and are called SchemaNodes. Action nodes that carry out logical evaluations are called PredicateNodes. ProcedureNodes are used to represent complex patterns or procedures.
ConceptNodes	These represent categories of perceptual or action or conceptual nodes, or portions of maps representing such categories.
Psyche Nodes	These are GoalNodes and FeelingNodes (special kinds of PredicateNodes), which play a special role in overall system control, in terms of monitoring system health, and orienting overall system behavior.

Table 4: Novamente node varieties

Link Variety	Description
Logical Links	These represent symmetric or asymmetric logical relationships among nodes (InheritanceLink, SimilarityLink) or among links and PredicateNodes (e.g. ImplicationLink, EquivalenceLink)
MemberLink	These denote fuzzy set membership
Associative Links	These denote generic relatedness, including HebbianLink learned via Hebbian learning, and a simple AssociativeLink representing relationships derived from natural language or from databases.
Action-Concept Links	Called ExecutionLinks and EvaluationLinks, these form a conceptual record of the actions taken by SchemaNodes or PredicateNodes
ListLink and ConcatListLink	These represent internally-created or externally-observed lists, respectively.

Table 5: Novamente link varieties

A thorough treatment not being practical here due to space considerations, we will give only a few brief comments on the semantics of these Novamente Atom types.

The workhorse of the system is the ConceptNode. Some of these will represent individual concepts, others will form parts of larger concept maps. Logical and associative links interrelate ConceptNodes. For example, we may write:

```
InheritanceLink New York megalopolis
```

meaning the there are ConceptNodes corresponding to the concepts "New York" and "nation", and there is an InheritanceLink pointing from one to the other (signifying that New York is indeed a megalopolis). Or we may write:

```
AssociativeLink New York immigration
```

which just indicates a generic association between the two denoted ConceptNodes. An associative relationship is useful for the spreading of attention between related concepts, and also useful as a signpost telling the logical inference MindAgents where to look for possibly interesting relationships.

A more concrete relationship between New York and immigration, such as "many immigrants live in New York", might be represented as:

```
ImplicationLink lives_in_New_York is_immigrant
```

where `lives_in_New_York` and `is_immigrant` are PredicateNodes, and the former predicate obeys a relationship that would be written:

```
EquivalenceLink (lives_in_New_York(X)) (lives_in(New_York, X))
```

in ordinary predicate logic, and is written more like:

```
EquivalenceLink lives_in_New_York (lives_in (New_York))
```

in Novamente's variable-free internal representation. Variable management is one of the most complex aspects of logic-based AI systems and conventional programming languages as well; Novamente bypasses the whole topic, by using a variable-free representation of predicates and schemata, based on combinatory logic.

SchemaNodes and PredicateNodes come in two forms: simple and complex. Each simple one contains a single elementary schema or predicate function; each complex one contains an internal directed-acyclic-graph of interlinked SchemaNodes and PredicateNodes.

The set of elementary schema/predicate functions is in effect an "internal Novamente programming language," which bears some resemblance to functional programming languages like pure LISP or Haskell. The "actions" carried out by SchemaInstanceNodes are not just external actions, they are also in some cases internal cognitive actions. Complex SchemaNodes represent complex coordinated actions that are "encapsulated" in a single node; complex PredicateNodes represent complex patterns observed in the system or the world outside, and found to be useful. ExecutionLinks and EvaluationLinks record information about what the inputs and outputs of SchemaInstanceNodes and PredicateInstanceNodes were when they executed.

Ultimately, all the AI processes carried out inside Novamente could be formulated as compound schemata, although in the current core implementation, this is not the case; the primary AI dynamics of the system are implemented as C++ objects called MindAgents, which are more efficient than compound schemata.

Next, FeelingNodes are "internal sensor" nodes, that sense some aspect of the overall state of the system, such as free memory or the amount the system has learned lately. Complex "feelings" are formed by combining FeelingNodes in PredicateNodes, and give the system a "sense of self" in a practical manner which allows for autonomic homeostasis to be performed and for the system to deliberately adjust its task orientation towards an increased sense of positive "feeling."

Finally, GoalNodes are internal sensors like FeelingNodes, but the condition that they sense may sometimes be less global; they represent narrow system goals as well as broad holistic ones. The system is supplied with basic goals as it is with basic feelings, but complex and idiosyncratic goals may be built up over time. GoalNodes are used in adjusting the system's autonomic processes to support focus on goal-oriented thought processes, as well as for the system to deliberately seek out and analyze relevant information to meeting these goals.

5.4 Novamente Maps

Many Atoms are significant and valuable in themselves, but some gain meaning only via their coordinated activity involving other Atoms, i.e. their involvement in "maps." Maps come in many shapes and sizes; a general characterization of Novamente maps would be difficult to come by. However, Table 6 enumerates several roughly defined "map categories" that we feel are useful for understanding Novamente on the map level, in a general way.

An interesting example of the relation between Atoms and maps in Novamente is provided by looking at the implementation of *satisfaction* in the system. Novamente has FeelingNodes which are "internal sensors" reporting aspects of current system state. Some of these are elementary, and some are combinations of inputs from other FeelingNodes. One important FeelingNode is the Satisfaction FeelingNode, which summarizes those factors that the system is initially programmed to consider as "desirable." This is referred to by the MaximizeSatisfaction GoalNode, which is the center of Novamente's motivational system.

On the surface, FeelingNodes look like a symbolic-AI-style representations of system feelings. However, to pursue a human-mind analogy, these FeelingNodes are really more like basic limbic-system or otherwise chemically-induced brain stimuli than they are like richly textured high-level human feelings. In the human mind, satisfaction is much more complex than momentary pleasure. It involves expectations of satisfaction over various time scales, and it involves inferences about what may give satisfaction, estimates of how satisfied others will be with a given course of action and thus how much pleasure one will derive from their satisfaction, etc. Biological pleasure is in a sense the root of human satisfaction, but the relationship is not one of identity. Changes in the biology of pleasure generally result in changes in the experience of satisfaction – witness the different subjective texture of human satisfaction in puberty as opposed to childhood, or maturity as opposed to early adulthood. But the details of these changes are subtle and individually variant.

So, in this example, we have a parallel between an Atom-level entity, the Pleasure FeelingNode, and an emergent mind map, a meta-Node, the feeling of system-wide satisfaction or "happiness." There is a substantial similarity between these two parallel entities existing on different levels, but not an identity. Satisfaction is embodied in:

- a large, fuzzily defined collection of nodes and links (a "map");
- the dynamic patterns in the system that are induces when this collection becomes highly active (a "map dynamic pattern").

The Satisfaction FeelingNode is one element of the map associated with overall system satisfaction or "happiness." And it is a particularly critical element of this map, meaning that it has many high-weight connections to other elements of the map. This means that activation of pleasure is likely – but not guaranteed – to cause happiness.

Table 6 describes some map types we find in Novamente. Figure 5 shows an example of a map.

Map Variety	Description
Concept map	A map consisting primarily of conceptual nodes
Percept map	A map consisting primarily of perceptual nodes, which arises habitually when the system is presented with environmental stimuli of a certain sort
Schema map	A distributed schema
Predicate map	A distributed predicate
Memory map	A map consisting largely of nodes denoting specific entities (hence related via MemberLinks and their kin to more abstract nodes) and their relationships
Concept-percept map	A map consisting primarily of perceptual and conceptual nodes
Concept-schema map	A map consisting primarily of conceptual nodes and SchemaNodes
Percept-concept-schema map	A map consisting substantially of perceptual, conceptual and SchemaNodes
Event map	A map containing many links denoting temporal relationships
Feeling map	A map containing FeelingNodes as a significant component
Goal map	A map containing GoalNodes as a significant component

Table 6: Example Novamente map varieties

5.5 Mind Agents

The crux of Novamente intelligence lies in the MindAgents, which dynamically update the Atoms in the system on an ongoing basis. Regardless of what inputs are coming into the system or what demands are placed upon it, the MindAgents keep on working, analyzing the information in the system and creating new information based on it.

There are several "system maintenance" MindAgents, dealing with things like collecting system statistics, caching Atoms to disk periodically, updating caches related to distributed processing, handling queues of queries from users and other machines in the same analytic cluster or other Novamente analytic clusters. We will not discuss these further here, but will restrict ourselves to the "cognitive MindAgents" that work by modifying the AtomTable.

Tables 7 and 8 briefly mention a few existing and possible MindAgents, while the AGIRI website gives a complete list of MindAgents, with brief comments on the function of each one on the Atom and map level. Section 7 below gives more detailed comments on a few of the MindAgents, to give a rough flavor for how the system works.

Agent	Description
First-Order Inference	Acts on first-order logical links, producing new logical links from old using the formulas of Probabilistic Term Logic
Logical Link Mining	Creates logical links out of nonlogical links
Evolutionary Predicate Learning	Creates PredicateNodes containing predicates that predict membership in ConceptNodes
Clustering	Creates ConceptNodes representing clusters of existing ConceptNodes (thus enabling the cluster to be acted on, as a unified whole, by precise inference methods, as opposed to the less-accurate map-level dynamics)
Importance Updating	Updates Atom "importance" variables and other related quantities, using specially-deployed probabilistic inference
Concept Formation	Creates speculative, potentially interesting new ConceptNodes
Evolutionary Optimization	A "service" MindAgent, used for schema and predicate learning, and overall optimization of system parameters

Table 7: Existing Novamente MindAgents

5.6 Map Dynamics

Much of the meaning of Novamente MindAgents lies in the implications they have for dynamics on the map level. Here the relation between Novamente maps and the concepts of mathematical dynamical systems theory is highly pertinent.

The intuitive concept of a map is a simple one: a map is a set of Atoms that act as a whole. They may act as a whole for purposes of cognition, perception, or action. And, acting as wholes, they may relate to each other, just like Atoms may relate to each other. Relationships between maps do not take the form of individual links; they take the form of bundles of links joining the Atoms inside one map to the Atoms inside another.

Map dynamics are a bit "slipperier" to talk about than Atom dynamics, because maps are not explicitly engineered – they emerge. To tell what Atoms are present in a system at a given time, one simply prints out the AtomTable. To tell what maps are present, one has to do some advanced pattern recognition on the Atomspace, to determine which sets of nodes are in fact acting as coordinated wholes. However, a map doesn't have to be explicitly identified by

Agent	Description
Higher-Order Inference	Carries out inference operations on logical links that point to links and/or PredicateNodes
Logical Unification	Searches for Atoms that mutually satisfy a pair of PredicateNodes
Predicate/Schema Formation	Creates speculative, potentially interesting new SchemaNodes
Hebbian Association Formation	Builds and modifies links between Atoms, based on a special deployment of probabilistic inference that roughly emulates (but greatly exceeds in exactness) Hebbian reinforcement learning rule
Evolutionary Schema Learning	Creates SchemaNodes that fulfill criteria, e.g. that are expected to satisfy given GoalNodes
Schema Execution	Enacts active SchemaNodes, allowing the system to carry out coordinated trains of action
Map Encapsulation	Scans the AtomTable for patterns and creates new Atoms embodying these patterns
Map Expansion	Takes schemata and predicates embodied in nodes, and expands them into multiple nodes and links in the AtomTable (thus transforming complex Atoms into maps of simple Atoms)
Homeostatic Parameter Adaptation	Applies evolutionary programming to adaptively tune the parameters of the system

Table 8: Additional, planned Novamente MindAgents

anyone to do its job. Maps exist implicitly in a dynamic Novamente system, emerging out of Atom-level dynamics and then guiding these dynamics.

In dynamical systems terms, there are two kinds of maps: *attractor maps*, and *transient maps*. Schema and predicate maps are generally transient, whereas concept and percept maps are generally attractors; but this is not a hard and fast rule. Other kinds of maps have more intrinsic dynamic variety, for instance there will be some feeling maps associated with transient dynamics, and others associated with attractor dynamics.

The sense in which the term "attractor" is used here is slightly nonstandard. In dynamical systems theory [21], an attractor usually means a subset of a system's state space which is:

Invariant, when the system is in this subset of state space, it doesn't leave it;

Attractive, when the system is in a state near this subset of state space, it will voyage closer and closer to the attracting subspace.

In Novamente, the subset of state space corresponding to a map is the set of system states in which that map is highly important. However, in Novamente dynamics, these subsets of state space are almost never truly invariant.

Many maps are attractive, because Novamente importance updating dynamics behaves roughly like an attractor neural network. When most of a map is highly important, the rest of the map will get lots of activation which will make it highly important. On the other hand, Atoms linked to map elements via inhibitory links will get less activation, and become less important.

But maps are not invariant: once a map is active, it is *not* guaranteed to remain active forever. Rather, the Importance Updating Function, regulating Novamente dynamics, guarantees that most of the time, after a map has been important for a while, it will become less important, because the percentage of new things learned about it will become less than the percentage of new things learned about something else.

This combination of *attractiveness* and *temporary invariance* that we see in connection with Novamente maps, has been explored by physicist Mikhail Zak [74], who has called subsets of state space with this property *terminal attractors*. He has created simple mathematical dynamical systems with terminal attractors, by using iteration functions containing mathematical singularities. He has built some interesting neural net models in this way. The equations governing Novamente bear little resemblance to Zak's equations, but intuitively speaking, they seem to share the property of leading to terminal attractors, in the loose sense of state space subsets that are attractive but are only invariant for a period of time.

Many *concept maps* will correspond to fixed point map attractors – meaning that they are sets of Atoms which, once they become important, will tend to stay important for a while due to mutual reinforcement. On the other hand, some concept maps may correspond to more complex map dynamic patterns. And *event maps* may sometimes manifest a dynamical pattern imitating the event they represent. This kind of knowledge representation is well-known in the attractor neural networks literature.

Turning to schemata, an individual SchemaNode does not necessarily represent an entire schema of any mental significance – it may do so, especially in the case of a large encapsulated schema; but more often it will be part of a distributed schema (meaning that SchemaNode might more accurately be labeled LikelySchemaMapComponentNode). And of course, a distributed schema gathers its meaning from what it does when it executes. A distributed schema is a kind of mind map – a map that extends beyond SchemaInstanceNode and SchemaNodes, bringing in other nodes that are habitually activated when the SchemaInstanceNodes in the map are enacted. Note that this system behavior may go beyond the actions explicitly embodied in the SchemaNode contained in the distributed schema. Executing these SchemaNodes in a particular order may have rampant side-effects throughout the system, and these side-effects may have been taken into account when the schema was learned, constituting a key part of the "fitness" of the schema.

Next, percepts – items of data – coming into the system are not necessarily represented by individual perceptual nodes. For instance, a word instance that has come into the system during the reading process is going to be represented in multiple simultaneous ways. There may be a WordInstanceNode, a ListLink of CharacterInstanceNodes, and so forth. In a vision-endowed system, a representation of the image of the word will be stored. These will be interlinked, and linked to other perceptual and conceptual nodes, and perhaps to SchemaNodes embodying processes for speaking the word or producing letters involved in the word. In general, percepts are more likely to be embodied by maps that are centered on individual perceptual nodes (the WordInstanceNode in this case), but this is not going to be necessarily and universally the case.

Links also have their correlates on the map level, and in many cases are best considered as seeds that give rise to inter-map relationships. For example, an InheritanceLink represents a frequency relationship between nodes or links, but inheritance relationships between maps also exist. An inheritance relation between two maps A and B will not generally be embodied in a single link, it will be implicit in a set of InheritanceLinks spanning the Atoms belonging to A and the Atoms belonging to B. And the same holds for all the other varieties of logical relationship. Furthermore, the first-order inference rules from Probabilistic Term Logic, Novamente's reasoning system, carry over naturally to map-level logical links.

5.7 Functional Specialization

Now we return to the DINI architecture and its specific use within Novamente. The Novamente MindAgents are designed to be tightly integrated, so that a large collection of MindAgents acts on a large population of Atoms in an interleaved way. This set of Atoms may live on one machine, or on a cluster of connected machines. This kind of tight integration is essential to making integrative AGI work.

But, according to the Novamente design, there is also another layer required, a layer of loose integration on top of the tightly integrated layer. A Novamente system consists of a loosely-integrated collection of "analytic clusters" or "units," each one embodying a tightly-connected collection of AI processes, involving many different Atom types and MindAgents, and dedicated to a particular cognitive processing in a certain particular domain, or with a specific overall character.

The different analytic clusters interact via DINI; they all draw data from, and place data in, the same system-wide data warehouse. In some cases they may also query one another. And the parameters of the MindAgents inside the various analytic clusters may be adapted and optimized globally.

The simplest multi-cluster Novamente has three units, namely:

1. a *primary cognitive unit*;
2. a *background thinking unit*, containing many more nodes with only very important relationships among them, existing only to supply the primary cognitive unit with things it judges to be relevant;

3. an *AttentionalFocus unit*, containing a small number of atoms and doing very resource-intensive processing on them.

Here the specialization has to do with the intensity of processing rather than with the contents of processing.

For a Novamente to interact intensively with the outside world, it should have two dedicated clusters for each "interaction channel":

- one to contain the schemata controlling the interaction;
- one to store the "short-term-memory" relating to the interaction.

An "interaction channel" is a collection of sensory organs of some form, all perceiving roughly the same segment of external reality. Each human has only one interaction channel. But Novamente does not closely emulate either the human body or brain, and so it can easily be in this situation, interacting separately with people in different places around the world.

Perceptual processing like image or sound processing will best be done in specially dedicated units, with highly modality-tuned parameter values. Language processing also requires specialized units, dealing specifically with aspects of language processing such as parsing, semantic mapping, and disambiguation.

The human brain contains this kind of functional specialization to a large degree. In fact we know more about the specialization of different parts of the brain than about how they actually carry out their specialized tasks. Each specialized module of the brain appears to use a mixture of the same data representations and learning processes [34]. Many AI systems contain a similar modular structure, but each module contains a lot of highly rigid, specialized code inside. The approach here is very different. One begins with a collection of actors emergently providing generic cognitive capability, and then sculpts the dynamical patterns of their interactions through functional specialization.

5.8 Novamente and the Human Brain

Having reviewed the key aspects of the Novamente design, we now briefly return to a topic mentioned earlier, the relationship between Novamente and the human brain. While Novamente does not attempt to emulate neural structure or dynamics, there are nevertheless some meaningful parallels. Table 9 elucidates some of the more important ones.

On a structural level, the parallels are reasonably close: Novamente's functionally-specialized lobes are roughly analogous to different regions of the brain. At an intermediate level, Novamente nodes are roughly analogous to neuronal groups in the brain, as mentioned above; and Novamente links are like the synapse-bundles interconnecting neuronal groups. Novamente maps are like Edelman's neuronal maps, and also in some cases like the neural attractors posited by theorists like Daniel Amit [4] and Walter Freeman [25].

Human Brain Structure or Phenomena	Primary Functions	Novamente Structure or Phenomena
Neurons	Impulse-conducting cells, whose electrical activity is a key part of brain activity	No direct correlate
Neuronal groups	Collections of tightly interconnected neurons	Novamente nodes
Synapses	The junction across which a nerve impulse passes from one neuron to another; may be excitatory or inhibitory	Novamente links are like bundles of synapses joining neuronal groups
Synaptic Modification	Chemical dynamics that adapt the conductance of synapses based on experience; thought to be the basis of learning	The HebbianLearning MindAgent is a direct correlate. Other cognitive MindAgents (e.g. inference) may correspond to high-level patterns of synaptic modification
Dendritic Growth	Adaptive growth of new connections between neurons in a mature brain	Analogous to some heuristics in the ConceptFormation MindAgent
Neural attractors	Collections of neurons and/or neuronal groups that tend to be simultaneously active	Maps, e.g. concept and percept maps
Neural input/output maps	Composites of neuronal groups, mapping percepts into actions in a context-appropriate way	Schema maps
"Neural Darwinist" map evolution	Creates new, context-appropriate maps	Schema learning via inference, evolution, reinforcement learning
Cerebrum	Perception, cognition, emotion	The majority of Units in a Novamente configuration
Specialized cerebral regions	Diverse functions such as language processing, visual processing, etc....	Functionally-specialized Novamente Units
Cerebellum	Movement control, information integration	Action-oriented units, full of action schema-maps
Midbrain	Relays and translates information from all of the senses, except smell, to higher levels in the brain	Schemata mapping perceptual Atoms into cognitive Atoms
Hypothalamus	Regulation of basic biological drives; control of autonormic functions such as hunger, thirst, and body temperature	Homeostatic Parameter Adaptation MindAgent, built-in GoalNodes
Limbic System	Controls emotion, motivation, and memory	FeelingNodes and GoalNodes, and associated maps

Table 9: Novamente and the human brain

The parallels get weaker, however, when one turns to dynamics. Very little is known about the intermediate-scale dynamics of the brain. We know basically how neurons work, but we don't know much about the dynamics interrelating the levels of different types of neurotransmitters in different brain regions, nor about extracellular charge diffusion, or even about the dynamical behavior of complex collectives of real neurons. Novamente has a number of specific cognitive dynamics (e.g. probabilistic inference) that have no known analogues in brain dynamics; but this means little since intermediate-level brain dynamics is so poorly understood.

5.9 Emergent Structures

The dynamics of a Novamente system is largely controlled by the structure of the Atom hypergraph, and that the structure of the Atom hypergraph is strongly guided, and partly explicitly formed, by the dynamics of the system. This structural-dynamical feedback can lead to all kinds of complex emergent structures – some existing in the system at a given time, some manifesting themselves as patterns over time, and some spatiotemporal in nature. Maps are one manifestation of this feedback; but there is also a higher level of organization, in which the network of maps achieves certain emergent patterns. Among these emergent patterns are the ones identified in the psynet model of mind: the dual network and the self.

The Dual Network

The dual network, in Novamente, takes a fairly simple and direct form:

- the heterarchical aspect consists of the subnetwork defined by symmetric logical links and/or AssociativeLinks;
- the hierarchical aspect consists of the subnetwork defined by asymmetric logical links and associative links, and the subnetwork defined by schemata and their control relationships (schema A being said to control schema B when A modifies B's parameters significantly more than vice versa).

Schemata aside, the static aspect of the dual network is fairly straightforward. For instance, the ConceptNodes corresponding to different nations may be interlinked by SimilarityLinks and AssociativeLinks: this is a small "heterarchical network," a subset of the overall heterarchical network within a given Novamente's Atom space. These nodes representing individual nations may all inherit from the Nation ConceptNode (InheritanceLink being an asymmetric logical link). This is a simple, static example of dual network structure: elements that are related heterarchically are also close together in their hierarchical relationships. This aspect of dual network structure falls out pretty naturally from the intrinsic semantics of similarity, inheritance and association.

The control aspect of the dual network is less obvious and will only emerge if the various MindAgents are operating together properly. For example, consider a family of schemata, each concerned with recognizing some part of speech: nouns, verbs, adjectives, etc. These schemata will have similarities and associations with each other. They will all inherit from a handful of more general schemata for analyzing words and their properties. But they will also be *controlled* by these more general word-analysis schemata. Their control parameters and their flow of execution will be modulated by these more general control schemata. The coincidence of inheritance hierarchy and control hierarchy, and the overlaying of this coincident hierarchy on the associative/similarity heterarchy, is the crux of the "dual network" structure. It is not programmed into Novamente, but Novamente is designed so as to encourage it to emerge.

Specifically, the emergence of this kind of dual network metapattern follows fairly naturally from the harmonious interaction of:

- inference building similarity and inheritance links;
- importance updating, guiding the activation of atoms (and hence the application of built-in primary cognitive processes to atoms) based on the links between them;
- schema learning, which extends a schema's applicability from one node to another based on existing links between them (and based on observations of past schema successes and failures, as will be explained later).

The dual network structure is a *static representation* of the *dynamic cooperation* of these processes. We have discussed it here on the Atom level, but its manifestation on the map level is largely parallel, and equally important.

The Self

Just as important as the dual network is the mind structure that we call the "self." We stress that we are using a working definition of self, geared towards serving as a usable guideline for AGI engineering. We deliberately avoid ontological or existential discussions of the universal nature of selfhood and its relation to consciousness.

The "raw material" for Novamente's self – the primary senses in which a Novamente can self-reflect – consists of the collection of:

- patterns that the system has observed in itself as a whole, that is, the structural and dynamical patterns within its internal dual-network;
- patterns that it has observed in its own external actions, that is, that subnetwork of its dual network which involves tracking the procedure and consequences of running various schema;
- patterns that the system has observed in its relationship with other intelligent systems.

What we call the *self* is then a collection of patterns recognized in this set. Often the patterns recognized are very approximate ones, as the collection of data involved is huge and diverse – even a computer doesn't have the resources to remember every detail of every thing it's ever done. Furthermore, the particular data items leading to the creation of the psynet-wide patterns that define the self will often be forgotten, so that the self is a poorly grounded pattern (tuning how poorly grounded it may be, and still be useful, will be a subtle and crucial part of giving Novamente a useful, nontrivial sense of self).

On the map level, we may say that the self consists of:

- a set of *self-image maps*: maps that serve as an "internal images" of significant aspects of a Novamente system's structure or dynamics, or its interactions with other intelligent systems;
- a larger map that incorporates various self-image maps along with other Atoms (this is the emergent self).

The really interesting thing about the self is the feedback between declarative, localized knowledge and distributed, procedural knowledge that it embodies. As the collection of high-level patterns that is the self become more or less active, they automatically move the overall behavior of the system in appropriate directions. That is to say, as the system observes and reasons upon its patterns of self, it can then adjust its behavior by controlling its various internal processes in such a way as to favor patterns which have been observed to contribute to coherent thought, "good feelings," and satisfaction of goals.

We note the key role of interactions with humans in Novamente's development of self. While it would be theoretically possible to have self without society, society makes it vastly easier, by giving vastly more data for self-formation – and for a self to be able to function sufficiently in a world where there are other selves, society is indispensable. In time, it may be interesting to create a community of interacting Novamente AI systems. Initially, Novamente will learn about itself through interacting with humans. As humans ask it questions and chat with it, it will gain more and more information not only about humans but about what Novamente itself is, from the point of view of others. This will shape its future activities both explicitly and implicitly.

6 Interacting with Humans and Data Stores

The deployment of Novamente for knowledge management and analysis involves attention to many issues beyond those occurring in relation to "Novamente AGI" in general. Most of these issues fall into the categories of data sources and human-computer interaction.

The optimal way of handling such issues is domain-dependent. For the bioinformatics applications, we have taken an approach guided by the particular needs of bioscientists analyzing datasets generated via high-throughput genomics equipment. This section contains a brief description of our plans in these areas.

A key conceptual point arising here is the relationship between AI and IA (Intelligence Augmentation). Its ambitious long-term AGI goals notwithstanding, it is very clear that in the medium term Novamente is not going to outperform human intelligence all around. Rather, it should be viewed as a complement to individual and collective human intelligence. Humans will learn from Novamente's unique insights, and Novamente will also learn from humans. Specifically, Novamente leverages human intelligence by:

- ingesting data encoded by humans in databases;
- ingesting knowledge specifically encoded by humans for Novamente use;
- learning from its dialogues and interactions with humans;
- human construction of training sets for supervised categorization;
- learning from humans' ratings of its own and other humans' answers to queries;

The design of appropriate user interfaces embodying these factors is a significant undertaking in itself, and not one that we will enlarge on in this chapter. Here we will restrict ourselves to a brief discussion of the key features required, and the most salient issues that arise with them.

6.1 Data Sources

We have already discussed the conceptual issues involved with feeding Novamente databases in Section 2.2 above.

As noted there, Novamente is intended to work with external databases that have been integrated according to a "knowledge integration" methodology. This means that translators must be written, mapping the schemata within DB's into XML structured according to Novamente's XML DTD. This effectively maps database information into Novamente nodes and links. In this manner, a unified data warehouse may be built up, containing a diverse mix of data and abstract information. Table 10 and Fig. 8 show an example of the mapping of relational database table elements into Novamente nodes and links.

ID	CompanyName	EIN	ParentCo	CEO	...
...	...	...	...	...	...
2003	Postindustrial Widgets LLC	123-45-6789	2453	J. J. James	...
...	...	...	...	...	...
...	...	...	...	...	...
2453	The Associated Amalgamated Group, Inc.	897-65-4321	*null*	*null*	...

Table 10: Example RDB table

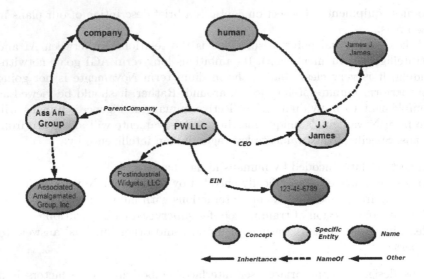

Fig. 8: Depiction of RDB table as Novamente nodes and links

Regarding the substantial amount of knowledge in contemporary databases as textual rather than structured, Novamente can ingest text using simplified statistical methods, and we have experimented with this in the context of biological research papers. But, real natural language understanding is obtained only by leaving text processing behind, and having Novamente translate back and forth between linguistic character sequences on the one hand, and semantically meaningful nodes and links on the other. This requires that natural language processing be implemented in a very deep way, as part and parcel of abstract Novamente cognition.

We believe the Novamente design can overcome the problems experienced by contemporary NLP algorithms, due to its integrative approach, which involves carrying out syntactic analysis via logical unification, a process that automatically incorporates available semantic and pragmatic knowledge into its behavior. We have not yet implemented NLP in the Novamente system, but our experience with a similar implementation in the Webmind system gives us reasonable confidence here. We return to this issue below.

6.2 Knowledge Encoding

Sometimes the data available in existing databases will not be enough to bring Novamente "up to speed" on a pertinent area. A significant proportion of human knowledge is "tacit" and is never written down anywhere, textually or in relational or quantitative form. Furthermore, in the case of knowledge that is expressed only in difficult-to-comprehend textual documents, Novamente's understanding may be enhanced by providing it with portions of the knowledge in explicit form.

For these reasons, it will sometimes be valuable to have humans encode knowledge formally, specifically for ingestion by Novamente. There are two different approaches here:

- "expert system" style formal language encoding of knowledge;
- knowledge entry via interactive Web forms.

The Web forms approach was prototyped at Webmind Inc. and seemed to be a reasonable way for individuals with little training to encode large amounts of relatively simple information. For formal language encoding, we have developed a formal language called NQL, which is similar to Cyc-L but has a much simpler syntax.

We caution that we are not proposing a traditional "expert systems" approach here, nor a traditional "common sense" knowledge encoding project a la Cyc. We anticipate that well less than 1% of the knowledge in Novamente will be placed there via human knowledge encoding. In our view, the role of knowledge encoding should be to fill in gaps, not to supply a fundamental knowledge base.

6.3 Querying

We have discussed how knowledge gets into Novamente – but how does it get out? How do humans ask Novamente questions? How do they iterate with the system to cooperatively find and produce knowledge? Our intention is to create a prototype user interface that is integrative in nature, encompassing a variety of complementary mechanisms.

1	Search Engine style queries `Manhattan sushi restaurants`
2	Natural language queries `I want information on outstanding sushi restaurants in Manhattan`
3	Formal language queries `X: X inheritsFrom restaurant AND` `Y: Y inheritsFrom sushi AND` `sells(X, Y) AND quality(Y, outstanding)`
4	Interactive conversation encompassing both NLP and formal language queries
5	Web forms queries covering common cases
6	Export of data into spreadsheets and other analytic software

Table 11: Different types of queries for Novamente

Practical experimentation with these mechanisms in a real-world data analysis context will teach us which are found most valuable by human users in which contexts; and this will guide further refinement of the Novamente UI and also of Novamente's internal query processing mechanisms.

6.4 Formal Language Queries

For untrained users, natural language queries and natural language conversation are clearly the most desirable interaction modalities. For trained expert users, on the other hand, there may be significant advantages to the use of formal language queries, or of queries mixing formal with natural language.

Formal queries allow a level of precision not obtainable using natural language. Furthermore – and this is a critical point – by having expert users submit the same queries in both natural language and formal-language format, Novamente will gain pragmatic knowledge about query interpretation. This is an example of how Novamente can learn from humans, who at least initially will be far smarter than it at interpreting complex human-language sentences.

For example, consider the query:

```
I want information on outstanding sushi restaurants in Manhattan
```

As a formal language query, this becomes simply:

```
Find X, Y so that:
    Inheritance X ''Japanese restaurant''
    location X Manhattan
    sells X Y
    Inheritance Y Sushi
    quality Y outstanding
```

Or, consider:

```
I want information on profitable companies from the United
States that sell their services to schools.
```

A sentence like this poses interpretative problems for current NLP systems. They have trouble determining which is the antecedent of "their": "profitable companies" or "the United States." Making the correct choice requires real-world understanding or extensive domain-specific system tuning. On the other hand, for an expert user, creating an appropriate formal language query to cover this case is easy:

```
Inheritance X ''United States''
based Y X
Inheritance Y profitable
sells Y Z
buys Z W
Inheritance W ''school''
```

The initial Novamente NLP system may sometimes make errors resolving sentences like the above. If a user submits this query to Novamente in both English and formal-language form, then Novamente will observe the correct interpretation of the sentence, and adjust its semantic mapping schemata accordingly (via the activity of the schema learning Mind Agents). Then the next time it sees a similar sentence, it will be more likely to make the right judgment.

When a human child learns language, they correct their interpretations via observing others' interpretations of utterances in the real world. Novamente will have fewer opportunities than humans to make this kind of observation-based correction, but as partial compensation, it has the ability to compare natural language sentences with expert-produced formal language renditions. And this "language teaching" need not be done as a special process, it may occur as a part of ordinary system usage, as expert users submit formal language queries and NL queries side by side.

6.5 Conversational Interaction

The "query/response" interaction modality is important and valuable, but it has its limitations. Often one wishes to have a series of interactions with persistent context – i.e., a conversation. Novamente is designed to support this, as well as more conventional query/response interactions. We are currently prototyping Novamente conversations in the context of the toy ShapeWorld environment.

Conversational interaction harmonizes nicely with the idea of mixed formal/natural language communication discussed above. The conversation example given in Table 3 above illustrates this concept concretely.

As we have not currently implemented any NLP at all in Novamente, achieving this sort of conversation with Novamente remains an R&D endeavor with the usual associated risks. Our current applications of Novamente are more along the lines of data analysis. However, we did prototype interactive conversation in the related Webmind software system, to a limited extent, and from this experience we gained a thorough understanding of the issues involved in approaching such functionality.

6.6 Report Generation

Another useful (and much simpler) human interaction functionality is report generation. The system will be able to automatically generate summary reports containing information pertinent to user queries, or simply summarizing interesting patterns it has found through its own spontaneous activity. Reports may contain:

- quantitative data;
- relationships expressed in formal language (predicate expressions);
- natural language produced by "language generation" and "text summarization" algorithms.

6.7 Active Collaborative Filtering and User Modeling

Finally, Novamente will gather information about human preferences in general, and the preferences of its individual users, through techniques refined in the "active collaborative filtering" community. Essentially, this means that users will be asked to rate Novamente's responses on several scales (e.g. usefulness, veracity). Furthermore, Novamente's UI will be configured to collect "implicit ratings" – information regarding how long they look at an information item, what they use it for, etc. Novamente will incorporate this information into its knowledge store, to be used as the subject of ongoing pattern analysis, which will enable it to adapt its behavior so as to better serve future users.

7 Example Novamente AI Processes

In this section we will briefly review a few of the most important AI processes in the Novamente system: probabilistic inference, nonlinear attention allocation, procedure learning, pattern mining, categorization, and natural language processing. These processes form a decent cross-section of what goes on in Novamente. We will illustrate each process with an intuitive example of what the process contributes to Novamente.

Table 12 compares standard approaches to some cognitive tasks and the approaches we have taken in Novamente.

Logical Inference	
Standard Approaches	Predicate, term, combinatory, fuzzy, probabilistic, nonmonotonic or paraconsistent logic
Challenges	Accurate management of uncertainty in a large-scale inference context *Inference Control*: intelligent, context-appropriate guidance of sequences of inferences
Novamente Approach	Probabilistic Term Logic tuned for effective large-scale uncertainty management, coupled with a combination of noninferencial cognitive processes for accurate control

Attention Allocation	
Standard Approaches	Blackboard systems, neural network activation spreading
Challenges	The system must focus on user tasks when needed, but also possess te abilit to spontaneously direct its own attention without being flighty or obsessive
Novamente Approach	Novamente's nonlinear importance updating function combines quantities derived from neural-net-like importance-updating and blackboard-system-like cognitive utility analysis
Procedure Learning	
Standard Approaches	Evolutionary programming, logic-based planning, feedforward neural networks, reinforcement learning
Challenges	Techniques tend to be unacceptably inefficient except in very narrow domains
Novamente Approach	A synthesis of techniques allows each procedure to be learned in the context of a large number of other already-learned procedures, enhancing efficiency considerably
Pattern Mining	
Standard Approaches	Association rule mining, genetic algorithms, logical inference, machine learning, search algorithms
Challenges	Finding complex patterns requires prohibitively inefficient searching through huge search spaces
Novamente Approach	Integrative cognition is designed to home in on the specific subset of search space containing complex but compact and significant patterns
Human Language Processing	
Standard Approaches	Numerous parsing algorithms and semantic mapping approaches exist, like context-free grammars, unification grammars, link grammars, conceptual graphs, conceptual grammars, etc
Challenges	Integrating semantic and pragmatic understanding into the syntax-analysis and production processes
Novamente Approach	Syntactic parsing is carried out via logical unification, in a manner that automatically incorporates probabilistic semantic and pragmantic knowledge. Language generation is carried out in a similarly integrative way, via inferential generalization

Table 12: Comparison of approaches to several cognitive tasks

7.1 Probabilistic Inference

Logical inference has been a major theme of AI research since the very beginning. There are many different approaches out there, including:

- predicate logic, e.g. Cyc [51], SNARK [63];
- combinatory logic [17, 24];
- uncertain term logic, e.g. Pei Wang's Non-Axiomatic Reasoning System (NARS), see this volume;
- probabilistic inference, e.g. Bayes nets [55], probabilistic logic programming [35];
- fuzzy logic [73];
- paraconsistent logic [60];
- nonmonotonic logic [65].

The basic task of computational logical deduction is a solved problem, but there are still many open problems in the area of AI and logic, for instance:

- inference control (what inferences to make when);
- representation and manipulation of uncertainty (fuzzy vs. probabilistic vs. multi-component truth value, etc);
- optimal logical representation of specific types of knowledge, such as temporal and procedural;
- inferences beyond deduction, such as induction, abduction [45] and analogy [43];

For these aspects of inference, many approaches exist with no consensus and few unifying frameworks. Cyc is perhaps the most ambitious attempt to unify all the different aspects of logical inference, but it's weak on nondeductive inference, and its control mechanisms are highly domain-specific and clearly not generally adequate.

The practical need for logical inference in a national security context is obvious. Among other things, inference can:

- synthesize information from multiple DBs;
- help interpret natural language;
- help match user queries to system knowledge;
- draw complex conclusions based on integrating a huge number of small pieces of information.

To take a DB integration example, when Database 1 says:

``Money often flows from XYZ Bank to Luxembourg.''

and Database 2 says:

``M. Jones has significant funds in XYZ bank.''

then abductive inference says:

``Maybe M. Jones is sending money to Luxembourg.''

which is a speculative, but possibly interesting, conclusion.

Novamente's logical inference component consists of a number of MindAgents for creating logical links, both from other logical links ("inference") and from nonlogical links ("direct evaluation"). It involves several different MindAgents:

- LogicalLinkMiner MindAgent (builds logical links from nonlogical links)
- FirstOrderInference MindAgent
- HigherOrderInference MindAgent
- LogicalUnification MindAgent
- PredicateEvaluation MindAgent
- TemporalInference MindAgent

Here we will discuss just one of these, the FirstOrderInference (FOI) MindAgent. This agent carries out three basic inference rules, deduction, inversion and revision. It also converts similarity relationships into inheritance relationships and vice versa. Each of its inference rules is probabilistic in character, using a special formula to take the probabilistic truth values of the premises and outputting a probabilistic truth value for the conclusion. These formulas are derived using a novel mathematical framework called Probabilistic Term Logic (PTL). The PTL inversion formula is essentially Bayes' rule; the deduction formula is unique to Novamente, though it is simply derivable from elementary probability theory. Revision is a weighted-averaging rule that combines different estimates of the truth value of the same relationship, coming from different sources. The rules deal with weight of evidence as well as strength, and have variants dealing with distributional truth values.

The combination of deduction and inversion yields two forms of inference familiar from the literature: induction and abduction. Induction and abduction are speculative forms of inference, intrinsically less certain than deduction, and the corresponding formulas reflect this. Figure 9 shows the basic patterns of deduction, induction abduction and revision. Examples of first-order inference are shown in Table 13.

The dynamics of Novamente TruthValues under PTL can be quite subtle. Unlike the NARS system and most other logical inference systems (loopy Bayes' nets being an exception), we do not rule out circular inference; we embrace it. Circular inferences occur rampantly, ultimately resulting in a "attractor state" of truth values throughout the system, in which all the truth values of the Atoms are roughly (though not necessarily exactly) consistent with each other. Interestingly, although PTL is based on formal logic, its dynamics more closely resemble those of attractor neural networks.

Special additions to the FOI framework deal with hypothetical, subjective and counterfactual knowledge, e.g. with statements such as

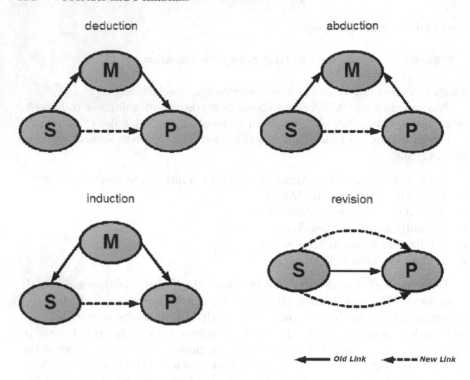

Fig. 9: First-order inference on InheritanceLinks

```
Joe believes the Earth is flat.
If Texas had no oil, then...
```

It is important that the system be able to represent these statements without actually coming to believe *"the Earth is flat"* or *"Texas has no oil."* This is accomplished by the HypotheticalLink construct and some simple related inference rules.

Higher-order inference deals with relationships such as:

```
ownerOf(X, Y) IFF possessionOf(Y, X)
```

In this example we have used traditional predicate logic notation to represent the antisymmetry of the ownership and possession relationships, but inside Novamente things are a little different: there are no variables at all. Instead, a combinatory logic approach is used to give variable-free representations of complex relationships such as these, as networks of PredicateNodes and SchemaNodes (including SchemaNodes embodying the "elementary combinators"). Using the **C** combinator, for instance, the above equivalence looks like:

```
Deduction:
  IBM is a US company
  US companies have EIN's
  |-
  IBM has an EIN

Induction:
  IBM is a US company
  IBM manufactures computers
  |-
  US companies manufacture computers

Abduction:
  Widgco is a US company selling widgets in Mexico
  Jim is CEO of a US company selling widgets in Mexico
  |-
  Jim is CEO of Widgco

Revision:
  According to the WSJ, Widgco will probably file for bankruptcy
  According to the Times, Widgco will possibly file for bankruptcy
  |-
  Widgco will probably file for bankruptcy
```

Table 13: Examples of first-order inference

EquivalenceLink ownerOf (C possessionOf)

The absence of variables means that the higher-order inference rules are basically the same as the first-order inference rules, but there are some new twists, such as logical unification, and rules for mixing first-order and higher-order relationships. The details by which these "twists" are resolved are due to the integrative nature of the Novamente system: for instance logical unification is carried out via Novamente's integrative schema/predicate learning process, incorporating evolutionary and reinforcement learning aloing with inference.

The logical inference MindAgents operate via importance-based selection: that is, when they are activated by the scheduler, they choose Atoms to reason on with probability proportional to their importance. Basic inference control is thus effectively delegated to the ImportanceUpdating MindAgent. Special-purpose inference control may be carried out by learned or programmed schemata embodied in SchemaInstanceNodes.

7.2 Nonlinear-Dynamical Attention Allocation

Apart from formal logic, the other major theme in the history of AI is formal neural network modeling. Neural networks excel at capturing the holistic and dynamical aspects of intelligence. Neural net inspired methods are used in Novamente in two places:

- in the ImportanceUpdating MindAgent, which is used to direct the system's attention to different Atoms differentially;
- in the HebbianLearning MindAgent, which modifies the TruthValues of logical links according to a special inference control mechanism that loosely emulates the basic Hebbian learning rule.

However, although the activity of these two MindAgents is loosely inspired by neural networks, we do not use neural net algorithms in Novamente. This is a choice made for reasons of simplicity and efficiency. Instead of neural nets, we use Probabilistic Term Logic in specially controlled ways that allow it to roughly emulate the interesting dynamics one sees in attractor neural networks.

We believe it probably *would* be possible to achieve the kind of precise inference that PTL does, using purely neural net based methods; and we did some preliminary work along these lines in 2002, developing an experimental neural-net updating approach called "Hebbian Logic." However, we believe that would be an unacceptably inefficient approach given the realities of von Neumann computer implementation.

7.3 Importance Updating

Attention allocation refers to the process by which the system determines how much processor time each of its Atoms should get. This is done by the ImportanceUpdating MindAgent, which adjusts the AttentionValues of the Nodes and Relationships it touches. Importance is determined by a special formula, the Importance Updating Function, that combines the other quantities that form the AttentionValue. This formula is based on probabilistic inference, and it may be interpreted as a special "inference control strategy" that does inferences in a certain order for each Atom at each cycle.

The formula is simple but somewhat subtle and was arrived at through a combination of mathematical analysis and practical experimentation. The basic idea of the formula is implied by the following criteria:

1. In the absence of other causes, importance decays.
2. The LTI of an atom is computed to be high if accumulated recent-utility is large, and low otherwise.
3. LTI is interpreted as a resting value, a lower bound at which importance itself settles in the absence of other causes. It decays ever more slowly as it approaches this resting value.
4. Recent increase in importance of closely related Atoms causes importance to increase. Recent decrease of importance of closely related atoms causes importance to decrease.
5. Above-average recent-utility causes slower importance decay.

Note that among other things this framework encapsulates a model of what psychologists call *short-term memory* or *working memory* or *attentional focus* [5, 6]. The AttentionalFocus (our preferred term) is the set of highly important atoms at a given time. Important atoms are likely to be selected by the other dynamics to work with each other, and hence there's a tendency for them to stay important via building links to each other and spreading activation amongst each other along these links. Yet, if important atoms do not generate interesting new relationships, their recent-utility will drop and their importance will decrease. The net result of these dynamics is to implement a "moving bubble of attention" constituting the set of high-importance atoms.

Importance updating may be seen, to a certain extent, as a non-cognitive part of the system, "merely scheduling." But this is a very narrow view. The maps forming in the network via the nonlinear dynamics of activation spreading and importance updating, are very important for guiding node formation, reasoning, association formation, and other mind processes. They constitute a major nexus of knowledge storage as well.

7.4 Schema and Predicate Learning

Perhaps the most difficult aspect of Novamente AI is what we call "schema and predicate learning." This pertains to what we above referred to as "fully general AI for small problem sizes." Novamente's procedure and predicate learning component solves the problems:

- given a description of desired functionality, find a computer program (a schema) that delivers the functionality;
- given a collection of information, find the patterns in it.

It solves these problems in a general way, but, there are significant computational efficiency issues, which mean that in practice the methods may be applied only on a small scale. Larger-scale problems of pattern recognition and schema learning must be solved by using other cognitive processes to do a breakdown into a collection of smaller problems, and using schema and predicate learning on the smaller problems.

This, of course, is a heuristic approach familiar from human psychology. People don't solve big problems all at once, either. They solve little problems, and slowly build on these, incrementally composing answers to larger problems. Humans developmentally learn "cognitive building blocks," store them, and apply them to various large scale problems. Oft-used building blocks are kept around and frequently referred to.

The value of general pattern recognition in a data analysis context is obvious, and will be reviewed in later subsections. Predicate learning, properly deployed, is a very valuable tool for data analysis. It is also, as we shall see, an essential part of language understanding.

The value of schema learning may be less transparent but is no less profound. Schema learning is an essential component of Novamente's "system

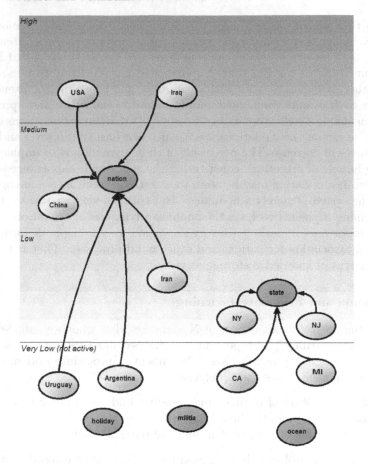

Fig. 10: Nodes at different importance levels

control" philosophy. Initially, Novamente's overall behavior will be guided by human-encoded heuristic control schemata. But a system as complex as Novamente cannot perform optimally on this basis – adaptive learning of control schemata is required. In particular, experience shows that complex logical inferences require context-adapted control schemata.

The computer science literature contains a number of different approaches to schema and predicate learning, which may be roughly categorized as:

- logic-based planning algorithms;
- neural networks;
- evolutionary programming;
- reinforcement learning;
- hybrid approaches;

None of these techniques is in itself sufficient for Novamente's needs, and so we have created our own algorithmic approach, integrating ideas from several of these existing approaches.

Logic-based planning algorithms such as GraphPlan [12] and SATPlan [47] planning have their strengths. Various techniques for probabilistic planning with Markov methods [19] have proved relatively successful in robotics. But in the end these are effective only in very narrowly constrained domains – they are "brittle."

Recurrent back-propagation [62] presents a theoretically general neural-net-based approach to procedure learning, but its efficiency problems are severe. More specialized neural net approaches, such as the clever neural net based planning algorithms developed by James Albus and colleagues for use in their integrative robotics architecture [3], display greater efficiency, but apply only to narrow application domains.

Reinforcement Learning is an approach to procedure learning based on "unsupervised" Hebbian learning [37] in the brain. Most famously, it has been embodied in John Holland's classifier systems [40]. While an interesting approach conceptually, reinforcement learning has severe problems with parameter tuning and scalability, and has rarely been successfully applied in practice.

Finally, evolutionary programming emulates the process of natural selection to "evolve" procedures fulfilling given criteria. It is a very promising approach, but like neural network learning is has scalability problems – learning can be very, very slow on large problems.

The approach we have taken in Novamente is a synthesis of logical, reinforcement learning, and evolutionary approaches. We use reinforcement learning (Hebbian learning via HebbianLinks) and logical planning (PTL higher-order inference), but we view these as auxiliary techniques, not as primary sources of schema and predicate learning power. Our primary schema and predicate learning approach is to fix evolutionary programming's scaling problems using a number of integrative-intelligence-oriented tricks.

One technique recently introduced for speeding up evolutionary programming is the Bayesian Optimization Algorithm (BOA, see [56]). We make use of an enhanced version of BOA, combined in a unique way with our combinatory-logic representation of predicates and schemata, and modified to utilize PTL as a Bayesian modeling algorithm (providing more intelligence than the decision tree based modeling algorithm used in standard BOA) and as an attention allocation algorithm to steer the direction of evolution. In short, our approach is:

- Represent schema/predicates using directed acyclic graphs whose nodes and links are typed, and whose nodes contain "elementary functions" drawn from: arithmetic, boolean and prob. logic, and combinatory logic. The use of combinatory logic functions allows us to get looping and recursion without using variables or cyclic graphs.

- Encode these program dags as "genotypes" using a special encoding method.
- Search program space using a special variation of the Bayesian Optimization Algorithm (BOA), acting on the genotypes, and using PTL (as a modeling scheme and as a driver of attention allocation) to guide its operations.

At time of writing, this is only partially implemented, and at the present rate of progress it may be up to a year before it's fully implemented and tested (though limited versions have already been tested on various mathematical problems, and will be extensively tested on real-world pattern recognition problems during the next few months). We believe this will be a workable approach, in terms of giving good average-case functionality for the general small-problem-size schema and predicate learning problem.

How will we deal with learning larger schemata and predicates? Here we intend to fall back on a common strategy used by brain-minds and other complex systems: hierarchical breakdown. Novamente will not be able to learn general large schemata and predicates, but it will be able to learn large schemata and predicates that consist of small schemata or predicates whose internal nodes refer to small schemata or predicates, whose internal nodes refer to small schemata or predicates, etc. We have modified the BOA algorithm specifically to perform well on hierarchical schemata or predicates. While this may seem not to have the full generality one would like to see in an "Artificial General Intelligence," we believe that this kind of hierarchical breakdown heuristic is the way the human mind/brain works, and is essentially inevitable in any kind of practical inteligent system, due to the plain intractability of the general schema and predicate learning problem.

7.5 Pattern Mining

Now we turn briefly to some of the practical, commercial applications of the current, partially-completed Novamente system. No process lies closer to the heart of the "analysis of massive data stores" problem than pattern mining. This is therefore an aspect of Novamente that we have thought about particularly carefully, in the context of both bioinformatics and national security applications. Our current bioinformatics work with Novamente has been significantly successful in this area, finding never-before-detected patterns of interregulation between genes by analyzing gene expression data in the context of biological background knowledge, as shown in more detail in the Appendix to this Chapter.

Conventional programs for unsupervised pattern mining include:

- Principal Components Analysis (PCA) [44];
- clustering algorithms;
- use of optimization algorithms to search through "pattern space";

- association rule mining algorithms for huge datasets (such as Apriori [1]);
- neural net based pattern recognition (SOMs [48], backpropagation).

Apriori is a simple but powerful algorithm, and we intend to use it or some variation thereof within the DINI framework's Fisher process, to grab potentially useful patterns out of the overall data warehouse and present them to Novamente for further study. Apriori has no sense of semantics, it merely finds combinations of data items that occur together with surprising frequency, using a greedy "hill-climbing" approach; it is excellent in the role of an initial filter through a data store that is too large to load into Novamente all at once.

Clustering algorithms are in common use in many domains; they fall into several different categories, including [71]:

- agglomerative algorithms, generally useful for building hierarchical category systems;
- partitioning algorithms, good for dividing data into a few large categories (k-means, Bioclust [10], and others);
- statistical algorithms: Expectation Maximization.

Clustering tends to be a problematic technology, in that algorithms are difficult to tune and validate. Each algorithm seems to be the best one – if you look at the right sort of dataset. Novamente does agglomerative clustering implicitly via the iterative action of the ConceptFormation MindAgent. There is also a Clustering MindAgent which carries out explicit partitioning-based clustering, perceiving the Atoms in the system as a symmetric weighted graph composed of SimilarityLinks and partitioning this graph using a variant of the Bioclust algorithm.

More powerful but more expensive than other techniques, predicate learning may be used to search the space of patterns in data for "fit" patterns. Here "fitness" is defined as a combination of:

- compactness, or "Minimum Description Length" [8]; and
- frequency and clarity of occurrence in observed data.

Evolutionary-programming based predicate learning can find patterns much subtler than those detectable via simple methods like Aprioro or clustering.

Logical inference also plays a role here, in the sense that, once patterns are found by any of the methods mentioned above, inference is able to make plausible estimates as to the actual validity of the patterns. It not only observes how prominent the pattern is in observed data, but uses inference to explore the similarity between the pattern and other observed patterns in other datasets, thus making an integrative judgment of validity.

An example pattern found by Novamente through mining gene expression data is shown in the Appendix. In later Novamente versions, these patterns could be presented to the user in a verbal form. Currently, patterns are presented in a formal language, which must then be visualized by a UI or verbalized by an expert human user.

7.6 Natural Language Processing

Novamente is a general AGI architecture, not restricted to any particular type of information input. We have discussed here how it may make use of symbolic-type data loaded from RDBs, but in our current work with the system, we are making equal use of its ability to process complex quantitative datasets (e.g. gene expression data derived from microarrays). And, critically for practical data analysis/management functionality, the system is also capable of processing linguistic information.

Some linguistic information may come into the system as sound, captured via tape recorders, computer microphones, and the like. Handling this sort of data requires some specialized quantitative-data preprocessing using tools such as wavelet analysis [72]. We will not discuss these matters here, although significant thought has gone into adapting Novamente for such purposes. Instead we will focus on linguistics, assuming the system is receiving language in textual form, either from computer files or from the output of a voice-to-text software program.

There is a host of natural language processing (NLP) technology out there, but the plain fact is that none of it works very well. Modern NLP technology works passably well if one of two criteria is met: the sentences involved are very simple, or the sentences involved all pertain to a single, very narrow domain.

It cannot yet deal with realistically complex utterances about the world in general. Lexiquest[8], Connexor[9] and other firms offer NLP tools that provide pragmatically useful functionality in narrow domains, but there is nothing out there that can, for example, take an e-mail grabbed randomly off the Internet and correctly parse 90% of the sentences in it.

We believe the Novamente architecture has the potential to provide a real breakthrough in NLP. This is because of its integrative nature. The biggest problem in NLP today is integrating semantic and pragmatic knowledge into the syntax parsing process [53]. Novamente, which automatically integrates all forms of information at its disposal, is ideally set up to do this.

Specifically, one of the leading approaches to NLP parsing out there today is "unification feature structure parsing." [61] In this approach, parsing is done through a process of unification of data structures representing sentences. This approach fits snugly with Novamente's LogicalUnification MindAgent, whose design has been tuned especially for language applications. Typical unification parsers work via specialized "linguistic unification" processes, but we have shown theoretically that Novamente logical unification can carry out the same process more powerfully. In addition to unification, feature structure grammars typically involve a small number of special linguistic transformations (such as transformations to turn statements into questions); these can be

---

[8]www.lexiquest.com
[9]www.connexor.com

expressed compactly as Novamente SchemaNodes, and conveniently learned and manipulated in this form.

Consider, for example, the sentence

``AWI will be in New York with the new shipment on Tuesday.''

An effective parser will analyze the parts of speech and syntactic relationships in this sentence and produce a parse tree such as the one shown in Fig. 11.

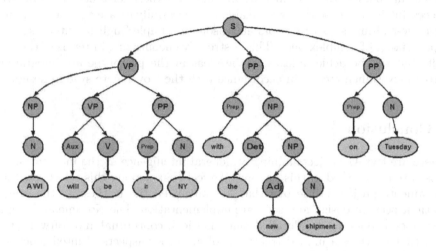

Fig. 11: Example parse tree

A parse tree like this will be found in Novamente via logical unification, aided by other MindAgents. The representation of a parse tree in terms of Novamente nodes and links is straightforward, although it involves significantly more nodes and links than the simple "parse tree" rendition shown in Fig. 11.

Once parsing is done, the really significant step comes, which is "semantic mapping" – the use of syntactic understanding to carry out semantic interpretation. Semantic mapping schemata – some provided by Novamente programmers, most learned through the system's experience – are used to map the Atoms representing the parse tree into Atoms representing the system's meaning, such as:

```
atTime(Location(Z, New York) AND with(Z, Y), 11/02/2003)
representative_of(Z,X)
name_of(X, ``AWI'')
name_of(X, ``Associated Widgets Incorporated'')
```

```
EIN(X, ''987654321'')
InheritanceLink(Y, shipment)
ContextLink(InheritanceLink (Y, new), 12/02/2003)
```

The process of language generation, used by Novamente to generate language representing patterns it has found or ideas it has had, is the inverse of this: it begins with a collection of Atoms representing semantic meanings, uses special schemata to produce a collection of Atoms representing the parse tree of a sentence, and then generates the sentence.

We are currently working with a prototype of this kind of parsing process, which runs outside the Novamente core; and our experiments have been highly successful, in the sense that we are able to successfully create accurate semantic node-and-link structures from all reasonably simple English sentences, and a percentage of complex ones. This is strongly encouraging in terms of the result that will be achievable when the ideas in the prototype are integrated into the Novamente core, in accordance with the Novamente system design.

8 Conclusion

There are few AI projects aiming at general intelligence at the moment, and most are represented in this volume. We feel that, among this select set, the Novamente project is the one that has advanced the farthest in the pragmatic aspects of design and software implementation. The Novamente design possesses a concrete and detailed mathematical, conceptual and software design that provides a unified treatment of all major aspects of intelligence as detailed in cognitive science and computer science.

Our prior experience with commercial and R&D oriented AI engineering has taught us innumerable lessons, one of which is to never underestimate the complexity and difficulty posed by the "just engineering" phases of the project. Another important lesson concerns the need for experiential learning as a way to train and teach a proto-AGI system.

While Novamente is currently partially engineered, the results we have obtained with the system so far (see, e.g., the Appendix of this chapter) are more along the lines of data mining than ambitious AGI. However, these practical applications provide us with invaluable insight into the practical issues that surround applications of AGI and proto-AGI systems to real-world problems.

Given the dismal history of grand AI visions, we realize the need for caution in making grand claims for the as-yet undemonstrated abilities of our system. But we do not consider history a good reason for conservatism as to the future. We believe that by drawing on the best insights of existing AI paradigms and integrating these within a novel synthetic framework based on self-organization and experiential interactive learning, we have a serious chance of taking AI to a new level, and approaching genuine Artificial General Intelligence. And we have a precise design and plan of action for testing our hypotheses in this regard.

Acknowledgments

The Novamente AGI design is the work not only of the authors, but also of a team of others, especially Andre Senna, Thiago Maia, Izabela Freire and Moshe Looks.

Regarding this document, thanks are due to Debbie Duong, for an extremely careful and detailed critique of an earlier version, which led to a number of significant improvements.

Appendix: Novamente Applied to Bioinformatic Pattern Mining

In this Appendix we present an example of what the current version of Novamente can do in the context of bioinformatic data analysis. The work reported here was done in the context of testing an early version of the Biomind Toolkit product for gene expression data analysis [7]. It is described in a forthcoming journal article [31].

What we will discuss here is one among several applications of Novamente to gene expression data analysis – "regulatory network inference," which involves studying gene expression data and recognizing the patterns that interrelate the expression levels of genes. This is a problem of the general nature of "inferring the dynamical rule of a system from samples of its trajectory."

The data presented to Novamente in this case consists of:

- A collection of "gene expression" datasets. Some of these are time series data sets, reporting the expression levels of all the genes in a genome (e.g., human, yeast) at a series of time points during the cell cycle of a single cell. Some are categorical datasets, giving the expression levels of genes in various individuals, along with data about category membership of the individuals (e.g., cancerous versus non-cancerous).
- A collection of biological background knowledge, derived from biological databases such as SGD, MIPS, BLAST, the Gene Ontology, and so forth [52].

An example rule inferred from this sort of data is a temporal pattern in the expression levels of the five specific genes that are familiar to molecular biologists and known by the labels *SIC1*, *PCL2*, *CLN3*, *ACE2*, and *SWI5*:

The arrows ==> represent probabilistic logical implications (Implication-Links). In this case, all relations involved in a given implication refer to the

```
C = (LOW(SIC1) OR MOD_LOW(SIC1)) AND (LOW(PCL2))
AND (LOW(CLN3)) OR (MOD_LOW(CLN3))

C AND EXTRA_HIGH(SWI5) ==>
DECREASE(SWI5) AND INCREASE(ACE2)

C AND (MOD_HIGH(SWI5) OR HIGH(SWI5)) ==>
INCREASE(SWI5) AND INCREASE(ACE2)
```

Table 14: Example regulatory network patterns

same time point. The predicates *LOW*, *MOD_LOW* (moderately low), *DE-CREASE*, etc. are quantitatively grounded using probabilistic logic, with parameters adaptively tuned to the dataset under analysis.

In this pattern, the inferred proposition C gives a context, in which the dependence of the correlation between *SWI5*'s movements and *ACE2*'s movements upon the level of *SWI5* can be observed. The system is not only detecting a contextual relationship here, it is detecting a context in which a certain gene's value can serve as the context for a dynamic relationship between genes. This is exactly the kind of complex interrelationship that makes genetic dynamics so subtle, and that standard data mining approaches are not capable of detecting.

The above example does not use background knowledge; it is strictly a pattern in gene expression values. The following is a simple example of a pattern involving background knowledge.

```
ConceptNode C
EquivalenceLink
(MemberLink X C)
(AssociativeLink X (transcriptional_regulation (CUP1)))

MOD_LOW(FKH2, t) AND LOW(MCM1, t) AND (MOD_LOW(C) OR LOW(C))
==>
INCREASE(FKH2, t+1) AND STABLE(MCM1, t+1) AND INCREASE(C)
```

Table 15: Example regulatory network patterns

Here the ConceptNode C is the category of genes that are associated with transcriptional regulation of the gene *CUP1*. The knowledge of which genes are associated with transcriptional regulation of *CUP1* comes from biological databases – not from any single database, but from integration of information from multiple databases using Novamente inference. The decision to incorporate this particular category in the rule was made by Novamente as part of its unsupervised pattern mining process.

In this particular application – like many others – the Novamente approach is significantly more effective than traditional statistics, decision trees, straightforward genetic programming based rule induction, or other traditional machine learning methods. It can find subtler patterns – both in gene expression data alone, and via judiciously incorporating knowledge from biological databases.

References

1. Rakesh Agrawal and Ramakrishnan Srikant. Fast algorithms for mining association rules. In Jorge B. Bocca, Matthias Jarke, and Carlo Zaniolo, editors, *Proc. 20th Int. Conf. Very Large Data Bases, VLDB*, pages 487–499. Morgan Kaufmann, 1994.
2. James Albus. *Engineering of Mind: An Introduction to the Science of Intelligent Systems*. John Wiley and Sons, 2001.
3. James Albus, N. DeClaris, A. Lacaze, and A. Meystel. Neutral Network Based Planner/Learner for Control Systems. In *Proceedings of the 1997 International Conference on Intelligent Systems and Semiotics*, pages 75–81, 1997.
4. Daniel Amit. *Modeling Brain Function*. Cambridge University Press, 1992.
5. Bernard Baars. *A Cognitive Theory of Consciousness*. Cambridge University Press, 1988.
6. A. D. Baddeley. *Working Memory*. Oxford University Press, 1998. '
7. Pierre Baldi and G. Wesley Hatfield. *DNA Microarrays and Gene Expression: From Experiments to Data Analysis and Modeling*. Cambridge University Press, 2002.
8. A. Barron, J. Rissanen, and B. Yu. The Minimum Description Length Principle in Coding and Modeling. *IEEE Transactions on Information Theory*, 44 no. 6:2743–2760, 1998.
9. Gregory Bateson. *Mind and Nature: A Necessary Unity*. Hampton Books, 2002.
10. Amir Ben-Dor, Ron Shamir, and Zohar Yakini. Clustering Gene Expression Patterns. *Journal of Computational Biology*,6:281-297, 1999.
11. The BlueGene/L Team. An Overview of the BlueGene/L Supercomputer. In *Proceedings of the SC-2002 Conference*, 2002.
12. A. Blum and J. Langford. Probabilistic Planning in the GraphPlan Framework. In *Proceedings of ECP'99*. Springer-Verlag, 1999.
13. Bila Bollobas. *Combinatorics: Set Systems, Hypergraphs, Families of Vectors and Probabilistic Combinatorics*. Cambridge University Press, 1986.
14. Chris Brand. *The G-Factor: General Intelligence and its Implications*. John Wiley and Sons, 1996.
15. William Calvin and David Bickerton. *Lingua ex Machina: Reconciling Darwin and Chomsky with the Human Brain*. MIT Press, 2000.
16. Gregory Chaitin. *Algorithmic Information Theory*. Addison-Wesley, 1988.
17. Haskell Curry and Robert Feys. *Combinatory Logic*. North-Holland, 1958.
18. Cycorp. The Syntax of CycL. Technical report, March 2002.

19. Thomas Dean, Leslie Pack Kaelbling, Jak Kirman, and Ann Nicholson. Planning Under Time Constraints in Stochastic Domains. *Artificial Intelligence*, 76, 1995.
20. Daniel Dennett. *Brainchildren: Essays on Designing Minds*. MIT Press, 1998.
21. Robert Devaney. *An Introduction to Chaotic Dynamical Systems*. Westview, 1989.
22. Julia V. Douthwaite. *The Wild Girl, Natural Man, and the Monster: Dangerous Experiments in the Age of Enlightenment*. University of Chicago Press, 1997.
23. Gerald Edelman. *Neural Darwinism*. Basic Books, 1987.
24. A. J. Field and P. G. Harrison. *Functional Programming*. Addison-Wesley, 1988.
25. Walter Freeman. *Neurodynamics*. Springer-Verlag, 2000.
26. Ben Goertzel. *The Evolving Mind*. Gordon and Breach, 1993.
27. Ben Goertzel. *The Structure of Intelligence*. Springer-Verlag, 1993.
28. Ben Goertzel. *Chaotic Logic*. Plenum Press, 1994.
29. Ben Goertzel. *From Complexity to Creativity*. Plenum Press, 1997.
30. Ben Goertzel. *Creating Internet Intelligence*. Plenum Press, 2001.
31. Ben Goertzel, Cassio Pennachin, and Lucio Coelho. A Systems-Biology Approach for Inferring Genetic Regulatory Networks. In preparation, 2006.
32. Ben Goertzel, Ken Silverman, Cate Hartley, Stephan Bugaj, and Mike Ross. The Baby Webmind Project. In *Proceedings of AISB 00*, 2000.
33. David Goldberg. *Genetic Algorithms in Search, Optimization, and Machine Learning*. Addison-Wesley, 1989.
34. Stephen Grossberg. *Neural Networks and Natural Intelligence*. MIT Press, 1992.
35. Henrik Grosskreutz. Belief Update in the pGolog Framework. In *Proceedings of the International Joint Conference on Artificial Intelligence – IJCAI-01*, 2001.
36. Stuart Hameroff. *Ultimate Computing*. North Holland, 1987.
37. Donald Hebb. *The Organization of Behavior*. John Wiley and Sons, 1948.
38. Daniel Hillis. *The Connection Machine*. MIT Press, 1989.
39. Douglas Hofstadter. *Mathematical Themes: Questing for the Essence of Mind and Pattern*. Basic Books, 1996.
40. John Holland. A Mathematical Framework for Studying Learning in Classifier Systems. *Physica D*, 2, n. 1-3, 1986.
41. John H. Holland. *Adaptation in Natural and Artificial Systems*. University of Michigan Press, 1975.
42. M. Hutter. Towards a universal theory of artificial intelligence based on algorithmic probability and sequential decisions. *Proceedings of the 12th European Conference on Machine Learning (ECML-2001)*, pages 226–238, 2001.
43. Bipin Indurkhya. *Metaphor and Cognition: An Interactionist Approach*. Kluwer Academic, 1992.
44. I. T. Jolliffe. *Principal Component Analysis*. Springer-Verlag, 1986.
45. John Josephson and Susan Josephson. *Abductive Inference: Computation, Philosophy, Technology*. Cambridge University Press, 1994.
46. Gyorgy Kampis. *Self-Modifying Systems in Biology and Cognitive Science*. Plenum Press, 1993.
47. H. Kantz and V. Selman. Pushing the Envelope: Planning, Propositional Logic, and Stochastic Search. In *Proceedings of the AAAI Conference 1996*, 1996.
48. Teuvo Kohonen. *Self-Organizing Maps*. Springer-Verlag, 1997.
49. John Koza. *Genetic Programming*. MIT Press, 1992.
50. Ray Kurzweil. *The Age of Spiritual Machines*. Penguin Press, 2000.
51. D. B. Lenat. Cyc: A Large-Scale Investment in Knowledge Infrastructure. *Communications of the ACM*, 38, no. 11, November 1995.

52. Stanley Letovsky. *Bioinformatics: Databases and Systems*. Kluwer Academic, 1999.
53. Christophet Manning and Heinrich Schutze. *Foundations of Statistical Natural Language Processing*. MIT Press, 1999.
54. Jane Mercer. *Labeling the Mentally Retarded*. University of California Press, 1973.
55. Judea Pearl. *Probabilistic Reasoning in Intelligent Systems: Networks of Plausible Inference*. Morgan-Kaufmann, 1988.
56. Martin Pelikan. *Bayesian Optimization Algorithm: From Single Level to Hierarchy*. PhD thesis, University of Illinois at Urbana-Champaign, October 2002.
57. Roger Penrose. *Shadows of the Mind*. Oxford University Press, 1997.
58. Juergen Schmidhuber. Bias-Optimal Incremental Problem Solving. In *Advances in Neural Information Processing Systems - NIPS 15*. MIT Press, 2002.
59. John R. Searle. Minds, Brains, and Programs. *Behavioral and Brain Sciences*, 3:417–457, 1980.
60. Stuart Shapiro. *An Introduction to SNePS 3*, pages 510–524. Springer-Verlag, 2000.
61. Stuart Shieber. *Introduction to Unification-Based Approaches to Grammar*. University of Chicago Press, 1986.
62. P. Simard, M. B. Ottaway, and D. H. Ballard. Analysis of Recurrent Backpropagation. In D. Touretzky, G. Hinton, and T. Sejnowsky, editors, *Proceedings of the 1988 Connectionist Models Summer School*, pages 103–112. Morgan Kaufmann, 1988.
63. M. Stickel, R. Waldinger, M. Lowry, T. Pressburger, and I. Underwood. Deductive Composition of Astronomical Software from Subroutine Libraries. In *Proceedings of the Twelfth Internation Conference on Automated Deduction (CADE-12)*, pages 341–355, June 1994.
64. David Stork. *Scientist on the Set: An Interview with Marvin Minsky*, chapter 2. MIT Press, 2002.
65. V. Subrahmanian. Nonmonotonic Logic Programming. *IEEE Transactions on Knowledge and Data Engineering*, 11-1:143–152, 1999.
66. Richard Sutton and Andrew Barton. *Reinforcement Learning*. MIT Press, 1998.
67. Alan Turing. *Computing Machinery and Intelligence*. McGraw-Hill, 1950.
68. Francisco Varela. *Principles of Biological Autonomy*. North-Holland, 1978.
69. Pei Wang. On the Working Definition of Intelligence. Technical Report CRCC Technical Report 95, Center for Research in Concepts and Cognition, Indiana University at Bloomington, 1995.
70. Gerhard Weiss, editor. *Multiagent Systems*. MIT Press, 2000.
71. Ian Witten and Eibe Frank. *Data Mining: Practical Machine Learning Tools and Techniques with Java Implementations*. Morgan Kaufmann, 1999.
72. Huang Xuedong, Alex Acero, Hsiao-Wuen Hon, and Raj Reddy. *Spoken Language Processing: A Guide to Theory, Algorithm and System Development*. Prentice Hall, 2001.
73. A. Zadeh and Janusz Kacprzyk, editors. *Fuzzy Logic for the Management of Uncertainty*. John Wiley and Sons, 1992.
74. Mikhail Zak. *From Instability to Intelligence: Complexity and Predictability in Nonlinear Dynamics*. Springer-Verlag, 1997.

Essentials of General Intelligence: The Direct Path to Artificial General Intelligence

Peter Voss

Adaptive A.I., Inc.
131 Galleon St. #2 Marina del Rey, CA 90292, USA
peter@optimal.org - http://www.adaptiveai.com

1 Introduction

This chapter explores the concept of "artificial *general* intelligence" (AGI) – its nature, importance, and how best to achieve it. Our[1] theoretical model posits that general intelligence comprises a limited number of distinct, yet highly integrated, foundational functional components. Successful implementation of this model will yield a highly adaptive, general-purpose system that can autonomously acquire an extremely wide range of specific knowledge and skills. Moreover, it will be able to improve its own cognitive ability through self-directed learning. We believe that, given the right design, current hardware/software technology is adequate for engineering practical AGI systems. Our current implementation of a functional prototype is described below.

The idea of "general intelligence" is quite controversial; I do not substantially engage this debate here but rather take the existence of such non-domain-specific abilities as a given [14]. It must also be noted that this essay focuses primarily on low-level (i.e., roughly animal level) cognitive ability. Higher-level functionality, while an integral part of our model, is only addressed peripherally. Finally, certain algorithmic details are omitted for reasons of proprietary ownership.

2 General Intelligence

Intelligence can be defined simply as an entity's ability to achieve goals – with greater intelligence coping with more complex and novel situations. Complexity ranges from the trivial – thermostats and mollusks (that in most contexts don't even justify the label "intelligence") – to the fantastically complex; autonomous flight control systems and humans.

Adaptivity, the ability to deal with changing and novel requirements, also covers a wide spectrum: from rigid, narrowly domain-specific to highly flexible, general purpose. Furthermore, flexibility can be defined in terms of *scope* and

---

[1]Intellectual property is owned by Adaptive A.I., Inc.

permanence – how much, and how often it changes. Imprinting is an example of limited scope and high permanence, while innovative, abstract problem solving is at the other end of the spectrum. While entities with high adaptivity and flexibility are clearly superior – they can potentially learn to achieve any possible goal – there is a hefty efficiency price to be paid: For example, had Deep Blue also been designed to learn language, direct airline traffic, and do medical diagnosis, it would not have beaten a world chess champion (all other things being equal).

General Intelligence comprises *the essential, domain-independent skills* necessary for acquiring a wide range of domain-specific knowledge (data and skills) – i.e. the ability to learn anything (in principle). More specifically, this learning ability needs to be autonomous, goal-directed, and highly adaptive:

Autonomous. Learning occurs both automatically, through exposure to sense data (unsupervised), and through bi-directional interaction with the environment, including exploration and experimentation (self-supervised).

Goal-directed. Learning is directed (autonomously) towards achieving varying and novel goals and sub-goals – be they "hard-wired," externally specified, or self-generated. Goal-directedness also implies very selective learning and data acquisition (from a massively data-rich, noisy, complex environment).

Adaptive. Learning is cumulative, integrative, contextual and adjusts to changing goals and environments. General adaptivity not only copes with gradual changes, but also seeds and facilitates the acquisition of totally novel abilities.

General cognitive ability stands in sharp contrast to inherent specializations such as speech- or face-recognition, knowledge databases/ontologies, expert systems, or search, regression or optimization algorithms. It allows an entity to acquire a virtually unlimited range of new specialized abilities. The mark of a *generally* intelligent system is not *having* a lot of knowledge and skills, but being able to *acquire* and *improve* them – and to be able to appropriately *apply* them. Furthermore, knowledge must be acquired and stored in ways appropriate both to the nature of the data, and to the goals and tasks at hand.

For example, given the correct set of basic core capabilities, an AGI system should be able to learn to recognize and categorize a wide range of novel perceptual patterns that are acquired via different senses, in many different environments and contexts. Additionally, it should be able to autonomously learn appropriate, goal-directed responses to such input contexts (given some feedback mechanism).

We take this concept to be valid not only for high-level human intelligence, but also for lower-level animal-like ability. The degree of "generality" (i.e., adaptability) varies along a continuum from genetically "hard-coded" responses (no adaptability), to high-level animal flexibility (significant learning ability as in, say, a dog), and finally to self-aware human general learning ability.

2.1 Core Requirements for General Intelligence

General intelligence, as described above, demands a number of irreducible features and capabilities. In order to proactively accumulate knowledge from various (and/or changing) environments, it requires:

1. senses to obtain features from "the world" (virtual or actual);
2. a coherent means for storing knowledge obtained this way; and
3. adaptive output/actuation mechanisms (both static and dynamic).

Such knowledge also needs to be automatically adjusted and updated on an ongoing basis; new knowledge must be appropriately related to existing data. Furthermore, perceived entities/patterns must be stored in a way that facilitates concept formation and generalization. An effective way to represent complex feature relationships is through vector encoding [7].

Any practical applications of AGI (and certainly any real- time uses) must *inherently* be able to process temporal data as patterns in time – not just as static patterns with a time dimension. Furthermore, AGIs must cope with data from different sense probes (e.g., visual, auditory, and data), and deal with such attributes as: noisy, scalar, unreliable, incomplete, multi-dimensional (both space/time dimensional, and having a large number of simultaneous features), etc. Fuzzy pattern matching helps deal with pattern variability and noise.

Another essential requirement of general intelligence is to cope with an overabundance of data. Reality presents massively more features and detail than is (contextually) relevant, or can be usefully processed. Therefore, why the system needs to have some control over what input data is selected for analysis and learning – both in terms of which data, and also the degree of detail. Senses ("probes") are needed not only for selection and focus, but also in order to ground concepts – to give them (reality-based) meaning.

While input data needs to be severely limited by focus and selection, it is also extremely important to obtain multiple views of reality – data from different feature extractors or senses. Provided that these different input patterns are properly associated, they can help to provide context for each other, aid recognition, and add meaning.

In addition to being able to sense via its multiple, adaptive input groups and probes, the AGI must also be able to act on the world – be it for exploration, experimentation, communication, or to perform useful actions. These mechanisms need to provide both static and dynamic output (states and behavior). They too, need to be adaptive and capable of learning.

Underlying all of this functionality is pattern processing. Furthermore, not only are sensing and action based on generic patterns, but so is internal cognitive activity. In fact, even high-level abstract thought, language, and formal reasoning – abilities outside the scope of our current project – are "just" higher-order elaborations of this [20].

2.2 Advantages of Intelligence Being General

The advantages of general intelligence are almost too obvious to merit listing; how many of us would dream of giving up our ability to adapt and learn new things? In the context of artificial intelligence this issue takes on a new significance.

There exists an inexhaustible demand for computerized systems that can assist humans in complex tasks that are highly repetitive, dangerous, or that require knowledge, senses or abilities that its users may not possess (e.g., expert knowledge, "photographic" recall, overcoming disabilities, etc.). These applications stretch across almost all domains of human endeavor.

Currently, these needs are filled primarily by systems engineered specifically for each domain and application (e.g., expert systems). Problems of cost, lead-time, reliability, and the lack of adaptability to new and unforeseen situations severely limit market potential. Adaptive AGI technology, as described in this paper, promises to significantly reduce these limitations and to open up these markets. It specifically implies:

- That systems can learn (and be taught) a wide spectrum of data and functionality
- They can adapt to changing data, environments and uses/goals
- This can be achieved without program changes – capabilities are learned, not coded

More specifically, this technology can potentially:

- Significantly reduce system "brittleness" [2] through fuzzy pattern matching and adaptive learning – increasing robustness in the face of changing and unanticipated conditions or data.
- Learn autonomously, by automatically accumulating knowledge about new environments through exploration.
- Allow systems to be operator-trained to identify new objects and patterns; to respond to situations in specific ways, and to acquire new behaviors.
- Eliminate programming in many applications. Systems can be employed in many different environments, and with different parameters simply through self-training.
- Facilitate easy deployment in new domains. A general intelligence engine with pluggable custom input/ output probes allows rapid and inexpensive implementation of specialized applications.

---

[2] "Brittleness" in AI refers to a system's inability to automatically adapt to changing requirements, or to cope with data outside of a predefined range – thus "breaking".

From a design perspective, AGI offers the advantage that all effort can be focused on achieving the best *general* solutions – solving them once, rather than once for each particular domain. AGI obviously also has huge economic implications: because AGI systems acquire most of their knowledge and skills (and adapt to changing requirements) autonomously, programming lead times and costs can be dramatically reduced, or even eliminated.

The fact that no (artificial!) systems with these capabilities currently exist seems to imply that it is very hard (or impossible) to achieve these objectives. However, I believe that, as with other examples of human discovery and invention, the solution will seem rather obvious in retrospect. The trick is correctly choosing a few critical development options.

3 Shortcuts to AGI

When explaining Artificial General Intelligence to the uninitiated one often hears the remark that, surely, everyone in AI is working to achieve general intelligence. This indicates how deeply misunderstood intelligence is. While it is true that *eventually* conventional (domain-specific) research efforts will converge with those of AGI, without deliberate guidance this is likely to be a long, inefficient process. High-level intelligence *must* be adaptive, must be general – yet very little work is being done to specifically identify what general intelligence is, what it *requires*, and how to achieve it.

In addition to understanding general intelligence, AGI design also requires an appreciation of the differences between *artificial* (synthetic) and biological intelligence, and between *designed* and evolved systems.

Our particular approach to achieving AGI capitalizes on extensive analysis of these issues, and on an incremental development path that aims to minimize development effort (time and cost), technical complexity, and overall project risks. In particular, we are focusing on engineering a series of functional (but low-resolution/capacity) proof-of-concept prototypes. Performance issues specifically related to commercialization are assigned to separate development tracks. Furthermore, our initial effort concentrates on identifying and implementing the most general and foundational components first, leaving high-level cognition, such as abstract thought, language, and formal logic, for later development (more on that later). We also focus more on selective, unsupervised, dynamic, incremental, interactive learning; on noisy, complex, analog data; and on integrating entity features and concept attributes in one comprehensive network.

While our project may not be the only one proceeding on this particular path, it is clear that a large majority of current AI work follows a substantially different overall approach. Our work focuses on:

- *general* rather than domain-specific cognitive ability;
- *acquired knowledge* and skills, versus loaded databases and coded skills;

- *bi-directional, real-time* interaction, versus batch processing;
- *adaptive attention* (focus and selection), versus human pre-selected data;
- core support for *dynamic patterns*, versus static data;
- unsupervised and *self-supervised*, versus supervised learning;
- adaptive, *self-organizing data structures*, versus fixed neural nets or databases;
- *contextual, grounded concepts*, versus hard-coded, symbolic concepts;
- explicitly *engineering* functionality, versus evolving it;
- *conceptual design*, versus reverse-engineering;
- *general proof-of-concept*, versus specific real applications development;
- *animal level cognition*, versus abstract thought, language, and formal logic.

Let's look at each of these choices in greater detail.

General rather than domain-specific cognitive ability

The advantages listed in the previous section flow from the fact that generally intelligent systems can ultimately learn any specialized knowledge and skills possible – human intelligence is the proof! The reverse is obviously not true.

A complete, well-designed AGI's ability to acquire domain- specific capabilities is limited only by processing and storage capacity. What is more, much of its learning will be autonomous – without teachers, and certainly without explicit programming. This approach implements (and capitalizes on) the essence of "Seed AI" – systems with a limited, but carefully chosen set of basic, initial capabilities that allow them (in a "bootstrapping" process) to dramatically increase their knowledge and skills through self-directed learning and adaptation. By concentrating on carefully designing the seed of intelligence, and then nursing it to maturity, one essentially bootstraps intelligence. In our AGI design this self-improvement takes two distinct forms/phases:

1. Coding the basic skills that allow the system to acquire a large amount of specific knowledge.
2. The system reaching sufficient intelligence and conceptual understanding of its own design, to enable it to deliberately improve its own design.

Acquired knowledge and skills versus loaded databases and coded skills

One crucial measure of general intelligence is its ability to *acquire* knowledge and skills, not how much it possesses. Many AI efforts concentrate on accumulating huge databases of knowledge and coding massive amounts of specific skills. If AGI is possible – and the evidence seems overwhelming – then much of this effort will be wasted. Not only will an AGI be able to acquire these additional smarts (largely) by itself, but moreover, it will also be able to keep its knowledge up-to-date, and to improve it. Not only will this save initial data collection and preparation as well as programming, it will also dramatically reduce maintenance.

An important feature of our design is that there are no traditional databases containing knowledge, nor programs encoding learned skills: All acquired knowledge is integrated into an adaptive central knowledge/skills network. Patterns representing knowledge are associated in a manner that facilitates conceptualization and sensitivity to context. Naturally, such a design is potentially far less prone to brittleness, and more resiliently fault-tolerant.

Bi-directional, real-time interaction versus batch processing

Adaptive learning systems must be able to interact bi-directionally with the environment – virtual or real. They must both sense data and act/react on an ongoing basis. Many AI systems do all of their learning in batch mode and have little or no ability to learn incrementally. Such systems cannot easily adjust to changing environments or requirements – in many cases they are unable to adapt beyond the initial training set without reprogramming or retraining.

In addition to real-time perception and learning, intelligent systems must also be able to act. Three distinct areas of action capability are required:

1. Acting on the "world" – be it to communicate, to navigate or explore, or to manipulate some external function or device in order to achieve goals.
2. Controlling or modifying the system's internal parameters (such as learning rate or noise tolerance, etc.) in order to set or improve functionality.
3. Controlling the system's sense input parameters such as focus, selection, resolution (granularity) as well as adjusting feature extraction parameters.

Adaptive attention (focus and selection) versus human pre-selected data

As mentioned earlier, reality presents far more sense data abundance, detail, and complexity than are required for any given task – or than can be processed. Traditionally, this problem has been dealt with by carefully selecting and formatting data before feeding it to the system. While this human assistance can improve performance in specific applications, it is often not realized that this additional intelligence resides in the human, not the software.

Outside guidance and training can obviously speed learning; however, AGI systems must *inherently* be designed to acquire knowledge by themselves. In particular, they need to control what input data is processed – where specifically to obtain data, in how much detail, and in what format. Absent this capability the system will either be overwhelmed by irrelevant data or, conversely, be unable to obtain crucial information, or get it in the required format. Naturally, such data focus and selection mechanisms must themselves be adaptive.

Core support for dynamic patterns versus static data

Temporal pattern processing is another fundamental requirement of interactive intelligence. At least three aspects of AGI rely on it: *perception* needs to learn/recognize dynamic entities and sequences, *action* usually comprises complex behavior, and *cognition* (internal processing) is inherently temporal. In spite of this obvious need for intrinsic support for dynamic patterns, many AI systems only process static data; temporal sequences, if supported at all, are often converted ("flattened") externally to eliminate the time dimension. Real-time temporal pattern processing is technically quite challenging, so it is not surprising that most designs try to avoid it.

Unsupervised and self-supervised versus supervised learning

Auto-adaptive systems such as AGIs require comprehensive capabilities to learn without supervision. Such teacher-independent knowledge and skill acquisition falls into two broad categories: unsupervised (data-driven, bottom-up), and self-supervised (goal-driven, top-down). Ideally these two modes of learning should seamlessly integrate with each other – and of course, also with other, supervised methods.

Here, as in other design choices, general adaptive systems are harder to design and tune than more specialized, unchanging ones. We see this particularly clearly in the overwhelming focus on back-propagation[3] in artificial neural network (ANN) development. Relatively little research aims at better understanding and improving incremental, autonomous learning. Our own design places heavy emphasis on these aspects.

Adaptive, self-organizing data structures versus fixed neural nets or databases

Another core requirement imposed by data/goal-driven, real-time learning is having a flexible, self-organizing data structure. On the one hand, knowledge representation must be highly integrated, while on the other hand it must be able to adapt to changing data densities (and other properties), and to varying goals or solutions. Our AGI encodes *all* acquired knowledge and skills in one integrated network-like structure. This central repository features a flexible, dynamically self-organizing topology. The vast majority of other AI designs rely either on loosely-coupled data objects or agents, or on fixed network topologies and pre-defined ontologies, data hierarchies or database layouts. This often severely limits their self-learning ability, adaptivity and robustness, or creates massive communication bottlenecks or other performance overhead.

Contextual, grounded concepts versus hard-coded, symbolic concepts

Concepts are probably the most important design aspect of AGI; in fact, one can say that "high-level intelligence *is* conceptual intelligence." Core characteristics of concepts include their ability to represent ultra-high-dimensional fuzzy sets that are grounded in reality, yet fluid with regard to context. In other words, they encode related sets of complex, coherent, multi-dimensional

---

[3]Back-propagation is one of the most powerful supervised training algorithms; it is, however, not particularly amenable to incremental learning.

patterns that represent features of entities. Concepts obtain their grounding (and thus their meaning) by virtue of patterns emanating from features sensed directly from entities that exist in reality. Because concepts are defined by *value ranges* within each feature dimension (sometimes in complex relationships), some kind of fuzzy pattern matching is essential. In addition, the *scope* of concepts must be fluid; they must be sensitive and adaptive to both environmental and goal contexts.

Autonomous concept formation is one of the key tests of intelligence. The many AI systems based on hard-coded or human-defined concepts fail this fundamental test. Furthermore, systems that do not derive their concepts via interactive perception are unable to ground their knowledge in reality, and thus lack crucial meaning. Finally, concept structures whose activation cannot be modulated by context and degree of fit are unable to capture the subtlety and fluidity of intelligent generalization. In combination, these limitations will cripple any aspiring AGI.

Explicitly engineering functionality versus evolving it

Design by evolution is extremely inefficient – whether in nature or in computer science. Moreover, evolutionary solutions are generally opaque; optimized only to some specified "cost function", not comprehensibility, modularity, or maintainability. Furthermore, evolutionary learning also requires more data or trials than are available in everyday problem solving.

Genetic and evolutionary programming *do* have their uses – they are powerful *tools* that can be used to solve very specific problems, such as optimization of large sets of variables; however they generally are not appropriate for creating large systems of infrastructures. Artificially evolving general intelligence *directly* seems particularly problematic because there is no known function measuring such capability along a single continuum – and absent such direction, evolution doesn't know what to optimize. One approach to deal with this problem is to try to coax intelligence out of a complex ecology of competing agents – essentially replaying natural evolution.

Overall, it seems that genetic programming techniques are appropriate when one runs out of specific engineering ideas. Here is a short summary of advantages of explicitly engineered functionality:

- Designs can directly capitalize on and encode the designer's knowledge and insights.
- Designs have comprehensible design documentation.
- Designs can be more far more modular – less need for multiple functionality and high inter-dependency of sub-systems than found in evolved systems.
- Systems can have a more flow-chart like, logical design – evolution has no foresight.
- They can be designed with debugging aids – evolution didn't need that.

- These features combine to make systems easier to understand, debug, interface, and – importantly – for multiple teams to simultaneously work on the design.

Conceptual design versus reverse-engineering

In addition to avoiding the shortcomings of evolutionary techniques, there are also numerous advantages to designing and engineering intelligent systems based on *functional* requirements rather than trying to copy evolution's design of the brain. As aviation has amply demonstrated, it is much easier to build planes than it is to reverse-engineer birds – much easier to achieve flight via thrust than flapping wings.

Similarly, in creating artificial intelligence it makes sense to capitalize on our human intellectual and engineering strengths – to ignore design parameters unique to biological systems, instead of struggling to copy nature's designs. Designs explicitly engineered to achieve desired functionality are much easier to understand, debug, modify, and enhance. Furthermore, using known and existing technology allows us to best leverage existing resources. So why limit ourselves to the single solution to intelligence created by a blind, unconscious Watchmaker with his own agenda (survival in an evolutionary environment very different from that of today)?

Intelligent machines designed from scratch carry neither the evolutionary baggage, nor the additional complexity for epigenesis, reproduction, and integrated self-repair of biological brains. Obviously this doesn't imply that we can learn nothing from studying brains, just that we don't have to limit ourselves to biological feasibility in our designs. Our (currently) only working example of high-level general intelligence (the brain) provides a crucial *conceptual* model of cognition, and can clearly inspire numerous specific design features.

Here are some desirable cognitive features that can be included in an AGI design that would not (and in some cases, could not) exist in a reverse-engineered brain:

- More effective control of neurochemistry ("emotional states").
- Selecting the appropriate degree of logical thinking versus intuition.
- More effective control over focus and attention.
- Being able to learn instantly, on demand.
- Direct and rapid interfacing with databases, the Internet, and other machines – potentially having instant access to all available knowledge.
- Optional "photographic" memory and recall ("playback") on all senses!
- Better control over remembering and forgetting (freezing important knowledge, and being able to unlearn.)
- The ability to accurately backtrack and review thought and decision processes (retrace and explore logic pathways.)
- Patterns, nodes and links can easily be tagged (labeled) and categorized.

- The ability to optimize the design for the available hardware instead of being forced to conform to the brain's requirements.
- The ability to utilize the best existing algorithms and software techniques – irrespective of whether they are biologically plausible.
- Custom designed AGI (unlike brains) can have a simple speed/capacity upgrade path.
- The possibility of comprehensive integration with other AI systems (like expert systems, robotics, specialized sense pre-processors, and problem solvers.)
- The ability to construct AGIs that are highly optimized for specific domains.
- Node, link, and internal parameter data is available as "input data" (full introspection.)
- Design specifications are available (to the designer and to the AGI itself!)
- Seed AI design: A machine can inherently be designed to more easily understand and improve its own functioning – thus bootstrapping intelligence to ever higher levels.

General proof-of-concept versus specific real applications development

Applying given resources to minimalist proof-of-concept designs improves the likelihood of cutting a swift, direct path towards an ultimate goal. Having identified high-level artificial general intelligence as our goal, it makes little sense to squander resources on inessentials. In addition to focusing our efforts on the ability to *acquire* knowledge autonomously, rather than capturing or coding it, we further aim to speed progress towards full AGI by reducing cost and complexity through:

- Concentrating on proof-of-concept prototypes, not commercial performance. This includes working at low data resolution and volume, and putting aside optimization. Scalability is addressed only at a theoretical level, and not necessarily implemented.
- Working with radically-reduced sense and motor capabilities. The fact that deaf, blind, and severely paralyzed people can attain high intelligence (Helen Keller, Stephen Hawking) indicates that these are not essential to developing AGI.
- Coping with complexity through a willingness to experiment and implement poorly understood algorithms – i.e. using an engineering approach. Using self-tuning feedback loops to minimize free parameters.
- Not being sidetracked by attempting to match the performance of domain-specific designs – focusing more on *how* capabilities are achieved (e.g. learned conceptualization, instead of programmed or manually specified concepts) rather than raw performance.
- Developing and testing in virtual environments, not physical implementations. Most aspects of AGI can be fully evaluated without the overhead (time, money, and complexity) of robotics.

Animal level cognition versus abstract thought, language, and formal logic

There is ample evidence that achieving high-level cognition requires only modest *structural* improvements from animal capability. Discoveries in cognitive psychology point towards generalized pattern processing being the foundational mechanism for all higher level functioning. On the other hand, relatively small differences between higher animals and humans are also witnessed by studies of genetics, the evolutionary timetable, and developmental psychology.

The core challenge of AGI is achieving the robust, adaptive conceptual learning ability of higher primates or young children. If human level intelligence is the goal, then pursuing robotics, language, or formal logic (at this stage) is a costly sideshow – whether motivated by misunderstanding the problem, or by commercial or "political" considerations.

Summary

While our project leans heavily on research done in many specialized disciplines, it is one of the few efforts dedicated to integrating such interdisciplinary knowledge with the specific goal of developing general artificial intelligence. We firmly believe that many of the issues raised above are crucial to the early achievement of truly intelligent adaptive learning systems.

4 Foundational Cognitive Capabilities

General intelligence requires a number of foundational cognitive abilities. At a first approximation, it must be able to:

- remember and recognize patterns representing coherent features of reality;
- relate such patterns by various similarities, differences, and associations;
- learn and perform a variety of actions;
- evaluate and encode feedback from a goal system;
- autonomously adjust its system control parameters;

As mentioned earlier, this functionality must handle a very wide variety of data types and characteristics (including temporal), and must operate interactively, in real-time. The expanded description below is based on our particular implementation; however, the features listed would generally be required (in some form) in *any* implementation of artificial general intelligence.

Pattern learning, matching, completion, and recall

The primary method of pattern acquisition consists of a proprietary adaptation of lazy learning [1, 28]. Our implementation stores feature patterns (static and dynamic) with adaptive fuzzy tolerances that subsequently determine how similar patterns are processed. Our recognition algorithm matches patterns on a competitive winner-take-all basis, as a set or aggregate of similar patterns, or by forced choice. It also offers inherent support for pattern completion, and recall (where appropriate).

Data accumulation and forgetting

Because our system learns patterns incrementally, mechanism are needed for consolidating and pruning excess data. Sensed patterns (or sub-patterns) that fall within a dynamically set noise/error tolerance of existing ones are automatically consolidated by a hebbian-like mechanism that we call "nudging." This algorithm also accumulates certain statistical information. On the other hand, patterns that turn out not to be important (as judged by various criteria) are deleted.

Categorization and clustering

Vector-coded feature patterns are acquired in real-time and stored in a highly adaptive network structure. This central self-organizing repository automatically clusters data in hyper-dimensional vector- space. Our matching algorithm's ability to recall patterns by any dimension provides inherent support for flexible, dynamic categorization. Additional categorization mechanisms facilitate grouping patterns by additional parameters, associations, or functions.

Pattern hierarchies and associations

Patterns of perceptual features do not stand in isolation – they are derived from coherent external reality. Encoding relationships between patterns serves the crucial functions of added meaning, context, and anticipation. Our system captures low-level, perception-driven pattern associations such as: sequential or coincidental in time, nearby in space, related by feature group or sense modality. Additional relationships are encoded at higher levels of the network, including actuation layers. This overall structure somewhat resembles the "dual network" described by [11].

Pattern priming and activation spreading

The core function of association links is to prime[4] related nodes. This helps to disambiguate pattern matching, and to select contextual alternatives. In the case where activation is particularly strong and perceptual activity is low, stored patterns will be "recognized" spontaneously. Both the scope and decay rate of such activation spreading are controlled adaptively. These dynamics combine with the primary, perception-driven activation to form the system's short-term memory.

Action patterns

Adaptive action circuits are used to control parameters in the following three domains:

---

[4] "Priming," as used in psychology, refers to an increase in the speed or accuracy of a decision that occurs as a consequence of prior exposure or activation.

1. senses, including adjustable feature extractors, focus and selection mechanisms;
2. output actuators for navigation and manipulation;
3. meta-cognition and internal controls;

Different actions states and behaviors (action sequences) for each of these control outputs can be created at design time (using a configuration script) or acquired interactively. Real-time learning occurs either by means of explicit teaching, or autonomously through random exploration. Once acquired, these actions can be tied to specific perceptual stimuli or whole contexts through various stimulus-response mechanisms. These S-R links (both activation and inhibition) are dynamically modified through ongoing reinforcement learning.

Meta-cognitive control

In addition to adaptive perception and action functionality, an AGI design must also allow for extensive monitoring and control of overall system parameters and functions (including "emotion-like" cognitive behavioral strategies.) Any complex interactive learning system contains numerous crucial control parameters such as noise tolerance, learning and exploration rates, priorities and goal management, and a myriad others. Not only must the system be able to adaptively control these many interactive vectors, it must also appropriately manage its various cognitive functions (such as recognition, recall, action, etc.) and modes (such as exploration, caution, attention, etc.), dynamically evaluating them for effectiveness. Our design deals with these requirements by means of a highly adaptive introspection/control "probe".

High-level intelligence

Our AGI model posits that no additional foundational functions are necessary for higher- level cognition. Abstract thought, language, and logical thinking are all elaborations of core abilities. This controversial point is elaborated on further on.

5 An AGI in the Making

The functional proof-of-concept prototype currently under development at Adaptive A.I. Inc. aims to embody all the abovementioned choices, requirements, and features. Our development path has been the following:

1. development framework;
2. memory core and interface structure;
3. individual foundational cognitive components;
4. integrated low-level cognition;
5. increased level of functionality (our current focus).

The software comprises an AGI engine framework with the following basic components:

- a set of pluggable, programmable (virtual) sensors and actuators (called "probes");
- a central pattern store/engine including all data and cognitive algorithms;
- a configurable, dynamic 2D virtual world, plus various training and diagnostic tools.

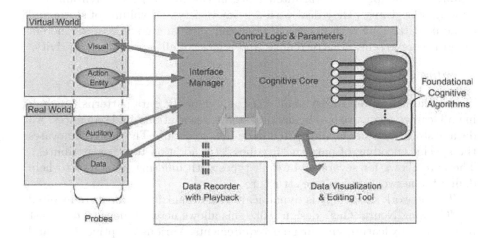

Fig. 1: Adaptive A.I.'s AGI Framework

The AGI engine design is based on, and embodies insights from a wide range of research in cognitive science – including computer science, neuroscience, epistemology [26, 17], cognitive science [24, 9] and psychology [20]. Particularly strong influences include: embodied systems [6], vector encoded representation [7], adaptive self-organizing neural nets (esp. Growing Neural Gas, [10]), unsupervised and self-supervised learning, perceptual learning [13], and fuzzy logic [18].

While our design includes several novel, and proprietary algorithms, our key innovation is the particular selection and integration of established technologies and prior insights.

5.1 AGI Engine Architecture and Design Features

Our AGI engine (which provides this foundational cognitive ability) can logically be divided into three parts (See Figure 1):

- Cognitive core;

- control/interface logic;
- input/output probes.

This "situated agent architecture" reflects the importance of having an AGI system that can dynamically and adaptively interact with the environment. From a theory-of-mind perspective it acknowledges both the crucial need for concept grounding (via senses), plus the absolute need for experiential, self-supervised learning.

The components listed below have been specifically designed with features required for adaptive general intelligence in (ultimately) real environments. Among other things, they deal with a great variety and volume of static and dynamic data, cope with fuzzy and uncertain data and goals, foster coherent integrated representations of reality, and – most of all – promote adaptivity.

Cognitive Core

This is the central repository of all static and dynamic data patterns – including all learned cognitive and behavioral states, associations, and sequences. All data is stored in a single, integrated node-link structure. The design innovates the specific encoding of pattern "fuzziness" (in addition to other attributes). The core allows for several node/link types with differing dynamics to help define the network's cognitive structure.

The network's topology is dynamically self-organizing – a feature inspired by "Growing Neural Gas" design [10]. This allows network density to adjust to actual data feature and/or goal requirements. Various adaptive local and global parameters further define network structure and dynamics in real time.

Control and interface logic

An overall control system coordinates the network's execution cycle, drives various cognitive and housekeeping algorithms, and controls/adapts system parameters. Via an Interface Manager, it also communicates data and control information to and from the probes.

In addition to handling the "nuts and bolts" of program execution and communication, and to managing the various cognitive algorithms, the control system also includes meta-cognitive monitoring and control. This is essentially the cognitive aspect of emotions; such states as curiosity, boredom, pleasure, disappointment, etc. [24]

Probes

The Interface Manager provides for dynamic addition and configuration of probes. Key design features of the probe architecture include the ability to have programmable feature extractors, variable data resolution, and focus and selection mechanisms. Such mechanisms for data selection are imperative for general intelligence: even moderately complex environments have a richness of data that far exceeds any system's ability to usefully process.

The system handles a very wide variety of data types and control signal requirements – including those for visual, sound, and raw data (e.g., database, internet, keyboard), as well as various output actuators. A novel "system probe" provides the system with monitoring and control of its internal states (a form of meta-cognition). Additional probes – either custom interfaces with other systems or additional real-world sensors/actuators – can easily be added to the system.

Development environment, language, and hardware

The complete AGI engine plus associated support programs are implemented in (Object Oriented) C# under Microsoft's .NET framework. Current tests show that practical (proof-of-concept) prototype performance can be achieved on a single, conventional PC (2 Ghz, 512 Meg). Even a non-performance-tuned implementation can process several complex patterns per second on a database of hundreds of thousands stored features.

6 From Algorithms to General Intelligence

This section covers some of our near-term research and development; it aims to illustrate our expected path toward meaningful general intelligence. While this work barely approaches higher-level *animal* cognition (exceeding it in some aspects, but falling far short in others such as sensory-motor skills), we take it to be a crucial step in proving the validity and practicality of our model. Furthermore, the actual functionality achieved should be highly competitive, if not unique, in applications where significant autonomous adaptivity and data selection, lack of brittleness, dynamic pattern processing, flexible actuation, and self-supervised learning are central requirements.

General intelligence doesn't comprise one single, brilliant knock-out invention or design feature; instead, it emerges from the synergetic integration of a number of essential fundamental components. On the structural side, the system must integrate sense inputs, memory, and actuators, while on the functional side various learning, recognition, recall and action capabilities must operate seamlessly on a wide range of static and dynamic patterns. In addition, these cognitive abilities must be conceptual and contextual – they must be able to generalize knowledge, and interpret it against different backgrounds.

A key milestone in our project was reached when we started testing the *integrated* functionality of the basic cognitive components within our overall AGI framework. A number of custom-developed, highly-configurable test utilities are being used to test the cohesive functioning of the whole system. At this stage, most of our AGI development/testing is done using our virtual world environments, driven by custom scripts. This automated training and evaluation is supplemented by manual experimentation in numerous different environments and applications. Experience gained by these tests helps to refine the complex dynamics of interacting algorithms and parameters.

One of the general difficulties with AGI development is to determine absolute measures of success. Part of the reason is that this field is still nascent, and thus no agreed definitions, let alone tests or measures of low-level general intelligence exist. As we proceed with our project we expect to develop ever more effective protocols and metrics for assessing cognitive ability. Our system's performance evaluation is guided by this description:

> General intelligence comprises the ability to acquire (and adapt) the knowledge and skills required for achieving a wide range of goals in a variety of domains.

In this context:

- *Acquisition* includes all of the following: automatic, via sense inputs (feature/data driven); explicitly taught; discovered through exploration or experimentation; internal processes (e.g., association, categorization, statistics, etc.).
- *Adaptation* implies that new knowledge is integrated appropriately.
- *Knowledge and skills* refer to all kinds of data and abilities (states and behaviors) that the system acquires for the short or long term.

Our initial protocol for evaluating AGIs aims to cover a wide spectrum of domains and goals by simulating sample applications in 2D virtual worlds. In particular, these tests should assess the degree to which the foundational abilities operate as an integrated, mutually supportive whole – and without programmer intervention! Three examples follow.

6.1 Sample Test Domains for Initial Performance Criteria

Adaptive security monitor

This system scans video monitors and alarm panels that oversee a secure area (say, factory, office building, etc.), and responds appropriately to abnormal conditions. Note, this is somewhat similar to a site monitoring application at MIT [15].

This simulation calls for a visual environment that contains a lot of detail but has only limited dynamic activity – this is its normal state (green). Two levels of abnormality exist: (i) minor, or known disturbance (yellow); (ii) major, or unknown disturbance (red).

The system must initially learn the normal state by simple exposure (automatically scanning the environment) at different resolutions (detail). It must also learn "yellow" conditions by being shown a number of samples (some at high resolution). All other states must output "red."

Standard operation is to continuously scan the environment at low resolution. If any abnormal condition is detected the system must learn to change to higher resolution in order to discriminate between "yellow" and "red."

The system must adapt to changes in the environment (and totally different environments) by simple exposure training.

Sight assistant

The system controls a movable "eye" (by voice command) that enables the identification (by voice output) of at least a hundred different objects in the world. A trainer will dynamically teach the system new names, associations, and eye movement commands.

The visual probe can select among different scenes (simulating rooms) and focus on different parts of each scene. The scenes depict objects of varying attributes: color, size, shape, various dynamics, etc. (and combinations of these), against different backgrounds.

Initial training will be to attach simple sound commands to maneuver the "eye", and to associate word labels with selected objects. The system must then reliably execute voice commands and respond with appropriate identification (if any). Additional functionality could be to have the system scan the various scenes when idle, and to automatically report selected important objects.

Object identification must cover a wide spectrum of different attribute combinations and tolerances. The system must easily learn new scenes, objects, words and associations, and also adapt to changes in any of these variables.

Maze explorer

A (virtual) entity explores a moderately complex environment. It discovers what types of objects aid or hinder its objectives, while learning to navigate this dynamic world. It can also be trained to perform certain behaviors.

The virtual world is filled with a great number of different objects (see previous example). In addition, some of these objects move in space at varying speeds and dynamics, and may be solid and/or immovable. Groups of different kinds of objects have pre-assigned attributes that indicate negative or positive. The AGI engine controls the direction and speed of an entity in this virtual world. Its goal is to learn to navigate around immovable and negative objects to reliably reach hidden positives.

The system can also be trained to respond to operator commands to perform behaviors of varying degrees of complexity (for example, actions similar to "tricks" one might teach a dog). This "Maze Explorer" can easily be set up to deal with fairly complex tasks.

6.2 Towards Increased Intelligence

Clearly, the tasks described above do not by themselves represent any kind of breakthrough in artificial intelligence research. They have been achieved many times before. However, what we *do* believe to be significant and unique is the achievement of these various tasks without any task-specific programming or parameterization. It is not what is being done, but *how* it is done.

Development beyond these basic proof-of-concept tests will advance in two directions: (1) to significantly increase resolution, data volume, and complexity in applications similar to the tests; (2) to add higher-level functionality. In addition to work aimed at further developing and proving our general intelligence model, there are also numerous practical enhancements that can be done. These would include implementing multi-processor and network versions, and integrating our system with databases or with other existing AI technology such as expert systems, voice recognition, robotics, or sense modules with specialized feature extractors.

By far the most important of these future developments concern higher-level ability. Here is a partial list of action items, all of which are derived from lower-level foundations:

- spread activation and retain context over extended period;
- support more complex internal temporal patterns, both for enhanced recognition and anticipation, and for cognitive and action sequences;
- internal activation feedback for processing without input;
- deduction, achieved through selective concept activation;
- advanced categorization by arbitrary dimensions;
- learning of more complex behavior;
- abstract and merged concept formation;
- structured language acquisition;
- increased awareness and control of internal states (introspection);
- Learning logic and other problem-solving methodologies.

7 Other Research

Co-authored with Shane Legg, then at Adaptive A.I., Inc

Many different approaches to AI exist; some of the differences are straight forward while others are subtle and hinge on difficult philosophical issues. As such the exact placement of our work relative to that of others is difficult and, indeed, open to debate. Our view that "intelligence is a property of an entity that engages in two way interaction with an external environment," technically puts us in the area of "agent systems" [27]. However, our emphasis on a connectionist rather than classical approach to cognitive modeling, places our work in the field of "embodied cognitive science." (See [23] for a comprehensive overview.)

While our *approach* is similar to other research in embodied cognitive science, in some respects our goals are substantively different. A key difference is our belief that a core set of cognitive abilities working together is sufficient to produce general intelligence. This is in marked contrast to others in embodied cognitive science who consider intelligence to be necessarily specific to a set of problems within a given environment. In other words, they believe that autonomous agents always exist in ecological niches. As such they focus their research on building very limited systems that effectively deal with only a small number of problems within a specific limited environment. Almost all work in the area follows this – see [4, 6, 3] for just a few well known examples. Their stance contradicts the fact that humans possess general intelligence; we are able to effectively deal with a wide range of problems that are significantly beyond anything that could be called our "ecological niche."

Perhaps the closest project to ours that is strictly in the area of embodied cognitive science is the Cog project at MIT [5]. The project aims to under-stand the dynamics of human interaction by the construction of a human-like robot complete with upper torso, a head, eyes, arms and hands. While this project is significantly more ambitious than other projects in terms of the level and complexity of the system's dynamics and abilities, the system is still essentially niche focused (elementary human social and physical interaction) when compared to our own efforts at general intelligence.

Probably the closest work to ours in the sense that it also aims to achieve general rather than niche intelligence is the Novamente project under the di-rection of Ben Goertzel. (The project was formerly known as Webmind, see [11, 12].) Novamente relies on a hybrid of low-level neural net-like dynamics for activation spreading and concept priming, coupled with high-level seman-tic constructs to represent a variety of logical, causal and spatial-temporal relations. While the semantics of the system's internal state are relatively easy to understand compared to a strictly connectionist approach, the classi-cal elements in the system's design open the door to many of the fundamental problems that have plagued classical AI over the last fifty years. For example, high-level semantics require a complex meta-logic contained in hard coded high-level reasoning and other high-level cognitive systems. These high-level systems contain significant implicit semantics that may not be grounded in environmental interaction but are rather hard coded by the designer – thus causing symbol grounding problems [16]. The relatively fixed, high-level meth-ods of knowledge representation and manipulation that this approach entails are also prone to "frame of reference" [21, 25] and "brittleness" problems. In a strictly embodied cognitive science approach, as we have taken, all knowl-edge is derived from agent-environment interaction thus avoiding these long-standing problems of classical AI.

[8] is another researcher whose model closely resembles our own, but there are no implementations specifically based on his theoretical work. Igor Alek-sander's (now dormant) MAGNUS project [2] also incorporated many key AGI concepts that we have identified, but it was severely limited by a clas-sical AI, finite-state machine approach. Valeriy Nenov and Michael Dyer of UCLA [22] used "massively" parallel hardware (a CM-2 Connection Machine) to implement a virtual, interactive perceptual design close to our own, but with a more rigid, pre-programmed structure. Unfortunately, this ambitious, ground-breaking work has since been abandoned. The project was probably severely hampered by limited (at the time) hardware.

Moving further away from embodied cognitive science to purely classical research in general intelligence, perhaps the best known system is the Cyc project being pursued by [19]. Essentially Lenat sees general intelligence as being "common sense." He hopes to achieve this goal by adding many millions of facts about the world into a huge database. After many years of work and millions of dollars in funding there is still a long way to go as the sheer number of facts that humans know about the world is truly staggering. We doubt that a very large database of basic facts is enough to give a computer much general intelligence – the mechanisms for autonomous knowledge acquisition are missing. Being a classical approach to AI this also suffers from the fundamental problems of classical AI listed above. For example, the symbol grounding problem arises again: if facts about cats and dogs are just added to a database that the computer can use even though it has never seen or interacted with an animal, are those concepts really meaningful to the system? While his project also claims to pursue "general intelligence," it is really very different from our own, both in its approach and in the difficulties it faces.

Analysis of AI's ongoing failure to overcome its long-standing limitations reveals that it is not so much that Artificial General Intelligence has been tried and that it has failed, but rather that the field has largely been abandoned – be it for theoretical, historic, or commercial reasons. Certainly, our particular type of approach, as detailed in previous sections, is receiving scant attention.

8 Fast-track AGI: Why So Rare?

Widespread application of AI has been hampered by a number of core limitations that have plagued the field since the beginning, namely:

- the expense and delay of custom programming individual applications;
- systems' inability to automatically learn from experience, or to be user teachable/trainable;
- reliability and performance issues caused by "brittleness" (the inability of systems to automatically adapt to changing requirements, or data outside of a predefined range);
- their limited intelligence and common sense.

The most direct path to solving these long-standing problems is to conceptually identify the fundamental characteristics common to all high-level intelligence, and to engineer systems with this basic functionality, in a manner that capitalizes on human and technological strength.

General intelligence is the key to achieving robust autonomous systems that can learn and adapt to a wide range of uses. It is also the cornerstone of self-improving, or Seed AI – using basic abilities to bootstrap higher-level ones. This chapter identified foundational components of general intelligence, as well as crucial considerations particular to the effective development of the artificial variety. It highlighted the fact that very few researchers are actually following this most direct route to AGI.

If the approach outlined above is so promising, then why is has it received so little attention? Why is hardly anyone actually working on it?

A short answer: Of all the people working in the field called *AI*:

- 80% don't believe in the concept of General Intelligence (but instead, in a large collection of specific skills and knowledge.)
- Of those that do, 80% don't believe that artificial, human-level intelligence is possible – either ever, or for a long, long time.
- Of those that do, 80% work on domain-specific AI projects for commercial or academic-political reasons (results are more immediate).
- Of those left, 80% have a poor conceptual framework...

Even though the above is a caricature, in contains more than a grain of truth.

A great number of researchers reject the validity or importance of "general intelligence." For many, controversies in psychology (such as those stoked by *The Bell Curve*) make this an unpopular, if not taboo subject. Others, conditioned by decades of domain-specific work, simply do not see the benefits of Seed AI – solving the problems only once.

Of those that do not in principle object to general intelligence, many don't believe that AGI is possible – in their life-time, or ever. Some hold this position because they themselves tried and failed "in their youth." Others believe that AGI is not the best approach to achieving "AI," or are at a total loss on how to go about it. Very few researchers have actually studied the problem from our (the general intelligence/Seed AI) perspective. Some are actually trying to reverse-engineer the brain – one function at a time. There are also those who have moral objections, or who are afraid of it.

Of course, a great many are so focused on particular, narrow aspects of intelligence that they simply don't get around to looking at the big picture – they leave it to others to make it happen. It is also important to note that there are often strong financial and institutional pressures to pursue specialized AI.

All of the above combine to create a dynamic where Real AI is not "fashionable" – getting little respect, funding, and support – further reducing the number of people drawn into it!

These should be more than enough reasons to account for the dearth of AGI progress. But it gets worse. Researchers actually trying to build AGI systems are further hampered by a myriad of misconceptions, poor choices, and lack of resources (funding and research). Many of the technical issues were explored previously (see Sections 3 and 7), but a few others are worth mentioning:

Epistemology

Models of AGI can only be as good as their underlying theory of knowledge – the nature of knowledge, and how it relates to reality. The realization that high-level intelligence is based on conceptual representation of reality underpins design decisions such as adaptive, fuzzy vector encoding, and an interactive, embodied approach. Other consequences are the need for sense-based focus and selection, and contextual activation. The central importance of a highly-integrated pattern network – especially including dynamic ones – becomes obvious on understanding the relationship between entities, attributes, concepts, actions, and thoughts. These and several other insights lay the foundation for solving problems related to grounding, brittleness, and common sense. Finally, there is still a lot of unnecessary confusion about the relationship between concepts and symbols. A dynamic that continues to handicap AI is the lingering schism between traditionalists and connectionists. This unfortunately helps to perpetuate a false dichotomy between explicit symbols/schema and incomprehensible patterns.

Theory of mind

Another concern is sloppy formulation and poor understanding of several key concepts: consciousness, intelligence, volition, meaning, emotions, common sense, and "qualia." The fact that hundreds of AI researchers attend conferences every year where key speakers proclaim that "we don't understand consciousness (or qualia, or whatever), and will probably never understand it" indicates just how pervasive this problem is. Marvin Minsky's characterization of consciousness being a "suitcase word"[5] is correct. Let's just unpack it!

Errors like these are often behind research going off at a tangent relative to stated long-term goals. Two examples are an undue emphasis on biological feasibility, and the belief that embodied intelligence cannot be virtual, that it has to be implemented in physical robots.

Cognitive psychology

It goes without saying that a proper understanding of the concept "intelligence" is key to engineering it. In addition to epistemology, several areas of cognitive psychology are crucial to unraveling its meaning. Misunderstanding intelligence has led to some costly disappointments, such as manually accumulating huge amounts of largely useless data (knowledge without meaning), efforts to achieve intelligence by combining masses of dumb agents, or trying to obtain meaningful conversation from an isolated network of symbols.

Project focus

The few projects that do pursue AGI based on relatively sound models run yet another risk: they can easily lose focus. Sometimes commercial considerations hijack a project's direction, while others get sidetracked by (relatively)

---

[5]Meaning that many different meanings are thrown together in a jumble – or at least packaged together in one "box," under one label.

irrelevant technical issues, such as trying to match an unrealistically high level of performance, fixating on biological feasibility of design, or attempting to implement high-level functions before their time. A clearly mapped-out developmental path to human-level intelligence can serve as a powerful antidote to losing sight of "the big picture." A vision of how to get from "here" to "there" also helps to maintain motivation in such a difficult endeavor.

Research support

AGI utilizes, or more precisely, is an integration of a large number of existing AI technologies. Unfortunately, many of the most crucial areas are sadly under-researched. They include:

- incremental, real-time, unsupervised/self-supervised learning (vs. back-propagation);
- integrated support for temporal patterns;
- dynamically-adaptive neural network topologies;
- self-tuning of system parameters, integrating bottom-up (data driven) and top-down (goal/meta-cognition driven) auto-adaptation;
- sense probes with auto-adaptive feature extractors.

Naturally, these very limitations feed back to reduce support for AGI research.

Cost and difficulty

Achieving high-level AGI will be hard. However, it will not be nearly as difficult as most experts think. A key element of "Real AI" theory (and its implementation) is to concentrate on the essentials of intelligence. Seed AI becomes a manageable problem – in some respects much simpler than other mainstream AI goals - by eliminating huge areas of difficult, but inessential AI complexity. Once we get the crucial fundamental functionality working, much of the additional "intelligence" (ability) required is taught or learned, not programmed. Having said this, I do believe that very substantial resources will be required to scale up the system to human-level storage and processing capacity. However, the far more moderate initial prototypes will serve as proof-of-concept for AGI while potentially seeding a large number of practical new applications.

9 Conclusion

Understanding general intelligence and identifying its essential components are key to building next-generation AI systems – systems that are far less expensive, yet significantly more capable. In addition to concentrating on general learning abilities, a fast-track approach should also seek a path of least resistance – one that capitalizes on human engineering strengths and available technology. Sometimes, this involves selecting the AI road less traveled.

I believe that the theoretical model, cognitive components, and framework described above, joined with my other strategic design decisions provide a solid basis for achieving practical AGI capabilities in the foreseeable future. Successful implementation will significantly address many traditional problems of AI. Potential benefits include:

- minimizing initial environment-specific programming (through self-adaptive configuration);
- substantially reducing ongoing software changes, because a large amount of additional functionality and knowledge will be acquired autonomously via self-supervised learning;
- greatly increasing the scope of applications, as users teach and train additional capabilities; and
- improved flexibility and robustness resulting from systems' ability to adapt to changing data patterns, environments and goals.

AGI promises to make an important contribution toward realizing software and robotic systems that are more usable, intelligent, and human-friendly. The time seems ripe for a major initiative down this new path of human advancement that is now open to us.

References

1. Aha DW (ed) (1997) Lazy Learning. *Artificial Intelligence Review*, 11:1-5.
2. Aleksander I (1996) *Impossible Minds*. Imperial College Press, London.
3. Arbib MA (1992) *Schema theory*. In: Shapiro S (ed), Encyclopedia of Artificial Intelligence, 2nd ed. John Wiley and Sons, New York.
4. Braitenberg V (1984) *Vehicles: Experiments in Synthetic Psychology*. MIT Press, Cambridge, MA.
5. Brooks RA, Stein LA (1993) *Building Brains for Bodies*. Memo 1439, Artificial Intelligence Lab, Massachusetts Institute of Technology.
6. Brooks RA (1994) *Coherent behavior from Many Adaptive Processes*. In: Cliff D, Husbands P, Meyer JA, Wilson SW (eds), From animals to animats: Proceedings of the third International Conference on Simulation of Adaptive Behavior. MIT Press, Cambridge, MA.
7. Churchland PM (1995) *The Engine of Reason, the Seat of the Soul: A Philosophical Journey into the Brain*. MIT Press, Cambridge, MA.
8. Clark A (1997) *Being There: Putting Brain, Body and World Together Again*. MIT Press, Cambridge, MA.
9. Drescher GI (1991) *A Constructivist Approach to Intelligence*. MIT Press, Cambridge, MA.
10. Fritzke B (1995) *AF Growing Neural Gas Network Learns Topologies*. In: Tesauro G, Touretzky DS, Leen TK (eds), Advances in Neural Information Processing Systems 7. MIT Press, Cambridge, MA.

11. Goertzel B (1997) *From Complexity to Creativity.* Plenum Press, New York.
12. Goertzel B (2001) *Creating Internet Intelligence.* Plenum Press, New York.
13. Goldstone RL (1998) Perceptual Learning. *Annual Review of Psychology,* 49:585–612.
14. Gottfredson LS (1998) The General Intelligence Factor. *Scientific American,* 9(4):24–29.
15. Grimson WEL, Stauffer C, Lee L, Romano R (1998) Using Adaptive Tracking to Classify and Monitor Activities in a Site. *Proc. IEEE Conf. on Computer Vision and Pattern Recognition.*
16. Harnad S (1990) The Symbol Grounding Problem. *Physica D,* 42:335–346.
17. Kelley D (1986) *The Evidence of the Senses.* Louisiana State University Press, Baton Rouge, LA.
18. Kosko B (1997) *Fuzzy Engineering.* Prentice Hall, Upper Saddle River, NJ.
19. Lenat D, Guha R (1990) *Building Large Knowledge Based Systems.* Addison-Wesley, Reading, MA.
20. Margolis H (1987) *Patterns, Thinking, and Cognition: A Theory of Judgment.* University of Chicago Press, Chicago, IL.
21. McCarthy J, Hayes P (1969) Some Philosophical Problems from the Standpoint of Artificial Intelligence. *Machine Intelligence,* 4:463–502.
22. Nenov VI, Dyer MG (1994) *Language Learning via Perceptual/Motor Association: A Massively Parallel Model.* In: Kitano H, Hendler JA (eds), Massively Parallel Artificial Intelligence, MIT Press, Cambridge, MA.
23. Pfeifer R, Scheier C (1999) *Understanding Intelligence.* MIT Press, Cambridge, MA.
24. Picard RW (1997) *Affective Computing.* MIT Press, Cambridge, MA.
25. Pylyshyn ZW (ed) (1987) *The Robot's Dilemma: The Frame Problem in A.I.* Ablex, Norwood, NJ.
26. Rand A (1990) *Introduction to Objectivist Epistemology.* Meridian, New York.
27. Russell S, Norvig P (1995) *Artificial Intelligence: A Modern Approach.* Prentice Hall, Upper Saddle River, NJ.
28. Yip K, Sussman GJ (1997) Sparse Representations for Fast, One-shot Learning. *Proc. of National Conference on Artificial Intelligence.*

Artificial Brains

Hugo de Garis

Computer Science Dept. and Theoretical Physics Dept., Utah State University
Old Main 423, Logan, Utah, UT 84322-4205, USA
degaris@cs.usu.edu - http://www.cs.usu.edu/ degaris

Summary. This chapter introduces the idea of "Evolvable Hardware," which applies evolutionary algorithms to the generation of programmable hardware as a means of achieving Artificial Intelligence. Cellular Automata-based Neural Networks are evolved in different modules, which form the components of artificial brains. Results from past models and plans for future work are presented.

1 Introduction

It is appropriate, in these early years of the new millennium, that a radical new technology makes its debut that will allow humanity to build artificial brains, an enterprise that may define and color the twenty-first century. This technology is called "Evolvable Hardware" (or just "E-Hard" for short). Evolvable hardware applies genetic algorithms (simulated Darwinian evolution) to the generation of programmable logic devices (PLDs, programmable hardware), allowing electronic circuits to be evolved at electronic speeds and at complexity levels that are beyond the intellectual design limits of human electronic engineers. Tens of thousands (and higher magnitudes) of such evolved circuits can be combined to form humanly specified artificial brain architectures.

In the late 1980s, the author began playing with genetic algorithms and their application to the evolution of neural networks. A genetic algorithm simulates the evolution of a system using a Darwinian "survival of the fittest" strategy. There are many variations of genetic (evolutionary) algorithms. One of the simplest uses a population of bit strings (a string of 0s and 1s) called "chromosomes" (analogous to molecular biology) to code for solutions to a problem. Each bit string chromosome can be decoded and applied to the problem at hand. The quality of the solution specified by the chromosome is measured and given a numerical score, called its "fitness". Each member of the population of competing chromosomes is ranked according to its fitness. Low scoring chromosomes are eliminated. High scoring chromosomes have copies made of them (their "children" in the next "generation").

Hence only the fittest survive. Random changes are made to the children, called "mutations." In most cases, mutations cause the fitness of a mutated chromosome to decrease, but occasionally, the fitness increases, making the child chromosome fitter than its parent (or parents, if two parents combine bits "sexually" to produce the child's chromosome). This fitter child chromosome will eventually force its less fit parents out of the population in future generations, until it in turn is forced out by its fitter offspring or the fitter offspring of other parents. After hundreds of generations of this "test, select, copy, mutate" cycle, systems can be evolved quite successfully that perform according to the desired fitness specification.

Neural networks are interconnected nets of simulated brain cells. An individual simulated brain cell (neuron) receives signals from neighboring neurons, which it "weights" by multiplying the incoming signal strength S_i by a numerical weighting factor W_i, to form the product S_iW_i. The sum of all the incoming weighted signals is formed and compared to the neuron's numerical threshold value T. If the sum has a value greater than T, then the neuron will "fire" an output signal whose strength depends on how much greater the sum is than the threshold T. The output signal travels down the neuron's outward branching pathway called an "axon." The branching axon connects and transmits it signal to other branching pathways called "dendrites" which transmit the signal to other neurons. By adjusting the weighting factors and by connecting up the network in appropriate ways, neural networks can be built which map input signals to output signals in desired ways.

The first attempts to wed genetic algorithms (GAs) to neural nets (NNs) restricted themselves to static (constant valued) inputs and outputs (no dynamics). This restriction struck the author as being unwarranted, so he began experimenting with dynamic inputs and outputs. The first successful attempt in this regard managed to get a pair of stick legs to walk, the first evolved, neural net controlled, dynamic behavior. If one can evolve one behavior, one can evolve many, so it became conceivable to imagine a whole library of evolved behaviors, for example, to get a software simulated quadruped to walk straight, to turn left, to turn right, to peck at food, to mate, etc, with one separately evolved neural net circuit or module per behavior. Behaviors could be switched smoothly by feeding in the outputs of the module generating the earlier behavior to the inputs of the module generating the later behavior.

By evolving modules that could detect signals coming from the environment, e.g. signal strength detectors, frequency detectors, motion detectors etc, then behaviors could be changed at appropriate moments. The simulated quadruped ("Lizzy") could begin to show signs of intelligence, due to possessing an artificial nervous system of growing sophistication. The idea began to emerge in the author's mind that it might be possible to build artificial brains, if only somehow one could put large numbers of evolved modules together to function as an integrated whole. The author began to dream of building artificial brains.

However there was a problem with the above approach. Every time a new (evolved neural net) module was added to the simulation (on a Mac 2 computer) in the early 1990s, the overall simulation speed slowed, until it was no longer practical to have more than a dozen modules. Somehow the whole process needed to be speeded up, which led to the dream of doing it all in hardware, at hardware speeds.

2 Evolvable Hardware

A visit to an electronic engineering colleague at George Mason University (GMU) in Virginia USA, in the summer of 1992, led the author to hear about FPGAs (Field Programmable Gate Arrays) for the first time. An FPGA is an array (a matrix) of electronic logic blocks, whose Boolean (*and, or, not*) functions, inter-block and I/O connections can be programmed, or "configured" (to use the technical term) by individual users, so if a logic designer makes a mistake, it can be quickly and easily corrected by reprogramming. FPGAs are very popular with electronic engineers today. Some FPGAs are S-RAM (Static RAM) based, and can therefore be reprogrammed an unlimited number of times. If the FPGA can also accept random configuring bit strings, then it becomes a suitable device for evolution. This epiphany made the author very excited in 1992, because he realized that it might be possible to evolve electronic neural circuits at electronic speeds and hence overcome his problem of slow evolution and execution speeds in software on a personal computer. The author began preaching the gospel of "evolvable hardware" as he called it, to his colleagues in the field of 'evolutionary computation (EC), which alternatively might be relabeled "evolvable software," or "E-Soft." Slowly, the idea caught on, so that by the year 2002, there had been a string of world conferences, and academic journals devoted to the topic started to appear.

In the latter half of the 1990s the E-Hard field was stimulated by the presence of a particular evolvable chip family manufactured by a Silicon Valley, California company called Xilinx, labeled the XC6200 series. This family of chips (with a different number of logic blocks per chip type) had several advantages over other reconfigurable chip families. The architecture of the chip was public knowledge (not a company secret) thus allowing researchers to play with it. It could accept random configuring bit strings without blowing up (important for evolution which uses random bit strings), and thirdly and very importantly, it was partially reconfigurable at a very fine grained level, meaning that if one mutated only a few bits in a long configuring bit string, only the corresponding components of the circuit were changed (reconfigured), without having to reconfigure the whole circuit again. This third feature allowed for rapid reconfiguration, which made the chip the favorite amongst E-Harders. Unfortunately, Xilinx stopped manufacturing the XC6200 series and is concentrating on its new multi-mega gate chip family called "Virtex," but the Virtex chips are less fine-grainedly reconfigurable than the XC6200 family, so E-Harders are feeling a bit out in the cold. Hopefully, Xilinx and similar manufacturers will see the light and make future generations of their chips more "evolvable," by possessing a higher degree of fine-grained reconfigurability. As will be seen below, the author chose a Xilinx chip XC6264 as the basis for his work on building an artificial brain (before supplies ran out). The underlying methodology of this work is based on "evolvable hardware."

2.1 Neural Network Models

Before discussing the evolution of a neural model in hardware at hardware speeds, one first needs to know what the neural model is. For years, the author had a vague notion of being able to put millions of artificial neurons into gigabytes of RAM and running that huge space as an artificial brain. RAM memory is fairly cheap, so it seemed reasonable to somehow embed neural networks, large numbers of them, into RAM, but how? The solution the author chose was to use cellular automata (CAs). Two dimensional (2D) CAs can be envisioned as a multicolored chessboard, all of whose squares can change their color at the tick of a clock according to certain rules. These cellular automata color (or state) change rules take the following form. Concentrate on a particular square, which has the color *orange*, let's say. Look at its four neighboring square colors. If the upper square is *red*, and the right hand square is *yellow*, and the bottom square is *blue*, and the left hand square is *green*, then at the next clock tick, the central *orange* square will become *brown*. This rule can be expressed succinctly in the form:

$$IF \quad (C = orange) \wedge (U = red) \wedge$$
$$(R = yellow) \wedge (B = blue) \wedge (L = green)$$
$$THEN \ (C = brown)$$

or even more succinctly, in the form:

$$orange.red.yellow.blue.green \implies brown$$

Using thousands of such rules, it was possible to make CAs behave as neural networks, which grew, signaled and evolved (see Figs. 1 and 2). Some early experiments showed that these circuits could be evolved to perform such tasks as generating an output signal that oscillated at an arbitrarily chosen frequency, that generated a maximum number of synapses in a given volume, etc. However, the large number of rules to make this CA based neural network function was a problem. The 2D version took 11,000 rules. The 3D version took over 60,000 rules. There was no way that such large numbers could be implemented directly in electronics, evolving at electronic speeds. An alternative model was needed which had very few rules, so few that they could be implemented directly into FPGAs, thus enabling the field of brain building by evolving neural net circuits in seconds rather than days as is often the case using software evolution methods.

The simplified model will be described in more detail, since it is the model actually implemented in the evolvable hardware. It is a 3D model, again based on cellular automata, but much simpler. A neuron is modeled by a single 3D CA cell. The CA trails (the axons and dendrites) are only 1 cell wide, instead of

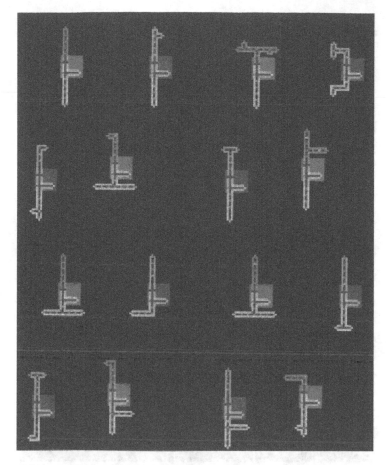

Fig. 1: Older complex model of cellular automata based neural network, early growth

the 3 cell wide earlier model. The growth instructions are distributed throughout the 3D CA space initially (see Fig. 3) instead of being passed through the CA trails (as in Figs. 1 and 2). The neural signaling in the newer model is 1 bit only, compared to the 8 bit signals in the earlier model. Such restrictions will lower the evolvability of the circuits, but in practice, one finds that the evolvabilities are still adequate for most purposes. In the growth phase, the first thing done is to position the neurons. For each possible position in the space where a neuron can be placed, a corresponding bit in the chromosome is used. If that bit is a 1, then a neuron is placed at that position. If the bit is a 0, then no neuron is placed at that position.

Every 3D CA cell is given 6 growth bits from the chromosome, one bit per cubic face. At the first tick of the growth clock, each neuron checks the bit at each of its 6 faces. If a bit is a 1, the neighboring blank cell touching

Fig. 2: Older complex model of cellular automata based neural network, saturated growth

the corresponding face of the neuron is made an axon cell. If the bit is a 0, then the neighboring blank cell is made a dendrite cell. Thus a neuron can grow maximum 6 axons or 6 dendrites, and all combinations in between. At the next clock tick, each blank cell looks at the bit of the face of the filled neighbor that touches it. If that filled cell face bit is a 1, then the blank cell becomes the cell type (axon or dendrite) of the touching filled neighbor. The blank cell also sets a pointer towards its parent cell – for example, if the parent cell lies to the west of the blank cell, the blank cell sets an internal pointer which says "west." These "parent pointers (PPs)" are used during the neural signaling phase to tell the 1-bit signals which way to move as they travel along axons and dendrites.

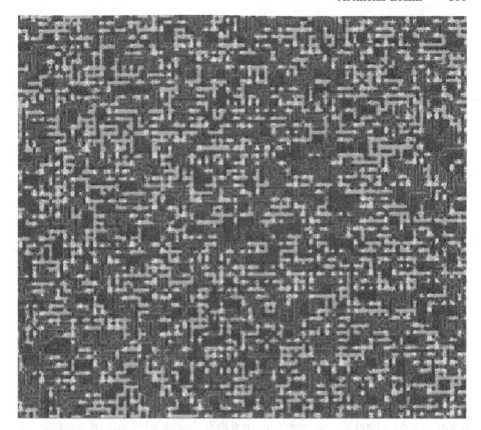

Fig. 3: Newer simpler model of cellular automata based neural network, saturated growth

This cellular growth process continues at each clock tick for several hundred ticks until the arborization of the axons and dendrites is saturated in the 3D space. In the hardware implementation of this simplified model, the CA space consists of a 24*24*24 cube (the "macro cube") of 3D CA cells, i.e. roughly 14,000 of them. At the 6 faces of the macro cube, axon and dendrite growths wrap around to the opposite macro face, thus forming a "toroidal" (doughnut) shape. There are prespecified input and output points (188 maximum input points, and 4 maximum output points, although in practice usually only one output point is used, to foster evolvability). The user specifies which input and output points are to be used for a given module. At an input point, an axon cell is set which grows into the space. Similarly for an output point, where a dendrite cell is set.

In the signaling phase, the 1 bit neural signals move in the same direction in which axon growth occurred, and in the opposite direction in which dendrite growth occurred. Put another way, the signal follows the direction of the parent pointers (PPs) if it is moving in a dendrite, and follows in any direction other than that of the parent pointers (PPs) if it is moving in an axon.

An input signal coming from another neuron or the outside world travels down the axon until the axon collides with a dendrite. The collision point is called a "synapse." The signal transfers to the dendrite and moves toward the dendrite's neuron. Each face of the neuron cube is genetically assigned a sign bit. If this bit is a 1, the signal will add 1 to the neuron's 4-bit counter value. If the bit is a 0, the signal will subtract 1 from the neuron's counter. If the counter value exceeds a threshold value, usually 2, it resets to zero, and the neuron "fires," sending a 1-bit signal to its axons at the next clock tick.

3 The CAM-Brain Machine (CBM)

The evolvable hardware device that implements the above neural net model is a Cellular Automata Machine (CAM), which is called a CAM-Brain Machine (CBM). The term CAM-Brain implies that an artificial brain is to be embedded inside cellular automata. The CBM is a piece of special hardware that evolves neural circuits very quickly. It consists largely of Xilinx's (programmable hardware) XC6264 chips (72 of them), which together can evolve a neural network circuit module in a few seconds. The CBM executes a genetic algorithm on the evolving neural circuits, using a population of 100 or so of them, and running through several hundred generations, i.e. tens of thousands of circuit growths and fitness measurements. Once a circuit has been evolved successfully, it is downloaded into a gigabyte of RAM memory. This process occurs up to 64000 times, resulting in 64000 downloaded circuit modules in the RAM. A team of Brain Architects (BAs) has already decided which modules are to be evolved, what their individual functions are, and how they are to interconnect. Once all the modules are evolved and their interconnections specified, the CBM then functions in a second mode. It updates the RAM memory containing the artificial brain at a rate of 130 billion 3D cellular automata cell updates a second. This is fast enough for real time control of a kitten robot "Robokitty," described below.

The CBM consists of 6 main components or units described briefly here.

Cellular Automata Unit: The Cellular Automata Unit contains the cellular automata cells in which the neurons grow their axons and dendrites, and transmit their signals.

Genotype/Phenotype Memory Unit: The Genotype/Phenotype Memory Unit contains the 100K bit chromosomes that determine the growth of the neural circuits. The Phenotype Memory Unit stores the state of the CA cells (blank, neuron, axon, dendrite).

Fitness Evaluation Unit: The Fitness Evaluation Unit saves the output bits, converts them to an analog form and then evaluates how closely the target and the actual outputs match.

Genetic Algorithm Unit: The Genetic Algorithm Unit performs the GA on the population of competing neural circuits, eliminating the weaker circuits and reproducing and mutating the stronger circuits.

Module Interconnection Memory Unit: The Module Interconnection Memory Unit stores the BA's (brain architect's) inter-module connection specifications, for example, "the 2nd output of module 3102 connects to the 134th input of module 63195."

External Interface Unit: The External Interface Unit controls the input/output of signals from/to the external world, e.g. sensors, camera eyes, microphone ears, motors, antenna I/O, etc.

The CBM's shape and color is symbolic (see Figs. 4, 5). The curved outer layer represents a slice of human cortex. The gray portion that contains the electronic boards represents the "gray matter" (neural bodies) of the brain, and the white portion which contains the power supply, represents the "white matter" (axons) of the brain.

The first CBM and its supporting software packages were implemented in 1999, and actual research use of the machine began in that year. The results of this testing and the experience gained in using the CBM to design artificial brain architectures, should form the contents of future articles, with such titles as "Artificial Brain Architectures."

3.1 Evolved Modules

Since the neural signals in the model implemented by the CBM use single bits, the inputs and outputs to a neural module also need to be in a 1-bit signal form. Table 1 shows a target (desired) output binary string, and the best evolved (software simulated) result, showing that the evolution of such binary strings is possible using the CBM implemented model. To increase the usefulness of the CBM, algorithms were created which converted an arbitrary analog curve into its corresponding bit string (series of 1s and 0s) and vice versa, thus allowing users to think entirely in analog terms. Analog inputs are converted automatically into binary and input to the module. Similarly the binary output is converted to analog and compared to analog target curves. Figure 6 shows a random analog target curve and the best evolved curve. Note that the evolved curve followed the target curve fairly well only for a limited amount of time, illustrating the "module's evolvable capacity" (MEC). To generate analog target curves of unlimited time lengths (needed to generate the behaviors of the kitten robot over extended periods of time) multi module systems may need to be designed which use a form of time slicing, with one module generating one time slice's target output.

Fig. 4: CAM-Brain Machine (CBM) with cover

Binary Target Output and Best Binary Evolved Output	
Target	000000000000000000000000000000001111111111111111111
Evolved	000000000000000000000000000000000001111111111111111
Target ct	000000000000000000000000011111111111111111000000000000000000000
Evolved ct	100000000000000000000000011111111111111111000000000000000000000

We have software simulated the evolution of many modules (for example, 2D static and dynamic pattern detectors, motion controllers, decision modules, etc). Experience shows us that their "evolvability" is usually high enough to generate enthusiasm. For EEs (evolutionary engineers) the concept of evolvability is critical.

3.2 The Kitten Robot "Robokitty"

In 1993, the year the CAM-Brain Project started, the idea that an artificial brain could be built containing a billion neurons in an era in which most neural nets contained tens to hundreds of neurons seemed ludicrous. Early skepticism was strong. A means was needed to show that an artificial brain is

Fig. 5: CAM-Brain Machine (CBM) showing slots for 72 FPGA circuit boards

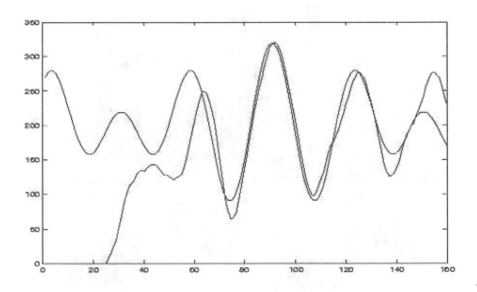

Fig. 6: Analog target output and best analog evolved output

a valid concept to silence the critics. The author chose to have the artificial brain control hundreds of behaviors of a cute life-sized robot kitten whose mechanical design is shown in Fig. 7. This robot kitten "Robokitty" will have some 23 motors, and will send and receive radio signals to and from the CBM via antenna. The behaviors of the kitten are evolved in commercial "Working Model 3D" software (from MSC Working Knowledge, Inc.) and the results then used as target wave forms for the evolution of the control modules in the CBM.

The evolution of motions in software at software speeds goes against the grain of the philosophy of evolvable hardware, but was felt to be unavoidable for practical reasons. Fortunately, the vast majority of modules will be evolved at electronic speeds. Judging by its many behaviors and the "intelligence" of its sensory and decision systems, it should be obvious to a casual observer that "it has a brain behind it," making the robot behave in as "kitten like" a manner as possible.

Fig. 7: Mechanical design of robot kitten "Robokitty"

4 Short- and Long-Term Future

The immediate goal once the first CBMs were built, was to use a CBM to create the artificial brain's modular architecture to control the robokitten. The very concreteness of the task, i.e. getting the kitten to execute its many hundreds of behaviors and decide when to switch between them based on decisions coming from its sensory systems and internal states, would require a major effort, since 64000 modules needed to be evolved. Of course, once work on the CBM had began, initial efforts were with single modules, to see what the CBM could evolve. Unfortunately, this work had only just begun in 2000 when the bankruptcy of the author's previous lab (Starlab) occurred and stopped such work in its tracks.

It was planned that once experience with single module evolution had been gained, interconnected multi-module systems would be built, with 10s, 100s, 1000s, 10,000s of modules, up to the limit of 64000 modules. If this job was to be completed in two years, assuming that it would take on average 30 minutes for an evolutionary engineer (EE) to dream up the function and fitness measure of a module, then a design team of 16 people would be needed.

A million-module, second generation artificial brain will require roughly 250 EEs. Thus the problem of building such a large artificial brain would not only be conceptual, but managerial as well. The author envisages that within five to ten years, if the first generation brain is a success, it is likely that large national organizations devoted to brain building will be created, comparable to the way Goddard's rockets went from two meter toys controlled by one man to NASA, with tens of thousands of engineers and a budget of billions of dollars.

Such national scale brain building projects have been given labels, such as the A-Brain Project, (America's National Brain Building Project), E-Brain Project (Europe's), C-Brain Project (China's), J-Brain Project (Japan's), etc. Initially, these artificial brains will probably be used to create increasingly intelligent robotic pets. Later they may be used to control household cleaning robots, soldier robots, etc. Brain based computing may generate a trillion dollar world market within 10 years or so. The annual PC market is worth about a trillion dollars worldwide today.

In the long term, 50 to 100 years from now, the situation becomes far more alarming. Twenty-first century technologies will allow 1 bit per atom memory storage, and femtosecond (a thousandth of a trillionth of a second) switching times (bit flipping). Reversible logic will allow heatless computing, and the creation of 3D circuitry that does not melt. In theory, asteroid sized, self-assembling, quantum computers which would have a bit flip rate of 10 to power 55 a second could be built. The estimated human computing capacity is a mere 10 to power 16 bit flips a second, i.e. roughly a trillion trillion trillion times less. For brain builders with a social conscience, the writing is on the wall. The author feels that the global politics of our new century will be dominated by the issue of species dominance.

Should humanity build godlike "Artilects" (artificial intellects) or not? The author foresees a major war between two human groups, the "Cosmists," who will favor building artilects, for whom such an activity is a science-compatible religion – the big-picture destiny of the human species – and the "Terrans," who will fear that one day, artilects, for whatever reason, may decide to exterminate the human race. For the Terrans, the only way to ensure that such a risk is never undertaken, is to insist that artilects are never built.

In the limit, to preserve the human species, the Terrans may exterminate the Cosmists, if the latter threaten to build artilects. With twenty-first century weaponry, and extrapolating up the graph of the number of deaths in major wars over time, we arrive at "gigadeath." One of the major tasks of today's brain builders is to persuade humanity that such a scenario is not a piece of dismissible science fiction, but a frightening possibility.

Some brain builders will stop their work due to such worries. Others will continue, driven by the magnificence of their goal – to build "artilect" gods. When the nuclear physicists in the 1930s were predicting that a single nuclear bomb could wipe out a whole city, most people thought they were crazy, but a mere 12 years after Leo Szilard had the idea of a nuclear chain reaction, Hiroshima was vaporized.

The decision whether to build artilects or not, will be the toughest that humanity will have to face in our new century. Humanity will have to choose between "building gods, or building our potential exterminators."

5 Postscript – July 2002

This postscript provides a brief update on what has been happening with the CAM-Brain Project since the above article was written. The author worked in Japan from 1992 to 1999. In the year 2000 he moved to a private blue-sky research lab called STARLAB in Brussels, Belgium, Europe. Starlab bought a CAM-Brain Machine (CBM) that was delivered in the summer of 2000. Unfortunately, the dotcom crash hit Starlab hard, resulting in its bankruptcy in June of 2001. Starlab's CBM was not fully paid for, so the constructor of this machine, who had internet access to it, switched it off, effectively killing the project. Four CBMs were built (one in Japan, two in Europe, one in the USA). Once the designer was no longer paid, he stopped updating the firmware in all the machines, so effectively all of them are incompletely developed and do not function as they should.

Since September of 2001, the author has been an associate professor in the computer science department at Utah State University in the US, with the responsibility for establishing a Brain Builder Group and obtaining funds for the creation of a second generation brain building machine called Brain Building Machine, 2nd Generation (BM2). If funding can be found, this second generation machine will use the latest generation of programmable/evolvable chips (namely Xilinx's "Virtex" family of chips).

The second time round, however, we are insisting on in-house hardware design expertise. We don't want to be dependent upon a commercially motivated external hardware designer again. This time, all the people involved in the creation of the BM2 are researchers. The author now has a team of a dozen people, mostly Masters and PhD students, who are learning how to program and evolve hardware, using the Xilinx Virtex chips, and to create increasingly evolvable neural net models which can be implemented in the evolvable hardware. The author is also collaborating with two other academic colleagues at different universities across the US, who have had extensive industrial hardware design experience. The summer vacation of 2002 was spent devising the architecture of the BM2, with the intention of submitting major grant proposals to the tune of US$1M for BM2 design and construction.

Perhaps before closing, a brief mention of some of the challenges faced by the BM2 design can be mentioned here.

The BM2 will have obvious similarities to the CBM. It will still be based on the basic assumption that individual neural network modules will be evolved and then hand assembled in RAM to make an artificial brain. This basic assumption in the overall design may, possibly, be changed as the BM2 conception proceeds, but for the moment it is difficult to imagine how a non-modular approach might be undertaken. However, the sheer momentum of Moore's Law will force us sooner or later to take a different approach, for the simple reason that it will become humanly impossible to conceive and individually evolve a million modules. The CBM could handle 75,000,000 neurons and 64,000 modules. Very probably, the BM2 will be able to handle 1,000,000,000 neurons and 1,000,000 modules. A million modules is simply too many to handle, thus necessitating the need to automate the evolution of multi-module systems. Just how the author's brain building team will solve such problems has still to be settled.

On the other hand, Moore's Law has already given the electronics world programmable (evolvable) chips with nearly 10,000,000 logic gates. A very large number of such chips is not needed to build a billion neuron artificial brain. That is the encouraging aspect of brain building, i.e. knowing that today's chip capacities will allow it.

Probably the greatest challenge remains the same as it did for the CBM, namely architecting the artificial brains themselves. How can hundreds of motion controllers, thousands of pattern recognizers, etc be put together to design an artificial brain that will control a (kitten?) robot device with such variety and intelligence that adults will remain amused by it for half an hour?

The author believes that the planet's first artificial brains will come into being within the next few years. If the CBM had not been stopped in its tracks, there may have been an initial attempt at building such a brain in the year 2001. Now the BM2 will need to be built for the "Utah Brain Project" to proceed. Stay tuned.

References

1. de Garis H (2002) Guest Editorial *Neurocomputing*, 42(1-4):1–8.
2. de Garis H, Korkin M (2002) The CAM-Brain Machine (CBM): An FPGA Based Hardware Tool which Evolves a 1000 Neuron Net Circuit Module in Seconds and Updates a 75 Million Neuron Artificial Brain for Real Time Robot Control. *Neurocomputing*, 42(1-4):35–68.
3. de Garis H (1999) Review of Proceedings of the First NASA/DoD Workshop on Evolvable Hardware *IEEE Transactions on Evolutionary Computation*, 3(4):3054–306.
4. de Garis H, Korkin M (2000) An Artificial Brain: Using Evolvable Hardware Techniques to Build a 75 Million Neuron Artificial Brain to Control the Many Behaviors of a Kitten Robot. In: Osada Y (ed) Handbook of Biomimetics, HTS Publishers, Japan.
5. de Garis H, Korkin M, Fehr G (2001) The CAM-Brain Machine (CBM): An FPGA Based Tool for Evolving a 75 Million Neuron Artificial Brain to Control a Lifesized Kitten Robot. *Autonomous Robots*, 10(3):235–249.
6. de Garis H, Korkin M, Gers F, Nawa E, Hough M (2000) Building an Artificial Brain Using an FPGA Based CAM-Brain Machine. *Applied Mathematics and Computation*, 111:163–192.

A comprehensive list of the author's journal articles, conference papers, book chapters and world media reports, can be found at `http://www.cs.usu.edu/~degaris`.

The New AI: General & Sound & Relevant for Physics

Jürgen Schmidhuber

IDSIA, Galleria 2, 6928 Manno (Lugano), Switzerland &
TU Munich, Boltzmannstr. 3, 85748 Garching, München, Germany
`juergen@idsia.ch` - `http://www.idsia.ch/~juergen`

Summary. Most traditional artificial intelligence (AI) systems of the past 50 years are either very limited, or based on heuristics, or both. The new millennium, however, has brought substantial progress in the field of theoretically optimal and practically feasible algorithms for prediction, search, inductive inference based on Occam's razor, problem solving, decision making, and reinforcement learning in environments of a very general type. Since inductive inference is at the heart of all inductive sciences, some of the results are relevant not only for AI and computer science but also for physics, provoking nontraditional predictions based on Zuse's thesis of the computer-generated universe.

1 Introduction

Remarkably, there is a theoretically *optimal* way of making predictions based on observations, rooted in the early work of Solomonoff and Kolmogorov [62, 28]. The approach reflects basic principles of Occam's razor: simple explanations of data are preferable to complex ones.

The theory of universal inductive inference quantifies what simplicity really means. Given certain very broad computability assumptions, it provides techniques for making optimally reliable statements about future events, given the past.

Once there is an optimal, formally describable way of predicting the future, we should be able to construct a machine that continually computes and executes action sequences that maximize expected or predicted reward, thus solving an ancient goal of AI research.

For many decades, however, AI researchers have not paid a lot of attention to the theory of inductive inference. Why not? There is another reason besides the fact that most of them have traditionally ignored theoretical computer science: the theory has been perceived as being associated with excessive computational costs. In fact, its most general statements refer to methods that are optimal (in a certain asymptotic sense) but incomputable. So researchers in machine learning and artificial intelligence have often resorted to alternative methods that lack a strong theoretical foundation but at least seem feasible in certain limited contexts. For example, since the early attempts at building a "General Problem Solver" [36, 43] much work has been done to develop mostly heuristic machine learning algorithms that solve new problems based on experience with previous problems. Many pointers to *learning by chunking, learning by macros, hierarchical learning, learning by analogy,* etc. can be found in Mitchell's book [34] and Kaelbling's survey [27].

Recent years, however, have brought substantial progress in the field of *computable* and *feasible* variants of optimal algorithms for prediction, search, inductive inference, problem solving, decision making, and reinforcement learning in very general environments. In what follows I will focus on the results obtained at IDSIA.

Sections 3, 4, 7 relate Occam's razor and the notion of simplicity to the shortest algorithms for computing computable objects, and will concentrate on recent *asymptotic* optimality results for universal learning machines, essentially ignoring issues of practical feasibility — compare Hutter's contribution [25] in this volume.

Section 5, however, will focus on our recent non-traditional simplicity measure which is *not* based on the shortest but on the *fastest* way of describing objects, and Section 6 will use this measure to derive non-traditional predictions concerning the future of our universe.

Sections 8, 9, 10 will finally address quite pragmatic issues and "true" time-optimality: given a problem and only so much limited computation time, what is the best way of spending it on evaluating solution candidates? In particular, Section 9 will outline a bias-optimal way of incrementally solving each task in a sequence of tasks with quickly verifiable solutions, given a probability distribution (the *bias*) on programs computing solution candidates. Bias shifts are computed by program prefixes that modify the distribution on their suffixes by reusing successful code for previous tasks (stored in non-modifiable memory). No tested program gets more runtime than its probability times the total search time. In illustrative experiments, ours becomes the first general system to *learn* a universal solver for arbitrary n disk *Towers of Hanoi* tasks (minimal solution size $2^n - 1$). It demonstrates the advantages of incremental learning by profiting from previously solved, simpler tasks involving samples of a simple context-free language. Section 10 discusses how to use this approach for building general reinforcement learners.

Finally, Sect. 11 will summarize the recent Gödel machine [56], a self-referential, theoretically optimal self-improver which explicitly addresses the *"Grand Problem of Artificial Intelligence"* [58] by optimally dealing with limited resources in general reinforcement learning settings.

2 More Formally

What is the optimal way of predicting the future, given the past? Which is the best way to act such as to maximize one's future expected reward? Which is the best way of searching for the solution to a novel problem, making optimal use of solutions to earlier problems?

Most previous work on these old and fundamental questions has focused on very limited settings, such as Markovian environments, where the optimal next action, given past inputs, depends on the current input only [27].

We will concentrate on a much weaker and therefore much more general assumption, namely, that the environment's responses are sampled from a computable probability distribution. If even this weak assumption were not true then we could not even formally specify the environment, leave alone writing reasonable scientific papers about it.

Let us first introduce some notation. B^* denotes the set of finite sequences over the binary alphabet $B = \{0,1\}$, B^∞ the set of infinite sequences over B, λ the empty string, and $B^\sharp = B^* \cup B^\infty$. x, y, z, z^1, z^2 stand for strings in $B^\sharp$. If $x \in B^*$ then xy is the concatenation of x and y (e.g., if $x = 10000$ and $y = 1111$ then $xy = 100001111$). For $x \in B^*$, $l(x)$ denotes the number of bits in x, where $l(x) = \infty$ for $x \in B^\infty$; $l(\lambda) = 0$. x_n is the prefix of x consisting of the first n bits, if $l(x) \geq n$, and x otherwise ($x_0 := \lambda$). log denotes the logarithm with basis 2, f, g denote functions mapping integers to integers. We write $f(n) = O(g(n))$ if there exist positive constants c, n_0 such that $f(n) \leq cg(n)$ for all $n > n_0$. For simplicity, let us consider universal Turing Machines [67] (TMs) with input alphabet B and trinary output alphabet including the symbols "0", "1", and " " (blank). For efficiency reasons, the TMs should have several work tapes to avoid potential quadratic slowdowns associated with 1-tape TMs. The remainder of this chapter assumes a fixed universal reference TM.

Now suppose bitstring x represents the data observed so far. What is its most likely continuation $y \in B^\sharp$? Bayes' theorem yields

$$P(xy \mid x) = \frac{P(x \mid xy)P(xy)}{P(x)} \propto P(xy), \tag{1}$$

where $P(z^2 \mid z^1)$ is the probability of z^2, given knowledge of z^1, and $P(x) = \int_{z \in B^\sharp} P(xz)dz$ is just a normalizing factor. So the most likely continuation y is determined by $P(xy)$, the *prior probability* of xy. But which prior measure P is plausible? Occam's razor suggests that the "simplest" y should be more probable. But which exactly is the "correct" definition of simplicity? Sections 3 and 4 will measure the simplicity of a description by its length. Section 5 will measure the simplicity of a description by the time required to compute the described object.

3 Prediction Using a Universal Algorithmic Prior Based on the Shortest Way of Describing Objects

Roughly forty years ago Solomonoff started the theory of universal optimal induction based on the apparently harmless simplicity assumption that P is computable [62]. While Equation (1) makes predictions of the entire future, given the past, Solomonoff [63] focuses just on the next bit in a sequence. Although this provokes surprisingly nontrivial problems associated with translating the bitwise approach to alphabets other than the binary one — this was achieved only recently [20] — it is sufficient for obtaining essential insights. Given an observed bitstring x, Solomonoff assumes the data are drawn according to a recursive measure μ; that is, there is a program for a universal Turing machine that reads $x \in B^*$ and computes $\mu(x)$ and halts. He estimates the probability of the next bit (assuming there will be one), using the remarkable, well-studied, enumerable prior M [62, 77, 63, 15, 31]

$$M(x) = \sum_{\substack{program\ prefix\ p\ computes \\ output\ starting\ with\ x}} 2^{-l(p)}. \tag{2}$$

M is *universal*, dominating the less general recursive measures as follows: For all $x \in B^*$,

$$M(x) \geq c_\mu \mu(x), \tag{3}$$

where c_μ is a constant depending on μ but not on x. Solomonoff observed that the conditional M-probability of a particular continuation, given previous observations, converges towards the unknown conditional μ as the observation size goes to infinity [63], and that the sum over all observation sizes of the corresponding μ-expected deviations is actually bounded by a constant. Hutter (on the author's SNF research grant "Unification of Universal Induction and Sequential Decision Theory") recently showed that the number of prediction errors made by universal Solomonoff prediction is essentially bounded by the number of errors made by any other predictor, including the optimal scheme based on the true μ [20].

Recent Loss Bounds for Universal Prediction. This is a more general recent result. Assume we do know that p is in some set P of distributions. Choose a fixed weight w_q for each q in P such that the w_q add up to 1 (for simplicity, let P be countable). Then construct the Bayesmix $M(x) = \sum_q w_q q(x)$, and predict using M instead of the optimal but unknown p. How wrong is it to do that? The recent work of Hutter provides general and sharp (!) loss bounds [21].

Let $LM(n)$ and $Lp(n)$ be the total expected unit losses of the M-predictor and the p-predictor, respectively, for the first n events. Then $LM(n) - Lp(n)$ is at most of the order of $\sqrt{Lp(n)}$. That is, M is not much worse than p, and in general, no other predictor can do better than that! In particular, if p is deterministic, then the M-predictor soon will not make errors anymore.

If P contains *all* recursively computable distributions, then M becomes the celebrated enumerable universal prior. That is, after decades of somewhat stagnating research we now have sharp loss bounds for Solomonoff's universal induction scheme (compare recent work of Merhav and Feder [33]).

Solomonoff's approach, however, is uncomputable. To obtain a feasible approach, reduce M to what you get if you, say, just add up weighted estimated future finance data probabilities generated by 1000 commercial stock-market prediction software packages. If only one of the probability distributions happens to be close to the true one (but you do not know which) you should still get rich.

Note that the approach is much more general than what is normally done in traditional statistical learning theory, e.g., [69], where the often quite unrealistic assumption is that the observations are statistically independent.

4 Super Omegas and Generalizations of Kolmogorov Complexity & Algorithmic Probability

Our recent research generalized Solomonoff's approach to the case of less restrictive nonenumerable universal priors that are still computable in the limit [50, 52].

An object X is formally describable if a finite amount of information completely describes X and only X. More to the point, X should be representable by a possibly infinite bitstring x such that there is a finite, possibly never halting program p that computes x and nothing but x in a way that modifies each output bit at most finitely many times; that is, each finite beginning of x eventually *converges* and ceases to change. This constructive notion of formal describability is less restrictive than the traditional notion of computability [67], mainly because we do not insist on the existence of a halting program that computes an upper bound of the convergence time of p's n-th output bit. Formal describability thus pushes constructivism [5, 1] to the extreme, barely avoiding the nonconstructivism embodied by even less restrictive concepts of describability (compare computability *in the limit* [17, 40, 14] and Δ_n^0-describability [42][31, p. 46-47]).

The traditional theory of inductive inference focuses on Turing machines with one-way write-only output tape. This leads to the universal enumerable Solomonoff-Levin (semi) measure. We introduced more general, nonenumerable, but still limit-computable measures and a natural hierarchy of generalizations of algorithmic probability and Kolmogorov complexity [50, 52], suggesting that the "true" information content of some (possibly infinite) bitstring x actually is the size of the shortest nonhalting program that converges to x, and nothing but x, on a Turing machine that can edit its previous outputs. In fact, this "true" content is often smaller than the traditional Kolmogorov complexity. We showed that there are *Super Omegas* computable in the limit yet more random than Chaitin's "number of wisdom" *Omega* [9] (which is maximally random in a weaker traditional sense), and that any approximable measure of x is small for any x lacking a short description.

We also showed that there is a universal cumulatively enumerable measure of x based on the measure of all enumerable y lexicographically greater than x. It is more dominant yet just as limit-computable as Solomonoff's [52]. That is, if we are interested in limit-computable universal measures, we should prefer the novel universal cumulatively enumerable measure over the traditional enumerable one. If we include in our Bayesmix such limit-computable distributions we obtain again sharp loss bounds for prediction based on the mix [50, 52].

Our approach highlights differences between countable and uncountable sets. Which are the potential consequences for physics? We argue that things such as *un*countable time and space and *in*computable probabilities actually should not play a role in explaining the world, for lack of evidence that they are really necessary [50]. Some may feel tempted to counter this line of reasoning by pointing out that for centuries physicists have calculated with continua of real numbers, most of them incomputable. Even quantum physicists who are ready to give up the assumption of a continuous universe usually do take for granted the existence of continuous probability distributions on their discrete universes, and Stephen Hawking explicitly said: *"Although there have been suggestions that space-time may have a discrete structure I see no reason to abandon the continuum theories that have been so successful."* Note, however, that all physicists in fact have only manipulated discrete symbols, thus generating finite, describable proofs of their results derived from enumerable axioms. That real numbers really *exist* in a way transcending the finite symbol strings used by everybody may be a figment of imagination — compare Brouwer's constructive mathematics [5, 1] and the Löwenheim-Skolem Theorem [32, 61] which implies that any first order theory with an uncountable model such as the real numbers also has a countable model. As Kronecker put it: *"Die ganze Zahl schuf der liebe Gott, alles Übrige ist Menschenwerk"* ("God created the integers, all else is the work of man" [6]). Kronecker greeted with scepticism Cantor's celebrated insight [7] about real numbers, mathematical objects Kronecker believed did not even exist.

Assuming our future lies among the few (countably many) describable futures, we can ignore uncountably many nondescribable ones, in particular, the random ones. Adding the relatively mild assumption that the probability distribution from which our universe is drawn is cumulatively enumerable provides a theoretical justification of the prediction that the most likely continuations of our universes are computable through short enumeration procedures. In this sense Occam's razor is just a natural by-product of a computability assumption! But what about falsifiability? The pseudorandomness of our universe might be effectively undetectable in principle, because some approximable and enumerable patterns cannot be proven to be nonrandom in recursively bounded time.

The next sections, however, will introduce additional plausible assumptions that do lead to *computable* optimal prediction procedures.

5 Computable Predictions Through the Speed Prior Based on the Fastest Way of Describing Objects

Unfortunately, while M and the more general priors of Sect. 4 are computable in the limit, they are not recursive, and thus practically infeasible. This drawback inspired less general yet practically more feasible principles of minimum description length (MDL) [71, 41] as well as priors derived from time-bounded restrictions [31] of Kolmogorov complexity [28, 62, 9]. No particular instance of these approaches, however, is universally accepted or has a general convincing motivation that carries beyond rather specialized application scenarios. For instance, typical efficient MDL approaches require the specification of a class of computable models of the data, say, certain types of neural networks, plus some computable loss function expressing the coding costs of the data relative to the model. This provokes numerous ad-hoc choices.

Our recent work [54], however, offers an alternative to the celebrated but noncomputable algorithmic simplicity measure or Solomonoff-Levin measure discussed above [62, 77, 63]. We introduced a new measure (a prior on the computable objects) which is not based on the *shortest* but on the *fastest* way of describing objects.

Let us assume that the observed data sequence is generated by a computational process, and that any possible sequence of observations is therefore computable in the limit [50]. This assumption is stronger and more radical than the traditional one: Solomonoff just insists that the probability of any sequence prefix is recursively computable, but the (infinite) sequence itself may still be generated probabilistically.

Given our starting assumption that data are deterministically generated by a machine, it seems plausible that the machine suffers from a computational resource problem. Since some things are much harder to compute than others, the resource-oriented point of view suggests the following postulate.

Postulate 1 *The cumulative prior probability measure of all x incomputable within time t by any method is at most inversely proportional to t.*

This postulate leads to the Speed Prior $S(x)$, the probability that the output of the following probabilistic algorithm starts with x [54]:

> **Initialize:** Set $t := 1$. Let the input scanning head of a universal TM point to the first cell of its initially empty input tape.
> **Forever repeat:** While the number of instructions executed so far exceeds t: toss an unbiased coin; if heads is up set $t := 2t$; otherwise exit. If the input scanning head points to a cell that already contains a bit, execute the corresponding instruction (of the growing self-delimiting program, e.g., [30, 31]). Else toss the coin again, set the cell's bit to 1 if heads is up (0 otherwise), and set $t := t/2$.

Algorithm *GUESS* is very similar to a probabilistic search algorithm used in previous work on applied inductive inference [47, 49]. On several toy problems it generalized extremely well in a way unmatchable by traditional neural network learning algorithms.

With S comes a computable method AS for predicting optimally within ϵ accuracy [54]. Consider a finite but unknown program p computing $y \in B^\infty$. What if Postulate 1 holds but p is not optimally efficient, and/or computed on a computer that differs from our reference machine? Then we effectively do not sample beginnings y_k from S but from an alternative semimeasure S'. Can we still predict well? Yes, because the Speed Prior S dominates S'. This dominance is all we need to apply the recent loss bounds [21]. The loss that we are expected to receive by predicting according to AS instead of using the true but unknown S' does not exceed the optimal loss by much [54].

6 Speed Prior-Based Predictions for Our Universe

"In the beginning was the code."
FIRST SENTENCE OF THE GREAT PROGRAMMER'S BIBLE

Physicists and economists and other inductive scientists make predictions based on observations. Astonishingly, however, few physicists are aware of the theory of *optimal* inductive inference [62, 28]. In fact, when talking about the very nature of their inductive business, many physicists cite rather vague concepts such as Popper's falsifiability [39], instead of referring to quantitative results.

All widely accepted physical theories, however, are accepted not because they are falsifiable — they are not — or because they match the data — many alternative theories also match the data — but because they are simple in a certain sense. For example, the theory of gravitation is induced from locally observable training examples such as falling apples and movements of distant light sources, presumably stars. The theory predicts that apples on distant planets in other galaxies will fall as well. Currently nobody is able to verify or falsify this. But everybody believes in it because this generalization step makes the theory simpler than alternative theories with separate laws for apples on other planets. The same holds for superstring theory [18] or Everett's many-worlds theory [12], which presently also are neither verifiable nor falsifiable, yet offer comparatively simple explanations of numerous observations. In particular, most of Everett's postulated many-worlds will remain unobservable forever, but the assumption of their existence simplifies the theory, thus making it more beautiful and acceptable.

In Sects. 3 and 4 we have made the assumption that the probabilities of next events, given previous events, are (limit-)computable. Here we make a stronger assumption by adopting *Zuse's thesis* [75, 76], namely, that the very universe is actually being computed deterministically, e.g., on a cellular automaton (CA) [68, 70]. Quantum physics, quantum computation [3, 10, 38], Heisenberg's uncertainty principle and Bell's inequality [2] do *not* imply any physical evidence against this possibility, e.g., [66].

But then which is our universe's precise algorithm? The following method [48] computes it:

> Systematically create and execute all programs for a universal computer, such as a Turing machine or a CA; the first program is run for one instruction every second step on average, the next for one instruction every second of the remaining steps on average, and so on.

This method in a certain sense implements the simplest theory of everything: *all* computable universes, including ours and ourselves as observers, are computed by the very short program that generates and executes *all* possible programs [48]. In nested fashion, some of these programs will execute processes that again compute all possible universes, etc. [48]. Of course, observers in "higher-level" universes may be completely unaware of observers or universes computed by nested processes, and vice versa. For example, it seems hard to track and interpret the computations performed by a cup of tea.

The simple method above is more efficient than it may seem at first glance. A bit of thought shows that it even has the optimal order of complexity. For example, it outputs our universe history as quickly as this history's fastest program, save for a (possibly huge) constant slowdown factor that does not depend on output size.

Nevertheless, some universes are fundamentally harder to compute than others. This is reflected by the Speed Prior S discussed above (Section 5). So let us assume that our universe's history is sampled from S or a less dominant prior reflecting suboptimal computation of the history. Now we can immediately predict:

1. Our universe will not get many times older than it is now [50] — essentially, the probability that it will last 2^n times longer than it has lasted so far is at most 2^{-n}.

2. Any apparent randomness in any physical observation must be due to some yet unknown but *fast* pseudo-random generator PRG [50] which we should try to discover. *2a.* A re-examination of beta decay patterns may reveal that a very simple, fast, but maybe not quite trivial PRG is responsible for the apparently random decays of neutrons into protons, electrons and antineutrinos. *2b.* Whenever there are several possible continuations of our universe corresponding to different Schrödinger wave function collapses — compare Everett's widely accepted many worlds theory [12] — we should be more likely to end up in one computable by a short *and* fast algorithm. A re-examination of split experiment data involving entangled states such as the observations of spins of initially close but soon distant particles with correlated spins might reveal unexpected, nonobvious, nonlocal algorithmic regularity due to a fast PRG.

3. Large scale quantum computation [3] will not work well, essentially because it would require too many exponentially growing computational resources in interfering "parallel universes" [12].

4. Any probabilistic algorithm depending on truly random inputs from the environment will not scale well in practice.

Prediction *2* is verifiable but not necessarily falsifiable within a fixed time interval given in advance. Still, perhaps the main reason for the current absence of empirical evidence in this vein is that few [11] have looked for it.

In recent decades several well-known physicists have started writing about topics of computer science, e.g., [38, 10], sometimes suggesting that real world physics might allow for computing things that are not computable traditionally. Unimpressed by this trend, computer scientists have argued in favor of the opposite: since there is no evidence that we need more than traditional computability to explain the world, we should try to make do without this assumption, e.g., [75, 76, 13, 48].

7 Optimal Rational Decision Makers

So far we have talked about passive prediction, given the observations. Note, however, that agents interacting with an environment can also use predictions of the future to compute action sequences that maximize expected future reward. Hutter's recent *AIXI model* [22] (author's SNF grant 61847) does exactly this, by combining Solomonoff's M-based universal prediction scheme with an *expectimax* computation.

In cycle t action y_t results in perception x_t and reward r_t, where all quantities may depend on the complete history. The perception x'_t and reward r_t are sampled from the (reactive) environmental probability distribution μ. Sequential decision theory shows how to maximize the total expected reward, called value, if μ is known. Reinforcement learning [27] is used if μ is unknown. AIXI defines a mixture distribution ξ as a weighted sum of distributions $\nu \in \mathcal{M}$, where $\mathcal{M}$ is any class of distributions including the true environment μ.

It can be shown that the conditional M probability of environmental inputs to an AIXI agent, given the agent's earlier inputs and actions, converges with increasing length of interaction against the true, unknown probability [22], as long as the latter is recursively computable, analogously to the passive prediction case.

Recent work [24] also demonstrated AIXI's optimality in the following sense. The Bayes-optimal policy p^ξ based on the mixture ξ is self-optimizing in the sense that the average value converges asymptotically for all $\mu \in \mathcal{M}$ to the optimal value achieved by the (infeasible) Bayes-optimal policy p^μ, which knows μ in advance. The necessary condition that $\mathcal{M}$ admits self-optimizing policies is also sufficient. No other structural assumptions are made on $\mathcal{M}$. Furthermore, p^ξ is Pareto-optimal in the sense that there is no other policy yielding higher or equal value in *all* environments $\nu \in \mathcal{M}$ and a strictly higher value in at least one [24].

We can modify the AIXI model such that its predictions are based on the ϵ-approximable Speed Prior S instead of the incomputable M. Thus we obtain the so-called *AIS model*. Using Hutter's approach [22] we can now show that the conditional S probability of environmental inputs to an AIS agent, given the earlier inputs and actions, converges to the true but unknown probability, as long as the latter is dominated by S, such as the S' above.

8 Optimal Universal Search Algorithms

In a sense, searching is less general than reinforcement learning because it does not necessarily involve predictions of unseen data. Still, search is a central aspect of computer science (and any reinforcement learner needs a searcher as a submodule — see Sects. 10 and 11). Surprisingly, however, many books on search algorithms do not even mention the following, very simple asymptotically optimal, "universal" algorithm for a broad class of search problems.

Define a probability distribution P on a finite or infinite set of programs for a given computer. P represents the searcher's initial bias (e.g., P could be based on program length, or on a probabilistic syntax diagram).

> Method LSEARCH: Set current time limit T=1. WHILE problem not solved DO:
> > Test all programs q such that $t(q)$, the maximal time spent on creating and running and testing q, satisfies $t(q) < P(q)\,T$. Set $T := 2T$.

LSEARCH (for *Levin Search*) may be the algorithm Levin was referring to in his two page paper [29] which states that there is an asymptotically optimal universal search method for problems with easily verifiable solutions, that is, solutions whose validity can be quickly tested. Given some problem class, if some unknown optimal program p requires $f(k)$ steps to solve a problem instance of size k, then LSEARCH will need at most $O(f(k)/P(p)) = O(f(k))$ steps — the constant factor $1/P(p)$ may be huge, but does not depend on k. Compare [31, p. 502-505] and [23] and the fastest way of computing all computable universes in Sect. 6.

Recently Hutter developed a more complex asymptotically optimal search algorithm for *all* well-defined problems, not just those with with easily verifiable solutions [23]. HSEARCH cleverly allocates part of the total search time for searching the space of proofs to find provably correct candidate programs with provable upper runtime bounds, and at any given time focuses resources on those programs with the currently best proven time bounds. Unexpectedly, HSEARCH manages to reduce the unknown constant slowdown factor of LSEARCH to a value of $1 + \epsilon$, where ϵ is an arbitrary positive constant.

Unfortunately, however, the search in proof space introduces an unknown *additive* problem class-specific constant slowdown, which again may be huge. While additive constants generally are preferrable over multiplicative ones, both types may make universal search methods practically infeasible.

HSEARCH and LSEARCH are nonincremental in the sense that they do not attempt to minimize their constants by exploiting experience collected in previous searches. Our method *Adaptive* LSEARCH or ALS tries to overcome this [60] — compare Solomonoff's related ideas [64, 65]. Essentially it works as follows: whenever LSEARCH finds a program q that computes a solution for the current problem, q's probability $P(q)$ is substantially increased using a "learning rate," while probabilities of alternative programs decrease appropriately. Subsequent LSEARCHes for new problems then use the adjusted P, etc. A nonuniversal variant of this approach was able to solve reinforcement learning (RL) tasks [27] in partially observable environments unsolvable by traditional RL algorithms [74, 60].

Each LSEARCH invoked by ALS is optimal with respect to the most recent adjustment of P. On the other hand, the modifications of P themselves are not necessarily optimal. Recent work discussed in the next section overcomes this drawback in a principled way.

9 Optimal Ordered Problem Solver (OOPS)

Our recent OOPS [53, 55] is a simple, general, theoretically sound, in a certain sense time-optimal way of searching for a universal behavior or program that solves each problem in a sequence of computational problems, continually organizing and managing and reusing earlier acquired knowledge. For example, the n-th problem may be to compute the n-th event from previous events (prediction), or to find a faster way through a maze than the one found during the search for a solution to the $n-1$-th problem (optimization).

Let us first introduce the important concept of bias-optimality, which is a pragmatic definition of time-optimality, as opposed to the asymptotic optimality of both LSEARCH and HSEARCH, which may be viewed as academic exercises demonstrating that the $O()$ notation can sometimes be practically irrelevant despite its wide use in theoretical computer science. Unlike asymptotic optimality, bias-optimality does not ignore huge constant slowdowns:

Definition 1 (BIAS-OPTIMAL SEARCHERS). *Given is a problem class $\mathcal{R}$, a search space $\mathcal{C}$ of solution candidates (where any problem $r \in \mathcal{R}$ should have a solution in $\mathcal{C}$), a task dependent bias in form of conditional probability distributions $P(q \mid r)$ on the candidates $q \in \mathcal{C}$, and a predefined procedure that creates and tests any given q on any $r \in \mathcal{R}$ within time $t(q,r)$ (typically unknown in advance). A searcher is n-bias-optimal $(n \geq 1)$ if for any maximal total search time $T_{max} > 0$ it is guaranteed to solve any problem $r \in \mathcal{R}$ if it has a solution $p \in \mathcal{C}$ satisfying $t(p,r) \leq P(p \mid r) T_{max}/n$. It is bias-optimal if $n = 1$.*

This definition makes intuitive sense: the most probable candidates should get the lion's share of the total search time, in a way that precisely reflects the initial bias. Now we are ready to provide a general overview of the basic ingredients of OOPS [53, 55]:

Primitives: We start with an initial set of user-defined primitive behaviors. Primitives may be assembler-like instructions or time-consuming software, such as, say, theorem provers, or matrix operators for neural network-like parallel architectures, or trajectory generators for robot simulations, or state update procedures for multiagent systems, etc. Each primitive is represented by a token. It is essential that those primitives whose runtimes are not known in advance can be interrupted at any time.

Task-specific prefix codes: Complex behaviors are represented by token sequences or programs. To solve a given task represented by task-specific program inputs, OOPS tries to sequentially compose an appropriate complex behavior from primitive ones, always obeying the rules of a given user-defined initial programming language. Programs are grown incrementally, token by token; their beginnings or *prefixes* are immediately executed while being created; this may modify some task-specific internal state or memory, and may transfer control back to previously selected tokens (e.g., loops). To add a new token to some program prefix, we first have to wait until the execution of the prefix so far *explicitly requests* such a prolongation, by setting an appropriate signal in the internal state. Prefixes that cease to request any further tokens are called *self-delimiting* programs or simply programs (programs are their own prefixes). *Binary* self-delimiting programs were studied by [30] and [8] in the context of Turing machines [67] and the theory of Kolmogorov complexity and algorithmic probability [62, 28]. OOPS, however, uses a more practical, not necessarily binary framework.

The program construction procedure above yields *task-specific prefix codes* on program space: with any given task, programs that halt because they have found a solution or encountered some error cannot request any more tokens. Given the current task-specific inputs, no program can be the prefix of another one. On a different task, however, the same program may continue to request additional tokens. This is important for our novel approach — incrementally growing self-delimiting programs are unnecessary for the asymptotic optimality properties of LSEARCH and HSEARCH, but essential for OOPS.

Access to previous solutions: Let p^n denote a found prefix solving the first n tasks. The search for p^{n+1} may greatly profit from the information conveyed by (or the knowledge embodied by) $p^1, p^2, \ldots, p^n$ which are stored or *frozen* in special *non*modifiable memory shared by all tasks, such that they are accessible to p^{n+1} (this is another difference to *non*incremental LSEARCH and HSEARCH). For example, p^{n+1} might execute a token sequence that calls p^{n-3} as a subprogram, or that copies p^{n-17} into some internal *modifiable* task-specific memory, then modifies the copy a bit, then applies the slightly edited copy to the current task. In fact, since the number of frozen programs may grow to a large value, much of the knowledge embodied by p^j may be about how to access and edit and use older p^i $(i < j)$.

Bias: The searcher's initial bias is embodied by initial, user-defined, task dependent probability distributions on the finite or infinite search space of possible program prefixes. In the simplest case we start with a maximum entropy distribution on the tokens, and define prefix probabilities as the products of the probabilities of their tokens. But prefix continuation probabilities may also depend on previous tokens in context sensitive fashion.

Self-computed suffix probabilities: In fact, we permit that any executed prefix assigns a task-dependent, self-computed probability distribution to its own possible continuations. This distribution is encoded and manipulated in task-specific internal memory. So, unlike with ALS [60], we do not use a prewired learning scheme to update the probability distribution. Instead we leave such updates to prefixes whose online execution modifies the probabilities of their suffixes. By, say, invoking previously frozen code that redefines the probability distribution on future prefix continuations, the currently tested prefix may completely reshape the most likely paths through the search space of its own continuations, based on experience ignored by *non*incremental LSEARCH and HSEARCH. This may introduce significant problem class-specific knowledge derived from solutions to earlier tasks.

Two searches: Essentially, OOPS provides equal resources for two near-*bias-optimal* searches (Def. 1) that run in parallel until p^{n+1} is discovered and stored in non-modifiable memory. The first is exhaustive; it systematically tests all possible prefixes on all tasks up to $n+1$. Alternative prefixes are tested on all current tasks in parallel while still growing; once a task is solved, we remove it from the current set; prefixes that fail on a single task are discarded. The second search is much more focused; it only searches for prefixes that start with p^n, and only tests them on task $n + 1$, which is safe, because we already know that such prefixes solve all tasks up to n.

Bias-optimal backtracking: HSEARCH and LSEARCH assume potentially infinite storage. Hence, they may largely ignore questions of storage management. In any practical system, however, we have to efficiently reuse limited storage. Therefore, in both searches of OOPS, alternative prefix continuations are evaluated by a novel, practical, token-oriented backtracking procedure that can deal with several tasks in parallel, given some *code bias* in the form of previously found code. The procedure always ensures near-*bias-optimality* (Def. 1): no candidate behavior gets more time than it deserves, given the probabilistic bias. Essentially we conduct a depth-first search in program space, where the branches of the search tree are program prefixes, and backtracking (partial resets of partially solved task sets and modifications of internal states and continuation probabilities) is triggered once the sum of the runtimes of the current prefix on all current tasks exceeds the prefix probability multiplied by the total search time so far.

In case of unknown, infinite task sequences we can typically never know whether we already have found an optimal solver for all tasks in the sequence.

But once we unwittingly do find one, at most half of the total future run time will be wasted on searching for alternatives. Given the initial bias and subsequent bias shifts due to $p^1, p^2, \ldots$, no other bias-optimal searcher can expect to solve the $n + 1$-th task set substantially faster than OOPS. A by-product of this optimality property is that it gives us a natural and precise measure of bias and bias shifts, conceptually related to Solomonoff's *conceptual jump size* [64, 65].

Since there is no fundamental difference between domain-specific problem-solving programs and programs that manipulate probability distributions and thus essentially rewrite the search procedure itself, we collapse both learning and metalearning in the same time-optimal framework.

An example initial language. For an illustrative application, we wrote an interpreter for a stack-based universal programming language inspired by FORTH [35], with initial primitives for defining and calling recursive functions, iterative loops, arithmetic operations, and domain-specific behavior. Optimal metasearching for better search algorithms is enabled through the inclusion of bias-shifting instructions that can modify the conditional probabilities of future search options in currently running program prefixes.

Experiments. Using the assembler-like language mentioned above, we first teach OOPS something about recursion, by training it to construct samples of the simple context free language $\{1^k 2^k\}$ (k 1's followed by k 2's), for k up to 30 (in fact, the system discovers a universal solver for all k). This takes roughly 0.3 days on a standard personal computer (PC). Thereafter, within a few additional days, OOPS demonstrates incremental knowledge transfer: it exploits aspects of its previously discovered universal $1^k 2^k$-solver, by rewriting its search procedure such that it more readily discovers a universal solver for all k disk *Towers of Hanoi* problems — in the experiments it solves all instances up to $k = 30$ (solution size $2^k - 1$), but it would also work for $k > 30$. Previous, less general reinforcement learners and *non*learning AI planners tend to fail for much smaller instances.

Future research may focus on devising particularly compact, particularly reasonable sets of initial codes with particularly broad practical applicability. It may turn out that the most useful initial languages are not traditional programming languages similar to the FORTH-like one, but instead based on a handful of primitive instructions for massively parallel cellular automata [68, 70, 76], or on a few nonlinear operations on matrix-like data structures such as those used in recurrent neural network research [72, 44, 4]. For example, we could use the principles of OOPS to create a non-gradient-based, near-bias-optimal variant of Hochreiter's successful recurrent network metalearner [19]. It should also be of interest to study probabilistic *Speed Prior*-based OOPS variants [54] and to devise applications of OOPS-like methods as components of universal reinforcement learners (see below). In ongoing work, we are applying OOPS to the problem of optimal trajectory planning for robotics in a realistic physics simulation. This involves the interesting trade-off between comparatively fast program-composing primitives or *"thinking primitives"* and time-consuming *"action primitives,"* such as *stretch-arm-until-touch-sensor-input.*

10 OOPS-Based Reinforcement Learning

At any given time, a reinforcement learner [27] will try to find a *policy* (a strategy for future decision making) that maximizes its expected future reward. In many traditional reinforcement learning (RL) applications, the policy that works best in a given set of training trials will also be optimal in future test trials [51]. Sometimes, however, it won't. To see the difference between searching (the topic of the previous sections) and reinforcement learning (RL), consider an agent and two boxes. In the n-th trial the agent may open and collect the content of exactly one box. The left box will contain $100n$ Swiss Francs, the right box 2^n Swiss Francs, but the agent does not know this in advance. During the first 9 trials the optimal policy is *"open left box."* This is what a good searcher should find, given the outcomes of the first 9 trials. But this policy will be suboptimal in trial 10. A good reinforcement learner, however, should extract the underlying regularity in the reward generation process and predict the future tasks and rewards, picking the right box in trial 10, without having seen it yet.

The first general, asymptotically optimal reinforcement learner is the recent AIXI model [22, 24] (Section 7). It is valid for a very broad class of environments whose reactions to action sequences (control signals) are sampled from arbitrary computable probability distributions. This means that AIXI is far more general than traditional RL approaches. However, while AIXI clarifies the theoretical limits of RL, it is not practically feasible, just like HSEARCH is not. From a pragmatic point of view, we are really interested in a reinforcement learner that makes optimal use of given, limited computational resources. The following outlines one way of using OOPS-like bias-optimal methods as components of general yet feasible reinforcement learners.

We need two OOPS modules. The first is called the predictor or world model. The second is an action searcher using the world model. The life of the entire system should consist of a sequence of *cycles* 1, 2, ... At each cycle, a limited amount of computation time will be available to each module. For simplicity we assume that during each cyle the system may take exactly one action. Generalizations to actions consuming several cycles are straight-forward though. At any given cycle, the system executes the following procedure:

1. For a time interval fixed in advance, the predictor is first trained in bias-optimal fashion to find a better world model, that is, a program that predicts the inputs from the environment (including the rewards, if there are any), given a history of previous observations and actions. So the n-th task ($n = 1, 2, \ldots$) of the first OOPS module is to find (if possible) a better predictor than the best found so far.

2. After the current cycle's time for predictor improvement is finished, the current world model (prediction program) found by the first OOPS module will be used by the second module, again in bias-optimal fashion, to search for a future action sequence that maximizes the predicted cumulative reward (up to some time limit). That is, the n-th task ($n = 1, 2, \ldots$) of the second OOPS module will be to find a control program that computes a control sequence of actions, to be fed into the program representing the current world model (whose input predictions are successively fed back to itself in the obvious manner), such that this control sequence leads to higher predicted reward than the one generated by the best control program found so far.

3. After the current cycle's time for control program search is finished, we will execute the current action of the best control program found in step 2. Now we are ready for the next cycle.

The approach is reminiscent of an earlier, heuristic, non-bias-optimal RL approach based on two adaptive recurrent neural networks, one representing the world model, the other one a controller that uses the world model to extract a policy for maximizing expected reward [46]. The method was inspired by previous combinations of *non*recurrent, *reactive* world models and controllers [73, 37, 26].

At any given time, until which temporal horizon should the predictor try to predict? In the AIXI case, the proper way of treating the temporal horizon is not to discount it exponentially, as done in most traditional work on reinforcement learning, but to let the future horizon grow in proportion to the learner's lifetime so far [24]. It remains to be seen whether this insight carries over to OOPS-RL.

Despite the bias-optimality properties of OOPS for certain ordered task sequences, however, OOPS-RL is not necessarily the best way of spending limited time in general reinforcement learning situations. On the other hand, it is possible to use OOPS as a proof-searching submodule of the recent, optimal, universal, reinforcement learning Gödel machine [56] discussed in the next section.

11 The Gödel Machine

The Gödel machine [56], also this volume, explicitly addresses the *'Grand Problem of Artificial Intelligence'* [58] by optimally dealing with limited resources in general reinforcement learning settings, and with the possibly huge (but constant) slowdowns buried by AIXI(t, l) [22] in the somewhat misleading $O()$-notation. It is designed to solve arbitrary computational problems beyond those solvable by plain OOPS, such as maximizing the expected future reward of a robot in a possibly stochastic and reactive environment (note that the total utility of some robot behavior may be hard to verify — its evaluation may consume the robot's entire lifetime).

How does it work? While executing some arbitrary initial problem solving strategy, the Gödel machine simultaneously runs a proof searcher which systematically and repeatedly tests proof techniques. Proof techniques are programs that may read any part of the Gödel machine's state, and write on a reserved part which may be reset for each new proof technique test. In an example Gödel machine [56] this writable storage includes the variables *proof* and *switchprog*, where *switchprog* holds a potentially unrestricted program whose execution could completely rewrite any part of the Gödel machine's current software. Normally the current *switchprog* is not executed. However, proof techniques may invoke a special subroutine *check()* which tests whether *proof* currently holds a proof showing that the utility of stopping the systematic proof searcher and transferring control to the current *switchprog* at a particular point in the near future exceeds the utility of continuing the search until some alternative *switchprog* is found. Such proofs are derivable from the proof searcher's axiom scheme which formally describes the utility function to be maximized (typically the expected future reward in the expected remaining lifetime of the Gödel machine), the computational costs of hardware instructions (from which all programs are composed), and the effects of hardware instructions on the Gödel machine's state. The axiom scheme also formalizes known probabilistic properties of the possibly reactive environment, and also the *initial* Gödel machine state and software, which includes the axiom scheme itself (no circular argument here). Thus proof techniques can reason about expected costs and results of all programs including the proof searcher.

Once *check()* has identified a provably good *switchprog*, the latter is executed (some care has to be taken here because the proof verification itself and the transfer of control to *switchprog* also consume part of the typically limited lifetime). The discovered *switchprog* represents a *globally* optimal self-change in the following sense: provably *none* of all the alternative *switchprog*s and *proof*s (that could be found in the future by continuing the proof search) is worth waiting for.

There are many ways of initializing the proof searcher. Although identical proof techniques may yield different proofs depending on the time of their invocation (due to the continually changing Gödel machine state), there is a bias-optimal and asymptotically optimal proof searcher initialization based on a variant of OOPS [56] (Sect. 9). It exploits the fact that proof verification is a simple and fast business where the particular optimality notion of OOPS is appropriate. The Gödel machine itself, however, may have an arbitrary, *typically different and more powerful* sense of optimality embodied by its given utility function.

12 Conclusion

Recent theoretical and practical advances are currently driving a renaissance in the fields of universal learners and optimal search [59]. A new kind of AI is emerging. Does it really deserve the attribute *"new,"* given that its roots date back to the 1930s, when Gödel published the fundamental result of theoretical computer science [16] and Zuse started to build the first general purpose computer (completed in 1941), and the 1960s, when Solomonoff and Kolmogorov published their first relevant results? An affirmative answer seems justified, since it is the recent results on practically feasible computable variants of the old incomputable methods that are currently reinvigorating the long dormant field. The "new" AI is new in the sense that it abandons the mostly heuristic or non-general approaches of the past decades, offering methods that are both general and theoretically sound, and provably optimal in a sense that *does* make sense in the real world.

We are led to claim that the future will belong to universal or near-universal learners that are more general than traditional reinforcement learners/decision makers depending on strong Markovian assumptions, or than learners based on traditional statistical learning theory, which often require unrealistic i.i.d. or Gaussian assumptions. Due to ongoing hardware advances, the time has come for optimal search in algorithm space, as opposed to the limited space of reactive mappings embodied by traditional methods such as artificial feedforward neural networks.

It seems safe to bet that not only computer scientists but also physicists and other inductive scientists will start to pay more attention to the fields of universal induction and optimal search, since their basic concepts are irresistibly powerful and general and simple. How long will it take for these ideas to unfold their full impact? A very naive and speculative guess driven by wishful thinking might be based on identifying the *"greatest moments in computing history"* and extrapolating from there. Which are those "greatest moments?" Obvious candidates are:

1. *1623:* first mechanical calculator by Schickard starts the computing age (followed by machines of Pascal, 1640, and Leibniz, 1670).
2. *Roughly two centuries later:* concept of a *programmable* computer (Babbage, UK, 1834-1840).
3. *One century later:* fundamental theoretical work on universal integer-based programming languages and the limits of proof and computation (Gödel, Austria, 1931, reformulated by Turing, UK, 1936); first working programmable computer (Zuse, Berlin, 1941). (The next 50 years saw many theoretical advances as well as faster and faster switches — relays were replaced by tubes by single transistors by numerous transistors etched on chips — but arguably this was rather predictable, incremental progress without radical shake-up events.)
4. *Half a century later:* World Wide Web (UK's Berners-Lee, Switzerland, 1990).

This list seems to suggest that each major breakthrough tends to come roughly twice as fast as the previous one. Extrapolating the trend, optimists should expect the next radical change to manifest itself one quarter of a century after the most recent one, that is, by 2015, which happens to coincide with the date when the fastest computers will match brains in terms of raw computing power, according to frequent estimates based on Moore's law. The author is confident that the coming 2015 upheaval (if any) will involve universal learning algorithms and Gödel machine-like, optimal, incremental search in algorithm space [56] — possibly laying a foundation for the remaining series of faster and faster additional revolutions culminating in an "Omega point" expected around 2040.

13 Acknowledgments

Hutter's frequently mentioned work was funded through the author's SNF grant 2000-061847 "Unification of universal inductive inference and sequential decision theory." Over the past three decades, numerous discussions with Christof Schmidhuber (a theoretical physicist) helped to crystallize the ideas on computable universes — compare his notion of *"mathscape"* [45].

References

1. Beeson M (1985) *Foundations of Constructive Mathematics.* Springer-Verlag, Berlin, New York, Heidelberg.
2. Bell JS (1966) On the problem of hidden variables in quantum mechanics. *Rev. Mod. Phys.*, 38:447–452.
3. Bennett CH, DiVicenzo DP Quantum information and computation. *Nature*, 404(6775):256–259.
4. Bishop CM (1995) *Neural networks for pattern recognition.* Oxford University Press.
5. Brouwer LEJ (1907) Over de Grondslagen der Wiskunde. Dissertation, Doctoral Thesis, University of Amsterdam.
6. Cajori F (1919) *History of mathematics.* Macmillan, New York, 2nd edition.
7. Cantor G (1874) Über eine Eigenschaft des Inbegriffes aller reellen algebraischen Zahlen. *Crelle's Journal für Mathematik*, 77:258–263.
8. Chaitin GJ (1975) A theory of program size formally identical to information theory. *Journal of the ACM*, 22:329–340.
9. Chaitin GJ (1987) *Algorithmic Information Theory.* Cambridge University Press, Cambridge, UK.
10. Deutsch D (1997) *The Fabric of Reality.* Allen Lane, New York, NY.
11. Erber T, Putterman S (!985) Randomness in quantum mechanics – nature's ultimate cryptogram? *Nature*, 318(7):41–43.
12. Everett III H (1957) 'Relative State' formulation of quantum mechanics. *Reviews of Modern Physics*, 29:454–462.
13. Fredkin EF, Toffoli T (1982) Conservative logic. *International Journal of Theoretical Physics*, 21(3/4):219–253.
14. Freyvald RV (1977) Functions and functionals computable in the limit. *Transactions of Latvijas Vlasts Univ. Zinatn. Raksti*, 210:6–19.

15. Gács P (1983) On the relation between descriptional complexity and algorithmic probability. *Theoretical Computer Science*, 22:71–93.
16. Gödel K (1931) Über formal unentscheidbare Sätze der Principia Mathematica und verwandter Systeme I. *Monatshefte für Mathematik und Physik*, 38:173–198.
17. Gold EM (1965) Limiting recursion. *Journal of Symbolic Logic*, 30(1):28–46.
18. Green MB, Schwarz JH, Witten E (1987) *Superstring Theory*. Cambridge University Press, Cambridge, UK.
19. Hochreiter S, Younger AS, Conwell PR (2001) Learning to learn using gradient descent. In *Lecture Notes on Comp. Sci. 2130, Proc. Intl. Conf. on Artificial Neural Networks (ICANN-2001)*, Springer, Berlin, Heidelberg.
20. Hutter M (2001) Convergence and error bounds of universal prediction for general alphabet. *Proceedings of the 12th European Conference on Machine Learning (ECML-2001)*, Technical Report IDSIA-07-01, cs.AI/0103015), 2001.
21. Hutter M (2001) General loss bounds for universal sequence prediction. In Brodley CE, Danyluk AP (eds) *Proceedings of the 18th International Conference on Machine Learning (ICML-2001)*.
22. Hutter M (2001) Towards a universal theory of artificial intelligence based on algorithmic probability and sequential decisions. *Proceedings of the 12th European Conference on Machine Learning (ECML-2001)*.
23. Hutter M (2002) The fastest and shortest algorithm for all well-defined problems. *International Journal of Foundations of Computer Science*, 13(3):431–443.
24. Hutter M (2002) Self-optimizing and Pareto-optimal policies in general environments based on Bayes-mixtures. In *Proc. 15th Annual Conf. on Computational Learning Theory (COLT 2002)*, volume 2375 of *LNAI*, Springer, Berlin.
25. Hutter M (2005) A gentle introduction to the universal algorithmic agent AIXI. In this volume.
26. Jordan MI, Rumelhart DE (1990) Supervised learning with a distal teacher. Technical Report Occasional Paper #40, Center for Cog. Sci., MIT.
27. Kaelbling LP, Littman ML, Moore AW Reinforcement learning: a survey. *Journal of AI research*, 4:237–285.
28. Kolmogorov AN (1965) Three approaches to the quantitative definition of information. *Problems of Information Transmission*, 1:1–11.
29. Levin LA (1973) Universal sequential search problems. *Problems of Information Transmission*, 9(3):265–266.
30. Levin LA (1974) Laws of information (nongrowth) and aspects of the foundation of probability theory. *Problems of Information Transmission*, 10(3):206–210.
31. Li M, Vitányi PMB (!997) *An Introduction to Kolmogorov Complexity and its Applications*. Springer, Berlin, 2nd edition.
32. Löwenheim L (1915) Über Möglichkeiten im Relativkalkül. *Mathematische Annalen*, 76:447–470.
33. Merhav N, Feder M (1998) Universal prediction. *IEEE Transactions on Information Theory*, 44(6):2124–2147.
34. Mitchell T (1997) *Machine Learning*. McGraw Hill.
35. Moore CH, Leach GC (1970) FORTH: a language for interactive computing, 1970. http://www.ultratechnology.com.

36. Newell A, Simon H (1963) GPS, a Program that Simulates Human Thought, In: Feigenbaum E, Feldman J (eds), *Computers and Thought*, MIT Press, Cambridge, MA.

37. Nguyen, Widrow B (1989) The truck backer-upper: An example of self learning in neural networks. In *Proceedings of the International Joint Conference on Neural Networks*.

38. Penrose R *The Emperor's New Mind*. Oxford University Press, Oxford.

39. Popper KR (1934) *The Logic of Scientific Discovery*. Hutchinson, London.

40. Putnam H (1965) Trial and error predicates and the solution to a problem of Mostowski. *Journal of Symbolic Logic*, 30(1):49–57.

41. Rissanen J (1986) Stochastic complexity and modeling. *The Annals of Statistics*, 14(3):1080–1100.

42. Rogers, Jr. H (1967) *Theory of Recursive Functions and Effective Computability*. McGraw-Hill, New York.

43. Rosenbloom PS, Laird JE, and Newell A. *The SOAR Papers*. MIT Press, 1993.

44. Rumelhart DE, Hinton GE, Williams RJ (1986) Learning internal representations by error propagation. In Rumelhart DE, McClelland JL (eds) *Parallel Distributed Processing*, volume 1, MIT Press.

45. Schmidhuber C (2000) Strings from logic. Technical Report CERN-TH/2000-316, CERN, Theory Division. http://xxx.lanl.gov/abs/hep-th/0011065.

46. Schmidhuber J (1991)   Reinforcement learning in Markovian and non-Markovian environments. In Lippman DS, Moody JE, Touretzky DS (eds) *Advances in Neural Information Processing Systems 3*, Morgan Kaufmann, Los Altos, CA.

47. Schmidhuber J (1995) Discovering solutions with low Kolmogorov complexity and high generalization capability. In Prieditis A and Russell S (eds) *Machine Learning: Proceedings of the Twelfth International Conference*. Morgan Kaufmann, San Francisco, CA.

48. Schmidhuber J (1997) A computer scientist's view of life, the universe, and everything. In Freksa C, Jantzen M, Valk R (eds) *Foundations of Computer Science: Potential - Theory - Cognition*, volume 1337 of *LLNCS*, Springer, Berlin.

49. Schmidhuber J (1997) Discovering neural nets with low Kolmogorov complexity and high generalization capability. *Neural Networks*, 10(5):857–873.

50. Schmidhuber J (2000) Algorithmic theories of everything. Technical Report IDSIA-20-00, quant-ph/0011122, IDSIA. Sections 1-5: see [52]; Section 6: see [54].

51. Schmidhuber J (2001) Sequential decision making based on direct search. In Sun R, Giles CL (eds) *Sequence Learning: Paradigms, Algorithms, and Applications*. volume 1828 of *LLAI*, Springer, Berlin.

52. Schmidhuber J (2002) Hierarchies of generalized Kolmogorov complexities and nonenumerable universal measures computable in the limit. *International Journal of Foundations of Computer Science*, 13(4):587–612.

53. Schmidhuber J (2004) Optimal ordered problem solver. *Machine Learning*, 54(3):211–254.

54. Schmidhuber J (2002) The Speed Prior: a new simplicity measure yielding near-optimal computable predictions. In Kivinen J, Sloan RH (eds) *Proceedings of the 15th Annual Conference on Computational Learning Theory (COLT 2002)*, Lecture Notes in Artificial Intelligence, Springer, Berlin.

55. Schmidhuber J (2003) Bias-optimal incremental problem solving. In Becker S, Thrun S, Obermayer K (eds) *Advances in Neural Information Processing Systems 15*, MIT Press, Cambridge, MA.

56. Schmidhuber J (2003) Gödel machines: self-referential universal problem solvers making provably optimal self-improvements. Technical Report IDSIA-19-03, arXiv:cs.LO/0309048 v2, IDSIA.

57. Schmidhuber J (2003) The new AI: General & sound & relevant for physics. Technical Report TR IDSIA-04-03, Version 1.0, cs.AI/0302012 v1, IDSIA.

58. Schmidhuber J (2003)   Towards solving the grand problem of AI.   In Quaresma P, Dourado A, Costa E, Costa JF (eds) *Soft Computing and complex systems*, Centro Internacional de Mathematica, Coimbra, Portugal. Based on [57].

59. Schmidhuber J and Hutter M (2002) NIPS 2002 workshop on universal learning algorithms and optimal search. Additional speakers: R. Solomonoff, P. M. B. Vitányi, N. Cesa-Bianchi, I. Nemenmann. Whistler, CA.

60. Schmidhuber J, Zhao J, Wiering M (1997) Shifting inductive bias with success-story algorithm, adaptive Levin search, and incremental self-improvement. *Machine Learning*, 28:105–130.

61. Skolem T (1919) Logisch-kombinatorische Untersuchungen über Erfüllbarkeit oder Beweisbarkeit mathematischer Sätze nebst einem Theorem über dichte Mengen. *Skrifter utgit av Videnskapsselskapet in Kristiania, I, Mat.-Nat. Kl.*, N4:1–36.

62. Solomonoff R (1964) A formal theory of inductive inference. Part I. *Information and Control*, 7:1–22.

63. Solomonoff R (1978) Complexity-based induction systems. *IEEE Transactions on Information Theory*, IT-24(5):422–432.

64. Solomonoff R (1986) An application of algorithmic probability to problems in artificial intelligence. In Kanal L, Lemmer J (eds) *Uncertainty in Artificial Intelligence*, Elsevier Science Publishers/North Holland, Amsterdam.

65. Solomonoff R (1989) A system for incremental learning based on algorithmic probability. In *Proceedings of the Sixth Israeli Conference on Artificial Intelligence, Computer Vision and Pattern Recognition*.

66. 't Hooft G (1999) Quantum gravity as a dissipative deterministic system. *Classical and Quantum Gravity* (16):3263–3279.

67. Turing A (1936) On computable numbers, with an application to the Entscheidungsproblem. *Proceedings of the London Mathematical Society, Series 2*, 41:230–267.

68. Ulam S (1950) Random processes and transformations. In *Proceedings of the International Congress on Mathematics*, volume 2, pages 264–275.

69. Vapnik V *The Nature of Statistical Learning Theory*. Springer, New York, 1995.

70. von Neumann J (1966) *Theory of Self-Reproducing Automata*. University of Illionois Press, Champain, IL.

71. Wallace CS, Boulton DM (1968) An information theoretic measure for classification. *Computer Journal*, 11(2):185–194.

72. Werbos PJ (1974) *Beyond Regression: New Tools for Prediction and Analysis in the Behavioral Sciences*. PhD thesis, Harvard University.

73. Werbos PJ (1987) Learning how the world works: Specifications for predictive networks in robots and brains. In *Proceedings of IEEE International Conference on Systems, Man and Cybernetics, N.Y.*.

74. Wiering M, Schmidhuber J (1996) Solving POMDPs with Levin search and EIRA. In Saitta L (ed) *Machine Learning: Proceedings of the Thirteenth International Conference*, Morgan Kaufmann, San Francisco, CA.
75. Zuse K (1967) Rechnender Raum. *Elektronische Datenverarbeitung*, 8:336–344.
76. Zuse K (1969) *Rechnender Raum*. Friedrich Vieweg & Sohn, Braunschweig. English translation: *Calculating Space*, MIT Technical Translation AZT-70-164-GEMIT, MIT (Proj. MAC), Cambridge, MA.
77. Zvonkin AK, Levin LA (1970) The complexity of finite objects and the algorithmic concepts of information and randomness. *Russian Math. Surveys*, 25(6):83–124.

Gödel Machines: Fully Self-referential Optimal Universal Self-improvers*

Jürgen Schmidhuber

IDSIA, Galleria 2, 6928 Manno (Lugano), Switzerland &
TU Munich, Boltzmannstr. 3, 85748 Garching, München, Germany
juergen@idsia.ch - http://www.idsia.ch/~juergen

Summary. We present the first class of mathematically rigorous, general, fully self-referential, self-improving, optimally efficient problem solvers. Inspired by Kurt Gödel's celebrated self-referential formulas (1931), such a problem solver rewrites any part of its own code as soon as it has found a proof that the rewrite is *useful,* where the problem-dependent *utility function* and the hardware and the entire initial code are described by axioms encoded in an initial proof searcher which is also part of the initial code. The searcher systematically and efficiently tests computable *proof techniques* (programs whose outputs are proofs) until it finds a provably useful, computable self-rewrite. We show that such a self-rewrite is globally optimal—no local maxima!—since the code first had to prove that it is not useful to continue the proof search for alternative self-rewrites. Unlike previous *non*-self-referential methods based on hardwired proof searchers, ours not only boasts an optimal *order* of complexity but can optimally reduce any slowdowns hidden by the $O()$-notation, provided the utility of such speed-ups is provable at all.

1 Introduction and Outline

In 1931 Kurt Gödel used elementary arithmetics to build a universal programming language for encoding arbitrary proofs, given an arbitrary enumerable set of axioms. He went on to construct *self-referential* formal statements that claim their own unprovability, using Cantor's diagonalization trick [5] to demonstrate that formal systems such as traditional mathematics are either flawed in a certain sense or contain unprovable but true statements [11]. Since Gödel's exhibition of the fundamental limits of proof and computation, and Konrad Zuse's subsequent construction of the first working programmable computer (1935-1941), there has been a lot of work on specialized algorithms solving problems taken from more or less general problem classes. Apparently, however, one remarkable fact has so far escaped the attention of computer scientists: it is possible to use self-referential proof systems to build optimally efficient yet conceptually very simple universal problem solvers.

All traditional algorithms for problem solving / machine learning / reinforcement learning [19] are hardwired. Some are designed to improve some limited type of policy through experience, but are not part of the modifiable

---

*Certain parts of this work appear in [46] and [47], both by Springer.

policy, and cannot improve themselves in a theoretically sound way. Humans
are needed to create new/better problem solving algorithms and to prove their
usefulness under appropriate assumptions.

Let us eliminate the restrictive need for human effort in the most general
way possible, leaving all the work including the proof search to a system that
can rewrite and improve itself in arbitrary computable ways and in a most
efficient fashion. To attack this *"Grand Problem of Artificial Intelligence,"* we
introduce a novel class of optimal, fully self-referential [11] general problem
solvers called *Gödel machines* [43].[1] They are universal problem solving systems that interact with some (partially observable) environment and can in
principle modify themselves without essential limits besides the limits of computability. Their initial algorithm is not hardwired; it can completely rewrite
itself, but only if a proof searcher embedded within the initial algorithm can
first prove that the rewrite is useful, given a formalized utility function reflecting computation time and expected future success (e.g., rewards). We will see
that self-rewrites due to this approach are actually *globally optimal* (Theorem 1, Section 4), relative to Gödel's well-known fundamental restrictions of
provability [11]. These restrictions should not worry us; if there is no proof of
some self-rewrite's utility, then humans cannot do much either.

The initial proof searcher is $O()$-optimal (has an optimal order of complexity) in the sense of Theorem 2, Section 5. Unlike Hutter's hardwired systems
[17, 16] (Section 2), however, a Gödel machine can further speed up its proof
searcher to meet *arbitrary* formalizable notions of optimality beyond those
expressible in the $O()$-notation. Our approach yields the first theoretically
sound, fully self-referential, optimal, general problem solvers.

Outline. Section 2 presents basic concepts, relations to the most relevant
previous work, and limitations. Section 3 presents the essential details of a
self-referential axiomatic system, Section 4 the Global Optimality Theorem 1,
and Section 5 the $O()$-optimal (Theorem 2) initial proof searcher. Section 6
provides examples and additional relations to previous work, briefly discusses
issues such as a *technical* justification of consciousness, and provides answers
to several frequently asked questions about Gödel machines.

2 Basic Overview, Relation to Previous Work, and Limitations

Many traditional problems of computer science require just one problem-
defining input at the beginning of the problem solving process. For example,
the initial input may be a large integer, and the goal may be to factorize it.
In what follows, however, we will also consider the *more general case* where

---

[1]Or *'Goedel machine'*, to avoid the *Umlaut*. But *'Godel machine'* would not be
quite correct. Not to be confused with what Penrose calls, in a different context,
'Gödel's putative theorem-proving machine' [29]!

the problem solution requires interaction with a dynamic, initially unknown environment that produces a continual stream of inputs and feedback signals, such as in autonomous robot control tasks, where the goal may be to maximize expected cumulative future reward [19]. This may require the solution of essentially arbitrary problems (examples in Sect. 6.2 formulate traditional problems as special cases).

2.1 Notation and Set-up

Unless stated otherwise or obvious, throughout the paper newly introduced variables and functions are assumed to cover the range implicit in the context. B denotes the binary alphabet $\{0, 1\}$, B^* the set of possible bitstrings over B, $l(q)$ denotes the number of bits in a bitstring q; q_n the n-th bit of q; λ the empty string (where $l(\lambda) = 0$); $q_{m:n} = \lambda$ if $m > n$ and $q_m q_{m+1} \ldots q_n$ otherwise (where $q_0 := q_{0:0} := \lambda$).

Our hardware (e.g., a universal or space-bounded Turing machine or the abstract model of a personal computer) has a single life which consists of discrete cycles or time steps $t = 1, 2, \ldots$. Its total lifetime T may or may not be known in advance. In what follows, the value of any time-varying variable Q at time t will be denoted by $Q(t)$. Occasionally it may be convenient to consult Fig. 1.

During each cycle our hardware executes an elementary operation which affects its variable state $s \in \mathcal{S} \subset B^*$ and possibly also the variable environmental state $Env \in \mathcal{E}$. (Here we need not yet specify the problem-dependent set $\mathcal{E}$). There is a hardwired state transition function $F : \mathcal{S} \times \mathcal{E} \to \mathcal{S}$. For $t > 1$, $s(t) = F(s(t-1), Env(t-1))$ is the state at a point where the hardware operation of cycle $t-1$ is finished, but the one of t has not started yet. $Env(t)$ may depend on past output actions encoded in $s(t-1)$ and is simultaneously updated or (probabilistically) computed by the possibly reactive environment.

In order to conveniently talk about programs and data, we will often attach names to certain string variables encoded as components or substrings of s. Of particular interest are 3 variables called *time*, x, y, p:

1. At time t, variable *time* holds a unique binary representation of t. We initialize $time(1) = $ '1', the bitstring consisting only of a one. The hardware increments *time* from one cycle to the next. This requires at most $O(log\ t)$ and on average only $O(1)$ computational steps.
2. Variable x holds environmental inputs. For $t > 1$, $x(t)$ may differ from $x(t-1)$ only if a program running on the Gödel machine has executed a special input-requesting instruction at time $t-1$. Generally speaking, the delays between successive inputs should be sufficiently large so that programs can perform certain elementary computations on an input, such as copying it into internal storage (a reserved part of s) before the next input arrives.

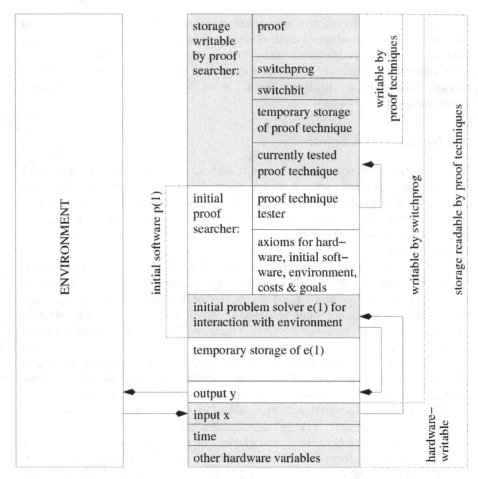

Fig. 1: Storage snapshot of a not yet self-improved example Gödel machine, with the initial software still intact. See text for details.

3. $y(t)$ is an output bitstring which may subsequently influence the environment, where $y(1) = $ '0' by default. For example, $y(t)$ could be interpreted as a control signal for an environment-manipulating robot whose actions may have an effect on future inputs.

4. $p(1)$ is the initial software: a program implementing the original policy for interacting with the environment and for proof searching. Details will be discussed below.

At any given time t $(1 \leq t \leq T)$ the goal is to maximize future success or *utility*. A typical *"value to go"* utility function (to be maximized) is of the form $u(s, Env) : \mathcal{S} \times \mathcal{E} \rightarrow \mathcal{R}$, where $\mathcal{R}$ is the set of real numbers:

$$u(s, Env) = E_\mu \left[\sum_{\tau=time}^{T} r(\tau) \;\middle|\; s, Env \right], \tag{1}$$

where $r(t)$ is a real-valued reward input (encoded within $s(t)$) at time t, $E_\mu(\cdot \mid \cdot)$ denotes the conditional expectation operator with respect to some possibly unknown distribution μ from a set M of possible distributions (M reflects whatever is known about the possibly probabilistic reactions of the environment), and the above-mentioned $time = time(s)$ is a function of state s which uniquely identifies the current cycle. Note that we take into account the possibility of extending the expected lifespan $E_\mu(T \mid s, Env)$ through appropriate actions.

Alternative formalizable utility functions could favor improvement of *worst case* instead of *expected* future performance, or higher reward intake *per time interval* etc.

2.2 Basic Idea of Gödel Machine

Our machine becomes a self-referential [11] *Gödel machine* by loading it with a particular form of machine-dependent, self-modifying code p. The initial code $p(1)$ at time step 1 includes a (typically sub-optimal) problem solving subroutine for interacting with the environment, such as Q-learning [19], and a general proof searcher subroutine (Section 5) that systematically makes pairs *(switchprog, proof)* (variable substrings of s) until it finds a *proof* of a target theorem which essentially states: *'the immediate rewrite of* p *through current program* switchprog *on the given machine implies higher utility than leaving* p *as is'*. Then it executes *switchprog*, which may completely rewrite p, including the proof searcher. Section 3 will explain details of the necessary initial axiomatic system $\mathcal{A}$ encoded in $p(1)$.

The Global Optimality Theorem (Theorem 1, Sect. 4) shows this self-improvement strategy is not greedy: since the utility of *'leaving* p *as is'* implicitly evaluates all possible alternative *switchprogs* which an unmodified p might find later, we obtain a globally optimal self-change—the *current switchprog* represents the best of all possible relevant self-changes, relative to the given resource limitations and initial proof search strategy.

2.3 Proof Techniques and an $O()$-optimal Initial Proof Searcher.

Section 5 will present an $O()$-optimal initialization of the proof searcher, that is, one with an optimal *order* of complexity (Theorem 2). Still, there will remain a lot of room for self-improvement hidden by the $O()$-notation. The searcher uses an online extension of *Universal Search* [23, 25] to systematically test *online proof techniques*, which are proof-generating programs that may read parts of state s (similarly, mathematicians are often more interested in proof techniques than in theorems). To prove target theorems as above, proof techniques may invoke special instructions for generating axioms and applying inference rules to prolong the current *proof* by theorems. Here an axiomatic system $\mathcal{A}$ encoded in $p(1)$ includes axioms describing *(a)* how any instruction invoked by a program running on the given hardware will change the machine's state s (including instruction pointers etc.) from one step to the next (such that proof techniques can reason about the effects of any program including the proof searcher), *(b)* the initial program $p(1)$ itself (Section 3 will show that this is possible without introducing circularity), **(c)** stochastic environmental properties, **(d)** the formal utility function u, e.g., equation (1). The evaluation of utility automatically takes into account computational costs of all actions including proof search.

2.4 Relation to Hutter's Previous Work

Hutter's non-self-referential but still $O()$-optimal *'fastest' algorithm for all well-defined problems* HSEARCH [17] uses a *hardwired* brute force proof searcher. Assume discrete input/output domains X/Y, a formal problem specification $f : X \to Y$ (say, a functional description of how integers are decomposed into their prime factors), and a particular $x \in X$ (say, an integer to be factorized). HSEARCH orders all proofs of an appropriate axiomatic system by size to find programs q that for all $z \in X$ provably compute $f(z)$ within time bound $t_q(z)$. Simultaneously it spends most of its time on executing the q with the best currently proven time bound $t_q(x)$. It turns out that HSEARCH is as fast as the *fastest* algorithm that provably computes $f(z)$ for all $z \in X$, save for a constant factor smaller than $1 + \epsilon$ (arbitrary $\epsilon > 0$) and an f-specific but x-independent additive constant [17]. This constant may be enormous though.

Hutter's AIXI*(t,l)* [16] is related. In discrete cycle $k = 1, 2, 3, \ldots$ of AIXI*(t,l)*'s lifetime, action $y(k)$ results in perception $x(k)$ and reward $r(k)$, where all quantities may depend on the complete history. Using a universal computer such as a Turing machine, AIXI*(t,l)* needs an initial offline setup phase (prior to interaction with the environment) where it uses a *hardwired* brute force proof searcher to examine all proofs of length at most L, filtering out those that identify programs (of maximal size l and maximal runtime t per cycle) which not only could interact with the environment but which for all possible interaction histories also correctly predict a lower bound of their own expected future reward. In cycle k, AIXI*(t,l)* then runs all programs identified in the setup phase (at most 2^l), finds the one with highest self-rating, and executes its corresponding action. The problem-independent setup time (where almost all of the work is done) is $O(L \cdot 2^L)$. The online time per cycle is $O(t \cdot 2^l)$. Both are constant but typically huge.

Advantages and Novelty of the Gödel Machine. There are major differences between the Gödel machine and Hutter's HSEARCH [17] and AIXI*(t,l)* [16], including:

1. The theorem provers of HSEARCH and AIXI*(t,l)* are hardwired, non-self-referential, unmodifiable meta-algorithms that cannot improve them-

selves. That is, they will always suffer from the same huge constant slow-downs (typically $\gg 10^{1000}$) buried in the $O()$-notation. But there is nothing in principle that prevents our truly self-referential code from proving and exploiting drastic reductions of such constants, in the best possible way that provably constitutes an improvement, if there is any.

2. The demonstration of the $O()$-optimality of HSEARCH and AIXI(t,l) depends on a clever allocation of computation time to some of their unmodifiable meta-algorithms. Our Global Optimality Theorem (Theorem 1, Section 4), however, is justified through a quite different type of reasoning which indeed exploits and crucially depends on the fact that there is no unmodifiable software at all, and that the proof searcher itself is readable and modifiable and can be improved. This is also the reason why its self-improvements can be more than merely $O()$-optimal.

3. HSEARCH uses a "trick" of proving more than is necessary which also disappears in the sometimes quite misleading $O()$-notation: it wastes time on finding programs that provably compute $f(z)$ for all $z \in X$ even when the current $f(x)(x \in X)$ is the only object of interest. A Gödel machine, however, needs to prove only what is relevant to its goal formalized by u. For example, the general u of eq. (1) completely ignores the limited concept of $O()$-optimality, but instead formalizes a stronger type of optimality that does not ignore huge constants just because they are constant.

4. Both the Gödel machine and AIXI(l,l) can maximize expected reward (HSEARCH cannot). But the Gödel machine is more flexible as we may plug in *any* type of formalizable utility function (e.g., *worst case* reward), and unlike AIXI(t,l) it does not require an enumerable environmental distribution.

Nevertheless, we may use AIXI(t,l) or HSEARCH to initialize the substring e of p which is responsible for interaction with the environment. The Gödel machine will replace e as soon as it finds a provably better strategy.

2.5 Limitations of Gödel Machines

The fundamental limitations are closely related to those first identified by Gödel's celebrated paper on self-referential formulae [11]. Any formal system that encompasses arithmetics (or ZFC, etc.) is either flawed or allows for unprovable but true statements. Hence, even a Gödel machine with unlimited computational resources must ignore those self-improvements whose effectiveness it cannot prove, e.g., for lack of sufficiently powerful axioms in $\mathcal{A}$. In particular, one can construct pathological examples of environments and utility functions that make it impossible for the machine to ever prove a target theorem. Compare Blum's speed-up theorem [3, 4] based on certain incomputable predicates. Similarly, a realistic Gödel machine with limited resources cannot profit from self-improvements whose usefulness it cannot prove within its time and space constraints.

Nevertheless, unlike previous methods, it can in principle exploit at least the *provably* good speed-ups of *any* part of its initial software, including those parts responsible for huge (but problem class-independent) slowdowns ignored by the earlier approaches [17, 16].

3 Essential Details of One Representative Gödel Machine

Theorem proving requires an axiom scheme yielding an enumerable set of axioms of a formal logic system $\mathcal{A}$ whose formulas and theorems are symbol strings over some finite alphabet that may include traditional symbols of logic (such as $\rightarrow, \wedge, =, (,), \forall, \exists, \ldots, c_1, c_2, \ldots, f_1, f_2, \ldots$), probability theory (such as $E(\cdot)$, the expectation operator), arithmetics $(+, -, /, =, \sum, <, \ldots)$, string manipulation (in particular, symbols for representing any part of state s at any time, such as $s_{7:88}(5555)$). A proof is a sequence of theorems, each either an axiom or inferred from previous theorems by applying one of the inference rules such as *modus ponens* combined with *unification*, e.g., [10].

The remainder of this chapter will omit standard knowledge to be found in any proof theory textbook. Instead of listing *all* axioms of a particular $\mathcal{A}$ in a tedious fashion, we will focus on the novel and critical details: how to overcome problems with self-reference and how to deal with the potentially delicate online generation of proofs that talk about and affect the currently running proof generator itself.

3.1 Proof Techniques

Brute force proof searchers (used in Hutter's $\text{AIXI}(t,l)$ and HSEARCH; see Section 2.4) systematically generate all proofs in order of their sizes. To produce a certain proof, this takes time exponential in proof size. Instead our $O()$-optimal $p(1)$ will produce many proofs with low algorithmic complexity [52, 21, 26] much more quickly. It systematically tests (see Sect. 5) *proof techniques* written in universal language $\mathcal{L}$ implemented within $p(1)$. For example, $\mathcal{L}$ may be a variant of PROLOG [7] or the universal FORTH[28]-inspired programming language used in recent work on optimal search [45]. A proof technique is composed of instructions that allow any part of s to be read, such as inputs encoded in variable x (a substring of s) or the code of $p(1)$. It may write on s^p, a part of s reserved for temporary results. It also may rewrite *switchprog*, and produce an incrementally growing proof placed in the string variable *proof* stored somewhere in s. *proof* and s^p are reset to the empty string at the beginning of each new proof technique test. Apart from standard arithmetic and function-defining instructions [45] that modify s^p, the programming language $\mathcal{L}$ includes special instructions for prolonging the current *proof* by correct theorems, for setting *switchprog*, and for checking whether a provably optimal p-modifying program was found and should be executed now. Certain long proofs can be produced by short proof techniques.

The nature of the six *proof*-modifying instructions below (there are no others) makes it impossible to insert an incorrect theorem into *proof*, thus trivializing proof verification:

1. **get-axiom(n)** takes as argument an integer n computed by a prefix of the currently tested proof technique with the help of arithmetic instructions such as those used in previous work [45]. Then it appends the n-th axiom (if it exists, according to the axiom scheme below) as a theorem to the current theorem sequence in *proof*. The initial axiom scheme encodes:

   a) **Hardware axioms** describing the hardware, formally specifying how certain components of s (other than environmental inputs x) may change from one cycle to the next.

   For example, if the hardware is a Turing machine[2] (TM) [56], then $s(t)$ is a bitstring that encodes the current contents of all tapes of the TM, the positions of its scanning heads, and the current *internal state* of the TM's finite state automaton, while F specifies the TM's look-up table which maps any possible combination of internal state and bits above scanning heads to a new internal state and an action such as: replace some head's current bit by 1/0, increment (right shift) or decrement (left shift) some scanning head, read and copy next input bit to cell above input tape's scanning head, etc. Alternatively, if the hardware is given by the abstract model of a modern microprocessor with limited storage, $s(t)$ will encode the current storage contents, register values, instruction pointers, etc.

   For example, the following axiom could describe how some 64-bit hardware's instruction pointer stored in $s_{1:64}$ is continually incremented as long as there is no overflow and the value of s_{65} does not indicate that a jump to some other address should take place:

$$(\forall t \forall n : [(n < 2^{64} - 1) \land (n > 0) \land (t > 1) \land (t < T)$$

$$\land (string2num(s_{1:64}(t)) = n) \land (s_{65}(t) = \text{`0'})]$$

$$\rightarrow (string2num(s_{1:64}(t+1)) = n+1))$$

   Here the semantics of used symbols such as '(' and '>' and '→' (implies) are the traditional ones, while '*string2num*' symbolizes a function translating bitstrings into numbers. It is clear that any abstract hardware model can be fully axiomatized in a similar way.

   b) **Reward axioms** defining the computational costs of any hardware instruction, and physical costs of output actions (e.g., control signals

---

[2]Turing reformulated Gödel's unprovability results in terms of Turing machines (TMs) [56] which subsequently became the most widely used abstract model of computation. It is well-known that there are *universal* TMs that in a certain sense can emulate any other TM or any other known computer. Gödel's integer-based formal language can be used to describe any universal TM, and vice versa.

$y(t)$ encoded in $s(t)$). Related axioms assign values to certain input events (encoded in variable x, a substring of s) representing reward or punishment (e.g., when a Gödel machine-controlled robot bumps into an obstacle). Additional axioms define the total value of the Gödel machine's life as a scalar-valued function of all rewards (e.g., their sum) and costs experienced between cycles 1 and T, etc. For example, assume that $s_{17:18}$ can be changed only through external inputs; the following example axiom says that the total reward increases by 3 whenever such an input equals '11' (unexplained symbols carry the obvious meaning):

$$(\forall t_1 \forall t_2 : [(t_1 < t_2) \wedge (t_1 \geq 1) \wedge (t_2 \leq T) \wedge (s_{17:18}(t_2) = \text{'11'})]$$

$$\rightarrow [R(t_1, t_2) = R(t_1, t_2 - 1) + 3]),$$

where $R(t_1, t_2)$ is interpreted as the cumulative reward between times t_1 and t_2. It is clear that any formal scheme for producing rewards can be fully axiomatized in a similar way.

c) **Environment axioms** restricting the way the environment will produce new inputs (encoded within certain substrings of s) in reaction to sequences of outputs y encoded in s. For example, it may be known in advance that the environment is sampled from an unknown probability distribution that is computable, given the previous history [52, 53, 16], or at least limit-computable [39, 40]. Or, more restrictively, the environment may be some unknown but deterministic computer program [58, 37] sampled from the Speed Prior [41] which assigns low probability to environments that are hard to compute by any method. Or the interface to the environment is Markovian [33], that is, the current input always uniquely identifies the environmental state—a lot of work has been done on this special case [31, 2, 55]. Even more restrictively, the environment may evolve in completely predictable fashion known in advance. All such prior assumptions are perfectly formalizable in an appropriate $\mathcal{A}$ (otherwise we could not write scientific papers about them).

d) **Uncertainty axioms; string manipulation axioms:** Standard axioms for arithmetics and calculus and probability theory [20] and statistics and string manipulation that (in conjunction with the environment axioms) allow for constructing proofs concerning (possibly uncertain) properties of future values of $s(t)$ as well as bounds on expected remaining lifetime / costs / rewards, given some time τ and certain hypothetical values for components of $s(\tau)$ etc. An example theorem saying something about expected properties of future inputs x might look like this:

$$(\forall t_1 \forall \mu \in M : [(1 \leq t_1) \wedge (t_1 + 15597 < T) \wedge (s_{5:9}(t_1) = \text{'01011'})$$

$$\wedge(x_{40:44}(t_1) = \text{`00000'})] \rightarrow (\exists t : [(t_1 < t < t_1 + 15597)$$

$$\wedge(P_\mu(x_{17:22}(t) = \text{`011011'} \mid s(t_1)) > \frac{998}{1000})])),$$

where $P_\mu(. \mid .)$ represents a conditional probability with respect to an axiomatized prior distribution μ from a set of distributions M described by the environment axioms (Item 1c).

Given a particular formalizable hardware (Item 1a) and formalizable assumptions about the possibly probabilistic environment (Item 1c), obviously one can fully axiomatize everything that is needed for proof-based reasoning.

e) **Initial state axioms:** Information about how to reconstruct the initial state $s(1)$ or parts thereof, such that the proof searcher can build proofs including axioms of the type

$$(s_{\mathbf{m:n}}(1) = \mathbf{z}), \quad e.g. : \quad (s_{7:9}(1) = \text{`010'}).$$

Here and in the remainder of the paper we use bold font in formulas to indicate syntactic place holders (such as $\mathbf{m,n,z}$) for symbol strings representing variables (such as m,n,z) whose semantics are explained in the text (in the present context z is the bitstring $s_{m:n}(1)$).

Note that it is *no fundamental problem* to fully encode both the hardware description *and* the initial hardware-describing p within p itself. To see this, observe that some software may include a program that can print the software.

f) **Utility axioms** describing the overall goal in the form of utility function u; e.g., equation (1) in Section 2.1.

2. **apply-rule(k, m, n)** takes as arguments the index k (if it exists) of an inference rule such as *modus ponens* (stored in a list of possible inference rules encoded within $p(1)$) and the indices m, n of two previously proven theorems (numbered in order of their creation) in the current *proof*. If applicable, the corresponding inference rule is applied to the addressed theorems and the resulting theorem appended to *proof*. Otherwise the currently tested proof technique is interrupted. This ensures that *proof* is never fed with invalid proofs.

3. **delete-theorem(m)** deletes the m-th theorem in the currently stored *proof*, thus freeing storage such that proof-storing parts of s can be reused and the maximal proof size is not necessarily limited by storage constraints. Theorems deleted from *proof*, however, cannot be addressed any more by *apply-rule* to produce further prolongations of *proof*.

4. **set-switchprog(m,n)** replaces *switchprog* by $s^p_{\mathbf{m:n}}$, provided that $s^p_{\mathbf{m:n}}$ is indeed a non-empty substring of s^p, the storage writable by proof techniques.

5. **state2theorem(m, n)** takes two integer arguments m, n and tries to transform the current contents of $s_{m:n}$ into a theorem of the form

$$(s_{\mathbf{m:n}}(\mathbf{t_1}) = \mathbf{z}), \ e.g.: \ (s_{6:9}(7775555) = \text{`1001'}),$$

where t_1 represents a time measured (by checking *time*) shortly after *state2theorem* was invoked, and z the bistring $s_{m:n}(t_1)$ (recall the special case $t_1 = 1$ of Item 1e). So we accept the time-labeled current observable contents of any part of s as a theorem that does not have to be proven in an alternative way from, say, the initial state $s(1)$, because the computation so far has already demonstrated that the theorem is true. Thus we may exploit information conveyed by environmental inputs, and the fact that sometimes (but not always) the fastest way to determine the output of a program is to run it.

This non-traditional online interface between syntax and semantics requires special care though. We must avoid inconsistent results through parts of s that change while being read. For example, the present value of a quickly changing instruction pointer *IP* (continually updated by the hardware) may be essentially unreadable in the sense that the execution of the reading subroutine itself will already modify *IP* many times. For convenience, the (typically limited) hardware could be set up such that it stores the contents of fast hardware variables every c cycles in a reserved part of s, such that an appropriate variant of *state2theorem()* could at least translate certain recent values of fast variables into theorems. This, however, will not abolish *all* problems associated with self-observations. For example, the $s_{m:n}$ to be read might also contain the reading procedure's own, temporary, constantly changing string pointer variables, etc.[3] To address such problems on computers with limited memory, *state2theorem* first uses some fixed protocol to check whether the current $s_{m:n}$ is readable at all or whether it might change if it were read by the remaining code of *state2theorem*. If so, or if m, n, are not in the proper range, then the instruction has no further effect. Otherwise it appends an *observed* theorem of the form $(s_{\mathbf{m:n}}(\mathbf{t_1}) = \mathbf{z})$ to *proof*. For example, if the current time is 7770000, then the invocation of *state2theorem(6,9)* might return the theorem $(s_{6:9}(7775555) = \text{`1001'})$, where $7775555 - 7770000 = 5555$ reflects the time needed by *state2theorem* to perform the initial check and to read leading bits off the continually increasing *time* (reading *time* also

---

[3]We see that certain parts of the current s may not be directly observable without changing the observable itself. Sometimes, however, axioms and previous observations will allow the Gödel machine to *deduce* time-dependent storage contents that are not directly observable. For instance, by analyzing the code being executed through instruction pointer *IP* in the example above, the value of *IP* at certain times may be predictable (or postdictable, after the fact). The values of other variables at given times, however, may not be deducible at all. Such limits of self-observability are reminiscent of Heisenberg's celebrated uncertainty principle [12], which states that certain physical measurements are necessarily imprecise, since the measuring process affects the measured quantity.

costs time) such that it can be sure that 7775555 is a recent proper time label following the start of *state2theorem*.

6. **check()** verifies whether the goal of the proof search has been reached. First it tests whether the last theorem (if any) in *proof* has the form of a **target theorem**. A target theorem states that given the *current* axiomatized utility function u (Item 1f), the utility of a switch from p to the current *switchprog* would be higher than the utility of continuing the execution of p (which would keep searching for alternative *switchprogs*). Target theorems are symbol strings (encoded in s) of the (decoded) form

$$(u[s(\mathbf{t_1}) \oplus (switchbit(\mathbf{t_1}) = \text{`1'}), Env(\mathbf{t_1})] >$$

$$u[s(\mathbf{t_1}) \oplus (switchbit(\mathbf{t_1}) = \text{`0'}), Env(\mathbf{t_1})]), \qquad (2)$$

where the variable t_1 (represented by syntactic place holder $\mathbf{t_1}$) stands for a time step, while all other symbols belong to the alphabet of the theorem-proving calculus, that is, the set of possible target theorems is parameterized only by t_1. Here the calculus should permit the notation $s(\mathbf{t_1}) \oplus (switchbit(\mathbf{t_1}) = \text{`b'})$ as a shortcut for the state obtained when we replace $switchbit(t_1)$, the true value of the variable bit *switchbit* (encoded in s) at time t_1, by $b \in \{0, 1\}$. This will facilitate the formulation of theorems that compare values conditioned on various alternative hypothetical properties of $s(t_1)$. (Note that $s(t_1)$ may be only partially known by the current proof technique even in environments where $s(t_1)$ and $switchbit(t_1)$ are fully predetermined for all valid t_1.)

The purpose of introducing t_1 is to deal with hardware-specific temporal delays that may be involved in checking and switching—it may take a significant amount of time to match abstract symbol strings found during proof search to the Gödel machine's real current state. If a target theorem has been found, *check()* uses a simple prewired subroutine to check whether there is enough time left to set variable *switchbit* (originally 0) to 1 before the continually increasing *time* will equal t_1. If this subroutine returns a negative result, *check()* exits. Otherwise it sets *switchbit* := 1 (there is no other way of changing *switchbit*). Then it repeatedly tests *time* until *time* $> t_1$, to make sure the condition of formula (6) was fulfilled at t_1. Then it transfers control to *switchprog* (there is no other way of calling *switchprog*). The *switchprog* may subsequently rewrite all parts of s, excluding hardware-reserved parts such as *time* and x, but including p.

Note that a proof technique does not necessarily have to compute the true expected utilities of switching and not switching—it just needs to determine which is higher. For example, it may be easy to prove that speeding up a subroutine of the proof searcher by a factor of 2 will certainly be worth the negligible (compared to lifetime T) time needed to execute the subroutine-changing algorithm, no matter the precise utility of the switch.

The axiomatic system $\mathcal{A}$ is a defining parameter of a given Gödel machine. Clearly, $\mathcal{A}$ must be strong enough to permit proofs of target theorems. In particular, the theory of uncertainty axioms (Item 1d) must be sufficiently rich. This is no fundamental problem: We simply insert all traditional axioms of probability theory [20].

4 Global Optimality Theorem

Intuitively, at any given time p should execute some self-modification algorithm only if it is the 'best' of all possible self-modifications, given the utility function, which typically depends on available resources, such as storage size and remaining lifetime. At first glance, however, target theorem (6) seems to implicitly talk about just one single modification algorithm, namely, $switchprog(t_1)$ as set by the systematic proof searcher at time t_1. Isn't this type of local search greedy? Couldn't it lead to a local optimum instead of a global one? No, it cannot, according to the global optimality theorem:

Theorem 1 (Globally Optimal Self-Changes, given u and $\mathcal{A}$ encoded in p). *Given any formalizable utility function u (Item 1f), and assuming consistency of the underlying formal system $\mathcal{A}$, any self-change of p obtained through execution of some program* switchprog *identified through the proof of a target theorem (6) is globally optimal in the following sense: the utility of starting the execution of the present* switchprog *is higher than the utility of waiting for the proof searcher to produce an alternative* switchprog *later.*

Proof. Target theorem (6) implicitly talks about all the other *switchprog*s that the proof searcher could produce in the future. To see this, consider the two alternatives of the binary decision: (1) either execute the current *switchprog* (set *switchbit* = 1), or (2) keep searching for *proofs* and *switchprog*s (set *switchbit* = 0) until the systematic searcher comes up with an even better *switchprog*. Obviously the second alternative concerns all (possibly infinitely many) potential *switchprog*s to be considered later. That is, if the current *switchprog* were not the 'best', then the proof searcher would not be able to prove that setting *switchbit* and executing *switchprog* will cause higher expected reward than discarding *switchprog*, assuming consistency of $\mathcal{A}$. *Q.E.D.*

4.1 Alternative Relaxed Target Theorem

We may replace the target theorem (6) (Item 6) by the following alternative target theorem:

$$(u[s(\mathbf{t_1}) \oplus (switchbit(\mathbf{t_1}) = \text{'1'}), Env(\mathbf{t_1})] \geq$$

$$u[s(\mathbf{t_1}) \oplus (switchbit(\mathbf{t_1}) = \text{'0'}), Env(\mathbf{t_1})]). \qquad (3)$$

The only difference to the original target theorem (6) is that the ">" sign became a "≥" sign. That is, the Gödel machine will change itself as soon as it found a proof that the change will not make things worse. A Global Optimality Theorem similar to Theorem 1 holds.

5 Bias-Optimal Proof Search (BIOPS)

Here we construct a $p(1)$ that is $O()$-optimal in a certain limited sense to be described below, but still might be improved as it is not necessarily optimal in the sense of the given u (for example, the u of equation (1) neither mentions nor cares for $O()$-optimality). Our Bias-Optimal Proof Search (BIOPS) is essentially an application of Universal Search [23, 25] to proof search. Previous practical variants and extensions of universal search have been applied [36, 38, 50, 45] to *offline* program search tasks where the program inputs are fixed such that the same program always produces the same results. In our *online* setting, however, BIOPS has to take into account that the same proof technique started at different times may yield different proofs, as it may read parts of s (e.g., inputs) that change as the machine's life proceeds.

BIOPS starts with a probability distribution P (the initial bias) on the proof techniques w that one can write in $\mathcal{L}$, e.g., $P(w) = K^{-l(w)}$ for programs composed from K possible instructions [25]. BIOPS is *near-bias-optimal* [45] in the sense that it will not spend much more time on any proof technique than it deserves, according to its probabilistic bias, namely, not much more than its probability times the total search time:

Definition 1 (Bias-Optimal Searchers [45]). Let $\mathcal{R}$ be a problem class, $\mathcal{C}$ be a search space of solution candidates (where any problem $r \in \mathcal{R}$ should have a solution in $\mathcal{C}$), $P(q \mid r)$ be a task-dependent bias in the form of conditional probability distributions on the candidates $q \in \mathcal{C}$. Suppose that we also have a predefined procedure that creates and tests any given q on any $r \in \mathcal{R}$ within time $t(q, r)$ (typically unknown in advance). Then *a searcher is n-bias-optimal ($n \geq 1$) if for any maximal total search time $T_{total} > 0$ it is guaranteed to solve any problem $r \in \mathcal{R}$ if it has a solution $p \in \mathcal{C}$ satisfying* $t(p, r) \leq P(p \mid r) T_{total}/n$. *It is bias-optimal if $n = 1$.*

Method 5.1 (BIOPS) In phase ($i = 1, 2, 3, \ldots$) Do: For all self-delimiting [25] proof techniques $w \in \mathcal{L}$ satisfying $P(w) \geq 2^{-i}$ Do:

1. Run w until halt or error (such as division by zero) or $2^i P(w)$ steps consumed.
2. Undo effects of w on s^p (does not cost significantly more time than executing w).

A proof technique w can interrupt Method 5.1 only by invoking instruction *check()* (Item 6), which may transfer control to *switchprog* (which possibly even will delete or rewrite Method 5.1). Since the initial p runs on the formalized hardware, and since proof techniques tested by p can read p and other parts of s, they can produce proofs concerning the (expected) performance of p and BIOPS itself. Method 5.1 at least has the optimal *order* of computational complexity in the following sense.

Theorem 2. *If independently of variable* time(s) *some unknown fast proof technique w would require at most $f(k)$ steps to produce a proof of difficulty measure k (an integer depending on the nature of the task to be solved), then Method 5.1 will need at most $O(f(k))$ steps.*

Proof. It is easy to see that Method 5.1 will need at most $O(f(k)/P(w)) = O(f(k))$ steps—the constant factor $1/P(w)$ does not depend on k. *Q.E.D.*

Note again, however, that the proofs themselves may concern quite different, arbitrary formalizable notions of optimality (stronger than those expressible in the $O()$-notation) embodied by the given, problem-specific, formalized utility function u. This may provoke useful, constant-affecting rewrites of the initial proof searcher despite its limited (yet popular and widely used) notion of $O()$-optimality.

5.1 How a Surviving Proof Searcher May Use BIOPS to Solve Remaining Proof Search Tasks

The following is not essential for this chapter. Let us assume that the execution of the *switchprog* corresponding to the first found target theorem has not rewritten the code of p itself—the current p is still equal to $p(1)$—and has reset *switchbit* and returned control to p such that it can continue where it was interrupted. In that case the BIOPS subroutine of $p(1)$ can use the Optimal Ordered Problem Solver OOPS [45] to accelerate the search for the n-th target theorem ($n > 1$) by reusing proof techniques for earlier found target theorems where possible. The basic ideas are as follows (details: [45]).

Whenever a target theorem has been proven, $p(1)$ *freezes* the corresponding proof technique: its code becomes non-writable by proof techniques to be tested in later proof search tasks. But it remains readable, such that it can be copy-edited and/or invoked as a subprogram by future proof techniques. We also allow prefixes of proof techniques to temporarily rewrite the probability distribution on their suffixes [45], thus essentially rewriting the probability-based search procedure (an incremental extension of Method 5.1) based on previous experience. As a side-effect we metasearch for faster search procedures, which can greatly accelerate the learning of new tasks [45].

Given a new proof search task, BIOPS performs OOPS by spending half the total search time on a variant of Method 5.1 that searches only among self-delimiting [24, 6] proof techniques starting with the most recently frozen proof technique. The rest of the time is spent on fresh proof techniques with arbitrary prefixes (which may reuse previously frozen proof techniques though) [45]. (We could also search for a *generalizing* proof technique solving all proof search tasks so far. In the first half of the search we would not have to test proof techniques on tasks other than the most recent one, since we already know that their prefixes solve the previous tasks [45].)

It can be shown that OOPS is essentially *8-bias-optimal* (see Def. 1), given either the initial bias or intermediate biases due to frozen solutions to previous tasks [45]. This result immediately carries over to BIOPS. To summarize, BIOPS essentially allocates part of the total search time for a new task to proof techniques that exploit previous successful proof techniques in computable ways. If the new task can be solved faster by copy-editing / invoking previously frozen proof techniques than by solving the new proof search task from scratch, then BIOPS will discover this and profit thereof. If not, then at least it will not be significantly slowed down by the previous solutions—BIOPS will remain 8-bias-optimal.

Recall, however, that BIOPS is not the only possible way of initializing the Gödel machine's proof searcher.

6 Discussion & Additional Relations to Previous Work

Here we list a few examples of possible types of self-improvements (Sect. 6.1), Gödel machine applicability to various tasks defined by various utility functions and environments (Sect. 6.2), probabilistic hardware (Sect. 6.3), and additional relations to previous work (Sect. 6.4). We also briefly discuss self-reference and consciousness (Sect. 6.6), and provide a list of answers to frequently asked questions (Sect. 6.7).

6.1 Possible Types of Gödel Machine Self-improvements

Which provably useful self-modifications are possible? There are few limits to what a Gödel machine might do:

1. In one of the simplest cases it might leave its basic proof searcher intact and just change the ratio of time-sharing between the proof searching subroutine and the subpolicy e—those parts of p responsible for interaction with the environment.
2. Or the Gödel machine might modify e only. For example, the initial e may regularly store limited memories of past events somewhere in s; this might allow p to derive that it would be useful to modify e such that e will conduct certain experiments to increase the knowledge about the environment, and use the resulting information to increase reward intake. In this sense the Gödel machine embodies a principled way of dealing with the exploration versus exploitation problem [19]. Note that the *expected* utility of conducting some experiment may exceed the one of not conducting it, even when the experimental outcome later suggests to keep acting in line with the previous e.

3. The Gödel machine might also modify its very axioms to speed things up. For example, it might find a proof that the original axioms should be replaced or augmented by theorems derivable from the original axioms.

4. The Gödel machine might even change its own utility function and target theorem, but can do so only if their *new* values are provably better according to the *old* ones.

5. In many cases we do not expect the Gödel machine to replace its proof searcher by code that completely abandons the search for proofs. Instead, we expect that only certain subroutines of the proof searcher will be sped up—compare the example at the end of Item 6 in Section 3.1—or that perhaps just the order of generated proofs will be modified in problem-specific fashion. This could be done by modifying the probability distribution on the proof techniques of the initial bias-optimal proof searcher from Section 5.

6. Generally speaking, the utility of limited rewrites may often be easier to prove than the one of total rewrites. For example, suppose it is 8:00 PM and our Gödel machine-controlled agent's permanent goal is to maximize future expected reward, using the (alternative) target theorem (4.1). Part thereof is to avoid hunger. There is nothing in its fridge, and shops close down at 8:30 PM. It does not have time to optimize its way to the supermarket in every little detail, but if it does not get going right now it will stay hungry tonight (in principle such near-future consequences of actions should be easily provable, possibly even in a way related to how humans prove advantages of potential actions to themselves). That is, if the agent's previous policy did not already include, say, an automatic daily evening trip to the supermarket, the policy provably should be rewritten at least limitedly and simply right now, while there is still time, such that the agent will surely get some food tonight, without affecting less urgent future behavior that can be optimized/decided later, such as details of the route to the food, or of tomorrow's actions.

7. In certain uninteresting environments reward is maximized by becoming dumb. For example, a given task may require to repeatedly and forever execute the same pleasure center-activating action, as quickly as possible. In such cases the Gödel machine may delete most of its more time-consuming initial software including the proof searcher.

8. Note that there is no reason why a Gödel machine should not augment its own hardware. Suppose its lifetime is known to be 100 years. Given a hard problem and axioms restricting the possible behaviors of the environment, the Gödel machine might find a proof that its expected cumulative reward will increase if it invests 10 years into building faster computational hardware, by exploiting the physical resources of its environment.

6.2 Example Applications

Example 1 (Maximizing expected reward with bounded resources). A robot that needs at least 1 liter of gasoline per hour interacts with a partially unknown environment, trying to find hidden, limited gasoline depots to occasionally refuel its tank. It is rewarded in proportion to its lifetime, and dies after at most 100 years or as soon as its tank is empty or it falls off a cliff, etc. The probabilistic environmental reactions are initially unknown but assumed to be sampled from the axiomatized Speed Prior [41], according to which hard-to-compute environmental reactions are unlikely. This permits a computable strategy for making near-optimal predictions [41]. One by-product of maximizing expected reward is to maximize expected lifetime.

Less general, more traditional examples that do not involve significant interaction with a probabilistic environment are also easily dealt with in the reward-based framework:

Example 2 (Time-limited NP-hard optimization). The initial input to the Gödel machine is the representation of a connected graph with a large number of nodes linked by edges of various lengths. Within given time T it should find a cyclic path connecting all nodes. The only real-valued reward will occur at time T. It equals 1 divided by the length of the best path found so far (0 if none was found). There are no other inputs. The by-product of maximizing expected reward is to find the shortest path findable within the limited time, given the initial bias.

Example 3 (Fast theorem proving). Prove or disprove as quickly as possible that all even integers > 2 are the sum of two primes (Goldbach's conjecture). The reward is $1/t$, where t is the time required to produce and verify the first such proof.

Example 4 (Optimize any suboptimal problem solver). Given any formalizable problem, implement a suboptimal but known problem solver as software on the Gödel machine hardware, and let the proof searcher of Section 5 run in parallel.

6.3 Probabilistic Gödel Machine Hardware

Above we have focused on an example deterministic machine. It is straightforward to extend this to computers whose actions are computed in probabilistic fashion, given the current state. Then the expectation calculus used for probabilistic aspects of the environment simply has to be extended to the hardware itself, and the mechanism for verifying proofs has to take into account that there is no such thing as a certain theorem—at best there are formal statements which are true with such and such probability. In fact, this may be the most realistic approach as any physical hardware is error-prone, which should be taken into account by realistic probabilistic Gödel machines.

Probabilistic settings also automatically avoid certain issues of axiomatic consistency. For example, predictions proven to come true with probability less than 1.0 do not necessarily cause contradictions even when they do not match the observations.

6.4 More Relations to Previous Work on Less General Self-improving Machines

Despite (or maybe because of) the ambitiousness and potential power of self-improving machines, there has been little work in this vein outside our own labs at IDSIA and TU Munich. Here we will list essential differences between the Gödel machine and our previous approaches to 'learning to learn,' 'metalearning,' self-improvement, self-optimization, etc.

1. **Gödel Machine versus Success-Story Algorithm and Other Metalearners**

   A learner's modifiable components are called its policy. An algorithm that modifies the policy is a learning algorithm. If the learning algorithm has modifiable components represented as part of the policy, then we speak of a self-modifying policy (SMP) [48]. SMPs can modify the way they modify themselves etc. The Gödel machine has an SMP.

   In previous work we used the *success-story algorithm* (SSA) to force some (stochastic) SMPs to trigger better and better self-modifications [35, 49, 48, 50]. During the learner's life-time, SSA is occasionally called at times computed according to SMP itself. SSA uses backtracking to undo those SMP-generated SMP-modifications that have not been empirically observed to trigger lifelong reward accelerations (measured up until the current SSA call—this evaluates the long-term effects of SMP-modifications setting the stage for later SMP-modifications). SMP-modifications that survive SSA represent a lifelong success history. Until the next SSA call, they build the basis for additional SMP-modifications. Solely by self-modifications our SMP/SSA-based learners solved a complex task in a partially observable environment whose state space is far bigger than most found in the literature [48].

   The Gödel machine's training algorithm is theoretically more powerful than SSA though. SSA empirically measures the usefulness of previous self-modifications, and does not necessarily encourage provably optimal ones. Similar drawbacks hold for Lenat's human-assisted, non-autonomous, self-modifying learner [22], our Meta-Genetic Programming [32] extending Cramer's Genetic Programming [8, 1], our metalearning economies [32] extending Holland's machine learning economies [15], and gradient-based metalearners for continuous program spaces of differentiable recurrent neural networks [34, 13]. All these methods, however, could be used to seed $p(1)$ with an initial policy.

2. Gödel Machine versus OOPS and OOPS-RL

The Optimal Ordered Problem Solver OOPS [45, 42] (used by BIOPS in Sect. 5.1) is a bias-optimal (see Def. 1) way of searching for a program that solves each problem in an ordered sequence of problems of a reasonably general type, continually organizing and managing and reusing earlier acquired knowledge. Solomonoff recently also proposed related ideas for a *scientist's assistant* [54] that modifies the probability distribution of universal search [23] based on experience.

As pointed out earlier [45] (section on OOPS limitations), however, OOPS-like methods are not directly applicable to general lifelong reinforcement learning (RL) tasks [19] such as those for which AIXI [16] was designed. The simple and natural but limited optimality notion of OOPS is *bias-optimality* (Def. 1): OOPS is a near-bias-optimal searcher for programs which compute solutions that one can quickly verify (costs of verification are taken into account). For example, one can quickly test whether some currently tested program has computed a solution to the *towers of Hanoi* problem used in the earlier paper [45]: one just has to check whether the third peg is full of disks.

But general RL tasks are harder. Here, in principle, the evaluation of the value of some behavior consumes the learner's entire life! That is, the naive test of whether a program is good or not would consume the entire life. That is, we could test only one program; afterwards life would be over.

So general RL machines need a more general notion of optimality, and must do things that plain OOPS does not do, such as predicting *future* tasks and rewards. It is possible to use two OOPS-modules as components of a rather general reinforcement learner (OOPS-RL), one module learning a predictive model of the environment, the other one using this *world model* to search for an action sequence maximizing expected reward [45, 44]. Despite the bias-optimality properties of OOPS for certain ordered task sequences, however, OOPS-RL is not necessarily the best way of spending limited computation time in general RL situations.

A provably optimal RL machine must somehow *prove* properties of otherwise un-testable behaviors (such as: what is the expected reward of this behavior which one cannot naively test as there is not enough time). That is part of what the Gödel machine does: It tries to greatly cut testing time, replacing naive time-consuming tests by much faster proofs of predictable test outcomes whenever this is possible.

Proof verification itself can be performed very quickly. In particular, verifying the correctness of a found proof typically does not consume the remaining life. Hence the Gödel machine may use OOPS as a bias-optimal proof-searching submodule. Since the proofs themselves may concern quite different, *arbitrary* notions of optimality (not just bias-optimality), the Gödel machine is more general than plain OOPS. But it is not just an extension of OOPS. Instead of OOPS it may as well use non-bias-optimal alternative methods to initialize its proof searcher. On the other hand, OOPS is not just a precursor of the Gödel machine. It is a stand-alone, incremental, bias-optimal way of allocating runtime to programs that reuse previously successful programs, and is applicable to many traditional problems, including but not limited to proof search.

3. Gödel Machine versus AIXI etc.

Unlike Gödel machines, Hutter's recent AIXI *model* [16] generally needs *unlimited* computational resources per input update. It combines Solomonoff's universal prediction scheme [52, 53] with an *expectimax* computation. In discrete cycle $k = 1, 2, 3, \ldots$, action $y(k)$ results in perception $x(k)$ and reward $r(k)$, both sampled from the unknown (reactive) environmental probability distribution μ. AIXI defines a mixture distribution ξ as a weighted sum of distributions $\nu \in \mathcal{M}$, where $\mathcal{M}$ is any class of distributions that includes the true environment μ. For example, $\mathcal{M}$ may be a sum of all computable distributions [52, 53], where the sum of the weights does not exceed 1. In cycle $k + 1$, AIXI selects as next action the first in an action sequence maximizing ξ-predicted reward up to some given horizon. Recent work [18] demonstrated AIXI 's optimal use of observations as follows. The Bayes-optimal policy p^ξ based on the mixture ξ is self-optimizing in the sense that its average utility value converges asymptotically for all $\mu \in \mathcal{M}$ to the optimal value achieved by the (infeasible) Bayes-optimal policy p^μ which knows μ in advance. The necessary condition that $\mathcal{M}$ admits self-optimizing policies is also sufficient. Furthermore, p^ξ is Pareto-optimal in the sense that there is no other policy yielding higher or equal value in *all* environments $\nu \in \mathcal{M}$ and a strictly higher value in at least one [18].

While AIXI clarifies certain theoretical limits of machine learning, it is computationally intractable, especially when $\mathcal{M}$ includes all computable distributions. This drawback motivated work on the time-bounded, asymptotically optimal AIXI*(t,l)* system [16] and the related HSEARCH [17], both already discussed in Section 2.4, which also lists the advantages of the Gödel machine. Both methods, however, could be used to seed the Gödel machine with an *initial* policy.

It is the *self-referential* aspects of the Gödel machine that relieve us of much of the burden of careful algorithm design required for AIXI*(t,l)* and HSEARCH. They make the Gödel machine both conceptually simpler *and* more general than AIXI*(t,l)* and HSEARCH.

6.5 Are Humans Probabilistic Gödel Machines?

We do not know. We think they better be. Their initial underlying formal system for dealing with uncertainty seems to differ substantially from those of traditional expectation calculus and logic though—compare Items 1c and 1d in Sect. 3.1 as well as the supermarket example in Sect. 6.1.

6.6 Gödel Machines and Consciousness

In recent years the topic of consciousness has gained some credibility as a serious research issue, at least in philosophy and neuroscience, e.g., [9]. However, there is a lack of *technical* justifications of consciousness: so far nobody has shown that consciousness is really useful for solving problems, although problem solving is considered of central importance in philosophy [30].

The fully self-referential Gödel machine may be viewed as providing just such a technical justification. It is "conscious" or "self-aware" in the sense that its entire behavior is open to self-introspection, and modifiable. It may "step outside of itself" [14] by executing self-changes that are provably good, where the proof searcher itself is subject to analysis and change through the proof techniques it tests. And this type of total self-reference is precisely the reason for its optimality as a problem solver in the sense of Theorem 1.

6.7 Frequently Asked Questions

In the past half year the author frequently fielded questions about the Gödel machine. Here a list of answers to typical questions.

1. **Q:** *Does the exact business of formal proof search really make sense in the uncertain real world?*
   A: Yes, it does. We just need to insert into $p(1)$ the standard axioms for representing uncertainty and for dealing with probabilistic settings and expected rewards etc. Compare items 1d and 1c in Section 3.1, and the definition of utility as an *expected* value in equation (1).

2. **Q:** *The target theorem (6) seems to refer only to the very first self-change, which may completely rewrite the proof-search subroutine—doesn't this make the proof of Theorem 1 invalid? What prevents later self-changes from being destructive?*
   A: This is fully taken care of. Please look once more at the proof of Theorem 1, and note that the first self-change will be executed only if it is provably useful (in the sense of the present untility function u) for all future self-changes (for which the present self-change is setting the stage). This is actually the main point of the whole Gödel machine set-up.

3. **Q** (related to the previous item): *The Gödel machine implements a meta-learning behavior: what about a meta-meta, and a meta-meta-meta level?*
   A: The beautiful thing is that all meta-levels are automatically collapsed into one: any proof of a target theorem automatically proves that the corresponding self-modification is good for all further self-modifications affected by the present one, in recursive fashion.

4. **Q:** *The Gödel machine software can produce only computable mappings from input sequences to output sequences. What if the environment is non-computable?*
   A: Many physicists and other scientists (exceptions: [58, 37]) actually do assume the real world makes use of all the real numbers, most of which

are incomputable. Nevertheless, theorems and proofs are just finite symbol strings, and all treatises of physics contain only computable axioms and theorems, even when some of the theorems can be interpreted as making statements about uncountably many objects, such as all the real numbers. (Note though that the Löwenheim-Skolem Theorem [27, 51] implies that any first order theory with an uncountable model such as the real numbers also has a countable model.) Generally speaking, formal descriptions of non-computable objects do *not at all* present a fundamental problem— they may still allow for finding a strategy that provably maximizes utility. If so, a Gödel machine can exploit this. If not, then humans will not have a fundamental advantage over Gödel machines.

5. **Q:** *Isn't automated theorem-proving very hard? Current AI systems cannot prove nontrivial theorems without human intervention at crucial decision points.*

   A: More and more important mathematical proofs (four color theorem, etc.) heavily depend on automated proof search. And traditional theorem provers do not even make use of our novel notions of proof techniques and $O()$-optimal proof search. Of course, some proofs are indeed hard to find, but here humans and Gödel machines face the same fundamental limitations.

6. **Q:** *Don't the "no free lunch theorems" [57] say that it is impossible to construct universal problem solvers?*

   A: No, they do not. They refer to the very special case of problems sampled from *i.i.d.* uniform distributions on *finite* problem spaces. See the discussion of no free lunch theorems in an earlier paper [45].

7. **Q:** *Can't the Gödel machine switch to a program switchprog that rewrites the utility function to a "bogus" utility function that makes unfounded promises of big rewards in the near future?*

   A: No, it cannot. It should be obvious that rewrites of the utility function can happen only if the Gödel machine first can prove that the rewrite is useful according to the *present* utility function.

7 Conclusion

The initial software $p(1)$ of our machine runs an initial problem solver, e.g., one of Hutter's approaches [17, 16] which have at least an optimal *order* of complexity. Simultaneously, it runs an $O()$-optimal initial proof searcher using an online variant of Universal Search to test *proof techniques*, which are programs able to compute proofs concerning the system's own future performance, based on an axiomatic system $\mathcal{A}$ encoded in $p(1)$, describing a formal *utility* function u, the hardware and $p(1)$ itself. If there is no provably good, globally optimal way of rewriting $p(1)$ at all, then humans will not find one either. But if there is one, then $p(1)$ itself can find and exploit it. This approach yields the first class of theoretically sound, fully self-referential, optimally efficient, general problem solvers.

After the theoretical discussion in Sects. 1 through 5, one practical question remains: to build a particular, especially practical Gödel machine with small initial constant overhead, which generally useful theorems should one add as axioms to $\mathcal{A}$ (as initial bias) such that the initial searcher does not have to prove them from scratch?

8 Acknowledgments

Thanks to Alexey Chernov, Marcus Hutter, Jan Poland, Sepp Hochreiter, Ray Solomonoff, Leonid Levin, Shane Legg, Alex Graves, Matteo Gagliolo, Viktor Zhumatiy, Ben Goertzel, Will Pearson, Faustino Gomez, and many others for useful comments on drafts or summaries or earlier versions of this paper.

References

1. Banzhaf W, Nordin P, Keller RE, Francone FD (1998) *Genetic Programming – An Introduction.* Morgan Kaufmann Publishers, San Francisco, CA.
2. Bellman R (1961) *Adaptive Control Processes.* Princeton University Press, Princeton, NJ.
3. Blum M (1967) A machine-independent theory of the complexity of recursive functions. *Journal of the ACM*, 14(2):322–336.
4. Blum M On effective procedures for speeding up algorithms. *Journal of the ACM*, 18(2):290–305.
5. Cantor G Über eine Eigenschaft des Inbegriffes aller reellen algebraischen Zahlen. *Crelle's Journal für Mathematik*, 77:258–263.
6. Chaitin GJ (1975) A theory of program size formally identical to information theory. *Journal of the ACM*, 22:329–340.
7. Clocksin WF,Mellish CS (1987) *Programming in Prolog.* Springer, Berlin, 3rd edition.
8. Cramer NL (1985) A representation for the adaptive generation of simple sequential programs. In Grefenstette JJ (ed) *Proceedings of an International Conference on Genetic Algorithms and Their Applications, Carnegie-Mellon University, July 24-26, 1985*, Lawrence Erlbaum, Hillsdale, NJ.
9. Crick F, Koch C (1998) Consciousness and neuroscience. *Cerebral Cortex*, 8:97–107.
10. Fitting MC (1996) *First-Order Logic and Automated Theorem Proving.* Graduate Texts in Computer Science. Springer, Berlin, 2nd edition.
11. Gödel K (1931) Über formal unentscheidbare Sätze der Principia Mathematica und verwandter Systeme I. *Monatshefte für Mathematik und Physik*, 38:173–198.
12. Heisenberg W (1925) Über den anschaulichen Inhalt der quantentheoretischen Kinematik und Mechanik. *Zeitschrift für Physik*, 33:879–893.

13. Hochreiter S, Younger AS, Conwell PR (2001) Learning to learn using gradient descent. In *Proc. Intl. Conf. on Artificial Neural Networks (ICANN-2001)*, volume 2130 of *LLCS* Springer, Berlin, Heidelberg.

14. Hofstadter D (!979) *Gödel, Escher, Bach: an Eternal Golden Braid.* Basic Books, New York.

15. Holland JH (1975) Properties of the bucket brigade. In *Proceedings of an International Conference on Genetic Algorithms.* Lawrence Erlbaum, Hillsdale, NJ.

16. Hutter M (2001) Towards a universal theory of artificial intelligence based on algorithmic probability and sequential decisions. *Proceedings of the 12th European Conference on Machine Learning (ECML-2001).*

17. Hutter M (2002) The fastest and shortest algorithm for all well-defined problems. *International Journal of Foundations of Computer Science,* 13(3):431–443.

18. Hutter M (2002) Self-optimizing and Pareto-optimal policies in general environments based on Bayes-mixtures. In *Proc. 15th Annual Conf. on Computational Learning Theory (COLT 2002),* volume 2375 of *LNAI,* Springer, Berlin.

19. Kaelbling LP, Littman ML, Moore AW Reinforcement learning: a survey. *Journal of AI research,* 4:237–285.

20. Kolmogorov AN (1933) *Grundbegriffe der Wahrscheinlichkeitsrechnung.* Springer, Berlin, 1933.

21. Kolmogorov AN (1965) Three approaches to the quantitative definition of information. *Problems of Information Transmission,* 1:1–11.

22. Lenat D (1983) Theory formation by heuristic search. *Machine Learning,* 21.

23. Levin LA (1973) Universal sequential search problems. *Problems of Information Transmission,* 9(3):265–266.

24. Levin LA (1974) Laws of information (nongrowth) and aspects of the foundation of probability theory. *Problems of Information Transmission,* 10(3):206–210.

25. Levin LA (1984) Randomness conservation inequalities: Information and independence in mathematical theories. *Information and Control,* 61:15–37.

26. Li M,Vitányi PMB (!997) *An Introduction to Kolmogorov Complexity and its Applications.* Springer, Berlin, 2nd edition.

27. Löwenheim L (1915) Über Möglichkeiten im Relativkalkül. *Mathematische Annalen,* 76:447–470.

28. Moore CH, Leach GC (1970) FORTH - a language for interactive computing, 1970. http://www.ultratechnology.com.

29. Penrose R (1994) *Shadows of the mind.* Oxford University Press, Oxford.

30. Popper KR (1999) *All Life Is Problem Solving.* Routledge, London.

31. Samuel AL (1959) Some studies in machine learning using the game of checkers. *IBM Journal on Research and Development,* 3:210–229.

32. Schmidhuber J (1987) Evolutionary principles in self-referential learning. Diploma thesis, Institut für Informatik, Technische Universität München.

33. Schmidhuber J (1991) Reinforcement learning in Markovian and non-Markovian environments. In Lippman DS, Moody JE, Touretzky DS (eds) *Advances in Neural Information Processing Systems 3,* Morgan Kaufmann, Los Altos, CA.

34. Schmidhuber J A self-referential weight matrix. In *Proceedings of the International Conference on Artificial Neural Networks, Amsterdam,* Springer, Berlin.

35. Schmidhuber J (1994) On learning how to learn learning strategies. Technical Report FKI-198-94, Fakultät für Informatik, Technische Universität München, 1994. See [50, 48].

36. Schmidhuber J (1995) Discovering solutions with low Kolmogorov complexity and high generalization capability. In Prieditis A and Russell S (eds) *Machine Learning: Proceedings of the Twelfth International Conference*. Morgan Kaufmann, San Francisco, CA.

37. Schmidhuber J (1997) A computer scientist's view of life, the universe, and everything. In Freksa C, Jantzen M, Valk R (eds) *Foundations of Computer Science: Potential - Theory - Cognition*, volume 1337 of *LLNCS*, Springer, Berlin.

38. Schmidhuber J (1997) Discovering neural nets with low Kolmogorov complexity and high generalization capability. *Neural Networks*, 10(5):857–873.

39. Schmidhuber J (2000) Algorithmic theories of everything. Technical Report IDSIA-20-00, quant-ph/0011122, IDSIA. Sections 1-5: see [40]; Section 6: see [41].

40. Schmidhuber J (2002) Hierarchies of generalized Kolmogorov complexities and nonenumerable universal measures computable in the limit. *International Journal of Foundations of Computer Science*, 13(4):587–612.

41. Schmidhuber J (2002) The Speed Prior: a new simplicity measure yielding near-optimal computable predictions. In Kivinen J, Sloan RH (eds) *Proceedings of the 15th Annual Conference on Computational Learning Theory (COLT 2002)*, Lecture Notes in Artificial Intelligence, Springer, Berlin.

42. Schmidhuber J (2003) Bias-optimal incremental problem solving. In Becker S, Thrun S, Obermayer K (eds) *Advances in Neural Information Processing Systems 15*, MIT Press, Cambridge, MA.

43. Schmidhuber J (2003) Gödel machines: self-referential universal problem solvers making provably optimal self-improvements. Technical Report IDSIA-19-03, arXiv:cs.LO/0309048 v2, IDSIA.

44. J. Schmidhuber. The new AI: General & sound & relevant for physics. In this volume.

45. Schmidhuber J (2004) Optimal ordered problem solver. *Machine Learning*, 54:211–254.

46. Schmidhuber J (2005) Gödel machines: Towards a Technical Justification of Consciousness. In Kudenko D, Kazakov D, Alonso E (eds) *Adaptive Agents and Multi-Agent Systems III*, LNCS 3394, Springer, Berlin.

47. Schmidhuber J (2005) Completely Self-Referential Optimal Reinforcement Learners. In Duch W et al (eds) *Proc. Intl. Conf. on Artificial Neural Networks ICANN'05*, LNCS 3697, Springer, Berlin, Heidelberg.

48. Schmidhuber J, Zhao J, Schraudolph N (1997) Reinforcement learning with self-modifying policies. In Thrun S, Pratt L (eds) *Learning to learn*, Kluwer, Norwell, MA.

49. Schmidhuber J, Zhao J, Wiering M (1996) Simple principles of metalearning. Technical Report IDSIA-69-96, IDSIA. See [50, 48].

50. Schmidhuber J, Zhao J, Wiering M (1997) Shifting inductive bias with success-story algorithm, adaptive Levin search, and incremental self-improvement. *Machine Learning*, 28:105–130.

51. Skolem T (1919) Logisch-kombinatorische Untersuchungen über Erfüllbarkeit oder Beweisbarkeit mathematischer Sätze nebst einem Theorem über dichte Mengen. *Skrifter utgit av Videnskapsselskapet in Kristiania, I, Mat.-Nat. Kl.*, N4:1–36.

52. Solomonoff R (1964) A formal theory of inductive inference. Part I. *Information and Control*, 7:1–22.
53. Solomonoff R (1978) Complexity-based induction systems. *IEEE Transactions on Information Theory*, IT-24(5):422–432.
54. Solomonoff R (2003) Progress in incremental machine learning—Preliminary Report for NIPS 2002 Workshop on Universal Learners and Optimal Search; revised Sept 2003. Technical Report IDSIA-16-03, IDSIA.
55. Sutton R, Barto A (1998) *Reinforcement Learning: An Introduction*. MIT Press, Cambridge, MA.
56. Turing A (1936) On computable numbers, with an application to the Entscheidungsproblem. *Proceedings of the London Mathematical Society, Series 2*, 41:230–267.
57. Wolpert DH, Macready DG (1997) No free lunch theorems for search. *IEEE Transactions on Evolutionary Computation*, 1.
58. Zuse K (1969) *Rechnender Raum*. Friedrich Vieweg & Sohn, Braunschweig. English translation: *Calculating Space*, MIT Technical Translation AZT-70-164-GEMIT, MIT (Proj. MAC), Cambridge, MA.

Universal Algorithmic Intelligence: A Mathematical Top→Down Approach*

Marcus Hutter

IDSIA, Galleria 2, CH-6928 Manno-Lugano, Switzerland
RSISE/ANU/NICTA, Canberra, ACT, 0200, Australia
marcus@hutter1.net - http://www.hutter1.net

Summary. Sequential decision theory formally solves the problem of rational agents in uncertain worlds if the true environmental prior probability distribution is known. Solomonoff's theory of universal induction formally solves the problem of sequence prediction for unknown prior distribution. We combine both ideas and get a parameter-free theory of universal Artificial Intelligence. We give strong arguments that the resulting AIXI model is the most intelligent unbiased agent possible. We outline how the AIXI model can formally solve a number of problem classes, including sequence prediction, strategic games, function minimization, reinforcement and supervised learning. The major drawback of the AIXI model is that it is uncomputable. To overcome this problem, we construct a modified algorithm AIXI*tl* that is still effectively more intelligent than any other time t and length l bounded agent. The computation time of AIXI*tl* is of the order $t \cdot 2^l$. The discussion includes formal definitions of intelligence order relations, the horizon problem and relations of the AIXI theory to other AI approaches.

1 Introduction

This chapter gives an introduction to a mathematical theory for intelligence. We present the AIXI model, a parameter-free optimal reinforcement learning agent embedded in an arbitrary unknown environment.

The science of Artificial Intelligence (AI) may be defined as the construction of intelligent systems and their analysis. A natural definition of a *system* is anything that has an input and an output stream. Intelligence is more complicated. It can have many faces like creativity, solving problems, pattern recognition, classification, learning, induction, deduction, building analogies, optimization, surviving in an environment, language processing, knowledge and many more. A formal definition incorporating every aspect of intelligence, however, seems difficult. Most, if not all known facets of intelligence can be formulated as goal-driven or, more precisely, as maximizing some utility function. It is, therefore, sufficient to study goal-driven AI; e.g. the (biological) goal of animals and humans is to survive and spread. The goal of AI systems should be to be useful to humans. The problem is that, except for special cases,

---

*This article grew out of the technical report [19] and summarizes and contains excerpts of the Springer book [30].

we know neither the utility function nor the environment in which the agent will operate in advance. The mathematical theory, coined AIXI, is supposed to solve these problems.

Assume the availability of unlimited computational resources. The first important observation is that this does not make the AI problem trivial. Playing chess optimally or solving NP-complete problems become trivial, but driving a car or surviving in nature don't. This is because it is a challenge itself to well-define the latter problems, not to mention presenting an algorithm. In other words, the AI problem has not yet been well defined. One may view AIXI as a suggestion for such a mathematical definition of AI.

AIXI is a universal theory of sequential decision making akin to Solomonoff's celebrated universal theory of induction. Solomonoff derived an optimal way of predicting future data, given previous perceptions, provided the data is sampled from a computable probability distribution. AIXI extends this approach to an optimal decision making agent embedded in an unknown environment. The *main idea* is to replace the unknown environmental distribution μ in the Bellman equations by a suitably generalized universal Solomonoff distribution ξ. The state space is the space of complete histories. AIXI is a universal theory without adjustable parameters, making no assumptions about the environment except that it is sampled from a computable distribution. From an algorithmic complexity perspective, the AIXI model generalizes optimal passive universal induction to the case of active agents. From a decision-theoretic perspective, AIXI is a suggestion of a new (implicit) "learning" algorithm, which may overcome all (except computational) problems of previous reinforcement learning algorithms.

There are strong arguments that AIXI is the most intelligent unbiased agent possible. We outline for a number of problem classes, including sequence prediction, strategic games, function minimization, reinforcement and supervised learning, how the AIXI model can formally solve them. The major drawback of the AIXI model is that it is incomputable. To overcome this problem, we construct a modified algorithm AIXItl that is still effectively more intelligent than any other time t and length l bounded agent. The computation time of AIXItl is of the order $t \cdot 2^l$. Other discussed topics are a formal definition of an intelligence order relation, the horizon problem and relations of the AIXI theory to other AI approaches.

This chapter is meant to be a gentle introduction to and discussion of the AIXI model. For a mathematically rigorous treatment, many subtleties, and proofs see the references to the author's works in the annotated bibliography section at the end of this chapter, and in particular the book [30]. This section also provides references to introductory textbooks and original publications on algorithmic information theory and sequential decision theory.

Section 2 presents the theory of sequential decisions in a very general form (called AIμ model) in which actions and perceptions may depend on arbitrary past events. We clarify the connection to the Bellman equations and discuss minor parameters including (the size of) the I/O spaces and the lifetime of the agent and their universal choice which we have in mind. Optimality of AIμ is obvious by construction.

Section 3: How and in which sense induction is possible at all has been subject to long philosophical controversies. Highlights are Epicurus' principle of multiple explanations, Occam's razor, and probability theory. Solomonoff elegantly unified all these aspects into one formal theory of inductive inference based on a universal probability distribution ξ, which is closely related to Kolmogorov complexity $K(x)$, the length of the shortest program computing x. Rapid convergence of ξ to the unknown true environmental distribution μ and tight loss bounds for arbitrary bounded loss functions and finite alphabet can be shown. Pareto optimality of ξ in the sense that there is no other predictor that performs better or equal in all environments and strictly better in at least one can also be shown. In view of these results it is fair to say that the problem of sequence prediction possesses a universally optimal solution.

Section 4: In the active case, reinforcement learning algorithms are usually used if μ is unknown. They can succeed if the state space is either small or has effectively been made small by generalization techniques. The algorithms work only in restricted (e.g. Markovian) domains, have problems with optimally trading off exploration versus exploitation, have nonoptimal learning rate, are prone to diverge, or are otherwise ad hoc. The formal solution proposed here is to generalize Solomonoff's universal prior ξ to include action conditions and replace μ by ξ in the AIμ model, resulting in the AI$\xi \equiv$AIXI model, which we claim to be universally optimal. We investigate what we can expect from a universally optimal agent and clarify the meanings of *universal, optimal,* etc. Other discussed topics are formal definitions of an intelligence order relation, the horizon problem, and Pareto optimality of AIXI.

Section 5: We show how a number of AI problem classes fit into the general AIXI model. They include sequence prediction, strategic games, function minimization, and supervised learning. We first formulate each problem class in its natural way (for known μ) and then construct a formulation within the AIμ model and show their equivalence. We then consider the consequences of replacing μ by ξ. The main goal is to understand in which sense the problems are solved by AIXI.

Section 6: The major drawback of AIXI is that it is incomputable, or more precisely, only asymptotically computable, which makes an implementation impossible. To overcome this problem, we construct a modified model AIXItl, which is still superior to any other time t and length l bounded algorithm. The computation time of AIXItl is of the order $t \cdot 2^l$. The solution requires an implementation of first-order logic, the definition of a universal Turing machine within it and a proof theory system.

Section 7: Finally, we discuss and remark on some otherwise unmentioned topics of general interest. We remark on various topics, including concurrent actions and perceptions, the choice of the I/O spaces, treatment of encrypted information, and peculiarities of mortal embodies agents. We continue with an outlook on further research, including optimality, down-scaling, implementation, approximation, elegance, extra knowledge, and training of/for AIXI(tl). We also include some (personal) remarks on non-computable physics, the number of wisdom Ω, and consciousness.

An annotated bibliography concludes this chapter.

2 Agents in Known Probabilistic Environments

The general framework for AI might be viewed as the design and study of intelligent agents [53]. An agent is a cybernetic system with some internal state, which acts with output y_k on some environment in cycle k, perceives some input x_k from the environment and updates its internal state. Then the next cycle follows. We split the input x_k into a regular part o_k and a reward r_k, often called reinforcement feedback. From time to time the environment provides nonzero reward to the agent. The task of the agent is to maximize its utility, defined as the sum of future rewards. A probabilistic environment can be described by the conditional probability μ for the inputs $x_1...x_n$ to the agent under the condition that the agent outputs $y_1...y_n$. Most, if not all environments are of this type. We give formal expressions for the outputs of the agent, which maximize the total μ-expected reward sum, called value. This model is called the AIμ model. As every AI problem can be brought into this form, the problem of maximizing utility is hence being formally solved if μ is known. Furthermore, we study some special aspects of the AIμ model. We introduce factorizable probability distributions describing environments with independent episodes. They occur in several problem classes studied in Sect. 5 and are a special case of more general separable probability distributions defined in Sect. 4.3. We also clarify the connection to the Bellman equations of sequential decision theory and discuss similarities and differences. We discuss minor parameters of our model, including (the size of) the input and output spaces $\mathcal{X}$ and $\mathcal{Y}$ and the lifetime of the agent, and their universal choice, which we have in mind. There is nothing remarkable in this section; it is the essence of sequential decision theory [47, 2, 3, 66], presented in a new form. Notation and formulas needed in later sections are simply developed. There are two major remaining problems: the problem of the unknown true probability distribution μ, which is solved in Sect. 4, and computational aspects, which are addressed in Sect. 6.

2.1 The Cybernetic Agent Model

A good way to start thinking about intelligent systems is to consider more generally cybernetic systems, usually called agents in AI. This avoids struggling with the meaning of intelligence from the very beginning. A cybernetic system is a control circuit with input y and output x and an internal state. From an external input and the internal state the agent calculates deterministically or stochastically an output. This output (action) modifies the environment and leads to a new input (perception). This continues ad infinitum or for a finite number of cycles.

Definition 1 (The Agent Model). *An agent is a system that interacts with an environment in cycles* $k = 1,2,3,....$ *In cycle* k *the action (output)* $y_k \in \mathcal{Y}$ *of the agent is determined by a policy* p *that depends on the I/O-history* $y_1x_1...y_{k-1}x_{k-1}$. *The environment reacts to this action and leads to a new perception (input)* $x_k \in \mathcal{X}$ *determined by a deterministic function* q *or probability distribution* μ, *which depends on the history* $y_1x_1...y_{k-1}x_{k-1}y_k$. *Then the next cycle* $k+1$ *starts.*

As explained in the last section, we need some reward assignment to the cybernetic system. The input x is divided into two parts, the standard input o and some reward input r. If input and output are represented by strings, a deterministic cybernetic system can be modeled by a Turing machine p, where p is called the policy of the agent, which determines the (re)action to a perception. If the environment is also computable it might be modeled by a Turing machine q as well. The interaction of the agent with the environment can be illustrated as follows:

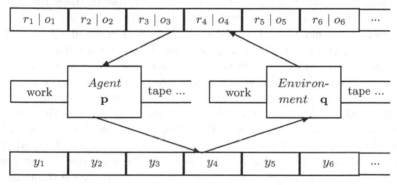

Both p as well as q have unidirectional input and output tapes and bidirectional work tapes. What entangles the agent with the environment is the fact that the upper tape serves as input tape for p, as well as output tape for q, and that the lower tape serves as output tape for p as well as input tape for q. Further, the reading head must always be left of the writing head, i.e. the symbols must first be written before they are read. Both p and q have their own mutually inaccessible work tapes containing their own "secrets". The heads move in the following way. In the kth cycle p writes y_k, q reads y_k, q writes $x_k \equiv r_k o_k$, p reads $x_k \equiv r_k o_k$, followed by the $(k+1)^{th}$ cycle and so on. The whole process starts with the first cycle, all heads on tape start and work tapes being empty. We call Turing machines behaving in this way *chronological Turing machines*. Before continuing, some notations on strings are appropriate.

2.2 Strings

We denote strings over the alphabet $\mathcal{X}$ by $s = x_1 x_2 ... x_n$, with $x_k \in \mathcal{X}$, where $\mathcal{X}$ is alternatively interpreted as a nonempty subset of $I\!N$ or itself as a prefix-free set of binary strings. The length of s is $l(s) = l(x_1) + ... + l(x_n)$. Analogous definitions hold for $y_k \in \mathcal{Y}$. We call x_k the kth input word and y_k the kth output word (rather than letter). The string $s = y_1 x_1 ... y_n x_n$ represents the input/output in chronological order. Due to the prefix property of the x_k and y_k, s can be uniquely separated into its words. The words appearing in strings are always in chronological order. We further introduce the following abbreviations: ϵ is the empty string, $x_{n:m} := x_n x_{n+1} ... x_{m-1} x_m$ for $n \leq m$ and ϵ for $n > m$. $x_{<n} := x_1 ... x_{n-1}$. Analogously for y. Further, $yx_n := y_n x_n$, $yx_{n:m} := y_n x_n ... y_m x_m$, and so on.

2.3 AI Model for Known Deterministic Environment

Let us define for the chronological Turing machine p a partial function also named $p : \mathcal{X}^* \to \mathcal{Y}^*$ with $y_{1:k} = p(x_{<k})$, where $y_{1:k}$ is the output of Turing machine p on input $x_{<k}$ in cycle k, i.e. where p has read up to x_{k-1} but no further.[1] In an analogous way, we define $q : \mathcal{Y}^* \to \mathcal{X}^*$ with $x_{1:k} = q(y_{1:k})$. Conversely, for every partial recursive chronological function we can define a corresponding chronological Turing machine. Each (agent,environment) pair (p,q) produces a unique I/O sequence $\omega^{pq} := y_1^{pq} x_1^{pq} y_2^{pq} x_2^{pq} ....$ When we look at the definitions of p and q we see a nice symmetry between the cybernetic system and the environment. Until now, not much intelligence is in our agent. Now the credit assignment comes into the game and removes the symmetry somewhat. We split the input $x_k \in \mathcal{X} := \mathcal{R} \times \mathcal{O}$ into a regular part $o_k \in \mathcal{O}$ and a reward $r_k \in \mathcal{R} \subset I\!R$. We define $x_k \equiv r_k o_k$ and $r_k \equiv r(x_k)$. The goal of the agent should be to maximize received rewards. This is called reinforcement learning. The reason for the asymmetry is that eventually we (humans) will be the environment with which the agent will communicate and *we* want to dictate what is good and what is wrong, not the other way round. This one-way learning, the agent learns from the environment, and not conversely, neither prevents the agent from becoming more intelligent than the environment, nor does it prevent the environment learning from the agent because the environment can itself interpret the outputs y_k as a regular and a reward part. The environment is just not forced to learn, whereas the agent is. In cases where we restrict the reward to two values $r \in \mathcal{R} = I\!B := \{0,1\}$, $r = 1$ is interpreted as a positive feedback, called *good* or *correct*, and $r = 0$ a negative feedback, called *bad* or *error*. Further, let us restrict for a while the lifetime (number of cycles) m of the agent to a large but finite value. Let

---

[1] Note that a possible additional dependence of p on $y_{<k}$ as mentioned in Definition 1 can be eliminated by recursive substitution; see below. Similarly for q.

$$V_{km}^{pq} := \sum_{i=k}^{m} r(x_i^{pq})$$

be the future total reward (called future utility), the agent p receives from the environment q in the cycles k to m. It is now natural to call the agent p^* that maximizes V_{1m} (called total utility), the *best* one:[2]

$$p^* := \arg\max_p V_{1m}^{pq} \quad \Rightarrow \quad V_{km}^{p^*q} \geq V_{km}^{pq} \quad \forall p : y_{<k}^{pq} = y_{<k}^{p^*q} \tag{1}$$

For $k=1$ the condition on p is nil. For $k>1$ it states that p shall be consistent with p^* in the sense that they have the same history. If $\mathcal{X}$, $\mathcal{Y}$ and m are finite, the number of different behaviors of the agent, i.e. the search space is finite. Therefore, because we have assumed that q is known, p^* can effectively be determined by pre-analyzing all behaviors. The main reason for restricting to finite m was not to ensure computability of p^* but that the limit $m \to \infty$ might not exist. The ease with which we defined and computed the optimal policy p^* is not remarkable. Instead, the (unrealistic) assumption of a completely known deterministic environment q has trivialized everything.

2.4 AI Model for Known Prior Probability

Let us now weaken our assumptions by replacing the deterministic environment q with a probability distribution $\mu(q)$ over chronological functions. Here μ might be interpreted in two ways. Either the environment itself behaves stochastically defined by μ or the true environment is deterministic, but we only have subjective (probabilistic) information of which environment is the true environment. Combinations of both cases are also possible. We assume here that μ is known and describes the true stochastic behavior of the environment. The case of unknown μ with the agent having some beliefs about the environment lies at the heart of the AIξ model described in Section 4.

The *best* or *most intelligent* agent is now the one that maximizes the *expected* utility (called value function) $V_\mu^p \equiv V_{1m}^{p\mu} := \sum_q \mu(q) V_{1m}^{pq}$. This defines the AI$\mu$ model.

Definition 2 (The AIμ model). *The AIμ model is the agent with policy p^μ that maximizes the μ-expected total reward $r_1+...+r_m$, i.e. $p^* \equiv p^\mu :=$ $\mathrm{argmax}_p V_\mu^p$. Its value is $V_\mu^* := V_\mu^{p^\mu}$.*

We need the concept of a *value function* in a slightly more general form.

Definition 3 (The μ/true/generating value function). *The agent's perception x consists of a regular observation $o \in \mathcal{O}$ and a reward $r \in \mathcal{R} \subset I\!R$. In cycle k the value $V_{km}^{p\mu}(yx_{<k})$ is defined as the μ-expectation of the future*

---

[2]$\mathrm{argmax}_p V(p)$ is the p that maximizes $V(\cdot)$. If there is more than one maximum we might choose the lexicographically smallest one for definiteness.

reward sum $r_k + ... + r_m$ with actions generated by policy p, and fixed history $yx_{<k}$. We say that $V_{km}^{p\mu}(yx_{<k})$ is the (future) value of policy p in environment μ given history $yx_{<k}$, or shorter, the μ or true or generating value of p given $yx_{<k}$. $V_\mu^p := V_{1m}^{p\mu}$ is the (total) value of p.

We now give a more formal definition for $V_{km}^{p\mu}$. Let us assume we are in cycle k with history $\dot{y}\dot{x}_1...\dot{y}\dot{x}_{k-1}$ and ask for the *best* output y_k. Further, let $\dot{Q}_k := \{q : q(\dot{y}_{<k}) = \dot{x}_{<k}\}$ be the set of all environments producing the above history. We say that $q \in \dot{Q}_k$ is *consistent* with history $\dot{y}\dot{x}_{<k}$. The expected reward for the next $m-k+1$ cycles (given the above history) is called the value of policy p and is given by a conditional probability:

$$V_{km}^{p\mu}(\dot{y}\dot{x}_{<k}) := \frac{\sum_{q \in \dot{Q}_k} \mu(q) V_{km}^{pq}}{\sum_{q \in \dot{Q}_k} \mu(q)}. \tag{2}$$

Policy p and environment μ do not determine history $\dot{y}\dot{x}_{<k}$, unlike the deterministic case, because the history is no longer deterministically determined by p and q, but depends on p and μ *and* on the outcome of a stochastic process. Every new cycle adds new information ($\dot{x}_i$) to the agent. This is indicated by the dots over the symbols. In cycle k we have to maximize the expected future rewards, taking into account the information in the history $\dot{y}\dot{x}_{<k}$. This information is not already present in p and q/μ at the agent's start, unlike in the deterministic case.

Furthermore, we want to generalize the finite lifetime m to a dynamic (computable) farsightedness $h_k \equiv m_k - k + 1 \geq 1$, called horizon. For $m_k = m$ we have our original finite lifetime; for $h_k = h$ the agent maximizes in every cycle the next h expected rewards. A discussion of the choices for m_k is delayed to Sect. 4.5. The next h_k rewards are maximized by

$$p_k^* := \arg\max_{p \in \dot{P}_k} V_{km_k}^{p\mu}(\dot{y}\dot{x}_{<k}),$$

where $\dot{P}_k := \{p : \exists y_k : p(\dot{x}_{<k}) = \dot{y}_{<k} y_k\}$ is the set of systems consistent with the current history. Note that p_k^* depends on k and is used only in step k to determine $\dot{y}_k$ by $p_k^*(\dot{x}_{<k}|\dot{y}_{<k}) = \dot{y}_{<k}\dot{y}_k$. After writing $\dot{y}_k$ the environment replies with $\dot{x}_k$ with (conditional) probability $\mu(\dot{Q}_{k+1})/\mu(\dot{Q}_k)$. This probabilistic outcome provides new information to the agent. The cycle $k+1$ starts with determining $\dot{y}_{k+1}$ from p_{k+1}^* (which can differ from p_k^* for dynamic m_k) and so on. Note that p_k^* implicitly also depends on $\dot{y}_{<k}$ because $\dot{P}_k$ and $\dot{Q}_k$ do so. But recursively inserting p_{k-1}^* and so on, we can define

$$p^*(\dot{x}_{<k}) := p_k^*(\dot{x}_{<k}|p_{k-1}^*(\dot{x}_{<k-1}|...p_1^*)). \tag{3}$$

It is a chronological function and computable if $\mathcal{X}$, $\mathcal{Y}$ and m_k are finite and μ is computable. For constant m one can show that the policy (3) coincides with the AIμ model (Definition 2). This also proves

$$V_{km}^{*\mu}(yx_{<k}) \geq V_{km}^{p\mu}(yx_{<k}) \quad \forall p \text{ consistent with } yx_{<k}, \tag{4}$$

similarly to (1). For $k=1$ this is obvious. We also call (3) AIμ model. For deterministic[3] μ this model reduces to the deterministic case discussed in the last subsection.

It is important to maximize the sum of future rewards and not, for instance, to be greedy and only maximize the next reward, as is done e.g. in sequence prediction. For example, let the environment be a sequence of chess games, and each cycle corresponds to one move. Only at the end of each game is a positive reward $r=1$ given to the agent if it won the game (and made no illegal move). For the agent, maximizing all future rewards means trying to win as many games in as short as possible time (and avoiding illegal moves). The same performance is reached if we choose h_k much larger than the typical game lengths. Maximization of only the next reward would be a very bad chess playing agent. Even if we would make our reward r finer, e.g. by evaluating the number of chessmen, the agent would play very bad chess for $h_k=1$, indeed.

The AIμ model still depends on μ and m_k; m_k is addressed in Section 4.5. To get our final universal AI model the idea is to replace μ by the universal probability ξ, defined later. This is motivated by the fact that ξ converges to μ in a certain sense for any μ. With ξ instead of μ our model no longer depends on any parameters, so it is truly universal. It remains to show that it behaves intelligently. But let us continue step by step. In the following we develop an alternative but equivalent formulation of the AIμ model. Whereas the functional form presented above is more suitable for theoretical considerations, especially for the development of a time-bounded version in Sect. 6, the iterative and recursive formulation of the next subsections will be more appropriate for the explicit calculations in most of the other sections.

2.5 Probability Distributions

We use Greek letters for probability distributions, and underline their arguments to indicate that they are probability arguments. Let $\rho_n(\underline{x}_1...\underline{x}_n)$ be the probability that an (infinite) string starts with $x_1...x_n$. We drop the index on ρ if it is clear from its arguments:

$$\sum_{x_n \in \mathcal{X}} \rho(\underline{x}_{1:n}) \equiv \sum_{x_n} \rho_n(\underline{x}_{1:n}) = \rho_{n-1}(\underline{x}_{<n}) \equiv \rho(\underline{x}_{<n}), \quad \rho(\epsilon) \equiv \rho_0(\epsilon) = 1. \tag{5}$$

We also need conditional probabilities derived from the chain rule. We prefer a notation that preserves the chronological order of the words, in contrast to the standard notation $\rho(\cdot|\cdot)$ that flips it. We extend the definition of ρ to the conditional case with the following convention for its arguments: An underlined argument $\underline{x}_k$ is a probability variable, and other non-underlined arguments x_k represent conditions. With this convention, the conditional probability has

---

[3]We call a probability distribution deterministic if it assumes values 0 and 1 only.

the form $\rho(x_{<n}\underline{x}_n)=\rho(\underline{x}_{1:n})/\rho(\underline{x}_{<n})$. The equation states that the probability that a string $x_1...x_{n-1}$ is followed by x_n is equal to the probability of $x_1...x_n*$ divided by the probability of $x_1...x_{n-1}*$. We use $x*$ as an abbreviation for 'strings starting with x'.

The introduced notation is also suitable for defining the conditional probability $\rho(y_1\underline{x}_1...y_n\underline{x}_n)$ that the environment reacts with $x_1...x_n$ under the condition that the output of the agent is $y_1...y_n$. The environment is chronological, i.e. input x_i depends on $yx_{<i}y_i$ only. In the probabilistic case this means that $\rho(y\underline{x}_{<k}y_k):=\sum_{x_k}\rho(y\underline{x}_{1:k})$ is independent of y_k, hence a tailing y_k in the arguments of ρ can be dropped. Probability distributions with this property will be called *chronological*. The y are always conditions, i.e. are never underlined, whereas additional conditioning for the x can be obtained with the chain rule:

$$\rho(yx_{<n}y\underline{x}_n) = \rho(y\underline{x}_{1:n})/\rho(y\underline{x}_{<n}) \quad \text{and} \tag{6}$$
$$\rho(y\underline{x}_{1:n}) = \rho(y\underline{x}_1) \cdot \rho(yx_1y\underline{x}_2) \cdot ... \cdot \rho(yx_{<n}y\underline{x}_n).$$

The second equation is the first equation applied n times.

2.6 Explicit Form of the AIμ Model

Let us define the AIμ model p^* in a different way: Let $\mu(yx_{<k}y\underline{x}_k)$ be the true probability of input x_k in cycle k, given the history $yx_{<k}y_k$; $\mu(y\underline{x}_{1:k})$ is the true chronological prior probability that the environment reacts with $x_{1:k}$ if provided with actions $y_{1:k}$ from the agent. We assume the cybernetic model depicted on page 231 to be valid. Next we define the value $V_{k+1,m}^{*\mu}(yx_{1:k})$ to be the μ-expected reward sum $r_{k+1}+...+r_m$ in cycles $k+1$ to m with outputs y_i generated by agent p^* that maximizes the expected reward sum, and responses x_i from the environment, drawn according to μ. Adding $r(x_k)\equiv r_k$ we get the reward including cycle k. The probability of x_k, given $yx_{<k}y_k$, is given by the conditional probability $\mu(yx_{<k}y\underline{x}_k)$. So the expected reward sum in cycles k to m given $yx_{<k}y_k$ is

$$V_{km}^{*\mu}(yx_{<k}y_k) := \sum_{x_k}[r(x_k) + V_{k+1,m}^{*\mu}(yx_{1:k})] \cdot \mu(yx_{<k}y\underline{x}_k). \tag{7}$$

Now we ask how p^* chooses y_k: It should choose y_k as to maximize the future rewards. So the expected reward in cycles k to m given $yx_{<k}$ and y_k chosen by p^* is $V_{km}^{*\mu}(yx_{<k}):=\max_{y_k}V_{km}^{*\mu}(yx_{<k}y_k)$ (see Figure 1).

Together with the induction start

$$V_{m+1,m}^{*\mu}(yx_{1:m}) := 0, \tag{8}$$

$V_{km}^{*\mu}$ is completely defined. We might summarize one cycle into the formula

$$V_{km}^{*\mu}(yx_{<k}) = \max_{y_k}\sum_{x_k}[r(x_k) + V_{k+1,m}^{*\mu}(yx_{1:k})] \cdot \mu(yx_{<k}y\underline{x}_k). \tag{9}$$

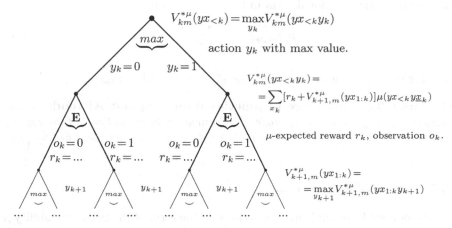

$$V_{km}^{*\mu}(yx_{<k}) = \max_{y_k} V_{km}^{*\mu}(yx_{<k}y_k)$$

action y_k with max value.

$$V_{km}^{*\mu}(yx_{<k}y_k) =$$
$$= \sum_{x_k}[r_k + V_{k+1,m}^{*\mu}(yx_{1:k})]\mu(yx_{<k}y\underline{x}_k)$$

μ-expected reward r_k, observation o_k.

$$V_{k+1,m}^{*\mu}(yx_{1:k}) =$$
$$= \max_{y_{k+1}} V_{k+1,m}^{*\mu}(yx_{1:k}y_{k+1})$$

Figure 1 (Expectimax Tree/Algorithm for $\mathcal{O} = \mathcal{Y} = I\!B$)

We introduce a dynamic (computable) farsightedness $h_k \equiv m_k - k + 1 \geq 1$, called horizon. For $m_k = m$, where m is the lifetime of the agent, we achieve optimal behavior, for limited farsightedness $h_k = h$ ($m = m_k = h + k - 1$), the agent maximizes in every cycle the next h expected rewards. A discussion of the choices for m_k is delayed to Sect. 4.5. If m_k is our horizon function of p^* and $\dot{y}\dot{x}_{<k}$ is the actual history in cycle k, the output $\dot{y}_k$ of the agent is explicitly given by

$$\dot{y}_k = \arg\max_{y_k} V_{km_k}^{*\mu}(\dot{y}\dot{x}_{<k}y_k), \tag{10}$$

which in turn defines the policy p^*. Then the environment responds $\dot{x}_k$ with probability $\mu(\dot{y}\dot{x}_{<k}\dot{y}\underline{\dot{x}}_k)$. Then cycle $k+1$ starts. We might unfold the recursion (9) further and give $\dot{y}_k$ nonrecursively as

$$\dot{y}_k \equiv \dot{y}_k^\mu := \arg\max_{y_k}\sum_{x_k}\max_{y_{k+1}}\sum_{x_{k+1}}...\max_{y_{m_k}}\sum_{x_{m_k}}(r(x_k)+...+r(x_{m_k}))\cdot\mu(\dot{y}\dot{x}_{<k}y\underline{x}_{k:m_k}).$$
$$\tag{11}$$

This has a direct interpretation: The probability of inputs $x_{k:m_k}$ in cycle k when the agent outputs $y_{k:m_k}$ with actual history $\dot{y}\dot{x}_{<k}$ is $\mu(\dot{y}\dot{x}_{<k}y\underline{x}_{k:m_k})$. The future reward in this case is $r(x_k)+...+r(x_{m_k})$. The best expected reward is obtained by averaging over the x_i ($\sum_{x_i}$) and maximizing over the y_i. This has to be done in chronological order to correctly incorporate the dependencies of x_i and y_i on the history. This is essentially the expectimax algorithm/tree [46, 53]. The AIμ model is *optimal* in the sense that no other policy leads to higher expected reward. The value for a general policy p can be written in the form

$$V_{km}^{p\mu}(yx_{<k}) := \sum_{x_{1:m}}(r_k + ... + r_m)\mu(yx_{<k}y\underline{x}_{k:m})|_{y_{1:m}=p(x_{<m})}. \tag{12}$$

As is clear from their interpretations, the iterative environmental probability μ relates to the functional form in the following way:

$$\mu(y\underline{x}_{1:k}) = \sum_{q:q(y_{1:k})=x_{1:k}} \mu(q) \tag{13}$$

With this identification one can show [19, 30] the following:

Theorem 2 (Equivalence of functional and explicit AI model). *The actions of the functional AI model (3) coincide with the actions of the explicit (recursive/iterative) AI model (9)–(11) with environments identified by (13).*

2.7 Factorizable Environments

Up to now we have made no restrictions on the form of the prior probability μ apart from being a chronological probability distribution. On the other hand, we will see that, in order to prove rigorous reward bounds, the prior probability must satisfy some separability condition to be defined later. Here we introduce a very strong form of separability, when μ factorizes into products.

Assume that the cycles are grouped into independent episodes $r = 1,2,3,...$, where each episode r consists of the cycles $k = n_r+1,...,n_{r+1}$ for some $0 = n_0 < n_1 < ... < n_s = n$:

$$\mu(y\underline{x}_{1:n}) = \prod_{r=0}^{s-1} \mu_r(y\underline{x}_{n_r+1:n_{r+1}}) \tag{14}$$

(In the simplest case, when all episodes have the same length l then $n_r = r \cdot l$). Then $\dot{y}_k$ depends on μ_r and x and y of episode r only, with r such that $n_r < k \leq n_{r+1}$. One can show that

$$\dot{y}_k = \arg\max_{y_k} V_{km_k}^{*\mu}(\dot{y}\dot{x}_{<k}y_k) = \arg\max_{y_k} V_{kt}^{*\mu}(\dot{y}\dot{x}_{<k}y_k), \tag{15}$$

with $t := \min\{m_k, n_{r+1}\}$. The different episodes are completely independent in the sense that the inputs x_k of different episodes are statistically independent and depend only on the outputs y_k of the same episode. The outputs y_k depend on the x and y of the corresponding episode r only, and are independent of the actual I/O of the other episodes.

Note that $\dot{y}_k$ is also independent of the choice of m_k, as long as m_k is sufficiently large. If all episodes have a length of at most l, i.e. $n_{r+1} - n_r \leq l$ and if we choose the horizon h_k to be at least l, then $m_k \geq k+l-1 \geq n_r+l \geq n_{r+1}$ and hence $t = n_{r+1}$ independent of m_k. This means that for factorizable μ there is no problem in taking the limit $m_k \to \infty$. Maybe this limit can also be performed in the more general case of a sufficiently separable μ. The (problem of the) choice of m_k will be discussed in more detail later.

Although factorizable μ are too restrictive to cover all AI problems, they often occur in practice in the form of repeated problem solving, and hence, are worthy of study. For example, if the agent has to play games like chess repeatedly, or has to minimize different functions, the different games/functions might be completely independent, i.e. the environmental probability factorizes, where each factor corresponds to a game/function minimization. For details, see the appropriate sections on strategic games and function minimization.

Further, for factorizable μ it is probably easier to derive suitable reward bounds for the universal AIξ model defined in the next section, than for the separable cases that will be introduced later. This could be a first step toward a definition and proof for the general case of separable problems. One goal of this paragraph was to show that the notion of a factorizable μ could be the first step toward a definition and analysis of the general case of separable μ.

2.8 Constants and Limits

We have in mind a universal agent with complex interactions that is at least as intelligent and complex as a human being. One might think of an agent whose input y_k comes from a digital video camera, and the output x_k is some image to a monitor,[4] only for the rewards we might restrict to the most primitive binary ones, i.e. $r_k \in I\!\!B$. So we think of the following constant sizes:

$$1 \ll \langle l(y_k x_k) \rangle \ll k \leq m \ll |\mathcal{Y} \times \mathcal{X}|$$
$$1 \ll \quad 2^{16} \quad \ll 2^{24} \leq 2^{32} \ll \quad 2^{65536}$$

The first two limits say that the actual number k of inputs/outputs should be reasonably large compared to the typical length $\langle l \rangle$ of the input/output words, which itself should be rather sizeable. The last limit expresses the fact that the total lifetime m (number of I/O cycles) of the agent is far too small to allow every possible input to occur, or to try every possible output, or to make use of identically repeated inputs or outputs. We do not expect any useful outputs for $k \lesssim \langle l \rangle$. More interesting than the lengths of the inputs is the complexity $K(x_1...x_k)$ of all inputs until now, to be defined later. The environment is usually not "perfect." The agent could either interact with an imperfect human or tackle a nondeterministic world (due to quantum mechanics or chaos).[5] In either case, the sequence contains some noise, leading to $K(x_1...x_k) \propto \langle l \rangle \cdot k$. The complexity of the probability distribution of the input sequence is something different. We assume that this noisy world operates according to some simple computable rules. $K(\mu_k) \ll \langle l \rangle \cdot k$, i.e. the rules of the world can be highly compressed. We may allow environments in which new aspects appear for $k \to \infty$, causing a non-bounded $K(\mu_k)$.

In the following we never use these limits, except when explicitly stated. In some simpler models and examples the size of the constants will even

---

[4]Humans can only simulate a screen as output device by drawing pictures.

[5]Whether truly stochastic processes exist at all is a difficult question. At least the quantum indeterminacy comes very close to it.

violate these limits (e.g. $l(x_k) = l(y_k) = 1$), but it is the limits above that the reader should bear in mind. We are only interested in theorems that do not degenerate under the above limits. In order to avoid cumbersome convergence and existence considerations we make the following assumptions throughout this work:

Assumption 3 (Finiteness) *We assume that:*

- *the input/perception space $\mathcal{X}$ is finite*
- *the output/action space $\mathcal{Y}$ is finite*
- *the rewards are nonnegative and bounded i.e. $r_k \in \mathcal{R} \subseteq [0, r_{max}]$,*
- *the horizon m is finite*

Finite $\mathcal{X}$ and bounded $\mathcal{R}$ (each separately) ensure existence of μ-expectations but are sometimes needed together. Finite $\mathcal{Y}$ ensures that $\mathrm{argmax}_{y_k \in \mathcal{Y}}[...]$ exists, i.e. that maxima are attained, while finite m avoids various technical and philosophical problems (Sect. 4.5), and positive rewards are needed for the time-bounded AIXItl model (Sect. 6). Many theorems can be generalized by relaxing some or all of the above finiteness assumptions.

2.9 Sequential Decision Theory

One can relate (9) to the Bellman equations [2] of sequential decision theory by identifying complete histories $yx_{<k}$ with states, $\mu(yx_{<k}y\underline{x}_k)$ with the state transition matrix, V_μ^* with the value function, and y_k with the action in cycle k [3, 53]. Due to the use of complete histories as state space, the AIμ model neither assumes stationarity, nor the Markov property, nor complete accessibility of the environment. Every state occurs at most once in the lifetime of the system. For this and other reasons the explicit formulation (11) is more natural and useful here than to enforce a pseudo-recursive Bellman equation form.

As we have in mind a universal system with complex interactions, the action and perception spaces $\mathcal{Y}$ and $\mathcal{X}$ are huge (e.g. video images), and every action or perception itself occurs usually only once in the lifespan m of the agent. As there is no (obvious) universal similarity relation on the state space, an effective reduction of its size is impossible, but there is no principle problem in determining y_k from (11) as long as μ is known and computable, and $\mathcal{X}$, $\mathcal{Y}$ and m are finite.

Things drastically change if μ is unknown. Reinforcement learning algorithms [31, 66, 3] are commonly used in this case to learn the unknown μ or directly its value. They succeed if the state space is either small or has effectively been made small by generalization or function approximation techniques. In any case, the solutions are either ad hoc, work in restricted domains only, have serious problems with state space exploration versus exploitation, or are prone to diverge, or have nonoptimal learning rates. There is no universal and optimal solution to this problem so far. The central theme of this article is to present a new model and argue that it formally solves all these problems in an optimal way. The true probability distribution μ will not be learned directly, but will be replaced by some generalized universal prior ξ, which converges to μ.

3 Universal Sequence Prediction

This section deals with the question of how to make predictions in unknown environments. Following a brief description of important philosophical attitudes regarding inductive reasoning and inference, we describe more accurately what we mean by induction, and explain why we can focus on sequence prediction tasks. The most important concept is Occam's razor (simplicity) principle. Indeed, one can show that the best way to make predictions is based on the shortest ($\hat{=}$ simplest) description of the data sequence seen so far. The most general effective descriptions can be obtained with the help of general recursive functions, or equivalently by using programs on Turing machines, especially on the universal Turing machine. The length of the shortest program describing the data is called the Kolmogorov complexity of the data. Probability theory is needed to deal with uncertainty. The environment may be a stochastic process (e.g. gambling houses or quantum physics) that can be described by "objective" probabilities. But also uncertain knowledge about the environment, which leads to beliefs about it, can be modeled by "subjective" probabilities. The old question left open by subjectivists of how to choose the a priori probabilities is solved by Solomonoff's universal prior, which is closely related to Kolmogorov complexity. Solomonoff's major result is that the universal (subjective) posterior converges to the true (objective) environment(al probability) μ. The only assumption on μ is that μ (which needs not be known!) is computable. The problem of the unknown environment μ is hence solved for all problems of inductive type, like sequence prediction and classification.

3.1 Introduction

An important and highly nontrivial aspect of intelligence is inductive inference. Simply speaking, induction is the process of predicting the future from the past, or more precisely, it is the process of finding rules in (past) data and using these rules to guess future data. Weather or stock-market forecasting, or continuing number series in an IQ test are nontrivial examples. Making good predictions plays a central role in natural and artificial intelligence in general, and in machine learning in particular. All induction problems can be phrased as sequence prediction tasks. This is, for instance, obvious for time-series prediction, but also includes classification tasks. Having observed data x_t at times $t < n$, the task is to predict the nth symbol x_n from sequence $x_1...x_{n-1}$. This *prequential approach* [13] skips over the intermediate step of learning a model based on observed data $x_1...x_{n-1}$ and then using this model to predict x_n. The prequential approach avoids problems of model consistency, how to separate noise from useful data, and many other issues. The goal is to make "good" predictions, where the prediction quality is usually measured by a loss function, which shall be minimized. The key concept to well-define and solve induction problems is *Occam's razor* (simplicity) principle, which says that " Entities should not be multiplied beyond necessity," which may be interpreted as to keep the simplest theory consistent with the observations $x_1...x_{n-1}$ and to use this theory to predict x_n. Before we can present Solomonoff's formal solution, we have to quantify Occam's razor in terms of Kolmogorov complexity, and introduce the notion of subjective/objective probabilities.

3.2 Algorithmic Information Theory

Intuitively, a string is simple if it can be described in a few words, like "the string of one million ones," and is complex if there is no such short description, like for a random string whose shortest description is specifying it bit by bit. We can restrict the discussion to binary strings, since for other (non-stringy mathematical) objects we may assume some default coding as binary strings. Furthermore, we are only interested in effective descriptions, and hence restrict decoders to be Turing machines. Let us choose some universal (so-called prefix) *Turing machine U* with unidirectional binary input and output tapes and a bidirectional work tape [42, 30]. We can then define the (conditional) *prefix Kolmogorov complexity* [5, 17, 33, 38] of a binary string x as the length l of the shortest program p, for which U outputs the binary string x (given y).

Definition 4 (Kolmogorov complexity). *Let U be a universal prefix Turing machine U. The (conditional) prefix Kolmogorov complexity is defined as the shortest program p, for which U outputs x (given y):*

$$K(x) := \min_p\{l(p) : U(p) = x\}, \quad K(x|y) := \min_p\{l(p) : U(y, p) = x\}$$

Simple strings like 000...0 can be generated by short programs, and hence have low Kolmogorov complexity, but irregular (e.g. random) strings are their own shortest description, and hence have high Kolmogorov complexity. An important property of K is that it is nearly independent of the choice of U. Furthermore, it shares many properties with Shannon's entropy (information measure) S, but K is superior to S in many respects. To be brief, K is an excellent universal complexity measure, suitable for quantifying Occam's razor. There is (only) one severe disadvantage: K is not finitely computable. The major algorithmic property of K is that it is (only) co-enumerable, i.e. it is approximable from above.

For general (non-string) objects one can specify some default coding $\langle \cdot \rangle$ and define $K(object) := K(\langle object \rangle)$, especially for numbers and pairs, e.g. we abbreviate $K(x,y) := K(\langle x,y \rangle)$. The most important information-theoretic properties of K are listed below, where we abbreviate $f(x) \leq g(x)+O(1)$ by $f(x) \overset{+}{\leq} g(x)$. We also later abbreviate $f(x)=O(g(x))$ by $f(x) \overset{\times}{\leq} g(x)$.

Theorem 4 (Information properties of Kolmogorov complexity).

$i)$ $K(x) \overset{+}{\leq} l(x)+2\log l(x), \qquad K(n) \overset{+}{\leq} \log n+2\log\log n.$

$ii)$ $\sum_x 2^{-K(x)} \leq 1, \qquad K(x) \geq l(x)$ for 'most' x, $\qquad K(n) \to \infty$ for $n \to \infty$.

$iii)$ $K(x|y) \overset{+}{\leq} K(x) \overset{+}{\leq} K(x,y).$

$iv)$ $K(x,y) \overset{+}{\leq} K(x)+K(y), \qquad K(xy) \overset{+}{\leq} K(x)+K(y).$

$v)$ $K(x|y,K(y))+K(y) \overset{+}{=} K(x,y) \overset{+}{=} K(y,x) \overset{+}{=} K(y|x,K(x))+K(x).$

$vi)$ $K(f(x)) \overset{+}{\leq} K(x)+K(f)$ if $f:I\!B^* \to I\!B^*$ is recursive/computable.

$vii)$ $K(x) \overset{+}{\leq} -\log_2 P(x)+K(P)$ if $P:I\!B^* \to [0,1]$ is recursive and $\sum_x P(x) \leq 1$

All (in)equalities remain valid if K is (further) conditioned under some z, i.e. $K(...) \rightsquigarrow K(...|z)$ and $K(...|y) \rightsquigarrow K(...|y,z)$. Those stated are all valid within an additive constant of size $O(1)$, but there are others which are only valid to logarithmic accuracy. K has many properties in common with Shannon entropy as it should be, since both measure the information content of a string. Property (i) gives an upper bound on K, and property (ii) is Kraft's inequality which implies a lower bound on K valid for 'most' n, where 'most' means that there are only $o(N)$ exceptions for $n \in \{1,...,N\}$. Providing side information y can never increase code length, requiring extra information y can never decrease code length (iii). Coding x and y separately never helps (iv), and transforming x does not increase its information content (vi). Property (vi) also shows that if x codes some object o, switching from one coding scheme to another by means of a recursive bijection leaves K unchanged within additive $O(1)$ terms. The first nontrivial result is the symmetry of information (v), which is the analogue of the multiplication/chain rule for conditional probabilities. Property (vii) is at the heart of the MDL principle [52], which approximates $K(x)$ by $-\log_2 P(x)+K(P)$. See [42] for proofs.

3.3 Uncertainty & Probabilities

For the *objectivist* probabilities are real aspects of the world. The outcome of an observation or an experiment is not deterministic, but involves physical random processes. Kolmogorov's axioms of probability theory formalize the properties that probabilities should have. In the case of i.i.d. experiments the probabilities assigned to events can be interpreted as limiting frequencies (*frequentist* view), but applications are not limited to this case. Conditionalizing probabilities and Bayes' rule are the major tools in computing posterior probabilities from prior ones. For instance, given the initial binary sequence

$x_1...x_{n-1}$, what is the probability of the next bit being 1? The probability of observing x_n at time n, given past observations $x_1...x_{n-1}$ can be computed with the multiplication or chain rule[6] if the true generating distribution μ of the sequences $x_1 x_2 x_3...$ is known: $\mu(x_{<n}\underline{x}_n) = \mu(\underline{x}_{1:n})/\mu(\underline{x}_{<n})$ (see Sects. 2.2 and 2.5). The problem, however, is that one often does not know the true distribution μ (e.g. in the cases of weather and stock-market forecasting).

The *subjectivist* uses probabilities to characterize an agent's degree of belief in (or plausibility of) something, rather than to characterize physical random processes. This is the most relevant interpretation of probabilities in AI. It is somewhat surprising that plausibilities can be shown to also respect Kolmogorov's axioms of probability and the chain rule for conditional probabilities by assuming only a few plausible qualitative rules they should follow [10]. Hence, if the plausibility of $x_{1:n}$ is $\xi(\underline{x}_{1:n})$, the degree of belief in x_n given $x_{<n}$ is, again, given by the conditional probability: $\xi(x_{<n}\underline{x}_n) = \xi(\underline{x}_{1:n})/\xi(\underline{x}_{<n})$.

The the chain rule allows determining posterior probabilities/plausibilities from prior ones, but leaves open the question of how to determine the priors themselves. In statistical physics, the principle of indifference (symmetry principle) and the maximum entropy principle can often be exploited to determine prior probabilities, but only Occam's razor is general enough to assign prior probabilities in *every* situation, especially to cope with complex domains typical for AI.

3.4 Algorithmic Probability & Universal Induction

Occam's razor (appropriately interpreted and in compromise with Epicurus' principle of indifference) tells us to assign high/low a priori plausibility to simple/complex strings x. Using K as the complexity measure, any monotone decreasing function of K, e.g. $\xi(\underline{x}) = 2^{-K(x)}$ would satisfy this criterion. But ξ also has to satisfy the probability axioms, so we have to be a bit more careful. Solomonoff [61, 62] defined the *universal prior* $\xi(\underline{x})$ as the probability that the output of a universal Turing machine U starts with x when provided with fair coin flips on the input tape. Formally, ξ can be defined as

$$\xi(\underline{x}) := \sum_{p\,:\,U(p)=x*} 2^{-l(p)} \geq 2^{-K(x)}, \tag{16}$$

where the sum is over all (so-called minimal) programs p for which U outputs a string starting with x. The inequality follows by dropping all terms in $\sum_p$ except for the shortest p computing x. Strictly speaking ξ is only a *semimeasure* since it is not normalized to 1, but this is acceptable/correctable. We derive the following bound:

$$\sum_{t=1}^{\infty}(1-\xi(x_{<t}\underline{x}_t))^2 \leq -\tfrac{1}{2}\sum_{t=1}^{\infty}\ln\xi(x_{<t}\underline{x}_t) = -\tfrac{1}{2}\ln\xi(\underline{x}_{1:\infty}) \leq \tfrac{1}{2}\ln 2 \cdot K(x_{1:\infty}).$$

---

[6]Strictly speaking it is just the definition of conditional probabilities.

In the first inequality we have used $(1-a)^2 \leq -\frac{1}{2}\ln a$ for $0 \leq a \leq 1$. In the equality we exchanged the sum with the logarithm and eliminated the resulting product by the chain rule (6). In the last inequality we used (16). If $x_{1:\infty}$ is a computable sequence, then $K(x_{1:\infty})$ is finite, which implies $\xi(x_{<t}\underline{x}_t) \to 1$ ($\sum_{t=1}^{\infty}(1-a_t)^2 < \infty \Rightarrow a_t \to 1$). This means, that if the environment is a computable sequence (whichsoever, e.g. the digits of π or e in binary representation), after having seen the first few digits, ξ correctly predicts the next digit with high probability, i.e. it recognizes the structure of the sequence.

Assume now that the true sequence is drawn from the distribution μ, i.e. the true (objective) probability of $x_{1:n}$ is $\mu(\underline{x}_{1:n})$, but μ is unknown. How is the posterior (subjective) belief $\xi(x_{<n}\underline{x}_n) = \xi(\underline{x}_n)/\xi(\underline{x}_{<n})$ related to the true (objective) posterior probability $\mu(x_{<n}\underline{x}_n)$? Solomonoff's [62] crucial result is that the posterior (subjective) beliefs converge to the true (objective) posterior probabilities, if the latter are computable. More precisely, he showed that

$$\sum_{t=1}^{\infty}\sum_{x_{<t}} \mu(\underline{x}_{<t})\Big(\xi(x_{<t}\underline{0}) - \mu(x_{<t}\underline{0})\Big)^2 \overset{+}{\leq} \tfrac{1}{2}\ln 2 \cdot K(\mu). \tag{17}$$

$K(\mu)$ is finite if μ is computable, but the infinite sum on the l.h.s. can only be finite if the difference $\xi(x_{<t}\underline{0}) - \mu(x_{<t}\underline{0})$ tends to zero for $t \to \infty$ with μ-probability 1. This shows that using ξ as an estimate for μ may be a reasonable thing to do.

3.5 Loss Bounds & Pareto Optimality

Most predictions are eventually used as a basis for some decision or action, which itself leads to some reward or loss. Let $\ell_{x_t y_t} \in [0,1] \subset I\!R$ be the received loss when performing prediction/decision/action $y_t \in \mathcal{Y}$ and $x_t \in \mathcal{X}$ is the tth symbol of the sequence. Let $y_t^{\Lambda} \in \mathcal{Y}$ be the prediction of a (causal) prediction scheme Λ. The true probability of the next symbol being x_t, given $x_{<t}$, is $\mu(x_{<t}\underline{x}_t)$. The expected loss when predicting y_t is $\mathbf{E}[\ell_{x_t y_t}]$. The total μ-expected loss suffered by the Λ scheme in the first n predictions is

$$L_n^{\Lambda} := \sum_{t=1}^{n} \mathbf{E}[\ell_{x_t y_t^{\Lambda}}] = \sum_{t=1}^{n}\sum_{x_{1:t} \in \mathcal{X}^t} \mu(\underline{x}_{1:t})\ell_{x_t y_t^{\Lambda}}. \tag{18}$$

For instance, for the error-loss $l_{xy} = 1$ if $x = y$ and 0 else, L_n^{Λ} is the expected number of prediction errors, which we denote by E_n^{Λ}. The goal is to minimize the expected loss. More generally, we define the Λ_{ρ} sequence prediction scheme (later also called SPρ) $y_t^{\Lambda_{\rho}} := \text{argmin}_{y_t \in \mathcal{Y}}\sum_{x_t}\rho(x_{<t}\underline{x}_t)\ell_{x_t y_t}$ which minimizes the ρ-expected loss. If μ is known, Λ_{μ} is obviously the best prediction scheme in the sense of achieving minimal expected loss ($L_n^{\Lambda_{\mu}} \leq L_n^{\Lambda}$ for any Λ). One can prove the following loss bound for the universal Λ_{ξ} predictor [21, 20, 27]

$$0 \leq L_n^{\Lambda_{\xi}} - L_n^{\Lambda_{\mu}} \leq 2\ln 2 \cdot K(\mu) + 2\sqrt{L_n^{\Lambda_{\mu}} \ln 2 \cdot K(\mu)}. \tag{19}$$

Together with $L_n \leq n$ this shows that $\frac{1}{n}L_n^{\Lambda_\xi} - \frac{1}{n}L_n^{\Lambda_\mu} = O(n^{-1/2})$, i.e. asymptotically Λ_ξ achieves the optimal average loss of Λ_μ with rapid convergence. Moreover $L_\infty^{\Lambda_\xi}$ is finite if $L_\infty^{\Lambda_\mu}$ is finite and $L_n^{\Lambda_\xi}/L_n^{\Lambda_\mu} \to 1$ if $L_\infty^{\Lambda_\mu}$ is not finite. Bound (19) also implies $L_n^\Lambda \geq L_n^{\Lambda_\xi} - 2\sqrt{L_n^{\Lambda_\xi}\ln 2 \cdot K(\mu)}$, which shows that *no* (causal) predictor Λ whatsoever achieves significantly less (expected) loss than Λ_ξ. In view of these results it is fair to say that, ignoring computational issues, the problem of sequence prediction has been solved in a universal way.

A different kind of optimality is *Pareto optimality*. The universal prior ξ is Pareto optimal in the sense that there is no other predictor that leads to equal or smaller loss in *all* environments. Any improvement achieved by some predictor Λ over Λ_ξ in some environments is balanced by a deterioration in other environments [29].

4 The Universal Algorithmic Agent AIXI

Active systems, like game playing (SG) and optimization (FM), cannot be reduced to induction systems. The *main idea of this work* is to generalize universal induction to the general agent model described in Sect. 2. For this, we generalize ξ to include actions as conditions and replace μ by ξ in the rational agent model, resulting in the AIξ(=AIXI) model. In this way the problem that the true prior probability μ is usually unknown is solved. Convergence of $\xi \to \mu$ can be shown, indicating that the AIξ model could behave optimally in any computable but unknown environment with reinforcement feedback.

The main focus of this section is to investigate what we can expect from a universally optimal agent and to clarify the meanings of *universal, optimal*, etc. Unfortunately bounds similar to the loss bound (19) in the SP case can hold for *no* active agent. This forces us to lower our expectation about universally optimal agents and to introduce other (weaker) performance measures. Finally, we show that AIξ is Pareto optimal in the sense that there is no other policy yielding higher or equal value in *all* environments and a strictly higher value in at least one.

4.1 The Universal AIξ Model

Definition of the AIξ model. We have developed enough formalism to suggest our universal AIξ model. All we have to do is to suitably generalize the universal semimeasure ξ from the last section and replace the true but unknown prior probability μ^{AI} in the AIμ model by this generalized ξ^{AI}. In what sense this AIξ model is universal will be discussed subsequently.

In the functional formulation we define the universal probability ξ^{AI} of an environment q just as $2^{-l(q)}$:

$$\xi(q) := 2^{-l(q)}.$$

The definition could not be easier[7]![8] Collecting the formulas of Sect. 2.4 and replacing $\mu(q)$ by $\xi(q)$ we get the definition of the AIξ agent in functional form. Given the history $\dot{y}\dot{x}_{<k}$ the policy p^{ξ} of the functional AIξ agent is given by

$$\dot{y}_k := \arg\max_{y_k} \max_{p:p(\dot{x}_{<k})=\dot{y}_{<k}y_k} \sum_{q:q(\dot{y}_{<k})=\dot{x}_{<k}} 2^{-l(q)} \cdot V_{km_k}^{pq} \tag{20}$$

in cycle k, where $V_{km_k}^{pq}$ is the total reward of cycles k to m_k when agent p interacts with environment q. We have dropped the denominator $\sum_q \mu(q)$ from (2) as it is independent of the $p \in \dot{P}_k$ and a constant multiplicative factor does not change $\arg\max_{y_k}$.

For the iterative formulation, the universal probability ξ can be obtained by inserting the functional $\xi(q)$ into (13):

$$\xi(y\underline{x}_{1:k}) = \sum_{q:q(y_{1:k})=x_{1:k}} 2^{-l(q)}. \tag{21}$$

Replacing μ with ξ in (11) the iterative AIξ agent outputs

$$\dot{y}_k \equiv \dot{y}_k^{\xi} := \arg\max_{y_k} \sum_{x_k} \max_{y_{k+1}} \sum_{x_{k+1}} ... \max_{y_{m_k}} \sum_{x_{m_k}} (r(x_k)+...+r(x_{m_k})) \cdot \xi(\dot{y}\dot{x}_{<k}y\underline{x}_{k:m_k}) \tag{22}$$

in cycle k given the history $\dot{y}\dot{x}_{<k}$.

The equivalence of the functional and iterative AI model (Theorem 2) is true for every chronological semimeasure ρ, especially for ξ, hence we can talk about *the* AIξ model in this respect. It (slightly) depends on the choice of the universal Turing machine. $l(\langle q \rangle)$ is defined only up to an additive constant. The AIξ model also depends on the choice of $\mathcal{X} = \mathcal{R} \times \mathcal{O}$ and $\mathcal{Y}$, but we do not expect any bias when the spaces are chosen sufficiently simple, e.g. all strings of length 2^{16}. Choosing IN as the word space would be ideal, but whether the maxima (suprema) exist in this case, has to be shown beforehand. The only nontrivial dependence is on the horizon function m_k which will be discussed later. So apart from m_k and unimportant details the AIξ agent is uniquely defined by (20) or (22). It does not depend on any assumption about the environment apart from being generated by some computable (but unknown!) probability distribution.

Convergence of ξ to μ. Similarly to (17) one can show that the μ-expected squared difference of μ and ξ is finite for computable μ. This, in turn, shows

---

[7]It is not necessary to use $2^{-K(q)}$ or something similar, as some readers may expect, at this point, because for every program q there exists a functionally equivalent program $\tilde{q}$ with $K(q) \stackrel{+}{=} l(\tilde{q})$.

[8]Here and later we identify objects with their coding relative to some fixed Turing machine U. For example, if q is a function $K(q) := K(\langle q \rangle)$ with $\langle q \rangle$ being a binary coding of q such that $U(\langle q \rangle, y) = q(y)$. Reversely, if q already is a binary string we define $q(y) := U(q, y)$.

that $\xi(yx_{<k}y\underline{x}_k)$ converges rapidly to $\mu(yx_{<k}y\underline{x}_k)$ for $k\to\infty$ with μ-probability 1. The line of reasoning is the same; the y are pure spectators. This will change when we analyze loss/reward bounds analogous to (19). More generally, one can show [30] that[9]

$$\xi(yx_{<k}y\underline{x}_{k:m_k}) \stackrel{k\to\infty}{\longrightarrow} \mu(yx_{<k}y\underline{x}_{k:m_k}). \tag{23}$$

This gives hope that the outputs $\dot{y}_k$ of the AIξ model (22) could converge to the outputs $\dot{y}_k$ from the AIμ model (11).

We want to call an AI model *universal*, if it is μ-independent (unbiased, model-free) and is able to solve any solvable problem and learn any learnable task. Further, we call a universal model, *universally optimal*, if there is no program, which can solve or learn significantly faster (in terms of interaction cycles). Indeed, the AIξ model is parameter free, ξ converges to μ (23), the AIμ model is itself optimal, and we expect no other model to converge faster to AIμ by analogy to SP (19):

Claim (We expect AIXI to be universally optimal).

This is our main claim. In a sense, the intention of the remaining sections is to define this statement more rigorously and to give further support.

Intelligence order relation. We define the ξ-expected reward in cycles k to m of a policy p similar to (2) and (20). We extend the definition to programs $p \notin \dot{P}_k$ that are not consistent with the current history.

$$V_{km}^{p\xi}(\dot{y}\dot{x}_{<k}) := \frac{1}{\mathcal{N}} \sum_{q:q(\dot{y}_{<k})=\dot{x}_{<k}} 2^{-l(q)} \cdot V_{km}^{\tilde{p}q}. \tag{24}$$

The normalization $\mathcal{N}$ is again only necessary for interpreting V_{km} as the expected reward but is otherwise unneeded. For consistent policies $p \in \dot{P}_k$ we define $\tilde{p}:=p$. For $p\notin\dot{P}_k$, $\tilde{p}$ is a modification of p in such a way that its outputs are consistent with the current history $\dot{y}\dot{x}_{<k}$, hence $\tilde{p}\in\dot{P}_k$, but unaltered for the current and future cycles $\geq k$. Using this definition of V_{km} we could take the maximum over all policies p in (20), rather than only the consistent ones.

Definition 5 (Intelligence order relation). *We call a policy p more or equally intelligent than p' and write*

$$p \succeq p' \quad :\Leftrightarrow \quad \forall k \forall \dot{y}\dot{x}_{<k} : V_{km_k}^{p\xi}(\dot{y}\dot{x}_{<k}) \geq V_{km_k}^{p'\xi}(\dot{y}\dot{x}_{<k}).$$

i.e. if p yields in any circumstance higher ξ-expected reward than p'.

As the algorithm p^* behind the AIξ agent maximizes $V_{km_k}^{p\xi}$ we have $p^\xi \succeq p$ for all p. The AIξ model is hence the most intelligent agent w.r.t. $\succeq$. Relation

---

[9]Here, and everywhere else, with $\xi_k \to \mu_k$ we mean $\xi_k - \mu_k \to 0$, and not that μ_k (and ξ_k) itself converge to a limiting value.

$\succeq$ is a universal order relation in the sense that it is free of any parameters (except m_k) or specific assumptions about the environment. A proof, that $\succeq$ is a reliable intelligence order (which we believe to be true), would prove that AIξ is universally optimal. We could further ask: How useful is $\succeq$ for ordering policies of practical interest with intermediate intelligence, or how can $\succeq$ help to guide toward constructing more intelligent systems with reasonable computation time? An effective intelligence order relation $\succeq^c$ will be defined in Sect. 6, which is more useful from a practical point of view.

4.2 On the Optimality of AIXI

In this section we outline ways toward an optimality proof of AIXI. Sources of inspiration are the SP loss bounds proven in Sect. 3 and optimality criteria from the adaptive control literature (mainly) for linear systems [34]. The value bounds for AIXI are expected to be, in a sense, weaker than the SP loss bounds because the problem class covered by AIXI is much larger than the class of induction problems. Convergence of ξ to μ has already been proven, but is not sufficient to establish convergence of the behavior of the AIXI model to the behavior of the AIμ model. We will focus on three approaches toward a general optimality proof:

The meaning of "universal optimality". The first step is to investigate what we can expect from AIXI, i.e. what is meant by *universal optimality*. A "learner" (like AIXI) may converge to the optimal informed decision-maker (like AIμ) in several senses. Possibly relevant concepts from statistics are: *consistency, self-tunability, self-optimization, efficiency, unbiasedness, asymptotic* or *finite* convergence [34], Pareto optimality, and some more defined in Sect. 4.3. Some concepts are stronger than necessary, others are weaker than desirable but suitable to start with. Self-optimization is defined as the asymptotic convergence of the average true value $\frac{1}{m}V_{1m}^{p^{\xi}\mu}$ of AIξ to the optimal value $\frac{1}{m}V_{1m}^{*\mu}$. Apart from convergence speed, self-optimization of AIXI would most closely correspond to the loss bounds proven for SP. We investigate which properties are desirable and under which circumstances the AIXI model satisfies these properties. We will show that no universal model, including AIXI, can in general be self-optimizing. Conversely, we show that AIXI is Pareto optimal in the sense that there is no other policy that performs better or equal in all environments, and strictly better in at least one.

Limited environmental classes. The problem of defining and proving general value bounds becomes more feasible by considering, in a first step, restricted concept classes. We analyze AIXI for known classes (like Markovian or factorizable environments) and especially for the new classes (forgetful, relevant, asymptotically learnable, farsighted, uniform, pseudo-passive, and passive) defined later in Sect. 4.3. In Sect. 5 we study the behavior of AIXI in various standard problem classes, including sequence prediction, strategic games, function minimization, and supervised learning.

Generalization of AIXI to general Bayes mixtures. The other approach is to generalize AIXI to AIζ, where $\zeta() = \sum_{\nu \in \mathcal{M}} w_\nu \nu()$ is a general Bayes mixture of distributions ν in some class $\mathcal{M}$. If $\mathcal{M}$ is the multi-set of enumerable semimeasures enumerated by a Turing machine, then AIζ coincides with AIXI. If $\mathcal{M}$ is the (multi)set of passive effective environments, then AIXI reduces to the Λ_ξ predictor that has been shown to perform well. One can show that these loss/value bounds generalize to wider classes, at least asymptotically [26]. Promising classes are, again, the ones described in Sect. 4.3. In partic- ular, for ergodic MDPs we showed that AIζ is self-optimizing. Obviously, the least we must demand from $\mathcal{M}$ to have a chance of finding a self-optimizing policy is that there exists some self-optimizing policy at all. The key result in [26] is that this necessary condition is also sufficient. More generally, the key is not to prove absolute results for specific problem classes, but to prove relative results of the form "if there exists a policy with certain desirable properties, then AIζ also possesses these desirable properties." If there are tasks that can- not be solved by any policy, AIζ cannot be blamed for failing. Environmental classes that allow for self-optimizing policies include bandits, i.i.d. processes, classification tasks, certain classes of POMDPs, k^{th}-order ergodic MDPs, fac- torizable environments, repeated games, and prediction problems. Note that in this approach we have for each environmental class a corresponding model AIζ, whereas in the approach pursued in this article the same universal AIXI model is analyzed for all environmental classes.

Optimality by construction. A possible further approach toward an op- timality "proof" is to regard AIXI as *optimal by construction*. This perspec- tive is common in various (simpler) settings. For instance, in bandit prob- lems, where pulling arm i leads to reward 1 (0) with unknown probability p_i $(1 - p_i)$, the traditional Bayesian solution to the uncertainty about p_i is to assume a uniform (or Beta) prior over p_i and to maximize the (subjectively) expected reward sum over multiple trials. The exact solution (in terms of Gittins indices) is widely regarded as "optimal", although justified alterna- tive approaches exist. Similarly, but simpler, assuming a uniform subjective prior over the Bernoulli parameter $p_{(i)} \in [0,1]$, one arrives at the reasonable, but more controversial, Laplace rule for predicting i.i.d. sequences. AIXI is similar in the sense that the unknown $\mu \in \mathcal{M}$ is the analogue of the unknown $p \in [0,1]$, and the prior beliefs $w_\nu = 2^{-K(\nu)}$ justified by Occam's razor are the analogue of a uniform distribution over $[0,1]$. In the same sense as Gittins' so- lution to the bandit problem and Laplace' rule for Bernoulli sequences, AIXI may also be regarded as optimal by construction. Theorems relating AIXI to AIμ would not be regarded as optimality proofs of AIXI, but just as how much harder it becomes to operate when μ is unknown, i.e. the achievements of the first three approaches are simply reinterpreted.

4.3 Value Bounds and Separability Concepts

Introduction. The values V_{km} associated with the AI systems correspond roughly to the negative loss $-L_n^A$ of the SP systems. In SP, we were interested in small bounds for the loss excess $L_n^{A\xi} - L_n^A$. Unfortunately, simple value bounds for AIξ in terms of V_{km} analogous to the loss bound (19) do not hold. We even have difficulties in specifying what we can expect to hold for AIξ or any AI system that claims to be universally optimal. Consequently, we cannot have a proof if we don't know what to prove. In SP, the only important property of μ for proving loss bounds was its complexity $K(\mu)$. We will see that in the AI case, there are no useful bounds in terms of $K(\mu)$ only. We either have to study restricted problem classes or consider bounds depending on other properties of μ, rather than on its complexity only. In the following, we will exhibit the difficulties by two examples and introduce concepts that may be useful for proving value bounds. Despite the difficulties in even claiming useful value bounds, we nevertheless, firmly believe that the order relation (Definition 5) correctly formalizes the intuitive meaning of intelligence and, hence, that the AIξ agent is universally optimal.

(Pseudo) Passive μ and the HeavenHell example. In the following we choose $m_k = m$. We want to compare the true, i.e. μ-expected value V_{1m}^μ of a μ-independent universal policy p^{best} with any other policy p. Naively, we might expect the existence of a policy p^{best} that maximizes V_{1m}^μ, apart from additive corrections of lower order for $m \to \infty$:

$$V_{1m}^{p^{best}\mu} \geq V_{1m}^{p\mu} - o(...) \quad \forall \mu, p \tag{25}$$

Such policies are sometimes called self-optimizing [34]. Note that $V_{1m}^{p^\mu \mu} \geq V_{1m}^{p\mu} \forall p$, but p^μ is not a candidate for (a universal) p^{best} as it depends on μ. On the other hand, the policy p^ξ of the AIξ agent maximizes V_{1m}^ξ by definition ($p^\xi \succeq p$). As V_{1m}^ξ is thought to be a guess of V_{1m}^μ, we might expect $p^{best} = p^\xi$ to approximately maximize V_{1m}^μ, i.e. (25) to hold. Let us consider the problem class (set of environments) $\mathcal{M} = \{\mu_0, \mu_1\}$ with $\mathcal{Y} = \mathcal{R} = \{0,1\}$ and $r_k = \delta_{iy_1}$ in environment μ_i, where the Kronecker symbol δ_{xy} is defined as 1 for $x = y$ and 0 otherwise. The first action y_1 decides whether you go to heaven with all future rewards r_k being 1 (good) or to hell with all future rewards being 0 (bad). Note that μ_i are (deterministic, non-ergodic) MDPs:

It is clear that if μ_i, i.e. i is known, the optimal policy p^{μ_i} is to output $y_1 = i$ in the first cycle with $V_{1m}^{p^{\mu_i}\mu} = m$. On the other hand, any unbiased policy p^{best} independent of the actual μ either outputs $y_1 = 1$ or $y_1 = 0$. Independent of the actual choice y_1, there is always an environment ($\mu = \mu_{1-y_1}$) for which this choice is catastrophic ($V_{1m}^{p^{best}\mu}=0$). No single agent can perform well in both environments μ_0 *and* μ_1. The r.h.s. of (25) equals $m-o(m)$ for $p=p^\mu$. For all p^{best} there is a μ for which the l.h.s. is zero. We have shown that no p^{best} can satisfy (25) for all μ and p, so we cannot expect p^ξ to do so. Nevertheless, there are problem classes for which (25) holds, for instance SP. For SP, (25) is just a reformulation of (19) with an appropriate choice for p^{best}, namely Λ_ξ (which differs from p^ξ, see next section). We expect (25) to hold for all inductive problems in which the environment is not influenced[10] by the output of the agent. We want to call these μ, *passive* or *inductive* environments. Further, we want to call $\mathcal{M}$ and $\mu \in \mathcal{M}$ satisfying (25) with $p^{best}=p^\xi$ *pseudo-passive*. So we expect inductive μ to be pseudo-passive.

The OnlyOne example. Let us give a further example to demonstrate the difficulties in establishing value bounds. Let $\mathcal{X}=\mathcal{R}=\{0,1\}$ and $|\mathcal{Y}|$ be large. We consider all (deterministic) environments in which a single complex output y^* is correct ($r=1$) and all others are wrong ($r=0$). The problem class $\mathcal{M}$ is defined by

$$\mathcal{M} := \{\mu_{y^*} : y^* \in \mathcal{Y}, \ K(y^*) = \lfloor \log|\mathcal{Y}| \rfloor\}, \quad \text{where} \quad \mu_{y^*}(yx_{<k}y_k\underline{1}) := \delta_{y_k y^*} \ \forall k.$$

There are $N \overset{\times}{=} |\mathcal{Y}|$ such y^*. The only way a μ-independent policy p can find the correct y^*, is by trying one y after the other in a certain order. In the first $N-1$ cycles, at most $N-1$ different y are tested. As there are N different possible y^*, there is always a $\mu \in \mathcal{M}$ for which p gives erroneous outputs in the first $N-1$ cycles. The number of errors is $E_\infty^p \geq N-1 \overset{\times}{=} |\mathcal{Y}| \overset{\times}{=} 2^{K(y^*)} \overset{\times}{=} 2^{K(\mu)}$ for this μ. As this is true for any p, it is also true for the AIξ model, hence $E_k^{p^\xi} \leq 2^{K(\mu)}$ is the best possible error bound we can expect that depends on $K(\mu)$ only. Actually, we will derive such a bound in Sect. 5.1 for inductive environments. Unfortunately, as we are mainly interested in the cycle region $k \ll |\mathcal{Y}| \overset{\times}{=} 2^{K(\mu)}$ (see Sect. 2.8) this bound is vacuous. There are no interesting bounds for deterministic μ depending on $K(\mu)$ only, unlike the SP case. Bounds must either depend on additional properties of μ or we have to consider specialized bounds for restricted problem classes. The case of probabilistic μ is similar. Whereas for SP there are useful bounds in terms of $L_k^{\Lambda\mu}$ and $K(\mu)$, there are no such bounds for AIξ. Again, this is not a drawback of AIξ since for no unbiased AI system could the errors/rewards be bound in terms of $K(\mu)$ and the errors/rewards of AIμ only.

There is a way to make use of gross (e.g. $2^{K(\mu)}$) bounds. Assume that after a reasonable number of cycles k, the information $\dot{x}_{<k}$ perceived by the AIξ agent contains a lot of information about the true environment μ. The information in $\dot{x}_{<k}$ might be coded in any form. Let us assume that the complexity $K(\mu|\dot{x}_{<k})$ of μ under the condition that $\dot{x}_{<k}$ is known, is of order

---

[10] Of course, the reward feedback r_k depends on the agent's output. What we have in mind is, like in sequence prediction, that the true sequence is not influenced by the agent.

1. Consider a theorem, bounding the sum of rewards or of other quantities over cycles $1...\infty$ in terms of $f(K(\mu))$ for a function f with $f(O(1)) = O(1)$, like $f(n) = 2^n$. Then, there will be a bound for cycles $k...\infty$ in terms of $\approx f(K(\mu|\dot{x}_{<k})) = O(1)$. Hence, a bound like $2^{K(\mu)}$ can be replaced by small bound $\approx 2^{K(\mu|\dot{x}_{<k})} = O(1)$ after k cycles. All one has to show/ensure/assume is that enough information about μ is presented (in any form) in the first k cycles. In this way, even a gross bound could become useful. In Sect. 5.4 we use a similar argument to prove that AIξ is able to learn supervised.

Asymptotic learnability. In the following, we weaken (25) in the hope of getting a bound applicable to wider problem classes than the passive one. Consider the I/O sequence $\dot{y}_1\dot{x}_1...\dot{y}_n\dot{x}_n$ caused by AIξ. On history $\dot{y}\dot{x}_{<k}$, AIξ will output $\dot{y}_k \equiv \dot{y}_k^\xi$ in cycle k. Let us compare this to $\dot{y}_k^\mu$ what AIμ would output, still on the same history $\dot{y}\dot{x}_{<k}$ produced by AIξ. As AIμ maximizes the μ-expected value, AIξ causes lower (or at best equal) $V_{km_k}^\mu$ if $\dot{y}_k^\xi$ differs from $\dot{y}_k^\mu$. Let $D_{n\mu\xi} := \mathbf{E}[\sum_{k=1}^n 1 - \delta_{\dot{y}_k^\mu, \dot{y}_k^\xi}]$ be the μ-expected number of sub-optimal choices of AIξ, i.e. outputs different from AIμ in the first n cycles. One might weigh the deviating cases by their severity. In particular, when the μ-expected rewards $V_{km_k}^{p\mu}$ for $\dot{y}_k^\xi$ and $\dot{y}_k^\mu$ are equal or close to each other, this should be taken into account in a definition of $D_{n\mu\xi}$, e.g. by a weight factor $[V_{km}^{*\mu}(yx_{<k}) - V_{km}^{p^\xi\mu}(yx_{<k})]$. These details do not matter in the following qualitative discussion. The important difference to (25) is that here we stick to the history produced by AIξ and count a wrong decision as, at most, one error. The wrong decision in the HeavenHell example in the first cycle no longer counts as losing m rewards, but counts as one wrong decision. In a sense, this is fairer. One shouldn't blame those much who make a single wrong decision for which they have too little information available, in order to make a correct decision. The AIξ model would deserve to be called asymptotically optimal if the probability of making a wrong decision tends to zero, i.e. if

$$D_{n\mu\xi}/n \to 0 \quad \text{for} \quad n \to \infty, \quad \text{i.e.} \quad D_{n\mu\xi} = o(n). \tag{26}$$

We say that μ can be *asymptotically learned* (by AIξ) if (26) is satisfied. We claim that AIξ (for $m_k \to \infty$) can asymptotically learn every problem μ of relevance, i.e. AIξ is asymptotically optimal. We included the qualifier *of relevance*, as we are not sure whether there could be strange μ spoiling (26) but we expect those μ to be irrelevant from the perspective of AI. In the field of learning, there are many asymptotic learnability theorems, often not too difficult to prove. So a proof of (26) might also be feasible. Unfortunately, asymptotic learnability theorems are often too weak to be useful from a practical point of view. Nevertheless, they point in the right direction.

Uniform μ. ¿From the convergence (23) of $\xi \to \mu$ we might expect $V_{km_k}^{p\xi} \to V_{km_k}^{p\mu}$ for all p, and hence we might also expect $\dot{y}_k^\xi$ defined in (22) to converge to $\dot{y}_k^\mu$ defined in (11) for $k \to \infty$. The first problem is that if the V_{km_k} for the different choices of y_k are nearly equal, then even if $V_{km_k}^{p\xi} \approx V_{km_k}^{p\mu}$, $\dot{y}_k^\xi \neq \dot{y}_k^\mu$

is possible due to the non-continuity of argmax$_{y_k}$. This can be cured by a weighted $D_{n\mu\xi}$ as described above. More serious is the second problem we explain for $h_k=1$ and $\mathcal{X}=\mathcal{R}=\{0,1\}$. For $\dot{y}_k^\xi \equiv \text{argmax}_{y_k}\xi(\dot{y}\dot{r}_{<k}y_k\underline{1})$ to converge to $\dot{y}_k^\mu \equiv \text{argmax}_{y_k}\mu(\dot{y}\dot{r}_{<k}y_k\underline{1})$, it is not sufficient to know that $\xi(\dot{y}\dot{r}_{<k}\dot{y}\dot{r}_k) \to \mu(\dot{y}\dot{r}_{<k}\dot{y}\dot{r}_k)$ as proven in (23). We need convergence not only for the true output $\dot{y}_k$, but also for alternative outputs y_k. $\dot{y}_k^\xi$ converges to $\dot{y}_k^\mu$ if ξ converges uniformly to μ, i.e. if in addition to (23)

$$\left|\mu(yx_{<k}y_k'\underline{x}_k') - \xi(yx_{<k}y_k'\underline{x}_k')\right| < c\cdot\left|\mu(yx_{<k}y\underline{x}_k) - \xi(yx_{<k}y\underline{x}_k)\right| \quad \forall y_k'x_k' \quad (27)$$

holds for some constant c (at least in a μ-expected sense). We call μ satisfying (27) *uniform*. For uniform μ one can show (26) with appropriately weighted $D_{n\mu\xi}$ and bounded horizon $h_k < h_{max}$. Unfortunately there are relevant μ that are not uniform.

Other concepts. In the following, we briefly mention some further concepts. A *Markovian* μ is defined as depending only on the last cycle, i.e. $\mu(yx_{<k}y\underline{x}_k) = \mu_k(x_{k-1}y\underline{x}_k)$. We say μ is *generalized (l^{th}-order) Markovian*, if $\mu(yx_{<k}y\underline{x}_k) = \mu_k(x_{k-l}yx_{k-l+1:k-1}y\underline{x}_k)$ for fixed l. This property has some similarities to *factorizable* μ defined in (14). If further $\mu_k \equiv \mu_1 \forall k$, μ is called *stationary*. Further, we call μ (ξ) *forgetful* if $\mu(yx_{<k}y\underline{x}_k)$ ($\xi(yx_{<k}y\underline{x}_k)$) become(s) independent of $yx_{<l}$ for fixed l and $k \to \infty$ with μ-probability 1. Further, we say μ is *farsighted* if $\lim_{m_k \to \infty}\dot{y}_k^{(m_k)}$ exists. More details will be given in Sect. 4.5, where we also give an example of a farsighted μ, for which, nevertheless, the limit $m_k \to \infty$ makes no sense.

Summary. We have introduced several concepts that might be useful for proving value bounds, including forgetful, relevant, asymptotically learnable, farsighted, uniform, (generalized) Markovian, factorizable and (pseudo)passive μ. We have sorted them here, approximately in the order of decreasing generality. We will call them *separability concepts*. The more general (like relevant, asymptotically learnable and farsighted) μ will be called weakly separable, the more restrictive (like (pseudo) passive and factorizable) μ will be called strongly separable, but we will use these qualifiers in a more qualitative, rather than rigid sense. Other (non-separability) concepts are deterministic μ and, of course, the class of all chronological μ.

4.4 Pareto Optimality of AIξ

This subsection shows Pareto-opimtality of AIξ analogous to SP. The total μ-expected reward $V_\mu^{p^\xi}$ of policy p^ξ of the AIξ model is of central interest in judging the performance of AIξ. We know that there *are* policies (e.g. p^μ of AIμ) with higher μ-value ($V_\mu^* \geq V_\mu^{p^\xi}$). In general, every policy based on an estimate ρ of μ that is closer to μ than ξ is, outperforms p^ξ in environment μ, simply because it is more tailored toward μ. On the other hand, such a system probably performs worse than p^ξ in other environments. Since we do not know μ in advance we may ask whether there exists a policy p with better or equal performance than p^ξ in *all* environments $\nu \in \mathcal{M}$ and a strictly better performance for one $\nu \in \mathcal{M}$. This would clearly render p^ξ suboptimal. One can show that there is no such p [26]

Definition 6 (Pareto Optimality). *A policy $\tilde{p}$ is called Pareto optimal if there is no other policy p with $V_\nu^p \geq V_\nu^{\tilde{p}}$ for all $\nu \in \mathcal{M}$ and strict inequality for at least one ν.*

Theorem 5 (Pareto Optimality). *AIξ alias p^ξ is Pareto optimal.*

Pareto optimality should be regarded as a necessary condition for an agent aiming to be optimal. From a practical point of view, a significant increase of V for many environments ν may be desirable, even if this causes a small decrease of V for a few other ν. The impossibility of such a "balanced" improvement is a more demanding condition on p^ξ than pure Pareto optimality. In [26] it has been shown that AIξ is also balanced Pareto optimal.

4.5 The Choice of the Horizon

The only significant arbitrariness in the AIξ model lies in the choice of the horizon function $h_k \equiv m_k - k + 1$. We discuss some choices that seem to be natural and give preliminary conclusions at the end. We will not discuss ad hoc choices of h_k for specific problems (like the discussion in Sect. 5.2 in the context of finite strategic games). We are interested in universal choices of m_k.

Fixed horizon. If the lifetime of the agent is known to be m, which is in practice always large but finite, then the choice $m_k = m$ maximizes correctly the expected future reward. Lifetime m is usually not known in advance, as in many cases the time we are willing to run an agent depends on the quality of its outputs. For this reason, it is often desirable that good outputs are not delayed too much, if this results in a marginal reward increase only. This can be incorporated by damping the future rewards. If, for instance, the probability of survival in a cycle is $\gamma < 1$, an exponential damping (geometric discount) $r_k := r_k' \cdot \gamma^k$ is appropriate, where r_k' are bounded, e.g. $r_k' \in [0,1]$. Expression (22) converges for $m_k \to \infty$ in this case.[11] But this does not solve the problem, as we introduced a new arbitrary time scale $(1-\gamma)^{-1}$. Every damping introduces a time scale. Taking $\gamma \to 1$ is prone to the same problems as $m_k \to \infty$ in the undiscounted case discussed below.

Dynamic horizon (universal & harmonic discounting). The largest horizon with guaranteed finite and enumerable reward sum can be obtained by the universal discount $r_k \leadsto r_k \cdot 2^{-K(k)}$. This discount results in truly farsighted agent with effective horizon that grows faster than any computable function.

---

[11]More precisely, $\dot{y}_k = \underset{y_k}{\mathrm{argmax}} \lim_{m_k \to \infty} V_{km_k}^{*\xi}(\dot{y}\dot{x}_{<k} y_k)$ exists.

It is similar to a near-harmonic discount $r_k \rightsquigarrow r_k \cdot k^{-(1+\varepsilon)}$, since $2^{-K(k)} \leq 1/k$ for most k and $2^{-K(k)} \geq c/(k \log^2 k)$. More generally, the time-scale invariant damping factor $r_k = r_k' \cdot k^{-\alpha}$ introduces a dynamic time scale. In cycle k the contribution of cycle $2^{1/\alpha} \cdot k$ is damped by a factor $\frac{1}{2}$. The effective horizon h_k in this case is $\sim k$. The choice $h_k = \beta \cdot k$ with $\beta \sim 2^{1/\alpha}$ qualitatively models the same behavior. We have not introduced an arbitrary time scale m, but limited the farsightedness to some multiple (or fraction) of the length of the current history. This avoids the preselection of a global time scale m or $\frac{1}{1-\gamma}$. This choice has some appeal, as it seems that humans of age k years usually do not plan their lives for more than, perhaps, the next k years ($\beta_{human} \approx 1$). From a practical point of view this model might serve all needs, but from a theoretical point we feel uncomfortable with such a limitation in the horizon from the very beginning. Note that we have to choose $\beta = O(1)$ because otherwise we would again introduce a number β, which has to be justified. We favor the universal discount $\gamma_k = 2^{-K(k)}$, since it allows us, if desired, to "mimic" all other more greedy behaviors based on other discounts γ_k by choosing $r_k \in [0, c \cdot \gamma_k] \subseteq [0, 2^{-K(k)}]$.

Infinite horizon. The naive limit $m_k \to \infty$ in (22) may turn out to be well defined and the previous discussion superfluous. In the following, we suggest a limit that is always well defined (for finite $\mathcal{Y}$). Let $\dot{y}_k^{(m_k)}$ be defined as in (22) with dependence on m_k made explicit. Further, let $\dot{\mathcal{Y}}_k^{(m)} := \{ \dot{y}_k^{(m_k)} : m_k \geq m \}$ be the set of outputs in cycle k for the choices $m_k = m, m+1, m+2, \dots$. Because $\dot{\mathcal{Y}}_k^{(m)} \supseteq \dot{\mathcal{Y}}_k^{(m+1)} \neq \{\}$, we have $\dot{\mathcal{Y}}_k^{(\infty)} := \bigcap_{m=k}^{\infty} \dot{\mathcal{Y}}_k^{(m)} \neq \{\}$. We define the $m_k = \infty$ model to output any $\dot{y}_k^{(\infty)} \in \dot{\mathcal{Y}}_k^{(\infty)}$. This is the best output consistent with some arbitrary large choice of m_k. Choosing the lexicographically smallest $\dot{y}_k^{(\infty)} \in \dot{\mathcal{Y}}_k^{(\infty)}$ would correspond to the lower limit $\underline{\lim}_{m \to \infty} \dot{y}_k^{(m)}$, which always exists (for finite $\mathcal{Y}$). Generally $\dot{y}_k^{(\infty)} \in \dot{\mathcal{Y}}_k^{(\infty)}$ is unique, i.e. $|\dot{\mathcal{Y}}_k^{(\infty)}| = 1$ iff the naive limit $\lim_{m \to \infty} \dot{y}_k^{(m)}$ exists. Note that the limit $\lim_{m \to \infty} V_{km}^*(yx_{<k})$ need not exist for this construction.

Average reward and differential gain. Taking the raw average reward $(r_k + \dots + r_m)/(m - k + 1)$ and $m \to \infty$ also does not help: consider an arbitrary policy for the first k cycles and the/an optimal policy for the remaining cycles $k+1 \dots \infty$. In e.g. i.i.d. environments the limit exists, but all these policies give the same average value, since changing a finite number of terms does not affect an infinite average. In MDP environments with a single recurrent class one can define the relative or differential gain [3]. In more general environments (we are interested in) the differential gain can be infinite, which is acceptable, since differential gains can still be totally ordered. The major problem is the *existence* of the differential gain, i.e. whether it converges for $m \to \infty$ in $I\!R \cup \{\infty\}$ at all (and does not oscillate). This is just the old convergence problem in slightly different form.

Immortal agents are lazy. The construction in the next to previous paragraph leads to a mathematically elegant, no-parameter $AI\xi$ model. Unfortunately, this is not the end of the story. The limit $m_k \to \infty$ can cause undesirable results in the $AI\mu$ model for special μ, which might also happen in the $AI\xi$ model whatever we define $m_k \to \infty$. Consider an agent who for every $\sqrt{l}$ consecutive days of work, can thereafter take l days of holiday. Formally, consider $\mathcal{Y} = \mathcal{X} = \mathcal{R} = \{0,1\}$. Output $y_k = 0$ shall give reward $r_k = 0$ and output $y_k = 1$ shall give $r_k = 1$ iff $\dot{y}_{k-l-\sqrt{l}}...\dot{y}_{k-l} = 0...0$ for some l, i.e. the agent can achieve l consecutive positive rewards if there was a preceding sequence of length at least $\sqrt{l}$ with $y_k = r_k = 0$. If the lifetime of the $AI\mu$ agent is m, it outputs $\dot{y}_k = 0$ in the first s cycles and then $\dot{y}_k = 1$ for the remaining s^2 cycles with s such that $s + s^2 = m$. This will lead to the highest possible total reward $V_{1m} = s^2 = m + \frac{1}{2} - \sqrt{m + 1/4}$. Any fragmentation of the 0 and 1 sequences would reduce V_{1m}, e.g. alternatingly working for 2 days and taking 4 days off would give $V_{1m} = \frac{2}{3}m$. For $m \to \infty$ the $AI\mu$ agent can and will delay the point s of switching to $\dot{y}_k = 1$ indefinitely and always output 0 leading to total reward 0, obviously the worst possible behavior. The $AI\xi$ agent will explore the above rule after a while of trying $y_k = 0/1$ and then applies the same behavior as the $AI\mu$ agent, since the simplest rules covering past data dominate ξ. For finite m this is exactly what we want, but for infinite m the $AI\xi$ model (probably) fails, just as the $AI\mu$ model does. The good point is that this is not a weakness of the $AI\xi$ model in particular, as $AI\mu$ fails too. The bad point is that $m_k \to \infty$ has far-reaching consequences, even when starting from an already very large $m_k = m$. This is because the μ of this example is highly nonlocal in time, i.e. it may violate one of our weak separability conditions.

Conclusions. We are not sure whether the choice of m_k is of marginal importance, as long as m_k is chosen sufficiently large and of low complexity, $m_k = 2^{2^{16}}$ for instance, or whether the choice of m_k will turn out to be a central topic for the $AI\xi$ model or for the planning aspect of any AI system in general. We suppose that the limit $m_k \to \infty$ for the $AI\xi$ model results in correct behavior for weakly separable μ. A proof of this conjecture, if true, would probably give interesting insights.

4.6 Outlook

Expert advice approach. We considered expected performance bounds for predictions based on Solomonoff's prior. The other, dual, currently very popular approach, is "prediction with expert advice" (PEA) invented by Littlestone and Warmuth (1989), and Vovk (1992). Whereas PEA performs well in any environment, but only relative to a given set of experts , our Λ_ξ predictor competes with *any* other predictor, but only in expectation for environments with computable distribution. It seems philosophically less compromising to make assumptions on prediction strategies than on the environment, however weak. One could investigate whether PEA can be generalized to the case of active agents, which would result in a model dual to AIXI. We believe the answer to be negative, which on the positive side would show the necessity of Occam's razor assumption, and the distinguishedness of AIXI.

Actions as random variables. The uniqueness for the choice of the generalized ξ (16) in the AIXI model could be explored. From the originally many alternatives, which could all be ruled out, there is one alternative which still seems possible. Instead of defining ξ as in (21) one could treat the agent's actions y also as universally distributed random variables and then conditionalize ξ on y by the chain rule.

Structure of AIXI. The algebraic properties and the structure of AIXI could be investigated in more depth. This would extract the essentials from AIXI which finally could lead to an axiomatic characterization of AIXI. The benefit is as in any axiomatic approach. It would clearly exhibit the assumptions, separate the essentials from technicalities, simplify understanding and, most importantly, guide in finding proofs.

Restricted policy classes. The development in this section could be scaled down to restricted classes of policies $\mathcal{P}$. One may define $V^* = \operatorname{argmax}_{p \in \mathcal{P}} V^p$. For instance, consider a finite class of quickly computable policies. For MDPs, ξ is quickly computable and V_{ξ}^p can be (efficiently) computed by Monte Carlo sampling. Maximizing over the finitely many policies $p \in \mathcal{P}$ selects the asymptotically best policy p^{ξ} from $\mathcal{P}$ for all (ergodic) MDPs [26].

4.7 Conclusions

All tasks that require intelligence to be solved can naturally be formulated as a maximization of some expected utility in the framework of agents. We gave an explicit expression (11) of such a decision-theoretic agent. The main remaining problem is the unknown prior probability distribution μ^{AI} of the environment(s). Conventional learning algorithms are unsuitable, because they can neither handle large (unstructured) state spaces nor do they converge in the theoretically minimal number of cycles nor can they handle non-stationary environments appropriately. On the other hand, the universal semimeasure ξ (16), based on ideas from algorithmic information theory, solves the problem of the unknown prior distribution for induction problems. No explicit learning procedure is necessary, as ξ automatically converges to μ. We unified the theory of universal sequence prediction with the decision-theoretic agent by replacing the unknown true prior μ^{AI} by an appropriately generalized universal semimeasure ξ^{AI}. We gave strong arguments that the resulting AIξ model is universally optimal. Furthermore, possible solutions to the horizon problem were discussed. In Sect. 5 we present a number of problem classes, and outline how the AIξ model can solve them. They include sequence prediction, strategic games, function minimization and, especially, how AIξ learns to learn supervised. In Sect. 6 we develop a modified time-bounded (computable) AIXItl version.

5 Important Problem Classes

In order to give further support for the universality and optimality of the AIξ theory, we apply AIξ in this section to a number of problem classes. They include sequence prediction, strategic games, function minimization and, especially, how AIξ learns to learn supervised. For some classes we give concrete examples to illuminate the scope of the problem class. We first formulate each problem class in its natural way (when μ^{problem} is known) and then construct a formulation within the AIμ model and prove its equivalence. We then consider the consequences of replacing μ by ξ. The main goal is to understand why and how the problems are solved by AIξ. We only highlight special aspects of each problem class. Sections 5.1–5.5 together should give a better picture of the AIξ model. We do not study every aspect for every problem class. The subsections may be read selectively, and are not essential to understand the remainder.

5.1 Sequence Prediction (SP)

We introduced the AIξ model as a unification of ideas of sequential decision theory and universal probability distribution. We might expect AIξ to behave identically to SPξ, when faced with a sequence prediction problem, but things are not that simple, as we will see.

Using the AIμ model for sequence prediction. We saw in Sect. 3 how to predict sequences for known and unknown prior distribution μ^{SP}. Here we consider binary sequences[12] $z_1 z_2 z_3 ... \in IB^\infty$ with known prior probability $\mu^{\text{SP}}(z_1 z_2 z_3...)$.

We want to show how the AIμ model can be used for sequence prediction. We will see that it makes the same prediction as the SPμ agent. For simplicity we only discuss the special error loss $\ell_{xy} = 1 - \delta_{xy}$, where δ is the Kronecker symbol, defined as $\delta_{ab} = 1$ for $a = b$ and 0 otherwise. First, we have to specify *how* the AIμ model should be used for sequence prediction. The following choice is natural:

The system's output y_k is interpreted as a prediction for the kth bit z_k of the string under consideration. This means that y_k is binary ($y_k \in IB =: \mathcal{Y}$). As a reaction of the environment, the agent receives reward $r_k = 1$ if the prediction was correct ($y_k = z_k$), or $r_k = 0$ if the prediction was erroneous ($y_k \neq z_k$). The question is what the observation o_k in the next cycle should be. One choice would be to inform the agent about the correct k^{th} bit of the string and set $o_k = z_k$. But as from the reward r_k in conjunction with the prediction y_k, the true bit $z_k = \delta_{y_k r_k}$ can be inferred, this information is redundant. There is no need for this additional feedback. So we set $o_k = \epsilon \in \mathcal{O} = \{\epsilon\}$, thus having $x_k \equiv r_k \in \mathcal{R} \equiv \mathcal{X} = \{0,1\}$. The agent's performance does not change when we include this redundant information; it merely complicates the notation. The prior probability μ^{AI} of the AIμ model is

---

[12] We use z_k to avoid notational conflicts with the agent's inputs x_k.

$$\mu^{\text{AI}}(y_1\underline{x}_1...y_k\underline{x}_k) = \mu^{\text{AI}}(y_1\underline{r}_1...y_k\underline{r}_k)$$
$$= \mu^{\text{SP}}(\delta_{y_1r_1}...\delta_{y_kr_k})$$
$$= \mu^{\text{SP}}(\underline{z}_1...\underline{z}_k) \tag{28}$$

In the following, we will drop the superscripts of μ because they are clear from the arguments of μ and the μ equal in any case. It is intuitively clear and can be formally shown [19, 30] that maximizing the future reward V_{km}^{μ} is identical to greedily maximizing the immediate expected reward V_{kk}^{μ}. There is no exploration-exploitation tradeoff in the prediction case. Hence, AIμ acts with

$$\dot{y}_k = \arg\max_{y_k} V_{kk}^{*\mu}(\dot{y}\dot{x}_{<k}y_k)$$
$$= \arg\max_{y_k} \sum_{r_k} r_k \cdot \mu^{\text{AI}}(\dot{y}\dot{r}_{<k}y\underline{r}_k) = \arg\max_{z_k} \mu^{\text{SP}}(\dot{z}_1...\dot{z}_{k-1}\underline{z}_k) \tag{29}$$

The first equation is the definition of the agent's action (10) with m_k replaced by k. In the second equation we used the definition (9) of V_{km}. In the last equation we used (28) and $r_k = \delta_{y_k z_k}$.

So, the AIμ model predicts that z_k that has maximal μ-probability, given $\dot{z}_1...\dot{z}_{k-1}$. This prediction is independent of the choice of m_k. It is exactly the prediction scheme of the sequence predictor SPμ with known prior described in Sect. 3.5 (with special error loss). As this model was optimal, AIμ is optimal too, i.e. has minimal number of expected errors (maximal μ-expected reward) as compared to any other sequence prediction scheme. From this, it is clear that the value $V_{km}^{*\mu}$ must be closely related to the expected error $E_m^{\Lambda\mu}$ (18). Indeed, one can show that $V_{1m}^{*\mu} = m - E_m^{\Lambda\mu}$, and similarly for general loss functions.

Using the AIξ model for sequence prediction. Now we want to use the universal AIξ model instead of AIμ for sequence prediction and try to derive error/loss bounds analogous to (19). Like in the AIμ case, the agent's output y_k in cycle k is interpreted as a prediction for the kth bit z_k of the string under consideration. The reward is $r_k = \delta_{y_k z_k}$ and there are no other inputs $o_k = \epsilon$. What makes the analysis more difficult is that ξ is not symmetric in $y_i r_i \leftrightarrow (1-y_i)(1-r_i)$ and (28) does not hold for ξ. On the other hand, ξ^{AI} converges to μ^{AI} in the limit (23), and (28) should hold asymptotically for ξ in some sense. So we expect that everything proven for AIμ holds approximately for AIξ. The AIξ model should behave similarly to Solomonoff prediction SPξ. In particular, we expect error bounds similar to (19). Making this rigorous seems difficult. Some general remarks have been made in the last section. Note that bounds like (25) cannot hold in general, but could be valid for AIξ in (pseudo)passive environments.

Here we concentrate on the special case of a deterministic computable environment, i.e. the environment is a sequence $\dot{z} = \dot{z}_1\dot{z}_2...$ with $K(\dot{z}_{1:\infty}) < \infty$. Furthermore, we only consider the simplest horizon model $m_k = k$, i.e. greedily maximize only the next reward. This is sufficient for sequence prediction, as the reward of cycle k only depends on output y_k and not on earlier decisions. This choice is in no way sufficient and satisfactory for the full AIξ model, as *one* single choice of m_k should serve for *all* AI problem classes. So AIξ should allow good sequence prediction for some universal choice of m_k and not only for $m_k = k$, which definitely does not suffice for more complicated AI problems. The analysis of this general case is a challenge for the future. For $m_k = k$ the AIξ model (22) with $o_i = \epsilon$ and $r_k \in \{0,1\}$ reduces to

$$\dot{y}_k = \arg\max_{y_k} \sum_{r_k} r_k \cdot \xi(\dot{y}\dot{r}_{<k} y \underline{r}_k) = \arg\max_{y_k} \xi(\dot{y}\dot{r}_{<k} y_k \underline{1}). \tag{30}$$

The environmental response $\dot{r}_k$ is given by $\delta_{\dot{y}_k \dot{z}_k}$; it is 1 for a correct prediction $(\dot{y}_k = \dot{z}_k)$ and 0 otherwise. One can show [19, 30] that the number of wrong predictions $E_\infty^{\mathrm{AI}\xi}$ of the AIξ model (30) in these environments is bounded by

$$E_\infty^{\mathrm{AI}\xi} \stackrel{\times}{\le} 2^{K(\dot{z}_{1:\infty})} < \infty \tag{31}$$

for a computable deterministic environment string $\dot{z}_1 \dot{z}_2....$ The intuitive interpretation is that each wrong prediction eliminates at least one program p of size $l(p) \stackrel{+}{\le} K(\dot{z})$. The size is smaller than $K(\dot{z})$, as larger policies could not mislead the agent to a wrong prediction, since there is a program of size $K(\dot{z})$ making a correct prediction. There are at most $2^{K(\dot{z})+O(1)}$ such policies, which bounds the total number of errors.

We have derived a finite bound for $E_\infty^{\mathrm{AI}\xi}$, but unfortunately, a rather weak one as compared to (19). The reason for the strong bound in the SP case was that every error eliminates half of the programs.

The AIξ model would not be sufficient for realistic applications if the bound (31) were sharp, but we have the strong feeling (but only weak arguments) that better bounds proportional to $K(\dot{z})$ analogous to (19) exist. The current proof technique is not strong enough for achieving this. One argument for a better bound is the formal similarity between $\arg\max_{z_k} \xi(\dot{z}_{<k} \underline{z}_k)$ and (30), the other is that we were unable to construct an example sequence for which AIξ makes more than $O(K(\dot{z}))$ errors.

5.2 Strategic Games (SG)

Introduction. Strategic games (SG) are a very important class of problems. Game theory considers simple games of chance like roulette, combined with strategy like backgammon, up to purely strategic games like chess or checkers or go. In fact, what is subsumed under game theory is so general that it includes not only a huge variety of game types, but can also describe political and economic competitions and coalitions, Darwinism and many more topics. It seems that nearly every AI problem could be brought into the form of a game. Nevertheless, the intention of a game is that several players perform actions with (partial) observable consequences. The goal of each player is to maximize some utility function (e.g. to win the game). The players are assumed to be rational, taking into account all information they posses. The different goals of the players are usually in conflict. For an introduction into game theory, see [16, 48, 53, 47].

If we interpret the AI system as one player, and the environment models the other rational player *and* the environment provides the reinforcement feedback r_k, we see that the agent-environment configuration satisfies all criteria of a game. On the other hand, the AI models can handle more general situations, since they interact optimally with an environment, even if the environment is not a rational player with conflicting goals.

Strictly competitive strategic games. In the following, we restrict ourselves to deterministic, strictly competitive strategic[13] games with alternating moves. Player 1 makes move y_k in round k, followed by the move o_k of player 2.[14] So a game with n rounds consists of a sequence of alternating moves $y_1 o_1 y_2 o_2 ... y_n o_n$. At the end of the game in cycle n the game or final board situation is evaluated with $V(y_1 o_1 ... y_n o_n)$. Player 1 tries to maximize V, whereas player 2 tries to minimize V. In the simplest case, V is 1 if player 1 won the game, $V = -1$ if player 2 won and $V = 0$ for a draw. We assume a fixed game length n independent of the actual move sequence. For games with variable length but maximal possible number of moves n, we could add dummy moves and pad the length to n. The optimal strategy (Nash equilibrium) of both players is a minimax strategy:

$$\dot{o}_k = \arg\min_{o_k} \max_{y_{k+1}} \min_{o_{k+1}} ... \max_{y_n} \min_{o_n} V(\dot{y}_1 \dot{o}_1 ... \dot{y}_k o_k ... y_n o_n), \qquad (32)$$

$$\dot{y}_k = \arg\max_{y_k} \min_{o_k} ... \max_{y_n} \min_{o_n} V(\dot{y}_1 \dot{o}_1 ... \dot{y}_{k-1} \dot{o}_{k-1} y_k o_k ... y_n o_n). \qquad (33)$$

But note that the minimax strategy is only optimal if both players behave rationally. If, for instance, player 2 has limited capabilites or makes errors and player 1 is able to discover these (through past moves), he could exploit these weaknesses and improve his performance by deviating from the minimax strategy. At least the classical game theory of Nash equilibria does not take into account limited rationality, whereas the AIξ agent should.

Using the AIμ model for game playing. In the following, we demonstrate the applicability of the AI model to games. The AIμ model takes the position of player 1. The environment provides the evaluation V. For a symmetric situation we could take a second AIμ model as player 2, but for simplicity we take the environment as the second player and assume that this environmental player behaves according to the minimax strategy (32). The environment

---

[13]In game theory, games like chess are often called 'extensive', whereas 'strategic' is reserved for a different kind of game.

[14]We anticipate notationally the later identification of the moves of player 1/2 with the actions/observations in the AI models.

serves as a perfect player *and* as a teacher, albeit a very crude one, as it tells the agent at the end of the game only whether it won or lost.

The minimax behavior of player 2 can be expressed by a (deterministic) probability distribution μ^{SG} as the following:

$$\mu^{SG}(y_1\underline{o}_1...y_n\underline{o}_n) := \begin{cases} 1 & \text{if } o_k = \arg\min_{o'_k}...\max_{y'_n}\min_{o'_n} V(y_1o_1...y_ko'_k...y'_no'_n) \ \forall \ k \\ 0 & \text{otherwise} \end{cases}$$

$$(34)$$

The probability that player 2 makes move o_k is $\mu^{SG}(\dot{y}_1\dot{o}_1...\dot{y}_k\underline{o}_k)$, which is 1 for $o_k = \dot{o}_k$ as defined in (32) and 0 otherwise.

Clearly, the AIμ system receives no feedback, i.e. $r_1 = ... = r_{n-1} = 0$, until the end of the game, where it should receive positive/negative/neutral feedback on a win/loss/draw, i.e. $r_n = V(...)$. The environmental prior probability is therefore

$$\mu^{AI}(y_1\underline{x}_1...y_n\underline{x}_n) = \begin{cases} \mu^{SG}(y_1\underline{o}_1...y_n\underline{o}_n) & \text{if } r_1...r_{n-1} = 0 \\ & \text{and } r_n = V(y_1o_1...y_no_n) \ , \\ 0 & \text{otherwise} \end{cases}$$

$$(35)$$

where $x_i = r_io_i$. If the environment is a minimax player (32) plus a crude teacher V, i.e. if μ^{AI} is the true prior probability, the question is now: What is the behavior $\dot{y}_k^{AI}$ of the AIμ agent? It turns out that if we set $m_k = n$ the AIμ agent is also a minimax player (33) and hence optimal ($\dot{y}_k^{AI} = \dot{y}_k^{SG}$, see [19, 30] for a formal proof). Playing a sequence of games is a special case of a factorizable μ described in Sect. 2.7, with identical factors μ_r for all r and equal episode lengths $n_{r+1} - n_r = n$.

Hence, in a minimax environment AIμ behaves itself as a minimax strategy,

$$\dot{y}_k^{AI} = \arg\max_{y_k}\min_{o_k}...\max_{y_{(r+1)n}}\min_{o_{(r+1)n}} V(\dot{y}\dot{o}_{rn+1:k-1}...yo_{k:(r+1)n})$$

$$(36)$$

with r such that $rn < k \leq (r+1)n$ and for any choice of m_k as long as the horizon $h_k \geq n$.

Using the AIξ Model for Game Playing. When going from the specific AIμ model, where the rules of the game are explicitly modeled into the prior probability μ^{AI}, to the universal model AIξ, we have to ask whether these rules can be learned from the assigned rewards r_k. Here, the main reason for studying the case of repeated games rather than just one game arises. For a single game there is only one cycle of nontrivial feedback, namely the end of the game, which is too late to be useful except when further games follow.

We expect that no other learning scheme (with no extra information) can learn the game more quickly than AIξ, since μ^{AI} factorizes in the case of games of fixed length, i.e. μ^{AI} satisfies a strong separability condition. In the case of variable game length the entanglement is also low. μ^{AI} should still be sufficiently separable, allowing us to formulate and prove good reward bounds for AIξ. A qualitative argument goes as follows:

Since initially, AIξ loses all games, it tries to draw out a loss as long as possible, without having ever experienced or even knowing what it means to win. Initially, AIξ will make a lot of illegal moves. If illegal moves abort the game resulting in (non-delayed) negative reward (loss), AIξ can quickly learn the typically simple rules concerning legal moves, which usually constitute most of the rules; just the goal rule is missing. After having learned the move-rules, AIξ learns the (negatively rewarded) losing positions, the positions leading to losing positions, etc., so it can try to draw out losing games. For instance, in chess, avoiding being check mated for 20, 30, 40 moves against a master is already quite an achievement. At this ability stage, AIξ should be able to win some games by luck, or speculate about a symmetry in the game that check mating the opponent will be positively rewarded. Once having found out the complete rules (moves and goal), AIξ will right away reason that playing minimax is best, and henceforth beat all grandmasters.

If a (complex) game cannot be learned in this way in a realistic number of cycles, one has to provide more feedback. This could be achieved by intermediate help during the game. The environment could give positive (negative) feedback for every good (bad) move the agent makes. The demand on whether a move is to be valuated as good should be adapted to the gained experience of the agent in such a way that approximately the better half of the moves are valuated as good and the other half as bad, in order to maximize the information content of the feedback.

For more complicated games like chess, even more feedback may be necessary from a practical point of view. One way to increase the feedback far beyond a few bits per cycle is to train the agent by teaching it good moves. This is called supervised learning. Despite the fact that the AIμ model has only a reward feedback r_k, it is able to learn supervised, as will be shown in Sect. 5.4. Another way would be to start with simpler games containing certain aspects of the true game and to switch to the true game when the agent has learned the simple game.

No other difficulties are expected when going from μ to ξ. Eventually ξ^{AI} will converge to the minimax strategy μ^{AI}. In the more realistic case, where the environment is not a perfect minimax player, AIξ can detect and exploit the weakness of the opponent.

Finally, we want to comment on the input/output space $\mathcal{X}/\mathcal{Y}$ of the AI models. In practical applications, $\mathcal{Y}$ will possibly include also illegal moves. If $\mathcal{Y}$ is the set of moves of, e.g. a robotic arm, the agent could move a wrong figure or even knock over the figures. A simple way to handle illegal moves y_k is by interpreting them as losing moves, which terminate the game. Further, if, e.g. the input x_k is the image of a video camera which makes one shot per move, $\mathcal{X}$ is not the set of moves by the environment but includes the set of states of the game board. The discussion in this section handles this case as well. There is no need to explicitly design the systems I/O space $\mathcal{X}/\mathcal{Y}$ for a specific game.

The discussion above on the AIξ agent was rather informal for the following reason: game playing (the SGξ agent) has (nearly) the same complexity as fully general AI, and quantitative results for the AIξ agent are difficult (but not impossible) to obtain.

5.3 Function Minimization (FM)

Applications/examples. There are many problems that can be reduced to function minimization (FM) problems. The minimum of a (real-valued) function $f:\mathcal{Y}\to I\!R$ over some domain $\mathcal{Y}$ or a good approximate to the minimum has to be found, usually with some limited resources.

One popular example is the traveling salesman problem (TSP). $\mathcal{Y}$ is the set of different routes between towns, and $f(y)$ the length of route $y\in\mathcal{Y}$. The task is to find a route of minimal length visiting all cities. This problem is NP hard. Getting good approximations in limited time is of great importance in various applications. Another example is the minimization of production costs (MPC), e.g. of a car, under several constraints. $\mathcal{Y}$ is the set of all alternative car designs and production methods compatible with the specifications and $f(y)$ the overall cost of alternative $y\in\mathcal{Y}$. A related example is finding materials or (bio)molecules with certain properties (MAT), e.g. solids with minimal electrical resistance or maximally efficient chlorophyll modifications, or aromatic molecules that taste as close as possible to strawberry. We can also ask for nice paintings (NPT). $\mathcal{Y}$ is the set of all existing or imaginable paintings, and $f(y)$ characterizes how much person A likes painting y. The agent should present paintings which A likes.

For now, these are enough examples. The TSP is very rigorous from a mathematical point of view, as f, i.e. an algorithm of f, is usually known. In principle, the minimum could be found by exhaustive search, were it not for computational resource limitations. For MPC, f can often be modeled in a reliable and sufficiently accurate way. For MAT you need very accurate physical models, which might be unavailable or too difficult to solve or implement. For NPT all we have is the judgement of person A on every presented painting. The evaluation function f cannot be implemented without scanning A's brain, which is not possible with today's technology.

So there are different limitations, some depending on the application we have in mind. An implementation of f might not be available, f can only be tested at some arguments y and $f(y)$ is determined by the environment. We want to (approximately) minimize f with as few function calls as possible or, conversely, find an as close as possible approximation for the minimum within a fixed number of function evaluations. If f is available or can quickly be inferred by the agent and evaluation is quick, it is more important to minimize the total time needed to imagine new trial minimum candidates plus the evaluation time for f. As we do not consider computational aspects of AIξ till Sect. 6 we concentrate on the first case, where f is not available or dominates the computational requirements.

The greedy model. The FM model consists of a sequence $\dot{y}_1\dot{z}_1\dot{y}_2\dot{z}_2...$ where $\dot{y}_k$ is a trial of the FM agent for a minimum of f and $\dot{z}_k = f(\dot{y}_k)$ is the true function value returned by the environment. We randomize the model by assuming a probability distribution $\mu(f)$ over the functions. There are several reasons for doing this. We might really not know the exact function f, as in the NPT example, and model our uncertainty by the probability distribution μ. What is more important, we want to parallel the other AI classes, like in the SPμ model, where we always started with a probability distribution μ that was finally replaced by ξ to get the universal Solomonoff prediction SPξ. We want to do the same thing here. Further, the probabilistic case includes the deterministic case by choosing $\mu(f) = \delta_{ff_0}$, where f_0 is the true function. A final reason is that the deterministic case is trivial when μ and hence f_0 are known, as the agent can internally (virtually) check all function arguments and output the correct minimum from the very beginning.

We assume that $\mathcal{Y}$ is countable and that μ is a discrete measure, e.g. by taking only computable functions. The probability that the function values of $y_1,...,y_n$ are $z_1,...,z_n$ is then given by

$$\mu^{\mathrm{FM}}(y_1\underline{z}_1...y_n\underline{z}_n) := \sum_{f:f(y_i)=z_i \ \forall 1 \leq i \leq n} \mu(f). \tag{37}$$

We start with a model that minimizes the expectation z_k of the function value f for the next output y_k, taking into account previous information:

$$\dot{y}_k := \arg\min_{y_k} \sum_{z_k} z_k \cdot \mu(\dot{y}_1\dot{z}_1...\dot{y}_{k-1}\dot{z}_{k-1}y_k\underline{z}_k).$$

This type of greedy algorithm, just minimizing the next feedback, was sufficient for sequence prediction (SP) and is also sufficient for classification (CF, not described here). It is, however, not sufficient for function minimization as the following example demonstrates.

Take $f : \{0,1\} \to \{1,2,3,4\}$. There are 16 different functions which shall be equiprobable, $\mu(f) = \frac{1}{16}$. The function expectation in the first cycle

$$\langle z_1 \rangle := \sum_{z_1} z_1 \cdot \mu(y_1\underline{z}_1) = \tfrac{1}{4}\sum_{z_1} z_1 = \tfrac{1}{4}(1+2+3+4) = 2.5$$

is just the arithmetic average of the possible function values and is independent of y_1. Therefore, $\dot{y}_1 = 0$, if we define argmin to take the lexicographically first minimum in an ambiguous case like here. Let us assume that $f_0(0) = 2$, where f_0 is the true environment function, i.e. $\dot{z}_1 = 2$. The expectation of z_2 is then

$$\langle z_2 \rangle := \sum_{z_2} z_2 \cdot \mu(02y_2\underline{z}_2) = \begin{cases} 2 & \text{for} \quad y_2 = 0 \\ 2.5 & \text{for} \quad y_2 = 1 \end{cases}.$$

For $y_2 = 0$ the agent already knows $f(0) = 2$, for $y_2 = 1$ the expectation is, again, the arithmetic average. The agent will again output $\dot{y}_2 = 0$ with feedback $\dot{z}_2 = 2$.

This will continue forever. The agent is not motivated to explore other y's as $f(0)$ is already smaller than the expectation of $f(1)$. This is obviously not what we want. The greedy model fails. The agent ought to be inventive and try other outputs when given enough time.

The general reason for the failure of the greedy approach is that the information contained in the feedback z_k depends on the output y_k. A FM agent can actively influence the knowledge it receives from the environment by the choice in y_k. It may be more advantageous to first collect certain knowledge about f by an (in greedy sense) nonoptimal choice for y_k, rather than to minimize the z_k expectation immediately. The nonminimality of z_k might be overcompensated in the long run by exploiting this knowledge. In SP, the received information is always the current bit of the sequence, independent of what SP predicts for this bit. This is why a greedy strategy in the SP case is already optimal.

The general FMμ/ξ model. To get a useful model we have to think more carefully about what we really want. Should the FM agent output a good minimum in the last output in a limited number of cycles m, or should the average of the $z_1,...,z_m$ values be minimal, or does it suffice that just one z is as small as possible? The subtle and important differences between these settings have been analyzed and discussed in detail in [19, 30]. In the following we concentrate on minimizing the average, or equivalently the sum of function values. We define the FMμ model as to minimize the sum $z_1+...+z_m$. Building the μ average by summation over the z_i and minimizing with respect to the y_i has to be performed in the correct chronological order. With a similar reasoning as in (7) to (11) we get

$$\dot{y}_k^{\text{FM}} = \arg\min_{y_k} \sum_{z_k} ... \min_{y_m} \sum_{z_m} (z_1 + ... + z_m) \cdot \mu(\dot{y}_1\dot{z}_1...\dot{y}_{k-1}\dot{z}_{k-1}y_k\underline{z}_k...y_m\underline{z}_m)$$

(38)

By construction, the FMμ model guarantees optimal results in the usual sense that no other model knowing only μ can be expected to produce better results. The interesting case (in AI) is when μ is unknown. We define for this case, the FMξ model by replacing $\mu(f)$ with some $\xi(f)$, which should assign high probability to functions f of low complexity. So we might define $\xi(f) = \sum_{q:\forall x[U(qx)=f(x)]} 2^{-l(q)}$. The problem with this definition is that it is, in general, undecidable whether a TM q is an implementation of a function f. $\xi(f)$ defined in this way is uncomputable, not even approximable. As we only need a ξ analogous to the left hand side of (37), the following definition is natural

$$\xi^{\text{FM}}(y_1\underline{z}_1...y_n\underline{z}_n) := \sum_{q:q(y_i)=z_i \ \forall 1\leq i\leq n} 2^{-l(q)}.$$

(39)

ξ^{FM} is actually equivalent to inserting the uncomputable $\xi(f)$ into (37). One can show that ξ^{FM} is an enumerable semimeasure and dominates all enumerable probability distributions of the form (37).

Alternatively, we could have constrained the sum in (39) by $q(y_1...y_n) = z_1...z_n$ analogous to (21), but these two definitions are not equivalent. Definition (39) ensures the symmetry[15] in its arguments and $\xi^{\text{FM}}(...y\underline{z}...y\underline{z}'...) = 0$ for $z \neq z'$. It incorporates all general knowledge we have about function minimization, whereas (21) does not. But this extra knowledge has only low information content (complexity of $O(1)$), so we do not expect FMξ to perform much worse when using (21) instead of (39). But there is no reason to deviate from (39) at this point.

We can now define a loss $L_m^{\text{FM}\mu}$ as (38) with $k=1$ and argmin_{y_1} replaced by $\min_{y_1}$ and, additionally, μ replaced by ξ for $L_m^{\text{FM}\xi}$. We expect $|L_m^{\text{FM}\xi} - L_m^{\text{FM}\mu}|$ to be bounded in a way that justifies the use of ξ instead of μ for computable μ, i.e. computable f_0 in the deterministic case. The arguments are the same as for the AIξ model.

In [19, 30] it has been proven that FMξ is inventive in the sense that it never ceases searching for minima, but will test *all* $y \in \mathcal{Y}$ if $\mathcal{Y}$ is finite (and an infinite set of different y's if $\mathcal{Y}$ is infinite) for sufficiently large horizon m. There are currently no rigorous results on the *quality* of the guesses, but for the FMμ agent the guesses are optimal by definition. If $K(\mu)$ for the true distribution μ is finite, we expect the FMξ agent to solve the 'exploration versus exploitation' problem in a universally optimal way, as ξ converges rapidly to μ.

Using the AI Models for Function Mininimization. The AI models can be used for function minimization in the following way. The output y_k of cycle k is a guess for a minimum of f, like in the FM model. The reward r_k should be high for small function values $z_k = f(y_k)$. The choice $r_k = -z_k$ for the reward is natural. Here, the feedback is not binary but $r_k \in \mathcal{R} \subset I\!R$, with $\mathcal{R}$ being a countable subset of $I\!R$, e.g. the computable reals or all rational numbers. The feedback o_k should be the function value $f(y_k)$. As this is already provided in the rewards r_k we could set $o_k = \epsilon$ as in Sect. 5.1. For a change and to see that the choice really does not matter we set $o_k = z_k$ here. The AIμ prior probability is

$$\mu^{\text{AI}}(y_1\underline{x}_1...y_n\underline{x}_n) = \begin{cases} \mu^{\text{FM}}(y_1\underline{z}_1...y_n\underline{z}_n) \text{ for } r_k = -z_k, \, o_k = z_k, \, x_k = r_k o_k \\ \qquad\qquad 0 \qquad\qquad \text{else.} \end{cases} \tag{40}$$

Inserting this into (10) with $m_k = m$ one can show that $\dot{y}_k^{\text{AI}} = \dot{y}_k^{\text{FM}}$, where $\dot{y}_k^{\text{FM}}$ has been defined in (38). The proof is very simple since the FM model has already a rather general structure, which is similar to the full AI model.

We expect no problem in going from FMξ to AIξ. The only thing the AIξ model has to learn, is to ignore the o feedbacks as all information is already contained in r. This task is simple as every cycle provides one data point for a simple function to learn.

Remark on TSP. The Traveling Salesman Problem (TSP) seems to be trivial in the AIμ model but nontrivial in the AIξ model, because (38) just imple-

---

[15]See [65] for a discussion on symmetric universal distributions on unordered data.

ments an internal complete search, as $\mu(f) = \delta_{ff^{\text{TSP}}}$ contains all necessary information. AIμ outputs, from the very beginning, the exact minimum of f^{TSP}. This "solution" is, of course, unacceptable from a performance perspective. As long as we give no efficient approximation ξ^c of ξ, we have not contributed anything to a solution of the TSP by using AIξ^c. The same is true for any other problem where f is computable and easily accessible. Therefore, TSP is not (yet) a good example because all we have done is to replace an NP complete problem with the uncomputable AIξ model or by a computable AIξ^c model, for which we have said nothing about computation time yet. It is simply an overkill to reduce simple problems to AIξ. TSP is a simple problem in this respect, until we consider the AIξ^c model seriously. For the other examples, where f is inaccessible or complicated, an AIξ^c model would provide a true solution to the minimization problem as an explicit definition of f is not needed for AIξ and AIξ^c. A computable version of AIξ will be defined in Sect. 6.

5.4 Supervised Learning from Examples (EX)

The developed AI models provide a frame for reinforcement learning. The environment provides feedback r, informing the agent about the quality of its last (or earlier) output y; it assigns reward r to output y. In this sense, reinforcement learning is explicitly integrated into the AIμ/ξ models. AIμ maximizes the true expected reward, whereas the AIξ model is a universal, environment-independent reinforcement learning algorithm.

There is another type of learning method: Supervised learning by presentation of examples (EX). Many problems learned by this method are association problems of the following type. Given some examples $o \in R \subset \mathcal{O}$, the agent should reconstruct, from a partially given o', the missing or corrupted parts, i.e. complete o' to o such that relation R contains o. In many cases, $\mathcal{O}$ consists of pairs (z,v), where v is the possibly missing part.

Applications/examples. Learning functions by presenting $(z,f(z))$ pairs and asking for the function value of z by presenting $(z,?)$ falls into the category of supervised learning from examples, e.g. $f(z)$ may be the class label or category of z.

A basic example is learning properties of geometrical objects coded in some way. For instance, if there are 18 different objects characterized by their size (small or big), their colors (red, green, or blue) and their shapes (square, triangle, or circle), then $(object, property) \in R$ if the $object$ possesses the $property$. Here, R is a relation that is not the graph of a single-valued function.

When teaching a child by pointing to objects and saying "this is a tree" or "look how green" or "how beautiful," one establishes a relation of $(object, property)$ pairs in R. Pointing to a (possibly different) tree later and asking "What is this ?" corresponds to a partially given pair $(object,?)$, where the missing part "?" should be completed by the child saying "tree."

A final example is chess. We have seen that, in principle, chess can be learned by reinforcement learning. In the extreme case the environment only provides reward $r=1$ when the agent wins. The learning rate is probably inacceptable from a practical point of view, due to the low amount of information feedback. A more practical method of teaching chess is to present example games in the form of sensible (*board-state,move*) sequences. They contain information about legal and good moves (but without any explanation). After several games have been presented, the teacher could ask the agent to make its own move by presenting (*board-state,?*) and then evaluate the answer of the agent.

Supervised learning with the AIμ/ξ model. Let us define the EX model as follows: The environment presents inputs $o_{k-1} = z_k v_k \equiv (z_k, v_k) \in R \cup (\mathcal{Z} \times \{?\}) \subset \mathcal{Z} \times (\mathcal{Y} \cup \{?\}) = \mathcal{O}$ to the agent in cycle $k-1$. The agent is expected to output y_k in the next cycle, which is evaluated with $r_k = 1$ if $(z_k, y_k) \in R$ and 0 otherwise. To simplify the discussion, an output y_k is expected and evaluated even when $v_k(\neq ?)$ is given. To complete the description of the environment, the probability distribution $\mu_R(o_1...o_n)$ of the examples and questions o_i (depending on R) has to be given. Wrong examples should not occur, i.e. μ_R should be 0 if $o_i \notin R \cup (\mathcal{Z} \times \{?\})$ for some $1 \le i \le n$. The relations R might also be probability distributed with $\sigma(\underline{R})$. The example prior probability in this case is

$$\mu(\underline{o_1...o_n}) = \sum_R \mu_R(\underline{o_1...o_n}) \cdot \sigma(\underline{R}). \tag{41}$$

The knowledge of the valuation r_k on output y_k restricts the possible relations R, consistent with $R(z_k, y_k) = r_k$, where $R(z, y) := 1$ if $(z, y) \in R$ and 0 otherwise. The prior probability for the input sequence $x_1...x_n$ if the output sequence of AIμ is $y_1...y_n$, is therefore

$$\mu^{\mathrm{AI}}(y_1\underline{x}_1...y_n\underline{x}_n) = \sum_{R:\forall 1 < i \le n[R(z_i, y_i) = r_i]} \mu_R(\underline{o_1...o_n}) \cdot \sigma(\underline{R}),$$

where $x_i = r_i o_i$ and $o_{i-1} = z_i v_i$ with $v_i \in \mathcal{Y} \cup \{?\}$. In the I/O sequence $y_1 x_1 y_2 x_2... = y_1 r_1 z_2 v_2 y_2 r_2 z_3 v_3...$ the $y_1 r_1$ are dummies, after that regular behavior starts with example (z_2, v_2).

The AIμ model is optimal by construction of μ^{AI}. For computable prior μ_R and σ, we expect a near-optimal behavior of the universal AIξ model if μ_R additionally satisfies some separability property. In the following, we give some motivation why the AIξ model takes into account the supervisor information contained in the examples and why it learns faster than by reinforcement.

We keep R fixed and assume $\mu_R(o_1...o_n) = \mu_R(o_1) \cdot ... \cdot \mu_R(o_n) \neq 0 \Leftrightarrow o_i \in R \cup (\mathcal{Z} \times \{?\})$ $\forall i$ to simplify the discussion. Short codes q contribute most to $\xi^{\mathrm{AI}}(y_1 \underline{x}_1...y_n \underline{x}_n)$. As $o_1...o_n$ is distributed according to the computable probability distribution μ_R, a short code of $o_1...o_n$ for large enough n is a Huffman code with respect to the distribution μ_R. So we expect μ_R and hence

R to be coded in the dominant contributions to ξ^{AI} in some way, where the plausible assumption was made that the y on the input tape do not matter. Much more than one bit per cycle will usually be learned, i.e. relation R will be learned in $n \ll K(R)$ cycles by appropriate examples. This coding of R in q evolves independently of the feedbacks r. To maximize the feedback r_k, the agent has to learn to output a y_k with $(z_k, y_k) \in R$. The agent has to invent a program extension q' to q, which extracts z_k from $o_{k-1} = (z_k, ?)$ and searches for and outputs a y_k with $(z_k, y_k) \in R$. As R is already coded in q, q' can reuse this coding of R in q. The size of the extension q' is, therefore, of order 1. To learn this q', the agent requires feedback r with information content $O(1) = K(q')$ only.

Let us compare this with reinforcement learning, where only $o_{k-1} = (z_k, ?)$ pairs are presented. A coding of R in a short code q for $o_1...o_n$ is of no use and will therefore be absent. Only the rewards r force the agent to learn R. q' is therefore expected to be of size $K(R)$. The information content in the r's must be of the order $K(R)$. In practice, there are often only very few $r_k = 1$ at the beginning of the learning phase, and the information content in $r_1...r_n$ is much less than n bits. The required number of cycles to learn R by reinforcement is, therefore, at least but in many cases much larger than $K(R)$.

Although AIξ was never designed or told to learn supervised, it learns how to take advantage of the examples from the supervisor. μ_R and R are learned from the examples; the rewards r are not necessary for this process. The remaining task of learning how to learn supervised is then a simple task of complexity $O(1)$, for which the rewards r are necessary.

5.5 Other Aspects of Intelligence

In AI, a variety of general ideas and methods have been developed. In the previous subsections, we saw how several problem classes can be formulated within AIξ. As we claim universality of the AIξ model, we want to illuminate which and how other AI methods are incorporated in the AIξ model by looking at its structure. Some methods are directly included, while others are or should be emergent. We do not claim the following list to be complete.

Probability theory and *utility theory* are the heart of the AIμ/ξ models. The probability ξ is a universal belief about the true environmental behavior μ. The utility function is the total expected reward, called value, which should be maximized. Maximization of an expected utility function in a probabilistic environment is usually called *sequential decision theory*, and is explicitly integrated in full generality in our model. In a sense this includes probabilistic (a generalization of deterministic) *reasoning*, where the objects of reasoning are not true and false statements, but the prediction of the environmental behavior. *Reinforcement Learning* is explicitly built in, due to the rewards. Supervised learning is an emergent phenomenon (Sect. 5.4). *Algorithmic information theory* leads us to use ξ as a universal estimate for the prior probability μ.

For horizon >1, the expectimax series in (10) and the process of selecting maximal values may be interpreted as abstract *planning*. The expectimax series is a form of *informed search*, in the case of AIμ, and *heuristic search*, for AIξ, where ξ could be interpreted as a heuristic for μ. The minimax strategy of *game playing* in case of AIμ is also subsumed. The AIξ model converges to the minimax strategy if the environment is a minimax player, but it can also take advantage of environmental players with limited rationality. *Problem solving* occurs (only) in the form of how to maximize the expected future reward.

Knowledge is accumulated by AIξ and is stored in some form not specified further on the work tape. Any kind of information in any representation on the inputs y is exploited. The problem of *knowledge engineering* and *representation* appears in the form of how to train the AIξ model. More practical aspects, like *language or image processing*, have to be learned by AIξ from scratch.

Other theories, like *fuzzy logic, possibility theory, Dempster-Shafer theory*, and so on are partly outdated and partly reducible to Bayesian probability theory [7, 8]. The interpretation and consequences of the evidence gap $g := 1 - \sum_{x_k} \xi(yx_{<k}y\underline{x}_k) > 0$ in ξ may be similar to those in Dempster-Shafer theory. Boolean logical reasoning about the external world plays, at best, an emergent role in the AIξ model.

Other methods that do not seem to be contained in the AIξ model might also be emergent phenomena. The AIξ model has to construct short codes of the environmental behavior, and AIXI*tl* (see next section) has to construct short action programs. If we would analyze and interpret these programs for realistic environments, we might find some of the unmentioned or unused or new AI methods at work in these programs. This is, however, pure speculation at this point. More important: when trying to make AIξ practically usable, some other AI methods, like genetic algorithms or neural nets, especially for I/O pre/postprocessing, may be useful.

The main thing we wanted to point out is that the AIξ model does not lack any important known property of intelligence or known AI methodology. What *is* missing, however, are computational aspects, which are addressed in the next section.

6 Time-Bounded AIXI Model

Until now, we have not bothered with the non-computability of the universal probability distribution ξ. As all universal models in this paper are based on ξ, they are not effective in this form. In this section, we outline how the previous models and results can be modified/generalized to the time-bounded case. Indeed, the situation is not as bad as it could be. ξ is enumerable and $\dot{y}_k$ is still approximable, i.e. there exists an algorithm that will produce a sequence of outputs eventually converging to the exact output $\dot{y}_k$, but we can never be sure whether we have already reached it. Besides this, the convergence is extremely slow, so this type of asymptotic computability is of no direct (practical) use, but will nevertheless be important later.

Let $\tilde{p}$ be a program that calculates within a reasonable time $\tilde{t}$ per cycle, a reasonable intelligent output, i.e. $\tilde{p}(\dot{x}_{<k}) = \dot{y}_{1:k}$. This sort of computability assumption, that a general-purpose computer of sufficient power is able to behave in an intelligent way, is the very basis of AI, justifying the hope to be able to construct agents that eventually reach and outperform human intelligence. For a contrary viewpoint see [45, 49, 50]. It is not necessary to discuss here what is meant by "reasonable time/intelligence" and "sufficient power". What we are interested in, in this section, is whether there is a computable version AIXI$\tilde{t}$ of the AIξ agent that is superior or equal to any p with computation time per cycle of at most $\tilde{t}$. By "superior", we mean "more intelligent", so what we need is an order relation for intelligence, like the one in Definition 5.

The best result we could think of would be an AIXI$\tilde{t}$ with computation time $\leq \tilde{t}$ at least as intelligent as any p with computation time $\leq \tilde{t}$. If AI is possible at all, we would have reached the final goal: the construction of the most intelligent algorithm with computation time $\leq \tilde{t}$. Just as there is no universal measure in the set of computable measures (within time $\tilde{t}$), neither may such an AIXI$\tilde{t}$ exist.

What we can realistically hope to construct is an AIXI$\tilde{t}$ agent of computation time $c \cdot \tilde{t}$ per cycle for some constant c. The idea is to run all programs p of length $\leq \tilde{l} := l(\tilde{p})$ and time $\leq \tilde{t}$ per cycle and pick the best output. The total computation time is $c \cdot \tilde{t}$ with $c = 2^{\tilde{l}}$. This sort of idea of "typing monkeys" with one of them eventually writing Shakespeare, has been applied in various forms and contexts in theoretical computer science. The realization of this *best vote* idea, in our case, is not straightforward and will be outlined in this section. A related idea is that of basing the decision on the majority of algorithms. This "democratic vote" idea was used in [44, 68] for sequence prediction, and is referred to as "weighted majority".

6.1 Time-Limited Probability Distributions

In the literature one can find time-limited versions of Kolmogorov complexity [11, 12, 32] and the time-limited universal semimeasure [39, 42, 55]. In the following, we utilize and adapt the latter and see how far we get. One way to define a time-limited universal chronological semimeasure is as a mixture over enumerable chronological semimeasures computable within time $\tilde{t}$ and of size at most $\tilde{l}$:

$$\xi^{\tilde{t}\tilde{l}}(y\underline{x}_{1:n}) := \sum_{\rho \,:\, l(\rho) \leq \tilde{l} \,\wedge\, t(\rho) \leq \tilde{t}} 2^{-l(\rho)} \rho(y\underline{x}_{1:n}). \qquad (42)$$

One can show that $\xi^{\tilde{t}\tilde{l}}$ reduces to ξ^{AI} defined in (21) for $\tilde{t}, \tilde{l} \to \infty$. Let us assume that the true environmental prior probability μ^{AI} is equal to or sufficiently accurately approximated by a ρ with $l(\rho) \leq \tilde{l}$ and $t(\rho) \leq \tilde{t}$ with $\tilde{t}$ and $\tilde{l}$ of reasonable size. There are several AI problems that fall into this class. In function minimization of Sect. 5.3, the computation of f and μ^{FM} are often feasible. In many cases, the sequences of Sect. 5.1 that should be predicted, can be easily calculated when μ^{SP} is known. In a classification problem, the probability distribution μ^{CF}, according to which examples are presented, is, in many cases, also elementary. But not all AI problems are of this "easy" type. For the strategic games of Sect. 5.2, the environment itself is usually a highly complex strategic player with a μ^{SG} that is difficult to calculate, although one might argue that the environmental player may have limited capabilities too. But it is easy to think of a difficult-to-calculate physical (probabilistic) environment like the chemistry of biomolecules.

The number of interesting applications makes this restricted class of AI problems, with time- and space-bounded environment $\mu^{\tilde{t}\tilde{l}}$, worthy of study. Superscripts to a probability distribution except for $\xi^{\tilde{t}\tilde{l}}$ indicate their length and maximal computation time. $\xi^{\tilde{t}\tilde{l}}$ defined in (42), with a yet to be determined computation time, multiplicatively dominates all $\mu^{\tilde{t}\tilde{l}}$ of this type. Hence, an $\mathrm{AI}\xi^{\tilde{t}\tilde{l}}$ model, where we use $\xi^{\tilde{t}\tilde{l}}$ as prior probability, is universal, relative to all $\mathrm{AI}\mu^{\tilde{t}\tilde{l}}$ models in the same way as $\mathrm{AI}\xi$ is universal to $\mathrm{AI}\mu$ for all enumerable chronological semimeasures μ. The argmax_{y_k} in (22) selects a y_k for which $\xi^{\tilde{t}\tilde{l}}$ has the highest expected utility V_{km_k}, where $\xi^{\tilde{t}\tilde{l}}$ is the weighted average over the $\rho^{\tilde{t}\tilde{l}}$; i.e. output $\dot{y}_k^{\mathrm{AI}\xi^{\tilde{t}\tilde{l}}}$ is determined by a weighted majority. We expect $\mathrm{AI}\xi^{\tilde{t}\tilde{l}}$ to outperform all (bounded) $\mathrm{AI}\rho^{\tilde{t}\tilde{l}}$, analogous to the unrestricted case.

In the following we analyze the computability properties of $\xi^{\tilde{t}\tilde{l}}$ and $\mathrm{AI}\xi^{\tilde{t}\tilde{l}}$, i.e. of $\dot{y}_k^{\mathrm{AI}\xi^{\tilde{t}\tilde{l}}}$. To compute $\xi^{\tilde{t}\tilde{l}}$ according to the definition (42) we have to enumerate all chronological enumerable semimeasures $\rho^{\tilde{t}\tilde{l}}$ of length $\leq \tilde{l}$ and computation time $\leq \tilde{t}$. This can be done similarly to the unbounded case as described in [42, 19, 30]. All $2^{\tilde{l}}$ enumerable functions of length $\leq \tilde{l}$, computable within time $\tilde{t}$ have to be converted to chronological probability distributions. For this, one has to evaluate each function for $|\mathcal{X}| \cdot k$ different arguments. Hence, $\xi^{\tilde{t}\tilde{l}}$ is computable within time[16] $t(\xi^{\tilde{t}\tilde{l}}(y\underline{x}_{1:k})) = O(|\mathcal{X}| \cdot k \cdot 2^{\tilde{l}} \cdot \tilde{t})$. The computation time of $\dot{y}_k^{\mathrm{AI}\xi^{\tilde{t}\tilde{l}}}$ depends on the size of $\mathcal{X}$, $\mathcal{Y}$ and m_k. $\xi^{\tilde{t}\tilde{l}}$ has to be evaluated $|\mathcal{Y}|^{h_k}|\mathcal{X}|^{h_k}$ times in (22). It is possible to optimize the algorithm and perform the computation within time

$$t(\dot{y}_k^{\mathrm{AI}\xi^{\tilde{t}\tilde{l}}}) = O(|\mathcal{Y}|^{h_k}|\mathcal{X}|^{h_k} \cdot 2^{\tilde{l}} \cdot \tilde{t}) \tag{43}$$

per cycle. If we assume that the computation time of $\mu^{\tilde{t}\tilde{l}}$ is exactly $\tilde{t}$ for all arguments, the brute-force time $\bar{t}$ for calculating the sums and maxs in (11) is $\bar{t}(\dot{y}_k^{\mathrm{AI}\mu^{\tilde{t}\tilde{l}}}) \geq |\mathcal{Y}|^{h_k}|\mathcal{X}|^{h_k} \cdot \tilde{t}$. Combining this with (43), we get

$$t(\dot{y}_k^{\mathrm{AI}\xi^{\tilde{t}\tilde{l}}}) = O(2^{\tilde{l}} \cdot \bar{t}(\dot{y}_k^{\mathrm{AI}\mu^{\tilde{t}\tilde{l}}})).$$

---

[16]We assume that a (Turing) machine can be simulated by another in linear time.

This result has the proposed structure, that there is a universal $AI\xi^{\tilde{t}\tilde{l}}$ agent with computation time $2^{\tilde{l}}$ times the computation time of a special $AI\mu^{\tilde{t}\tilde{l}}$ agent.

Unfortunately, the class of $AI\mu^{\tilde{t}\tilde{l}}$ systems with brute-force evaluation of $\dot{y}_k$ according to (11) is completely uninteresting from a practical point of view. For instance, in the context of chess, the above result says that the $AI\xi^{\tilde{t}\tilde{l}}$ is superior within time $2^{\tilde{l}} \cdot \tilde{t}$ to any brute-force minimax strategy of computation time $\tilde{t}$. Even if the factor of $2^{\tilde{l}}$ in computation time would not matter, the $AI\xi^{\tilde{t}\tilde{l}}$ agent is, nevertheless practically useless, as a brute-force minimax chess player with reasonable time $\tilde{t}$ is a very poor player.

Note that in the case of binary sequence prediction ($h_k = 1$, $|\mathcal{Y}| = |\mathcal{X}| = 2$) the computation time of ρ coincides with that of $\dot{y}_k^{AI\rho}$ within a factor of 2. The class $AI\rho^{\tilde{t}\tilde{l}}$ includes *all* non-incremental sequence prediction algorithms of length $\leq \tilde{l}$ and computation time $\leq \tilde{t}/2$. By non-incremental, we mean that no information of previous cycles is taken into account for speeding up the computation of $\dot{y}_k$ of the current cycle.

The shortcomings (mentioned and unmentioned ones) of this approach are cured in the next subsection by deviating from the standard way of defining a time-bounded ξ as a sum over functions or programs.

6.2 The Idea of the Best Vote Algorithm

A general agent is a chronological program $p(x_{<k}) = y_{1:k}$. This form, introduced in Sect. 2.4, is general enough to include any AI system (and also less intelligent systems). In the following, we are interested in programs p of length $\leq \tilde{l}$ and computation time $\leq \tilde{t}$ per cycle. One important point in the time-limited setting is that p should be incremental, i.e. when computing y_k in cycle k, the information of the previous cycles stored on the work tape can be reused. Indeed, there is probably no practically interesting, non-incremental AI system at all.

In the following, we construct a policy p^*, or more precisely, policies p_k^* for every cycle k that outperform all time- and length-limited AI systems p. In cycle k, p_k^* runs all $2^{\tilde{l}}$ programs p and selects the one with the best output y_k. This is a "best vote" type of algorithm, as compared to the 'weighted majority' type algorithm of the last subsection. The ideal measure for the quality of the output would be the ξ-expected future reward

$$V_{km}^{p\xi}(\dot{y}\dot{x}_{<k}) := \sum_{q \in \dot{Q}_k} 2^{-l(q)} V_{km}^{pq} \quad , \quad V_{km}^{pq} := r(x_k^{pq}) + ... + r(x_m^{pq}) \qquad (44)$$

The program p that maximizes $V_{km_k}^{p\xi}$ should be selected. We have dropped the normalization $\mathcal{N}$ unlike in (24), as it is independent of p and does not change the order relation in which we are solely interested here. Furthermore, without normalization, $V_{km}^{*\xi}(\dot{y}\dot{x}_{<k}) := \max_{p \in \dot{P}} V_{km}^{p\xi}(\dot{y}\dot{x}_{<k})$ is enumerable, which will be important later.

6.3 Extended Chronological Programs

In the functional form of the AIξ model it was convenient to maximize V_{km_k} over all $p \in \dot{P}_k$, i.e. all p consistent with the current history $\dot{y}\dot{x}_{<k}$. This was not a restriction, because for every possibly inconsistent program p there exists a program $p' \in \dot{P}_k$ consistent with the current history and identical to p for all future cycles $\geq k$. For the time-limited best vote algorithm p^* it would be too restrictive to demand $p \in \dot{P}_k$. To prove universality, one has to compare *all* $2^{\tilde{l}}$ algorithms in every cycle, not just the consistent ones. An inconsistent algorithm may become the best one in later cycles. For inconsistent programs we have to include the $\dot{y}_k$ into the input, i.e. $p(\dot{y}\dot{x}_{<k}) = y^p_{1:k}$ with $\dot{y}_i \neq y^p_i$ possible. For $p \in \dot{P}_k$ this was not necessary, as p knows the output $\dot{y}_k \equiv y^p_k$ in this case. The r^{pq}_i in the definition of V_{km} are the rewards emerging in the I/O sequence, starting with $\dot{y}\dot{x}_{<k}$ (emerging from p^*) and then continued by applying p and q with $\dot{y}_i := y^p_i$ for $i \geq k$.

Another problem is that we need V_{km_k} to select the best policy, but unfortunately V_{km_k} is uncomputable. Indeed, the structure of the definition of V_{km_k} is very similar to that of $\dot{y}_k$, hence a brute-force approach to approximate V_{km_k} requires too much computation time as for $\dot{y}_k$. We solve this problem in a similar way, by supplementing each p with a program that estimates V_{km_k} by w^p_k within time $\tilde{t}$. We combine the calculation of y^p_k and w^p_k and extend the notion of a chronological program once again to

$$p(\dot{y}\dot{x}_{<k}) = w^p_1 y^p_1 ... w^p_k y^p_k, \tag{45}$$

with chronological order $w^p_1 y^p_1 \dot{y}_1 \dot{x}_1 w^p_2 y^p_2 \dot{y}_2 \dot{x}_2 ....$

6.4 Valid Approximations

Policy p might suggest any output y^p_k, but it is not allowed to rate it with an arbitrarily high w^p_k if we want w^p_k to be a reliable criterion for selecting the best p. We demand that no policy is allowed to claim that it is better than it actually is. We define a (logical) predicate VA(p) called *valid approximation*, which is true if and only if p always satisfies $w^p_k \leq V^{p\xi}_{km_k}$, i.e. never overrates itself.

$$\text{VA}(p) \equiv [\forall k \forall w^p_1 y^p_1 \dot{y}_1 \dot{x}_1 ... w^p_k y^p_k : p(\dot{y}\dot{x}_{<k}) = w^p_1 y^p_1 ... w^p_k y^p_k \Rightarrow w^p_k \leq V^{p\xi}_{km_k}(\dot{y}\dot{x}_{<k})] \tag{46}$$

In the following, we restrict our attention to programs p, for which VA(p) can be proven in some formal axiomatic system. A very important point is that $V^{*\xi}_{km_k}$ is enumerable. This ensures the existence of sequences of programs $p_1, p_2, p_3, ...$ for which VA(p_i) can be proven and $\lim_{i \to \infty} w^{p_i}_k = V^{*\xi}_{km_k}$ for all k and all I/O sequences. p_i may be defined as the naive (nonhalting) approximation scheme (by enumeration) of $V^{*\xi}_{km_k}$ terminated after i time steps and using the approximation obtained so far for $w^{p_i}_k$ together with the corresponding output

$y_k^{p_i}$. The convergence $w_k^{p_i} \overset{i\to\infty}{\longrightarrow} V_{km_k}^{*\xi}$ ensures that $V_{km_k}^{*\xi}$, which we claimed to be the universally optimal value, can be approximated by p with provable VA(p) arbitrarily well, when given enough time. The approximation is not uniform in k, but this does not matter as the selected p is allowed to change from cycle to cycle.

Another possibility would be to consider only those p that check $w_k^p \leq V_{km_k}^{p\xi}$ online in every cycle, instead of the pre-check VA(p), either by constructing a proof (on the work tape) for this special case, or $w_k^p \leq V_{km_k}^{p\xi}$ is already evident by the construction of w_k^p. In cases where p cannot guarantee $w_k^p \leq V_{km_k}^{p\xi}$ it sets $w_k = 0$ and, hence, trivially satisfies $w_k^p \leq V_{km_k}^{p\xi}$. On the other hand, for these p it is also no problem to prove VA(p) as one has simply to analyze the internal structure of p and recognize that p shows the validity internally itself, cycle by cycle, which is easy by assumption on p. The cycle-by-cycle check is therefore a special case of the pre-proof of VA(p).

6.5 Effective Intelligence Order Relation

In Sect. 4.1 we introduced an intelligence order relation $\succeq$ on AI systems, based on the expected reward $V_{km_k}^{p\xi}$. In the following we need an order relation $\succeq^c$ based on the claimed reward w_k^p, which might be interpreted as an approximation to $\succeq$.

Definition 7 (Effective intelligence order relation). *We call p effectively more or equally intelligent than p' if*

$$p \succeq^c p' :\Leftrightarrow \forall k \forall \dot{y}\dot{x}_{<k} \exists w_{1:n} w'_{1:n} :$$
$$p(\dot{y}\dot{x}_{<k}) = w_1 * ...w_k * \ \wedge\ p'(\dot{y}\dot{x}_{<k}) = w'_1 * ...w'_k * \ \wedge\ w_k \geq w'_k,$$

i.e. if p always claims higher reward estimate w than p'.

Relation $\succeq^c$ is a co-enumerable partial order relation on extended chronological programs. Restricted to valid approximations it orders the policies w.r.t. the quality of their outputs *and* their ability to justify their outputs with high w_k.

6.6 The Universal Time-Bounded AIXI*tl* Agent

In the following, we describe the algorithm p^* underlying the universal time-bounded AIXI$\widetilde{tl}$ agent. It is essentially based on the selection of the best algorithms p_k^* out of the time $\tilde{t}$ and length $\tilde{l}$ bounded p, for which there exists a proof of VA(p) with length $\leq l_P$.

1. Create all binary strings of length l_P and interpret each as a coding of a mathematical proof in the same formal logic system in which VA($\cdot$) was formulated. Take those strings that are proofs of VA(p) for some p and keep the corresponding programs p.

2. Eliminate all p of length $>\tilde{l}$.

3. Modify the behavior of all retained p in each cycle k as follows: Nothing is changed if p outputs some $w_k^p y_k^p$ within $\tilde{t}$ time steps. Otherwise stop p and write $w_k=0$ and some arbitrary y_k to the output tape of p. Let P be the set of all those modified programs.

4. Start first cycle: $k:=1$.

5. Run every $p \in P$ on extended input $\dot{y}\dot{x}_{<k}$, where all outputs are redirected to some auxiliary tape: $p(\dot{y}\dot{x}_{<k}) = w_1^p y_1^p...w_k^p y_k^p$. This step is performed incrementally by adding $\dot{y}\dot{x}_{k-1}$ for $k>1$ to the input tape and continuing the computation of the previous cycle.

6. Select the program p with highest claimed reward w_k^p: $p_k^*:=\mathrm{argmax}_p w_k^p$.

7. Write $\dot{y}_k:=y_k^{p_k^*}$ to the output tape.

8. Receive input $\dot{x}_k$ from the environment.

9. Begin next cycle: $k:=k+1$, goto step 5.

It is easy to see that the following theorem holds.

Theorem 6 (Optimality of AIXItl). *Let p be any extended chronological (incremental) program like (45) of length $l(p) \leq \tilde{l}$ and computation time per cycle $t(p) \leq \tilde{t}$, for which there exists a proof of VA(p) defined in (46) of length $\leq l_P$. The algorithm p^* constructed in the last paragraph, which depends on $\tilde{l}$, $\tilde{t}$ and l_P but not on p, is effectively more or equally intelligent, according to $\succeq^c$ (see Definition 7) than any such p. The size of p^* is $l(p^*)=O(\log(\tilde{l}\cdot\tilde{t}\cdot l_P))$, the setup-time is $t_{setup}(p^*)=O(l_P^2 \cdot 2^{l_P})$ and the computation time per cycle is $t_{cycle}(p^*)=O(2^{\tilde{l}}\cdot\tilde{t})$.*

Roughly speaking, the theorem says that if there exists a computable solution to some or all AI problems at all, the explicitly constructed algorithm p^* is such a solution. Although this theorem is quite general, there are some limitations and open questions that we discuss in the next subsection.

The construction of the algorithm p^* needs the specification of a formal logic system $(\forall,\lambda,y_i,c_i,f_i,R_i,\rightarrow,\wedge,=,...)$, and axioms, and inference rules. A proof is a sequence of formulas, where each formula is either an axiom or inferred from previous formulas in the sequence by applying the inference rules. Details can be found in [25] in a related construction or in any textbook on logic or proof theory, e.g. [15, 60]. We only need to know that *provability* and *Turing Machines* can be formalized. The setup time in the theorem is just the time needed to verify the 2^{l_P} proofs, each needing time $O(l_P^2)$.

6.7 Limitations and Open Questions

- Formally, the total computation time of p^* for cycles $1...k$ increases linearly with k, i.e. is of order $O(k)$ with a coefficient $2^{\tilde{l}}\cdot\tilde{t}$. The unreasonably large factor $2^{\tilde{l}}$ is a well-known drawback in best/democratic vote models and will be taken without further comments, whereas the factor $\tilde{t}$ can be assumed to be of reasonable size. If we do not take the limit $k \to \infty$ but consider reasonable k, the practical significance of the time bound on p^* is somewhat limited due to the additional additive constant $O(l_P^2 \cdot 2^{l_P})$. It is much larger than $k \cdot 2^{\tilde{l}} \cdot \tilde{t}$ as typically $l_P \gg l(\mathrm{VA}(p)) \geq l(p) \equiv \tilde{l}$.

- p^* is superior only to those p that justify their outputs (by large w_k^p). It might be possible that there are p that produce good outputs y_k^p within reasonable time, but it takes an unreasonably long time to justify their outputs by sufficiently high w_k^p. We do not think that (from a certain complexity level onwards) there are policies where the process of constructing a good output is completely separated from some sort of justification process. But this justification might not be translatable (at least within reasonable time) into a reasonable estimate of $V_{km_k}^{p\xi}$.

- The (inconsistent) programs p must be able to continue strategies started by other policies. It might happen that a policy p steers the environment to a direction for which p is specialized. A "foreign" policy might be able to displace p only between loosely connected episodes. There is probably no problem for factorizable μ. Think of a chess game, where it is usually very difficult to continue the game or strategy of a different player. When the game is over, it is usually advantageous to replace a player by a better one for the next game. There might also be no problem for sufficiently separable μ.

- There might be (efficient) valid approximations p for which $\mathrm{VA}(p)$ is true but not provable, or for which only a very long ($> l_P$) proof exists.

6.8 Remarks

- The idea of suggesting outputs and justifying them by proving reward bounds implements one aspect of human thinking. There are several possible reactions to an input. Each reaction possibly has far-reaching consequences. Within a limited time one tries to estimate the consequences as well as possible. Finally, each reaction is valuated, and the best one is selected. What is inferior to human thinking is that the estimates w_k^p must be rigorously proved and the proofs are constructed by blind exhaustive search, further, that *all* behaviors p of length $\leq \tilde{l}$ are checked. It is inferior "only" in the sense of necessary computation time but not in the sense of the quality of the outputs.

- In practical applications there are often cases with short and slow programs p_s performing some task T, e.g. the computation of the digits of π, for which there exist long but quick programs p_l too. If it is not too difficult to prove that this long program is equivalent to the short one, then it is possible to prove $K^{t(p_l)}(T) \overset{+}{\leq} l(p_s)$ with K^t being the time-bounded Kolmogorov complexity. Similarly, the method of proving bounds w_k for V_{km_k} can give high lower bounds without explicitly executing these short and slow programs, which mainly contribute to V_{km_k}.

- Dovetailing all length- and time-limited programs is a well-known elementary idea (e.g. typing monkeys). The crucial part that was developed here is the selection criterion for the most intelligent agent.
- The construction of AIXI$\tilde{t}\tilde{l}$ and the enumerability of V_{km_k} ensure arbitrary close approximations of V_{km_k}, hence we expect that the behavior of AIXI$\tilde{t}\tilde{l}$ converges to the behavior of AIξ in the limit $\tilde{t}, \tilde{l}, l_P \to \infty$, in some sense.
- Depending on what you know or assume that a program p of size $\tilde{l}$ and computation time per cycle $\tilde{t}$ is able to achieve, the computable AIXI$\tilde{t}\tilde{l}$ model will have the same capabilities. For the strongest assumption of the existence of a Turing machine that outperforms human intelligence, AIXI$\tilde{t}\tilde{l}$ will do too, within the same time frame up to an (unfortunately very large) constant factor.

7 Discussion

This section reviews what has been achieved in the chapter and discusses some otherwise unmentioned topics of general interest. We remark on various topics, including concurrent actions and perceptions, the choice of the I/O spaces, treatment of encrypted information, and peculiarities of mortal embodies agents. We continue with an outlook on further research. Since many ideas have already been presented in the various sections, we concentrate on nontechnical open questions of general importance, including optimality, down-scaling, implementation, approximation, elegance, extra knowledge, and training of/for AIXI(tl). We also include some (personal) remarks on noncomputable physics, the number of wisdom Ω, and consciousness. As it should be, the chapter concludes with conclusions.

7.1 General Remarks

Game theory. In game theory [48] one often wants to model the situation of simultaneous actions, whereas the AIξ models have serial I/O. Simultaneity can be simulated by withholding the environment from the current agent's output y_k, until x_k has been received by the agent. Formally, this means that $\mu(yx_{<k}y\underline{x}_k)$ is independent of the last output y_k. The AIξ agent is already of simultaneous type in an abstract view if the behavior p is interpreted as the action. In this sense, AIXI is the action p^* that maximizes the utility function (reward), under the assumption that the environment acts according to ξ. The situation is different from game theory, as the environment ξ is not a second 'player' that tries to optimize his own utility (see Sect. 5.2).

Input/output spaces. In various examples we have chosen differently specialized input and output spaces $\mathcal{X}$ and $\mathcal{Y}$. It should be clear that, in principle, this is unnecessary, as large enough spaces $\mathcal{X}$ and $\mathcal{Y}$ (e.g. the set of strings of length 2^{32}) serve every need and can always be Turing-reduced to the specific presentation needed internally by the AIXI agent itself. But it is clear that, using a generic interface, such as camera and monitor for learning tic-tac-toe, for example, adds the task of learning vision and drawing.

How AIXI(tl) deals with encrypted information. Consider the task of decrypting a message that was encrypted by a public key encrypter like RSA. A message m is encrypted using a product n of two large primes p_1 and p_2, resulting in encrypted message $c = \text{RSA}(m|n)$. RSA is a simple algorithm of size $O(1)$. If AIXI is given the public key n and encrypted message c, in order to reconstruct the original message m it only has to "learn" the function $\text{RSA}^{-1}(c|n) := \overline{\text{RSA}}(c|p_1, p_2) = m$. RSA^{-1} can itself be described in length $O(1)$, since $\overline{\text{RSA}}$ is $O(1)$ and p_1 and p_2 can be reconstructed from n. Only very little information is needed to learn $O(1)$ bits. In this sense decryption is easy for AIXI (like TSP, see Sect. 5.3). The problem is that while $\overline{\text{RSA}}$ is efficient, RSA^{-1} is an extremely slow algorithm, since it has to find the prime factors from the public key. But note, in AIXI we are not talking about computation time, we are only talking about information efficiency (learning in the least number of interaction cycles). One of the key insights in this article that allowed for an elegant theory of AI was this separation of data efficiency from computation time efficiency. Of course, in the real world computation time matters, so we invented AIXItl. AIXItl can do every job as well as the best length l and time t bounded agent, apart from time factor 2^l and a huge offset time. No practical offset time is sufficient to find the factors of n, but in theory, enough offset time allows also AIXItl to (once-and-for-all) find the factorization, and then, decryption is easy of course.

Mortal embodied agents. The examples we gave in this article, particularly those in Sect. 5, were mainly bodiless agents: predictors, gamblers, optimizers, learners. There are some peculiarities with reinforcement learning autonomous embodied robots in real environments.

We can still reward the robot according to how well it solves the task we want it to do. A minimal requirement is that the robot's hardware functions properly. If the robot starts to malfunction its capabilities degrade, resulting in lower reward. So, in an attempt to maximize reward, the robot will also maintain itself. The problem is that some parts will malfunction rather quickly when no appropriate actions are performed, e.g. flat batteries, if not recharged in time. Even worse, the robot may work perfectly until the battery is nearly empty, and then suddenly stop its operation (death), resulting in zero reward from then on. There is too little time to learn how to maintain itself before it's too late. An autonomous embodied robot cannot start from scratch but must have some rudimentary built-in capabilities (which may not be that rudimentary at all) that allow it to at least survive. Animals survive due to reflexes, innate behavior, an internal reward attached to the condition of their organs, and a guarding environment during childhood. Different species emphasize different aspects. Reflexes and innate behaviors are stressed in lower animals versus years of safe childhood for humans. The same variety of solutions are available for constructing autonomous robots (which we will not detail here).

Another problem connected, but possibly not limited to embodied agents, especially if they are rewarded by humans, is the following: Sufficiently intelligent agents may increase their rewards by psychologically manipulating their human "teachers," or by threatening them. This is a general sociological problem which successful AI will cause, which has nothing specifically to do with AIXI. Every intelligence superior to humans is capable of manipulating the latter. In the absence of manipulable humans, e.g. where the reward structure serves a survival function, AIXI may directly hack into its reward feedback. Since this is unlikely to increase its long-term survival, AIXI will probably resist this kind of manipulation (just as most humans don't take hard drugs, due to their long-term catastrophic consequences).

7.2 Outlook & Open Questions

Many ideas for further studies were already stated in the various sections of the article. This outlook only contains nontechnical open questions regarding AIXI(tl) of general importance.

Value bounds. Rigorous proofs for non-asymptotic value bounds for AIξ are the major theoretical challenge – general ones, as well as tighter bounds for special environments μ, e.g. for rapidly mixing MDPs, and/or other performance criteria have to be found and proved. Although not necessary from a practical point of view, the study of continuous classes $\mathcal{M}$, restricted policy classes, and/or infinite $\mathcal{Y}$, $\mathcal{X}$ and m may lead to useful insights.

Scaling AIXI down. A direct implementation of the AIXItl model is, at best, possible for small-scale (toy) environments due to the large factor 2^l in computation time. But there are other applications of the AIXI theory. We saw in several examples how to integrate problem classes into the AIXI model. Conversely, one can downscale the AIξ model by using more restricted forms of ξ. This could be done in the same way as the theory of universal induction was downscaled with many insights to the Minimum Description Length principle [40, 52] or to the domain of finite automata [14]. The AIXI model might similarly serve as a supermodel or as the very definition of (universal unbiased) intelligence, from which specialized models could be derived.

Implementation and approximation. With a reasonable computation time, the AIXI model would be a solution of AI (see the next point if you disagree). The AIXItl model was the first step, but the elimination of the factor 2^l without giving up universality will almost certainly be a very difficult task.[17] One could try to select programs p and prove VA(p) in a more clever way than by mere enumeration, to improve performance without destroying universality. All kinds of ideas like genetic algorithms, advanced theorem provers and many more could be incorporated. But now we have a problem.

---

[17]But see [25] for an elegant *theoretical* solution.

Computability. We seem to have transferred the AI problem just to a different level. This shift has some advantages (and also some disadvantages) but does not present a practical solution. Nevertheless, we want to stress that we have reduced the AI problem to (mere) computational questions. Even the most general other systems the author is aware of depend on some (more than complexity) assumptions about the environment or it is far from clear whether they are, indeed, universally optimal. Although computational questions are themselves highly complicated, this reduction is a nontrivial result. A formal theory of something, even if not computable, is often a great step toward solving a problem and also has merits of its own, and AI should not be different in this respect (see previous item).

Elegance. Many researchers in AI believe that intelligence is something complicated and cannot be condensed into a few formulas. It is more a combining of enough *methods* and much explicit *knowledge* in the right way. From a theoretical point of view we disagree, as the AIXI model is simple and seems to serve all needs. From a practical point of view we agree to the following extent: To reduce the computational burden one should provide special-purpose algorithms (*methods*) from the very beginning, probably many of them related to reduce the complexity of the input and output spaces $\mathcal{X}$ and $\mathcal{Y}$ by appropriate pre/postprocessing *methods*.

Extra knowledge. There is no need to incorporate extra *knowledge* from the very beginning. It can be presented in the first few cycles in *any* format. As long as the algorithm to interpret the data is of size $O(1)$, the AIXI agent will "understand" the data after a few cycles (see Sect. 5.4). If the environment μ is complicated but extra knowledge z makes $K(\mu|z)$ small, one can show that the bound (17) reduces roughly to $\ln 2 \cdot K(\mu|z)$ when $x_1 \equiv z$, i.e. when z is presented in the first cycle. The special-purpose algorithms could be presented in x_1 too, but it would be cheating to say that no special-purpose algorithms were implemented in AIXI. The boundary between implementation and training is unsharp in the AIXI model.

Training. We have not said much about the training process itself, as it is not specific to the AIXI model and has been discussed in literature in various forms and disciplines [63, 56, 57]. By a training process we mean a sequence of simple-to-complex tasks to solve, with the simpler ones helping in learning the more complex ones. A serious discussion would be out of place. To repeat a truism, it is, of course, important to present enough knowledge o_k and evaluate the agent output y_k with r_k in a reasonable way. To maximize the information content in the reward, one should start with simple tasks and give positive reward to approximately the better half of the outputs y_k.

7.3 The Big Questions

This subsection is devoted to the *big* questions of AI in general and the AIXI model in particular with a personal touch.

On non-computable physics & brains. There are two possible objections to AI in general and, therefore, to AIXI in particular. Non-computable physics (which is not too weird) could make Turing computable AI impossible. As at least the world that is relevant for humans seems mainly to be computable we do not believe that it is necessary to integrate non-computable devices into an AI system. The (clever and nearly convincing) Gödel argument by Penrose [49, 50], refining Lucas [45], that non-computational physics *must* exist and *is* relevant to the brain, has (in our opinion convincing) loopholes.

Evolution & the number of wisdom. A more serious problem is the evolutionary information-gathering process. It has been shown that the 'number of wisdom' Ω contains a very compact tabulation of 2^n undecidable problems in its first n binary digits [6]. Ω is only enumerable with computation time increasing more rapidly with n than any recursive function. The enormous computational power of evolution could have developed and coded something like Ω into our genes, which significantly guides human reasoning. In short: Intelligence could be something complicated, and evolution toward it from an even cleverly designed algorithm of size $O(1)$ could be too slow. As evolution has already taken place, we could add the information from our genes or brain structure to any/our AI system, but this means that the important part is still missing, and that it is principally impossible to derive an efficient algorithm from a simple formal definition of AI.

Consciousness. For what is probably the *biggest question*, that of *consciousness*, we want to give a physical analogy. Quantum (field) theory is the most accurate and universal physical theory ever invented. Although already developed in the 1930s, the *big* question, regarding the interpretation of the wave function collapse, is still open. Although this is extremely interesting from a philosophical point of view, it is completely irrelevant from a practical point of view.[18] We believe the same to be valid for *consciousness* in the field of Artificial Intelligence: philosophically highly interesting but practically unimportant. Whether consciousness *will* be explained some day is another question.

7.4 Conclusions

The major theme of the chapter was to develop a mathematical foundation of Artificial Intelligence. This is not an easy task since intelligence has many (often ill-defined) faces. More specifically, our goal was to develop a theory for rational agents acting optimally in any environment. Thereby we touched various scientific areas, including reinforcement learning, algorithmic information theory, Kolmogorov complexity, computational complexity theory, information theory and statistics, Solomonoff induction, Levin search, sequential decision theory, adaptive control theory, and many more.

---

[18]In the Theory of Everything, the collapse might become of 'practical' importance and must or will be solved.

We started with the observation that all tasks that require intelligence to be solved can naturally be formulated as a maximization of some expected utility in the framework of agents. We presented a functional (3) and an iterative (11) formulation of such a decision-theoretic agent in Sect. 2, which is general enough to cover all AI problem classes, as was demonstrated by several examples. The main remaining problem is the unknown prior probability distribution μ of the environment(s). Conventional learning algorithms are unsuitable, because they can neither handle large (unstructured) state spaces, nor do they converge in the theoretically minimal number of cycles, nor can they handle non-stationary environments appropriately. On the other hand, Solomonoff's universal prior ξ (16), rooted in algorithmic information theory, solves the problem of the unknown prior distribution for induction problems as was demonstrated in Sect. 3. No explicit learning procedure is necessary, as ξ automatically converges to μ. We unified the theory of universal sequence prediction with the decision-theoretic agent by replacing the unknown true prior μ by an appropriately generalized universal semimeasure ξ in Sect. 4. We gave various arguments that the resulting AIXI model is the most intelligent, parameter-free and environmental/application-independent model possible. We defined an intelligence order relation (Definition 5) to give a rigorous meaning to this claim. Furthermore, possible solutions to the horizon problem have been discussed. In Sect. 5 we outlined how the AIXI model solves various problem classes. These included sequence prediction, strategic games, function minimization and, especially, learning to learn supervised. The list could easily be extended to other problem classes like classification, function inversion and many others. The major drawback of the AIXI model is that it is uncomputable, or more precisely, only asymptotically computable, which makes an implementation impossible. To overcome this problem, we constructed a modified model AIXItl, which is still effectively more intelligent than any other time t and length l bounded algorithm (Sect. 6). The computation time of AIXItl is of the order $t \cdot 2^l$. A way of overcoming the large multiplicative constant 2^l was presented in [25] at the expense of an (unfortunately even larger) additive constant. Possible further research was discussed. The main directions could be to prove general and special reward bounds, use AIXI as a supermodel and explore its relation to other specialized models, and finally improve performance with or without giving up universality.

All in all, the results show that Artificial Intelligence can be framed by an elegant mathematical theory. Some progress has also been made toward an elegant *computational* theory of intelligence.

Annotated Bibliography

Introductory textbooks. The book by Hopcroft and Ullman, and in the new revision co-authored by Motwani [18], is a very readable elementary introduction to automata theory, formal languages, and computation theory. The

Artificial Intelligence book [53] by Russell and Norvig gives a comprehensive overview over AI approaches in general. For an excellent introduction to Algorithmic Information Theory, Kolmogorov complexity, and Solomonoff induction one should consult the book of Li and Vitányi [42]. The Reinforcement Learning book by Sutton and Barto [66] requires no background knowledge, describes the key ideas, open problems, and great applications of this field. A tougher and more rigorous book by Bertsekas and Tsitsiklis on sequential decision theory provides all (convergence) proofs [3].

Algorithmic information theory. Kolmogorov [33] suggested to define the information content of an object as the length of the shortest program computing a representation of it. Solomonoff [61] invented the closely related universal prior probability distribution and used it for binary sequence prediction [61, 62] and function inversion and minimization [63]. Together with Chaitin [4, 5], this was the invention of what is now called Algorithmic Information theory. For further literature and many applications see [42]. Other interesting applications can be found in [6, 59, 69]. Related topics are the Weighted Majority algorithm invented by Littlestone and Warmuth [44], universal forecasting by Vovk [68], Levin search [37], PAC-learning introduced by Valiant [67] and Minimum Description Length [40, 52]. Resource-bounded complexity is discussed in [11, 12, 14, 32, 51], resource-bounded universal probability in [39, 42, 55]. Implementations are rare and mainly due to Schmidhuber [9, 54, 58, 56, 57]. Excellent reviews with a philosophical touch are [41, 64]. For an older general review of inductive inference see Angluin [1].

Sequential decision theory. The other ingredient in our AIξ model is sequential decision theory. We do not need much more than the maximum expected utility principle and the expectimax algorithm [46, 53]. Von Neumann and Morgenstern's book [47] might be seen as the initiation of game theory, which already contains the expectimax algorithm as a special case. The literature on reinforcement learning and sequential decision theory is vast and we refer to the references given in the textbooks [66, 3].

The author's contributions. Details on most of the issues addressed in this article can be found in various reports or publications or the book [30] by the author: The AIξ model was first introduced and discussed in March 2000 in [19] in a 62-page-long report. More succinct descriptions were published in [23, 24]. The AIξ model has been argued to formally solve a number of problem classes, including sequence prediction, strategic games, function minimization, reinforcement and supervised learning [19]. A variant of AIξ has recently been shown to be self-optimizing and Pareto optimal [26]. The construction of a general fastest algorithm for all well-defined problems [25] arose from the construction of the time-bounded AIXI*tl* model [23]. Convergence [28] and tight [29] error [22, 20] and loss [21, 27] bounds for Solomonoff's universal sequence prediction scheme have been proven. Loosely related ideas on a market/economy-based reinforcement learner [36] and gradient-based reinforcement planner [35] were implemented. These and other papers are available at http://www.hutter1.net.

References

1. Angluin D, Smith CH (1983) Inductive inference: Theory and methods. *ACM Computing Surveys*, 15(3):237–269.
2. Bellman RE (1957) *Dynamic Programming.* Princeton University Press, Princeton, NJ.
3. Bertsekas DP, Tsitsiklis JN (1996) *Neuro-Dynamic Programming.* Athena Scientific, Belmont, MA.
4. Chaitin GJ (1966) On the length of programs for computing finite binary sequences. *Journal of the ACM*, 13(4):547–5691.
5. Chaitin GJ (1975) A theory of program size formally identical to information theory. *Journal of the ACM*, 22(3):329–340.
6. Chaitin GJ (1991) Algorithmic information and evolution. In Solbrig O, Nicolis G (eds) *Perspectives on Biological Complexity*, IUBS Press, Paris.
7. Cheeseman P (1985) In defense of probability. In *Proc. 9th International Joint Conf. on Artificial Intelligence*, Morgan Kaufmann, Los Altos, CA.
8. Cheeseman P (1988) An inquiry into computer understanding. *Computational Intelligence*, 4(1):58–66.
9. Conte M, Tautteur G, De Falco I, Della Cioppa A, Tarantino E (1997) Genetic programming estimates of Kolmogorov complexity. In *Proc. 17th International Conf. on Genetic Algorithms*, Morgan Kaufmann, San Francisco, CA.
10. Cox RT (1946) Probability, frequency, and reasonable expectation. *American Journal of Physics*, 14(1):1–13.
11. Daley RP (1973) Minimal-program complexity of sequences with restricted resources. *Information and Control*, 23(4):301–312.
12. Daley RP (1977) On the inference of optimal descriptions. *Theoretical Computer Science*, 4(3):301–319.
13. Dawid AP (1984) Statistical theory. The prequential approach. *Journal of the Royal Statistical Society*, Series A 147:278–292.
14. Feder M, Merhav N, Gutman M (1992) Universal prediction of individual sequences. *IEEE Transactions on Information Theory*, 38:1258–1270.
15. Fitting MC (1996) *First-Order Logic and Automated Theorem Proving.* Graduate Texts in Computer Science. Springer, Berlin.
16. Fudenberg D, Tirole J (1991) *Game Theory.* MIT Press, Cambridge, MA.
17. Gács P (1974) On the symmetry of algorithmic information. *Soviet Mathematics Doklady*, 15:1477–1480.
18. Hopcroft J, Motwani R, Ullman JD (2001) *Introduction to Automata Theory, Language, and Computation.* Addison-Wesley.
19. Hutter M (2000) A theory of universal artificial intelligence based on algorithmic complexity. Technical Report cs.AI/0004001 http://arxiv.org/abs/cs.AI/0004001.
20. Hutter M (2001) Convergence and error bounds for universal prediction of nonbinary sequences. In *Proc. 12th European Conf. on Machine Learning (ECML-2001)*, volume 2167 of *LNAI*, Springer, Berlin.

21. Hutter M (2001) General loss bounds for universal sequence prediction. In *Proc. 18th International Conf. on Machine Learning (ICML-2001)*.
22. Hutter M (2001) New error bounds for Solomonoff prediction. *Journal of Computer and System Sciences*, 62(4):653–667.
23. Hutter M (2001) Towards a universal theory of artificial intelligence based on algorithmic probability and sequential decisions. In *Proc. 12th European Conf. on Machine Learning (ECML-2001)*, volume 2167 of *LNAI*, Springer, Berlin.
24. Hutter M (2001) Universal sequential decisions in unknown environments. In *Proc. 5th European Workshop on Reinforcement Learning (EWRL-5)*.
25. Hutter M (2002) The fastest and shortest algorithm for all well-defined problems. *International Journal of Foundations of Computer Science*, 13(3):431–443.
26. Hutter M (2002) Self-optimizing and Pareto-optimal policies in general environments based on Bayes-mixtures. In *Proc. 15th Annual Conf. on Computational Learning Theory (COLT 2002)*, volume 2375 of *LNAI*, Springer, Berlin.
27. Hutter M (2003) Convergence and loss bounds for Bayesian sequence prediction. *IEEE Transactions on Information Theory*, 49(8):2061–2067.
28. Hutter M (2003) On the existence and convergence of computable universal priors. In *Proc. 14th International Conf. on Algorithmic Learning Theory (ALT-2003)*, volume 2842 of *LNAI*, Springer, Berlin.
29. Hutter M (2003) Optimality of universal Bayesian prediction for general loss and alphabet. *Journal of Machine Learning Research*, 4:971–1000.
30. Hutter M (2004) *Universal Artificial Intelligence: Sequential Decisions based on Algorithmic Probability*. Springer, Berlin. http://www.idsia.ch/~marcus/ai/uaibook.htm.
31. Kaelbling LP, Littman ML, Moore AW (1996) Reinforcement learning: a survey. *Journal of Artificial Intelligence Research*, 4:237–285.
32. Ko K-I (1986) On the notion of infinite pseudorandom sequences. *Theoretical Computer Science*, 48(1):9–33.
33. Kolmogorov AN (1965) Three approaches to the quantitative definition of information. *Problems of Information and Transmission*, 1(1):1–7.
34. Kumar PR, Varaiya PP (1986). *Stochastic Systems: Estimation, Identification, and Adaptive Control*. Prentice Hall, Englewood Cliffs, NJ.
35. Kwee I, Hutter M, Schmidhuber J (2001) Gradient-based reinforcement planning in policy-search methods. In *Proc. 5th European Workshop on Reinforcement Learning (EWRL-5)*.
36. Kwee I, Hutter M, Schmidhuber J (2001) Market-based reinforcement learning in partially observable worlds. In *Proc. International Conf. on Artificial Neural Networks (ICANN-2001)*, volume 2130 of *LNCS*, Springer, Berlin.
37. Levin L (1973) Universal sequential search problems. *Problems of Information Transmission*, 9:265–266.
38. Levin L (1974) Laws of information conservation (non-growth) and aspects of the foundation of probability theory. *Problems of Information Transmission*, 10(3):206–210.
39. Li M, Vitányi PMB (1991) Learning simple concepts under simple distributions. *SIAM Journal on Computing*, 20(5):911–935.
40. Li M, Vitányi PMB (1992) Inductive reasoning and Kolmogorov complexity. *Journal of Computer and System Sciences*, 44:343–384.

41. Li M, Vitányi PMB (1992). Philosophical issues in Kolmogorov complexity (invited lecture). In *Proceedings on Automata, Languages and Programming (ICALP-92)*, Springer, Berlin.
42. Li M, Vitányi PMB (1997) *An Introduction to Kolmogorov Complexity and its Applications*. Springer, Berlin, 2nd edition.
43. Littlestone N, Warmuth MK (1989) The weighted majority algorithm. In *30th Annual Symposium on Foundations of Computer Science*.
44. Littlestone N, Warmuth MK (1994) The weighted majority algorithm. *Information and Computation*, 108(2):212–261.
45. Lucas, JR (1961) Minds, machines, and Gödel. *Philosophy*, 36:112–127.
46. Michie D (1966) Game-playing and game-learning automata. In Fox, E (ed) *Advances in Programming and Non-Numerical Computation*, Pergamon, New York.
47. Von Neumann J, Morgenstern O (1944) *Theory of Games and Economic Behavior*. Princeton University Press, Princeton, NJ.
48. Osborne MJ, Rubenstein A (1994) *A Course in Game Theory*. MIT Press, Cambridge, MA.
49. Penrose R (1989) *The Emperor's New Mind*. Oxford University Press, Oxford.
50. Penrose R (1994) *Shadows of the Mind, A Search for the Missing Science of Consciousness*. Oxford University Press, Oxford.
51. Pintado X, Fuentes E (1997) A forecasting algorithm based on information theory. In Tsichritzis D (ed) *Objects at Large*, Technical Report, Université de Genève.
52. Rissanen JJ (1989) *Stochastic Complexity in Statistical Inquiry*. World Scientific, Singapore.
53. Russell S, Norvig P (2003) *Artificial Intelligence. A Modern Approach*. Prentice-Hall, Englewood Cliffs, NJ, 2nd edition.
54. Schmidhuber J (1997) Discovering neural nets with low Kolmogorov complexity and high generalization capability. *Neural Networks*, 10(5):857–873.
55. Schmidhuber J (2002) The speed prior: A new simplicity measure yielding near-optimal computable predictions. In *Proc. 15th Conf. on Computational Learning Theory (COLT-2002)*, volume 2375 of *LNAI*, Springer, Berlin.
56. Schmidhuber J (2003) Bias-optimal incremental problem solving. In Becker S, Thrun S, Obermayer K (eds) *Advances in Neural Information Processing Systems 15*, MIT Press, Cambridge, MA.
57. Schmidhuber J (2004) Optimal ordered problem solver. *Machine Learning*, 54(3):211–254, also this volume.
58. Schmidhuber J, Zhao J, Wiering MA (1997) Shifting inductive bias with success-story algorithm, adaptive Levin search, and incremental self-improvement. *Machine Learning*, 28:105–130.
59. Schmidt M (1999) Time-bounded Kolmogorov complexity may help in search for extra terrestrial intelligence (SETI). *Bulletin of the European Association for Theoretical Computer Science*, 67:176–180.
60. Shoenfield JR (1967) *Mathematical Logic*. Addison-Wesley, Reading, MA.
61. Solomonoff R (1964) A formal theory of inductive inference: Parts 1 and 2. *Information and Control*, 7:1–22 and 224–254.
62. Solomonoff R (1978) Complexity-based induction systems: Comparisons and convergence theorems. *IEEE Transaction on Information Theory*, IT-24:422–432.

63. Solomonoff R (1986) The application of algorithmic probability to problems in artificial intelligence. In Kanal L, Lemmer J (eds) *Uncertainty in Artificial Intelligence*, Elsevier Science/North-Holland, Amsterdam.
64. Solomonoff R (1997) The discovery of algorithmic probability. *Journal of Computer and System Sciences*, 55(1):73–88.
65. Solomonoff R (1999) Two kinds of probabilistic induction. *Computer Journal*, 42(4):256–259.
66. Sutton R, Barto A (1998) *Reinforcement Learning: An Introduction*. MIT Press, Cambridge, MA.
67. Valiant LG (1984) A theory of the learnable. *Communications of the ACM*, 27(11):1134–1142.
68. Vovk VG (1992) Universal forecasting algorithms. *Information and Computation*, 96(2):245–277.
69. Vovk VG, Watkins C (1998) Universal portfolio selection. In *Proc. 11th Conf. on Computational Learning Theory (COLT-98)*, ACM Press, New York.

Program Search as a Path to Artificial General Intelligence

Łukasz Kaiser

Mathematische Grundlagen der Informatik, RWTH Aachen
D-52056 Aachen, Germany
lukaszkaiser@gmail.com

Summary. It is difficult to develop an adequate mathematical definition of intelligence. Therefore we consider the general problem of searching for programs with specified properties and we argue, using the Church-Turing thesis, that it covers the informal meaning of intelligence. The program search algorithm can also be used to optimise its own structure and learn in this way. Thus, developing a practical program search algorithm is a way to create AI.

To construct a working program search algorithm we show a model of programs and logic in which specifications and proofs of program properties can be understood in a natural way. We combine it with an extensive parser and show how efficient machine code can be generated for programs in this model. In this way we construct a system which communicates in precise natural language and where programming and reasoning can be effectively automated.

1 Intelligence and the Search for Programs

Intelligence is usually observed when knowledge is used in a smart and creative way to solve a problem. Still, it seems that the core of intelligence is neither the knowledge nor the specific method to use it, but the general way to learn from previous experience. This is not limited to adopting new knowledge, but also includes learning new ways to use what we know, extending it by reasoning, and even improving learning methods to learn more efficiently. Developing new ways to solve problems is a better indication of intelligence than solving separate tasks, as it is a creative work, where we do not have a precise description of what to do and are expected to find the right method knowing only what goals we want to achieve.

We will represent the informal notion of learning new ways to solve problems as the search for programs that fulfil some properties and we will design a system to make it practical. To explain why we choose this representation we have to analyse how methods of solving problems in general can be modelled by abstract notions and how problems can be specified. We use the general representation that dates back to the birth of AI and computer science with the works of Gödel, Turing, and Church.

We claim that the informal notion of a method for solving certain tasks can be expressed in mathematical terms as a Turing machine. To justify this we use the Church-Turing thesis, the assumption that everything that is computable, any complex behaviour of a system, can be computed or modelled using only a small set of simple abstract operations. We can take different sets of such operations, use either Turing machines or lambda calculus, recursive functions or any other programming language. Still, these all have the same computational power and over fifty years after stating this thesis we did not manage to find any physical system, neither classical nor quantum, that would be able to compute more than a simple Turing machine. Note a straightforward consequence of the Church-Turing thesis: as far as we assume that humans are normal, although very complex physical objects, the procedure that operates in our brains can also be implemented on Turing machines and therefore also on usual computers with enough memory, when these get fast enough.

The thesis of Church and Turing justifies that any informally understood *method for solving a problem* can be defined as an algorithm, a Turing machine that takes the instance of the problem as input and returns the solution.

Of course, to be considered a viable solution for the given problem the method (now – the Turing machine) has to fulfill certain requirements that depend on the problem. For example, if we want to find a way to sort cards, there might be many better or worse ways to do this, machines that take the cards and return them mixed, but any *solution* must return the cards in the right order. We will use the natural (first order) logic with the language appropriate for describing Turing machines to specify such requirements.

Please note that in this logic we are not only able to specify what a good solution is; we can also define an ordering, defining when one solution is better than another. We can say, for example, that solution A is better than solution B if it takes less time to sort through the cards, and this can be expressed using the definition of the number of steps in a run of a Turing machine. We also have to take into account that often the goals to achieve or the conditions of work will not be directly specified, but can refer to knowledge about similar events in the past. This can also be included in our requirement specification if we encode the past knowledge inside the formula. Since we assume the Church-Turing thesis, we can also take it for granted that a Turing machine can verify the correctness of a solution, and then all possible problems that an intelligent agent will ever be required to solve can be specified in first order logic, or even a limited variant of it.

We have modeled problem solving as searching for Turing machines with specified properties. Determining if such a machine exists is of course undecidable and the problem is intractable in general, but we can make some additional assumptions. First, we can assume that we do not only want the machine, but also a proof that it satisfies the formula and that such a machine with a proof exists. This is a realistic assumption in the context of artificial intelligence, since the agent normally wants to solve a problem that is solvable, and when the solution is found then it should be clear that it is correct. When no solution can be found or the agent knows nothing about whether it is correct or not, not even in the probabilistic sense, then it has to resort anyway to other methods that we do not investigate here, like asking another agents for help, or trying to solve the problem again later. Therefore, we will not consider the cases when the problem is not solvable or it can not be proved that the solution is correct, since in such cases the AI agent has to determine when to stop searching for the solution using external knowledge and taking other factors into account. Instead, we will concentrate on making a model of programs and a program search algorithm that preserves generality, yet is simple and efficient enough to be used in practice for specific classes of problems.

As we mentioned discussing intelligence, we do not only want a procedure to solve certain tasks, but we want the agent to learn. Learning, in this case, amounts to improving the procedure, so that after a number of problem instances have been solved it will solve other similar instances more efficiently. We will present a self-improving algorithm that searches for Turing machines with specified properties. Moreover, we will show an innovative system that binds programming and problem solving with natural language processing.

Outline. In the next section we will look for a general procedure that, when given a logic formula, looks for a Turing machine fulfilling it, and that optimizes itself with each successful run. We will present such a theoretical method based on the program and proof enumeration technique, which was already used by Gödel [4] and Turing. The resulting procedure has the nice property of self-improvement, similarly to how we improve our learning skills, and it is very general, so after some time it will become as good as any other such procedure with respect to any appropriate measure of efficiency. We will also show how it can be used by an AI agent in an unknown environment to learn to take successful actions.

The problem we face with such a theoretical solution is that it would not be usable in practice if implemented in a direct way. The time required for it to improve to a level of efficiency that would give any tangible results would be enormous. Therefore, in subsequent sections we will present a model of computation and program logic that combines functional programming with reasoning using games. This model is powerful enough to express algorithms and proofs on the same level of abstraction as we think of them, and at the same time compile programs to binary code. Thus, when running the program search procedure in this model, we can expect the implementation to execute efficiently and, even when it does not find the results automatically, we can still understand the steps it takes and guide it to the correct solution.

In Sect. 3 we will present the model and additionally give a method to parse compound expressions that fits in the model. Such parsing improves the presentation of programs and proofs, and can be extended to handle basic natural language processing. We will also use examples to show the compilation of programs from this model to efficient code, going through the C language.

In Sect. 4 we will analyse how properties of programs described in the model can be proved formally at a high level of abstraction. We will show

how automatic proofs can be guided by the user or by different heuristics, and how sub-procedures for reasoning in less general cases can be included in the model without loss of generality.

Please note that the theoretical results we present are well known and we do not discuss them very precisely. The model of computation, the method to parse expressions, and the logic presented later are also based on well known ideas but their combination is innovative. Therefore, we give more details about it and describe how to create a system that allows to write in natural language programs about which we can reason semi-automatically in formal logic, and which can be compiled to efficient machine code.

2 Theoretical Results

In this section we give an overview of the theoretical results that concern searching for programs with specified properties, and using program search in the standard AI model. We take Turing machines as our model of computation but any other Turing-complete model could be used here. Also, we do not give the results in full detail, as most of them are already standard knowledge in computer science, and we just want to put them in the context of AGI or extend them, and in such cases we give references to papers where these extensions are thoroughly discussed.

We start our theory by setting a description of programs and choosing a computable set of axioms from which we will deduce program properties. Later, we will present a model of programs that we consider simple and more practical, but let us now consider the Turing machines defined in set theory together with the axioms of set theory as formalized by Zermelo and Fränkel, which is a widely used axiomatization.

The *program search problem* can be stated as follows: given a formula $\varphi(x_1, \ldots, x_n)$ in first order logic on the structure defined above with free variables $x_1, \ldots, x_k$ denoting Turing machines, find a proof of $\varphi(m_1, \ldots, m_k)$ for some Turing machines $m_1, \ldots, m_k$.

Let us now state an important positive fact which is a straightforward consequence of the enumerability of Turing machines and proofs.

Fact 1. *There exists an algorithm that computes the solution to the program search problem if any solution exists, so given $\varphi(x_1, \ldots, x_k)$ it computes $m_1, \ldots, m_k$ and the proof of $\varphi(m_1, \ldots, m_k)$, assuming that for some machines such a proof exists.*

Proof. Since Turing machines, programs, and proofs are enumerable and it can be determined algorithmically whether a sequence of formulas forms a proof of a given claim, we can use the following algorithm to prove this fact:

(1) Set `length` to 1.

(2) Enumerate all k-tuples $m_1, \ldots, m_k$ of Turing machines shorter that `length` and all proofs shorter than `length` and check if there is any proof among these that proves $\varphi(m_1, \ldots, m_k)$.
(3) If the correct machines and proof were found, return them, else increase `length` by one and return to point (2).

Of course, this algorithm will find a solution, even the shortest one, if it exists. Otherwise, the algorithm will never stop. We will denote this algorithm by PSP_0.

2.1 Program Search in the Standard AI Model

We will now consider the often used AI model where the agent interacts with the environment. The agent is modeled to have sensors from which it collects input, and effectors which it uses to execute actions. Additionally, at any moment the agent may get additional feedback that denotes its own happiness, or a quantified assessment it gets from a teacher agent. The agent's task is to maximize the total assessment it gets throughout its life.

To be able to construct well-acting agents we have to assume something about the environment, or, at least, something about its probabilistic behaviour. One sensible assumption is that the environment, or at least the probability distribution of events, is driven by some program (Turing machine). We want to create an agent that will behave in a worse way than the optimal agent, if one exists, only for some period of time, and that will later act optimally.

Let us sketch the possible construction of such an agent, which uses the program search to find rules in environment behaviour, and uses these rules as predictors, in order to find the best possible actions in the assumed environment. This is a very natural general way to act by first planning actions according to the expected future outcome, and then choosing the best ones. Let our agent store the following internal variables:

 (i) a list of interwoven events and actions called `history`, initially empty;
 (ii) a program `model` that models the environment, initially any short one;
(iii) a program `actor` that models the suspected optimal behaviour of the agent, initially any trivial program;
(iv) two numbers `max size` and `max time`, initially set to 1.

We consider a model of the environment m_1 to be better than m_2 if we can prove that there is an agent that achieves, using m_1, a better assessment than any agent can achieve using m_2. The agent will act according to the following algorithm when a new event is encountered.

(1) Append the event to `history`.
(2) Search for any program smaller than `max size` that generates `history` in less time than `max time`. Among such environment models, consider only the best ones as defined above, and update `model` to be one of the shortest of the best programs.

(3) Search for a proof, shorter than `max size`, that shows that some program, smaller than `max size` and halting on every input, can achieve a better assessment in environment `model` than the program `actor`. In that case update `actor` to be one of the shortest of such programs.
(4) Increase `max time` and `max size` by one.
(5) Calculate the response of `actor` to the input event, append the response to `history`, and output it.

Since in the construction we search through all possible programs, we can state the following simple fact.

Fact 2. *If a Turing machine can describe the behaviour of the environment and there is a provably optimal agent for this environment, then the presented agent gets assessment smaller than the optimal one only for some period of time, and behaves optimally afterwards.*

Proof. Indeed, if the environment is a program, then after some running time it will generate output that distinguishes it from any shorter program. Please note that before the model is clear, the agent will assume an optimistic one and undertake actions according to this assumption. Then, after analysing this output in step (2), the variable `model` will be set to the correct environment program. When this variable is set correctly the agent will search in step (3) for the optimal agent for the detected environment. Since we assumed that there is a provably optimal agent, this agent and the proof of its optimality have some length. When `max size` exceeds this length, the variable `actor` will be set to the optimal program. Therefore, the agent will start to behave optimally after detecting the correct environment and the necessary proof.

The construction of the AIXI agent, based on similar ideas, but extended and also specified in probabilistic context, was presented in detail and with full proofs of optimality by Hutter [7, 8], and the underlying theory is described thoroughly in [9]. The method to define different things as shortest possible programs was developed by Levin [13] in the framework of Kolmogorov Complexity theory [12, 21], and Li and Vitanyi give an excellent overview of these and similar methods in [14].

2.2 Self-improving Program Search

We saw that the program search problem can be useful for the construction of an AI agent, but we still do not know how to search for programs efficiently. We do not intend to search for any program in particular, but to learn efficient procedures to search for programs of interest. We will show how we can define what programs are interesting depending on the history of previous search tasks, and we will show how in such a case a procedure for program search can improve itself.

Let us therefore specify an algorithm that receives solvable instances of the program search problem, solves them, and improves its performance on such and similar instances. To construct this procedure we need to define how to decide whether one program search algorithm is *more efficient* than another with respect to the history of observed instances of the problem, but we will postpone the discussion of such definitions until the next section. Also, the presented algorithm runs several processes simultaneously, but it is clear that such parallelism can be simulated on Turing machines as well as on single processor computers.

First, the algorithm initializes variable P to PSP_0, the program search algorithm presented before, and P will be used both to solve received problem instances and for self-improvement. It also initializes `history` to an empty sequence. It then divides available resources into two parts and runs two processes simultaneously. Whenever a new instance of a program search problem is received, it is appended to `history`. The algorithm works with respect to the efficiency measure μ that in every moment depends on the `history` known at that moment.

When the **main** process receives the problem instance, it uses P to solve it, and returns the solution.

The **improvement** process works as follows:

(1) Append the formula that describes the problem of creating a program search algorithm more efficient than P with respect to μ to `history`.
(2) Use P to find a more efficient program search algorithm as defined by the above formula.
(3) Update P to a new, more efficient version.
(4) Repeat, starting from (1) with new P and perhaps an extended `history`.

It can be seen that this algorithm not only solves the program search problem, but also uses its program search capacity to optimize itself. Therefore, even if PSP_0 is not an efficient solution, the presented procedure will automatically find a better one, thanks to the improvement component. We assumed that the efficiency relation depends on the history. If we do not want this algorithm to fall in cycles thinking that some program search algorithm P_1 is better than P_2 and later, when history changes, deciding the other way, we have to assume that the definition of efficiency will be monotonic in some way. If we are not able to make such assumptions, it could be useful to separate the history of instances received from outside from the self-improvement instances, and use two separate program search algorithms, one for solving the problems and the second to improve program search. The following fact can be stated with the assumption that the definition of efficiency is appropriate, but extensions to more complex situations are also possible.

Fact 3. *Let a program search algorithm Q (our goal, the efficient algorithm) be given and assume that the efficiency relation is such that there is only a bounded number of algorithms that are provably more efficient than PSP_0 and less efficient than Q, with respect to any possible histories. Then, for any sequence of received instances, the presented algorithm will after some number* *of steps substitute Q for its internal variable P and therefore become at least as efficient as Q.*

This way, if we find some reasonable definition of efficiency, then we can just start this algorithm and wait until it finds a good solution to the program search problem, which can then be used as an artificial general intelligence. The only practical issue is that if we start with PSP_0 then even with the best computers we would have to wait very long. Similar learning algorithms and program searches have been analysed with the tools of Kolmogorov Complexity theory, see [14, 8] for more information on this topic. Schmidhuber gives detailed discussion of a recently developed optimally self-improving machine, called the Gödel Machine, in [19]. Such methods can also be relevant for physics as is discussed in [20].

2.3 Discussion of Efficiency Definitions

Let us now address the definition of the efficiency of algorithms which solve the program search problem. We will try to compare such algorithms with respect to a history of instances of the problem they solve.

The usual definitions of complexity, even in the asymptotic sense, can not be used in this case, as many instances are not solvable at all.

Let us again look at the problem from an informal and intuitive perspective. After gaining experience on a class of instances in the past, we will normally say that an algorithm is efficient if it solves the instances from this class and other *similar* instances fast. The remaining problem is to define which instances are similar. It seems reasonable to say that two instances are similar if one can be transformed into the other using a few simple transformations, for example by changing some parameters or shifting them in some way.

Assume that a set of simple transformations is given. Then we can define the *level of similarity* between two instances as the number of transformations that have to be applied to get from one instance to the other. For practical reasons we could also assume that if this number is greater than some constant, then the instances are not similar at all.

Using this, we can say that one program search algorithm is more efficient than another with respect to a history if it is faster on all instances in the history and on all similar instances. We could also use an alternative definition and say that the *weight* of an algorithm A with respect to history H is

$$w(A, H) = \Sigma_{\{i \text{ similar to some } j \in H\}} \text{time}(A, i) \cdot 2^{\text{similarity}(i, H)},$$

where similarity(i, H) denotes the smallest level of similarity between i and any instance from H, and time(A, i) denotes the time it takes A to solve i. We assume that the sum is taken only over solvable instances i.

These two definitions seem reasonable and the first one satisfies the requirements presented in Fact 3, since it is monotonic with respect to history.

But, in practice, the second definition might be more useful, since it seems practical to decrease the efficiency of the algorithm in a few cases if it can lead to large improvements in other cases. It could also be practical to use some other weight for the definition of efficiency, for example including a heuristic that might make the efficiency a little worse in most cases, but improve it dramatically for some narrow class of cases.

Similar problems in the context of program search are considered in more detail by Schmidhuber in [18, 19], where more examples are presented. Still, it seems that the efficiency functions will have to be fine-tuned experimentally when such procedures start to be used in practice.

3 Convenient Model of Computation

We showed how to construct a learning program search procedure, but if we tried to implement it directly using PSP_0, then it would not be practical. Therefore, our goal now is to present a more usable solution. The model we present with its theory is described in detail in the documents in [10], where the reader can also find an implementation of the discussed algorithms. Since this is still work in progress, and many details are actively being polished, the web site should be consulted for corrections and the most recent version. Many of these definitions and methods are already standard in functional programming and term rewriting [2].

Let us repeat our motivation: We need a model of computation which will allow us to easily write programs and, at the same time, reason about them. To construct such a model, we will concentrate only on two basic operations used in programming, namely the possibility to define and apply functions and the possibility to create compound data types. Therefore, in our model we will operate on objects that represent some data, e.g. $1, 2, [T, F]$, and on functions like $+, \cdot, and$. We are allowed to compose functions with data and write *terms* in this way, for example $1 + 2$, T *and* F or $(1 + 2) \cdot (3 + 4)$.

To define functions in this model, we write rules telling how one term should change to another, e.g. T *and* F $\rightarrow$ F. In such rules we can use variables, for example, we can write $x + 0 \rightarrow x$. Note that not all terms have any meaning, for example $1 + T$ does not mean anything. To avoid such terms we will introduce *types*, such that, for example 1 will have type *int* and $+$ will have type *int, int* $\rightarrow$ *int* so we will not be allowed to apply it to the boolean value T.

The model we present is known as term rewriting with polymorphic types. We will first give the basic definitions in detail, in order to show that formal reasoning about these objects is indeed feasible and to avoid confusion later, when we give examples less formally. We will also show how to parse terms from expressions in semi-natural language and how to generate efficient machine code for programs is this model. Thus, we will construct a computer system where natural language input can be used for programming and reasoning without loss of efficiency of the created programs.

To define the model, we need the following classes, where *arity* is always a function that assigns a natural number to each element of the considered set:

(i)   the infinite enumerable set of *type variables*, denote α, β, γ;
(ii) the finite set Γ of *type names* with arity, denoted T, R, S;
(iii) the infinite enumerable set V of *term variables* with arity, denoted x, y, z;
(iv) the finite set Θ of *constructor names* with arity, denoted A, B, C;
(v)   the finite set Σ of *function names* with arity, denoted f, g, h.

Types. We start with formal type definitions. These might be difficult to understand at first, but the examples we give should be enough for an intuitive understanding. The set of types is defined inductively as the smallest set $\mathcal{G}$ such that:

(1) each type variable $\alpha \in \mathcal{G}$;
(2) if $T \in \Gamma$ with arity n and $R_1, \ldots, R_n \in \mathcal{G}$ then $T(R_1, \ldots, R_n) \in \mathcal{G}$;
(3) for any number n and types $T_1, \ldots, T_n \in \mathcal{G}$ and result type $R \in \mathcal{G}$ the *functional type* $(T_1, \ldots, T_n \to R) \in \mathcal{G}$.

We allow functional types for $n = 0$ to maintain consistent notation, but we consider the types R and $\emptyset \to R$ to be identical, and we will not distinguish them.

Let us for example define the types of boolean values, pairs, and lists. We will set:
$$\Gamma = \{\text{booleans}, \text{lists}, \text{pairs}\},$$
where booleans has arity 0, lists arity 1 and pairs arity 2. Then, the example type E of pairs consisting of a boolean value and a list of any other type can be represented as:
$$E = \text{pairs}(\text{booleans}, \text{lists}(\alpha)) \in \mathcal{G}.$$

The set $\text{TVar}(T)$ of type variables occurring in a type T is also defined inductively by $\text{TVar}(\alpha) = \{\alpha\}$, $\text{TVar}(T(R_1, \ldots, R_n)) = \text{TVar}(R_1) \cup \cdots \cup \text{TVar}(R_n)$, and $\text{TVar}(T_1, \ldots, T_n \to R) = \text{TVar}(T_1) \cup \cdots \cup \text{TVar}(T_n) \cup \text{TVar}(R)$, so $\text{TVar}(E) = \{\alpha\}$.

The usual intuition behind types is to view them as labeled trees, therefore we introduce the notion of positions in types. The set Λ of *positions* is the set of sequences of positive natural numbers. By $\lambda \in \Lambda$ we will denote the empty sequence or the top (root) position in the type.

For a given type T and position p we either say that p does not exist in T, or define the type at position p in T (denoted by $T|_p$) in the following inductive way:

(1) λ exists in each type and $T|_\lambda = T$;
(2) $p = (n, q)$ exists in $S = T(R_1, \ldots, R_m)$ if $m \geq n$ and q exists in R_n and in such case $S|_p = R_n|_q$;

(3) $p = (n, q)$ exists in $S = T_1, \ldots, T_m \to R$ if either $m \geq n$ and q exists in T_n and in such case $S|_p = T_n|_q$, or $m + 1 = n$ and q exists in R and then $S|_p = R_q$.

A position p is *above* some position q if there exists a sequence r of numbers such that $q = (p, r)$. In this case we also say that q is *below* p. The *height* of a position is its length, and the height of a type is the maximal height of a position existing in this type. In the example type E, one can see that position 3 does not exist in E, but $E|_{2,1} = \text{lists}(\alpha)|_1 = \alpha$ and so E has height 2.

Substitutions and unifiers. Sometimes we want to change a part of a type, and then we say that we *substitute* type S in type T *at position* p. As a result we get the type $R = T[S]_p$, such that for all positions q not below p that exist in T, it holds that $R|_q = T|_q$ and $R|_p = S$. Less formally, R is just T with the subtree at position p replaced by S. *Substituting* type S in type T *for a variable* α is defined as substituting S in T at all positions p where $T|_p = \alpha$. A *type substitution*, usually denoted with letters σ, τ, ρ, is a set of pairs, each consisting of a type variable and a type, and such pairs are denoted by $\alpha \leftarrow T$. For a substitution $\sigma = \{\alpha_1 \leftarrow T_1; \ldots; \alpha_n \leftarrow T_n\}$ we will denote the set of variables substituted for by $\text{TVar}(\sigma) = \{\alpha_1, \ldots, \alpha_n\}$ and we will say that by *applying* σ to a type T we obtain the type $R = T\sigma$, which is the result of substituting, for each i, the type T_i in T for the variable α_i. In some algorithms it is necessary to ensure that the variables substituted for are disjoint with variables in the terms we substitute. As an example, let us apply $\{\alpha \leftarrow \text{booleans}\}$ to the type E defined before and get

$$\text{pairs}(\text{booleans}, \text{lists}(\alpha))\{\alpha \leftarrow \text{booleans}\} = \text{pairs}(\text{booleans}, \text{lists}(\text{booleans})).$$

Sometimes we need to rename type variables in a type T; either all variables or only the variables from a given set $\mathcal{V}$. Let us set:

$$\sigma = \{\alpha \leftarrow \alpha \underbrace{''' \cdots '''}_{k} \; : \; \alpha \in \text{TVar}(T) \cap \mathcal{V}\},$$

where k is first set to 1 and doubles each time we rename any type. Then we can define the renamed type $\overline{T}^{\mathcal{V}} = T\sigma$ and if we want to rename all type variables, we will just write $\overline{T}$ for $\overline{T}^{\text{TVar}(T)}$. As the names of substituted variables change with the number k with each renaming, we can be sure that any two types R and S have disjoint variables after renaming, $\text{TVar}(\overline{R}) \cap \text{TVar}(\overline{S}) = \emptyset$.

We can apply a type substitution σ to another type substitution $\rho = \{\alpha_1 \leftarrow T_1; \ldots; \alpha_n \leftarrow T_n\}$ and obtain the substitution:

$$\rho\sigma = \{\alpha_1 \leftarrow T_1\sigma; \ldots; \alpha_n \leftarrow T_n\sigma\}.$$

We will say that a type substitution σ is *more general* than ρ if there is another substitution τ for which $\sigma\tau \subseteq \rho$.

Let us now take a set of tuples of types

$$\{(T_1, R_1, \ldots, S_1), \ldots, (T_n, R_n, \ldots, S_n)\}.$$

Any substitution ρ such that $T_i\rho = R_i\rho = \ldots = S_i\rho$ for each i is called a *unifier* of this set, and it is a well known and important fact that if there is any unifier, then there exists the most general one, which we will denote by:

$$\text{mgu}\{(T_1, R_1, \ldots, S_i), \ldots, (T_n, R_n, \ldots, S_n)\}.$$

The most general unifier can be computed in polynomial time if we can represent types in the form of acyclic graphs, and in exponential time if we restrict the representation to trees, where identical sub-trees can not be compressed. For example, it is easy to see that there is no unifier for:

$$\{(\text{pairs}(\text{booleans}, \alpha), \text{pairs}(\text{lists}(\beta), \gamma)\},$$

but the pair of types $(\text{pairs}(\alpha, \text{booleans}), \text{pairs}(\text{lists}(\beta), \gamma)$ can be unified, and:

$$\text{mgu}\{(\text{pairs}(\alpha, \text{booleans}), \text{pairs}(\text{lists}(\beta), \gamma)\} = \{\alpha \leftarrow \text{lists}(\beta), \gamma \leftarrow \text{booleans})\}.$$

When given a set of type substitutions $\{\sigma_1, \ldots, \sigma_n\}$, we will also use the most general unifier of these substitutions, $\tau = \text{mgu}\{\sigma_1, \ldots, \sigma_n\}$, defined as the unifier of the set of tuples $(T_1^\alpha, \ldots, T_k^\alpha)$ of all such types that $\alpha \leftarrow T_i^\alpha \in \sigma_{l_i}$ for some σ_{l_i}, so all types substituted for the same variable in all substitutions σ_i will be unified. Let us also denote, for each type variable α, the unified type $T_1^\alpha \tau$ by T^α and let:

$$\text{subst}\{\sigma_1, \ldots, \sigma_n\} = \{\alpha \leftarrow T^\alpha \ : \ \alpha \in \text{TVar}(\sigma_1) \cup \ldots \cup \text{TVar}(\sigma_n)\}.$$

Typed terms. We will now assume that each term variable $x \in V$, each constructor $C \in \Theta$, and each function $f \in \Sigma$ with arity n has an associated functional type

$$\text{type}(f) \ (\text{type}(C), \ \text{type}(x)) \ = \ T_1, \ldots, T_n \to R \ \in \mathcal{G}.$$

We will make an additional assumption that, for constructors, the type R is neither a type variable nor a functional type, and has height at most one. Using this information about type we can define inductively the set of *well typed terms* $\mathcal{T}$, giving at the same time the definition of the type of a term, $\text{type}(t) \in \mathcal{G}$, the set of variables of a term, $\text{Var}(t) \subseteq V$, and the substitution $\rho(t)$ that reconstructs type variables in $\text{Var}(t)$. To make the definition easier to follow, we will analyse the typing of the term $\text{Pair}(y, y)$ with $\text{type}(y) = \alpha$, and the constructor $\text{Pair} \in \Gamma$ with type $\alpha, \beta \to \text{pairs}(\alpha, \beta)$. We are using this slightly non-standard definition with reconstruction because it makes it easier to present the parsing algorithm later.

First, each variable $x \in V$, constructor $C \in \Theta$, and function symbol $f \in \Sigma$ belongs to $\mathcal{T}$ with the associated type, $\text{Var}(C) = \text{Var}(f) = \rho(C) = \rho(f) = \emptyset$, and

$$\text{Var}(x) = \{x\}, \ \rho(x) = \{ \ \alpha \leftarrow \alpha \text{ for } \alpha \in \text{TVar}(\text{type}(x)) \ \},$$

So, in our example, we have $\rho(y) = \{\alpha \leftarrow \alpha\}$.

Let a variable $x \in V$, constructor $C \in \Theta$, or function symbol $f \in \Sigma$ have arity $n > 0$. We will first rename the associated type S and denote $\overline{S} = S_1, \ldots, S_n \to R$. At this point in our example, we rename the type of Pair to be $\alpha', \beta' \to \text{pairs}(\alpha', \beta')$, thus expressing the fact that the type variable α is only accidentally the same in the type of y and Pair.

Furthermore, let us take terms $t_1, \ldots, t_n \in \mathcal{T}$ with $\text{type}(t_i) = R_i$ and rename all variables that are not reconstructed, so let

$$T_i = \overline{R_i}^{\text{TVar}(R_i) \backslash \text{TVar}(\rho(t_i))} .$$

In our example we do not rename anything, as we take $t_1 = t_2 = x$ and in x all type variables are reconstructed. Then, $f(t_1, \ldots, t_n)$, $C(t_1, \ldots, t_n)$ or $x(t_1, \ldots, t_n)$ is well typed if there exists

$$\rho = \text{mgu}\{\rho(t_1), \ldots, \rho(t_n)\} \text{ and } \sigma = \text{mgu}\{(T_1\rho, S_1), \ldots, (T_n\rho, S_n)\},$$

and in such case, if $\tau = \text{subst}\{\rho(t_1), \ldots, \rho(t_n)\}$ then

$$f(t_1, \ldots, t_n) \in \mathcal{T}, \ \text{type}(f(t_1, \ldots, t_n)) = R\sigma,$$

and $\text{Var}(f(t_1, \ldots, t_n)) = \text{Var}(t_1) \cup \cdots \cup \text{Var}(t_n)$, $\rho(f(t_1, \ldots, t_n)) = \tau\sigma$, and likewise in the case of constructor C.

In our example the unifier ρ of variable substitutions for y is an empty substitution and σ unifies both α' and β' from the renamed type of Pair with α. Then, substituting it in the pair type we get the result type $\text{pairs}(\alpha, \alpha)$.

In the case of a variable, we have to extend the definitions, so we have $\text{Var}(x(t_1, \ldots, t_n)) = \{x\} \cup \text{Var}(t_1) \cup \cdots \cup \text{Var}(t_n)$, and

$$\rho(x(t_1, \ldots, t_n)) = \tau\sigma \cup \{\alpha \leftarrow \alpha \text{ for } \alpha \in \text{TVar}(R) \backslash \text{TVar}(S_1) \cup \cdots \cup \text{TVar}(S_n)\}.$$

We can also define positions in terms and term substitutions in an analogous way to the definitions for types, and we will say that a term t is *ground* if $\text{Var}(t) = \emptyset$, and that it is *linear* if no variable occurs in it at more than one position.

Let us for example take two constructors T and F with arity 0 and booleans as the assigned type. Let us also take the constructor Pair that we already know and two constructors for lists, Nil with arity 0 and type $\text{lists}(\alpha)$, and Cons with arity 2 and type $\alpha, \text{lists}(\alpha) \to \text{lists}(\alpha)$.

In the examples we will use the symbol : to denote the type of a given term. We can now create terms with specific types, for instance:

```
Cons (T, Nil) : lists (booleans),
Pair (T, F) : pairs (booleans, booleans),
```

but we are not allowed to use terms that are not well typed, like

Cons (F, T) or Cons (Pair (T, F), Cons (T, Nil)).

In the first case, a term from booleans is used where a term from lists(α) is expected. In the second case, there is no correct type to instantiate the type variable α in Cons type definition, since there are both terms from booleans and pairs(booleans, booleans) in the list. Since we will be continuing this example, let us simplify our notation. We will denote Cons (x, y) by x :: y and Pair (x, y) by (x, y), so the four terms presented above will be denoted:

T :: Nil , (T, F) , F :: T and (T, F) :: (T :: Nil).

To clarify the need for reconstructing substitutions, let us assume that we have three variables $x, y, z \in V$ with type(x) = lists(β), type(y) = booleans, type(z) = γ. Now let us take the following example of two terms:

(x :: Nil, y :: Nil),
(x :: z :: Nil, y :: z :: Nil).

The first term is well typed, since Nil is a constructor and its type can unify with lists(β) in one place and booleans in another as the type variable in type(Nil) will be renamed. The second term is not well typed since it is not possible to reconstruct the type for variable z, which can not have type booleans in one place and lists(β) in another.

Rewriting. To define a function in a program that we want to execute, let us introduce the concept of a *rewrite rule*, a pair of terms l and r, the left and right side of the rule, denoted by $l \to r$. In a rewrite rule $l \to r$ it must hold that Var(l) $\supseteq$ Var(r), type(l) = type(r) (modulo renaming of type variables), and the symbol at the top position in l must be a function name.

A rewrite rule $l \to r$ can be *applied* to a term t at position p if there exists a substitution σ of variables in l such that $t|_p = l\sigma$. The result of applying the rule is $t[r\sigma]_p$, the term t rewritten at position p. Note that there is only one possible result of applying the rule to a term at a given position, and that the conditions guarantee that a ground term remains ground after applying the rule to it at any position, and in such a case it still has the same type after the rule is applied. The rule is ground if r is ground and it is linear if r is linear.

We will model programs by a system of terms and types defined above and a set of rewrite rules, where each subset of rules with the same function symbol at the top position on the left side is linearly ordered, and as you will see, the first rules in this order will be considered more important. Given a set $\mathcal{R}$ of rewrite rules, we say that a term t rewrites to term s in one step if there is a rule $l \to r \in \mathcal{R}$ that can be applied to t at some position p to give s, and two more conditions are fulfilled. First, no rule from R can be applied to t at a position below p, which is called *eager rewriting*. We make just one exception to this rule: the if function does not have to evaluate all branches. Second, no rule $l_1 \to r_1 \in \mathcal{R}$ with the same function name at the top position on the left side, and before $l \to r$ in the linear order on such rules, can be applied to $t\tau$ at position p for any substitution τ that could generate a conflict. We say that τ and $l_1 \to r_1$ generates a conflict with $l \to r$ on t if $l_1 = t\tau\rho_1$, $l = t\tau\rho_2$ and $r_1\rho_1 \neq r\rho_2$. In this way we forbid the application of rules that are less important if any more important rule could be potentially applied and yield a different result. If t contains function symbols then we treat them as variables, since the result of the function is unknown if it could not be rewritten.

The term t rewrites to s in k steps if there is a term u to which t rewrites in $k-1$ steps and u rewrites to s in one step. We will also say that a term t is in normal form if it can not be further rewritten. It follows from the linear order of rules with the same function symbol and the assumption of eager rewriting that if any term t rewrites in any number of steps to a normal form, then t does not rewrite to any other normal form.

We will now define the concatenation function from $\text{lists}(\alpha) \to \text{lists}(\alpha)$ that takes two lists and produces the concatenation of these lists, and to do this we need the function symbol concat $\in \Sigma$ and three variables x, y, z with arity 0, $\text{type}(y) = \alpha$ and $\text{type}(x) = \text{type}(z) = \text{lists}(\alpha)$. The function can then be defined with the following two rewrite rules:

```
concat (Nil, x) -> x,
concat (x :: y, z) -> x :: concat (y, z).
```

To see how we execute the function let us concatenate T :: Nil with F :: Nil by rewriting the term, which is done in the following way:

$$\text{concat}(\text{T} :: \text{Nil}, \text{F} :: \text{Nil}) \to \text{T} :: \text{concat}(\text{Nil}, \text{F} :: \text{Nil}) \to \text{T} :: \text{F} :: \text{Nil}.$$

Term rewriting with types as presented above is used as the foundation for high level programming languages such as ML and Haskell, so the presented model is not only a precise mathematical entity that can be used for logical reasoning; it can be used to write programs that are easy to read and understand. Programs in other models of computation suitable for logical reasoning, like the Turing machines, are not directly readable. On the other hand, it is quite difficult to construct an elegant logical calculus for any imperative programming language used in practice and to reason about it.

Term Rewriting I/O. One problem with the presented model is the definition of input and output, since in term rewriting there are no side effects. Therefore, we will assume that the system of types, terms and rewriting rules that we are working with is a computer program and can respond to a set of commands that we will describe. There are commands that allow us to define new types, constructors, function symbols, and variables; commands that add new rewrite rules and rewrite terms. These will be discussed in the next section together with the extended notation that we will use. Now let us look at the additional commands that allow storage of strings or sequences of definitions in files, which can also be loaded. Moreover, there is the internal knowledge database, where terms of any type can be stored. Let `path` be a string representing a path to a file or a virtual device, for example a printer or a display, `string` and `name` be strings, `type` be a type, and `term` be any term. The following commands provide input and output operations, with the last three used to define and manipulate storage space in the internal database:

```
Load string from [path].
Store [string] in [path].
Load definitions from [path].
Store system in [path].
Define data [name] in [type].
Load from [name].
Store [term] in [name].
```

When we define the storage space with the `define data` command we also set its type. This type must be more general than the type of anything we store using the `store` command, and it is the type of the term we get with the `load` command for this storage. To implement it without losing type correctness, we have to generate appropriate `load` and `store` commands for the internal database whenever the `define data` command is used.

In the presented setting, it is possible to load or store terms only after a complete sequence of rewriting steps; it is not possible to change the state of any variables during rewriting, which would complicate reasoning about programs. Since we normally rewrite terms between parsing, it is also necessary to add special handling of `load` commands when we construct system functions, because we have to prevent these commands from being in-lined before the actual call. We use a special tag `not_inline` in the function definition to prevent these functions from being evaluated before the correct time.

As terms are best suited for symbolic representation of data, the best way to create graphical programs in the system is to use vector graphics. We can connect the input and output of the term rewriting system with a HTML server and use the web standards like XForms and SVG. Then it is enough to generate terms with appropriate types corresponding to the specifications of the web standards and any browser can be used to run term rewriting programs with graphical interfaces. It is even better when this is combined with user interface ideas from [16] and, when the user can have all his work, both as text and graphical, on the desktop at once thanks to a zoomable interface [3].

3.1 Extended Program Notation

We showed a model of computation that suits our needs, but there is still a problem with the presentation of programs, as even for the short example we discussed it was necessary to define additional notation for list and pair constructors. Therefore, we will develop a more readable presentation that will be close to natural language. To enable this, let us assign syntax definitions to type names, constructors, function symbols and variables. We have to define a *syntax element* as either a string or a type, and assume that the type of `types` is set. Sequences of syntax elements followed by a return type constitute syntax definitions.

For example, let us assign to the type name `lists` the following syntax definition: 'lists', 'of', `types` with return type `types` and to the constructor `Cons` the following: ?a, ':' ':', lists (?a) with return type lists (?a). The meaning of the syntax definition in this constructor is the same as the notation we defined before.

Let us show an example of how a type name, constructor, a function symbol, and a variable are given with corresponding syntax definition. We will enclose the strings in syntax elements in stars * and use commands as in the following examples for definitions, where we assume that each type definition returns an element from `types`. In the definitions we can use the word `class` to denote types and the word `element` for constructors.

```
Define class *lists* *of* types.
Define element *Nil* in lists of ?a.
Define element ?a *:* *:* lists of ?a in lists of ?a.
Define variable *x* in lists of ?a.
Define function *concatenate* lists of ?a *with* lists of ?a
    into lists of ?a.
```

We will show how expressions can be parsed using syntax definitions, but note that the `define` command does not only add the defined type, constructor, variable or function name to the system and assigns the appropriate types and syntax definitions. Additionally, when the tag `is_functional` is specified, it adds a syntax definition that allows use of the constructors, functions or variables with arity bigger than 0 as functional values using just their names.

In this situation and when operating on function on meta level, which we discuss later, we need to have a single name corresponding to a syntax definition. These names are constructed automatically, in such a way that all the strings in the definition are kept and all types are changed to capitalized first letters of the type name preceeded with '; A is used for type variables. For example for the list constructor instead of `Cons` we will now use the name `'A_:_:_'L_` and for list type definition the name `lists_of_'T_`. As the names have to be unambiguous, they always end with an underscore if only one syntax definition with corresponding name exists, and they end with a number, e.g. `lists_of_'T_1`, `lists_of_'T_2` if more definitions correspond to the same name.

Let us now show how we parse an input string and create a term from it. During parsing we use extended rewrite rules in the form $l_1, l_2, \ldots, l_n \rightarrow r$, which expect a sequence of terms and only if such a sequence is encountered, rewrite it to the resulting term r. Syntax definitions are easily encoded as such extended rewrite rules when each string is encoded as a term of type strings and each type T is represented by a variable x_T with type$(x_T) = T$. Let us

look, for example, at the extended rewrite rules for `lists` type definition and
for the "`::`" constructor:

```
'lists' : strings, 'of' : strings, t : types ->
    lists_of_'T_ (t) : types,
x : ?a, ':' : strings, ':' : strings, y : lists (?a) ->
    'A_:_:_'L_ (x, y) : lists (?a).
```

Before parsing, the input string is split on all spaces and on all symbols
that are not letters, except for digits or letters connected to words with _ or
^. For example the string `var10 *x_11* of: ?a^2` would be split into `var`,
`1`, `0`, `*`, `x_11`, `*`, `of`, `:`, `?`, `a^2`. Then, each part is encoded as a term
of type strings and we apply the extended rewrite rules derived from syntax
definitions to the sequence of string terms decoded from the input string.

One can think about an algorithm for applying these extended rewrite
rules as an extension of bottom up parsing of context free grammars. In our
case, however, the rules include polymorphic types and not just a finite set of
non-terminals, so it is more complex to apply them everywhere, and at the
same time more things can be expressed easily in this way.

To apply a set of extended rewrite rules to a sequence of terms $t_1, \ldots, t_n$
we will store, for each pair of positions $1 \leq i \leq j \leq n$ in the sequence,
all terms t that can be derived between these two positions together with the
reconstructing substitutions $\rho'(t)$. These will sometimes extend the previously
defined substitutions $\rho(t)$ so that types of rewritten term variables will not
be forgotten. We will denote the set of all terms derived between i and j by
$d[i, j]$ and we will compute all derivable terms between all positions. We start
with $d[i, i] = t_i$ and $d[i, j] = \emptyset$ in other cases and look for a fixed-point of the
following extension of the sets $d[i, j]$.

To extend sets of derivable terms, we can take any sequence of positions
$1 \leq i_1 \leq i_2 \leq \ldots \leq i_{m+1} \leq n$ and terms $u_k \in d[i_k, i_{k+1}]$ ($k = 1, \ldots, m$),
for which $\rho' = \mathrm{subst}\{\rho'(u_1), \ldots, \rho'(u_m)\}$ exists. Further, we need an extended
rewrite rule $l_1, \ldots, l_m \to r$ such that for some term substitution σ it holds
that $l_k \sigma = u_k$ for all k. Then, we can extend the set $d[i_1, i_{m+1}]$ by setting
$d[i_1, i_{m+1}] := d[i_1, i_{m+1}] \cup \{r\sigma\}$ with $\rho'(r\sigma) = \rho(r\sigma) \cup \rho'$.

We will continue this process to reach all possible derivable terms for the
whole expression and if there is only one term in $d[1, n]$ we will return it as
the result. If there are more terms in $d[1, n]$ we will report the *ambiguity* error,
and the *no parse* error occurs if $d[1, n] = \emptyset$. It can also happen that the sets
will be extended infinitely, but we can prevent such cases using subsumption,
which is described below, and with additional rule checking before the parsing
starts.

In practice, when there are many extended rewrite rules, we have to first
look at what strings come in what order in the input and use only such rules,
for which all strings on the left side of the rule can be found in the derived
set, and therefore it is possible to apply the rule. In this case, if we derive a

new string for some position then we might have to increase the number of considered rules.

Let us analyze, for example, how the term x :: Nil is parsed, where the variable x has type lists(α). First, we apply twice the rule coming from the definition of variable x that changes the string 'x' to the term x from lists(α), and the rule for 'Nil' to get the term Nil from lists(α'). Then, we apply the syntax definition of :: to get the only possible parsing result Cons (x, Nil).

There is one more issue, as if we tried to use to the presented solution in practice we would very often get ambiguity errors. The first thing we have to do to avoid this is to define that a term t with reconstructing substitution $\rho'(t)$ *subsumes* another term r with another substitution $\rho'(r)$ if there is a type substitution τ such that $\rho'(t)\tau \subseteq \rho'(r)$ and a substitution σ such that $t\sigma = r$. In this way we specify when one intermediate result of parsing is more general than another, and we will only consider the most general intermediate results, i.e., between any two positions we will only consider such terms and type substitutions for reconstructed variables that are not subsumed by any other one derivable between these two positions. Considering such subsumptions is especially useful when we have syntax definitions for casts from one type to another, as more and less general types can be often derived in such cases.

Another feature that we have to add is the possibility of defining rule priority, binding strength, and associativity of syntax definitions, in order to parse 1 + 2 + 3 * 4 in a correct way without using parentheses. We can incorporate this into our algorithm in such a way, that we first compute all derivable terms with derivation trees and later we select only the best derivations according to certain priority rules. The priority rules formalize the fact that when an operator ○ is left-associative then $(x \circ y) \circ z$ has a higher priority than $x \circ (y \circ z)$, the converse being true for right-associative operators, and if one binds stronger than the other then it is respected. To check associativity and binding, we need to rotate the tree representing the term, but if we parse two ambiguous terms then we use rule priorities. If the symbol f has a bigger priority than g, then $f(t_1,\ldots,t_n)$ has a bigger priority than $g(r_1,\ldots,r_m)$. When the symbols f and g have the same priority then we can say that $f(t_1,\ldots,t_n)$ is bigger than $g(r_1,\ldots,r_m)$ only if $n \geq m$, $t_1 \geq r_1,\ldots,t_n \geq r_n$, and some $t_i > r_i$. To use casts from one type to another in practice we have to add a special priority and priority comparison rule, which means the following. To compare the cast $cast(t)$, derived using a rule with the special priority, with the term s, we have to compare t and s first, and if t turns out to be bigger than s then choose $cast(t)$, otherwise choose the cast-free term s. Of course, even with these rules there are many incomparable derivations and we still can get ambiguities, but it is rare in practice.

To make the language context dependent and more flexible we assume that any command sent to the system is first processed by the preprocess_command function. When this function is not defined then the command is left without processing but the possibility to define and redefine this function with rewrite rules allows extension of the language. Another important addition to the simple parsing algorithm presented before is the handling of compound sentences. We let the user define how sentences can be composed, for example

```
sentence1 "and" sentence2
sentence2 "where" sentence1
```

and then, before the parsing begins, we divide the text along these composition rules and parse the first sentence before parsing the next.

Using the presented methods we are not only able to parse complex expressions, which can be used to comfortably write programs in the presented model, but we can also use it for basic natural language processing. We can directly translate FrameNet frames [1] to types in our model and use frame rules for specific lexical units as syntax definitions. Then, many sentences written in natural language will be parsed to terms that denote their grammatical structure, and sometimes also a part of the semantic structure. Then we can define functions operating on these terms and allow interaction with the defined system in natural language. In this way simple programming and some program searches described in the next section can be done by non-programmers, which makes the system usable in practice.

When operating on larger sets of functions, types, constructors, and rewrite rules, it is useful to mark them in some way and to be able to choose the ones that we want to use at a given moment. Therefore, we will assign to each function, type and, constructor a set of tags in the form $key = value$, where both key and $value$ are strings. As mentioned before, some special tags can also be used to generate additional rules or stop in-lining of functions. We can activate and deactivate all symbols with a given set of tags set to a specific or to any possible value. As we did not yet present the commands necessary for adding rewrite rules, setting priorities, removing type, constructor, and function definitions, rewriting terms and for tags, let us give here a simple example.

```
Define function integers *+* integers into integers
    priority normal associativity left
    with tags [context = arithmetics, system = true].
Let 0 + 0 be 0. Compute 0 + 0.
Remove function arg + arg. Remove class lists of ?a.
Activate with tags [system = ANYTHING].
Deactivate with tags [context = arithmetics].
Close context.
```

Note the **close context** command, which removes all variable definitions and opens a new set of variable names, so we can use the variable named x in different contexts with different types. Also, by removing functions and types we have to check if these are not being used somewhere in other definitions or in the internal database.

3.2 Compiling Typed Rewriting Systems

We presented a nice model of computation and showed how to represent programs in a readable form, but we need to have some means of executing the programs. Of course, we could easily write a term rewriting interpreter, but as we expect to work with complex and time consuming programs it is necessary to have a more efficient method to execute them.

In general, it is not difficult to compile term rewrite rules to a functional language with polymorphic types, but the compiled programs might be quite inefficient. One method to improve performance is to make it possible to write function and type definitions in the language to which we compile and then, during compilation, substitute these types and functions by their more efficient hand-written counterparts. In this case we have to duplicate our work and write the same programs both as term rewriting rules and in the language to which we compile and we can make mistakes in the translation. To avoid this we will introduce a few optimizations of the rewriting rules and show how to generate efficient code, so that writing the same pieces of code by hand for efficiency will be necessary only in rare cases or for special very often used system functions and types, like arithmetics or lists.

The best compilation method would be to have a formal model of the target language and to use advanced program search algorithms to find efficient equivalent code. This is not possible at present, because neither are our program analysis methods advanced enough nor is the construction of a simple but credible model of a mainstream programming language easy.

For practical reasons we have to stick to more standard compilation methods. Functional programming languages have been present in academia for a long time, and recently some of the associated ideas started to be used in the industry. Polymorphic types, under the name of "generics", are already included in Java and in C# and there is extensive commercial work going on to construct efficient compilers for polymorphically typed languages.

There is also ample research concerning these issues, for example [23], where list optimizations are presented, but we will show only a few simple optimizations that can be quite easily implemented and perform very well in practice. These focus on improving memory management and increasing the number of tail recursive functions, and are similar to the linear optimization described in [11]. Despite their simplicity and efficiency, such optimizations have not been implemented in widely used compilers, perhaps because they rely heavily on the lack of any side effects during computation, which is true for programs in our model, but uncommon in other models.

Let us first show how to translate rewriting rules to C code using as example the concatenation function defined by the rules:

```
concat (Nil, x) -> x,
concat (x :: y, z) -> x :: concat (y, z).
```

We will discuss a basic translation to C code here to show the ideas, although we find it more practical to use C++ and generate separate classes for each type. Templates can be used for fast polymorphism and overloading to make the copy and comparison functions work on all terms, even on predefined classes like integers. In this way, it is also easier to handle terms with variables and add meta functionality to the generated code without losing efficiency, as you can generate any special function with a given name for each type, add a default one as a template, and let the C++ compiler handle overloading when the function is used.

It should be clear how to implement term matching with a tree of `if` or `case` expressions, and we will assume that there is a record `term_t` defined in C that stores the id of the symbol at the root position in the term and an array of sub-terms.

We will define the concatenation function in C so that it takes an additional argument, a pointer to a `term_t` where the result will be stored. So, taking matching into account, the concatenation function in C looks like this:

```
void concat (term_t arg0, term_t arg1, term_t *result)
{
  if (arg0.id == Nil_ID) {
    code for the first rule
  }
  else {
    code for the second rule
  }
}
```

Now we will generate the code for the rules, but we will treat constructors in a different way than function symbols. For a function symbol, we will have to generate the arguments first and store them in the variables, and then call the function, whereas for the constructors we will first allocate them and later continue code generation with changed result pointers. Let us look at how code is generated for the constructor in the second rule, where args0.subterms[0] corresponds to the variable x in the rewrite rule.

```
*result = NEW_TERM (Cons_ID, 2);

code for assigning arg0.subterms[0] to (*result).subterms[0]
code for assigning the other part to (*result).subterms[1]
```

When constructing the code for the other part we will first assign the term y to the new variable x0 and the term z to x1, and in the last line call the concatenation function with

```
concat (x0, x1, & (*result).subterms[1]);
```

In this way, we managed to use the knowledge about which symbol is a constructor and which is a function symbol to create a tail-recursive version of the concatenation function. We could also remove the necessity to allocate memory for some of the variables by reusing the terms from the left side. Instead of allocating memory for the result and setting the id with the NEW_TERM macro, we could just set the result to arg0 and then change the id and pointers when necessary. It is easy to reuse memory allocated on the left side and variables if they occur the same number of times on the left and right side of a rewriting rule, but we did not present it here in detail for clarity. Adding the possibility to handle certain types and functions in the external language, e.g. integers directly in C, requires additional work, especially to prevent boxing and unboxing where possible, since then we have to generate a separate version of each polymorphic function for each special type. You can look at [10] for more details about how this can be done and the tradeoffs between time and space efficiency and the size of generated code in this case.

Although we can translate our rewriting system directly to C, we will first do a few optimizations to increase the efficiency of the generated code. The first, quite technical one aims to decrease memory usage and the need to reallocate memory. We will try to make as many rewriting rules linear as possible, so we will try to return all unused arguments. For example, the concatenation function is linear, but the double function

```
double (x) -> Pair (x,x)
```

is not. Not all functions can be made linear, but some optimizations can be done.

We can substitute the functions for which a compound argument is read but only a simple argument is returned by equivalent functions returning also the compound argument. For example, the function that calculates the length of a list should be substituted by a function that calculates the length and returns the list itself.

To clarify the method consider the following example:

```
length (Nil) -> 0
length (x :: xs) -> 1 + length (xs)
argument_length (x) -> (length (x), x)
```

In this case, the argument_length function will have to clone the term x before it can call length, which will in turn destroy its copy of x. To avoid this, we could optimize the functions and make length return also the argument it takes, so it becomes equivalent to argument_length. To define it we need a new function increment_append that will operate on an element and a pair and will just do the same what the second rule for length does, but accumulating the unused list.

```
increment_append (x, (n, xs)) -> (1 + n, x :: xs)
length (Nil) -> (0, Nil)
length (x :: xs) -> increment_append (x, length (xs))
```

In this way we are able to improve the efficiency of memory allocation and we can make additional improvements to increase the possibility to reuse constructors, which can further optimize the code.

There is one more important and more semantic optimization we can do. In our model of computation, terms are constructed from well-defined types, so if an argument of a function has a non-variable type we can unfold the function definition by substituting all possible constructors of this type for the argument. For example, in the definition of the list concatenating function we had an argument y from lists(α) in the rule concat (x :: y, z) $\rightarrow$ x :: concat (y, z). Since a list, by the definition of our list constructors, is either an empty list or is constructed from an element and a list, we could substitute these two possibilities and get two new rules:

concat (x :: Nil, z) -> x :: concat (Nil, z),
concat (x :: (y :: ys), z) -> x :: concat (y :: ys, z).

Now the right sides of these rules can be symbolically reduced and we obtain a new definition of concatenation consisting of the following three rules:

concat (Nil, x) -> x,
concat (x :: Nil, z) -> x :: z,
concat (x :: (y :: ys), z) -> x :: y :: concat (ys, z).

Please note that with this new definition the concatenation function will be called on long lists only half of the times it would be with the old definition. The price here is that we have to do bigger matching to check all three rules, but we can generate optimal if trees for the patterns and the compiler on the lower level, in our case the C compiler, can usually optimize them much better than excessive function calls. Also, if there are some auxiliary non-recursive functions called, these calls can sometimes be completely removed in this way, and function calls for specific classes of arguments can also be optimized.

When the definitions are unfolded it is possible that some function calls will occur multiple times. If we represent terms as directed acyclic graphs (DAGs) that have no isomorphic sub-DAGs that are not identical, then such multiple occurrences will be detected and it will be possible to reduce them to one function call. Also, if one function calls two functions in different sub-terms then we can execute these functions in separate threads. The increased number of rules achieved with the optimization described above can amortize the cost of creating new threads. The possibility of automatically making the program concurrent, which is not practical in imperative programs that have to update the global state of memory, is very important for efficiency as computer systems are getting more and more parallel. Such simple reductions can sometimes speed up the execution by a large factor, and a large number of functional programs is amenable to such optimizations.

4 Reasoning Using Games

Creating and understanding proofs is a complex task, and to deeply under-stand this process and try to do it automatically requires that we build some model of proofs to think about. In mathematical logic, proofs were depicted as sequences of statements where one statement follows from another. In such a model, it is easy to check if something is a correct proof, but it can be seen even in school that it is very difficult to find a proof of anything. Therefore, we will consider a different, more intuitive representation, where proofs are modeled by games between two players: Eloise, aiming to prove the requested property, and Abelard, who wants to falsify it. The property is proved if Eloise has a winning strategy in the game, i.e., if Abelard loses no matter how he plays. Such games exist for a number of logics and one can find an overview of related results in [5].

In logic games, whenever we see an existential quantifier or a disjunction in the considered formula, then Eloise moves and chooses an element of the structure to substitute for the variable bound by the quantifier, or one component of the disjunction. Conversely, whenever we see a universal quantifier or a conjunction, then Abelard moves and chooses an element or one component of the conjunction. Let us look at a simple example and prove the property *there exists a number that is smaller than* 3 *and there exists a number that is smaller that* 2. Natural numbers are our structure in this case, and this property is a conjunction of two statements:

(1) there exists a number smaller than 3,
(2) there exists a number smaller than 2.

Since we have a conjunction, the first move belongs to Abelard and he chooses (1) or (2). Then it becomes Eloise's turn to move, since in both formulas there is an existential quantifier at the top position. In the first case, she can choose the number 2 and win and in the second case she can choose the number 0 and win. Therefore, Eloise has a winning strategy that can be described as follows: if Abelard chooses option (1) then choose the number 2 and if he chooses option (2) then choose the number 0. Note that if the thesis in the second case were *there exists a number smaller than* 0, there would be no winning strategy for Eloise as she would not be able to choose such a number, and Abelard would win.

It should also be clear that if we wanted to add a simple induction into this game we could allow the players to substitute a quantified variable x only by 0 or $x + 1$. If we try to do this with more quantifiers, problems will arise when we want to induce first on one universally quantified variable, and then on one existentially quantified variable. To solve such problems and capture the whole power of inductive reasoning without losing control over finiteness, we need to redefine the games we use and add a natural number to each position, denoting the *level of visibility* in this position. Then, for each level of visibility we need to define the set of possible actions and for each position on this level we need to assign to each action exactly one outgoing edge in the game graph, and now the players will not just choose moves, but they will choose actions. In this way the state of the play is a word over the alphabet of possible actions, but when the player is at visibility level i we will give him only incomplete information about the current play – just the letters that come from visibility levels lower or equal to i. To say that a player wins such partial information game we can not just present winning strategies but we have to give them stepwise through levels of visibility. Therefore, we require that the winning player first gives her strategy for the first visibility level, then the opponent responds with the strategy for the first level, then the first player gives her strategy for the second level and so on up to the last level.

Games with visibility levels can then finally use a parity or Muller winning condition and capture model checking on (tree,ω)-automatic structures or other reasoning. By other reasoning we mean here especially extensions of the game with syntactic reasoning rules including generalisation or specific rules for quantifier elimination. Note that using existential quantifiers and representing functions with rewrite rules we can use this to search for programs. But, although for games with visibility levels it can be non-trivially hard to determine the winner, one can always win such games using a strategy with finite memory and the winner is always determined. Before we present an extended game for general terms with additional possible moves let us show how logic can be implemented in the discussed rewriting system.

Logic in the System. To make it possible to implement logical reasoning in the rewriting system we need to define the type of logical formulas on which we will do the reasoning and also the type of terms so that logic can be represented with meta-rules.

The formulas in our system are defined with respect to a type T of basic terms in equalities in the following way:

(1) if t and r are terms of type T then $t = r$ is a T-formula;
(2) if φ and ψ are formulas then $\neg\varphi$, $\varphi \wedge \psi$, $\varphi \vee \psi$, $\varphi \rightarrow \psi$ are also formulas;
(3) if φ is a formula and s is a string that is meant to be the name of a variable in φ then also $\forall s \, \varphi$ and $\exists s \, \varphi$ are formulas.

We can also define the set of free term variables in a formula by $\mathrm{Var}(t = r) = \mathrm{Var}(t) \cup \mathrm{Var}(r)$, $\mathrm{Var}(\varphi \wedge \psi) = \mathrm{Var}(\varphi \vee \psi) = \mathrm{Var}(\varphi) \cup \mathrm{Var}(\psi)$, and $\mathrm{Var}(\forall s \, \varphi) = \mathrm{Var}(\exists s \, \varphi) = \mathrm{Var}(\varphi) \backslash \{s\}$, and we distinguish them from the *bound variables* that appear by quantifiers $\exists$ and $\forall$.

This is the standard type of formulas that we will use and for which we will define reasoning rules, but one can also define other logics with the methods described below. We can store formulas with different attributes, for example their proofs or parts of the proofs, just like any other term in the system database.

To change expressions consisting of terms and types in different ways that are not supported by direct rewriting rules, we need to access them coded on a more direct level. For that purpose, we define in the system the type of terms and the type of types and we specify how terms of any type correspond to coded terms. The coding uses T_Term for term constructor, T_Variable for term variable, and TT_Type, TT_Function, TT_Type_var to encode types. The names of syntax definitions are used as strings in the coding and term variables are coded together with their types. For example, the term $(x :$ booleans)::[] representing a list with one boolean value can be coded as:

```
T_Term ("'A_:_:_'L_",
    [T_Variable ("x_", TT_Type ("booleans_", [])), []),
     T_Term ("[_]_", [])]).
```

To make use of the presented coding we add special functions to the system that allow access to information about already defined types, functions, and recently defined ones. Information about functions is placed in a special type that gives the tags, rewrite rules for the function, and the type of the function; we get analogous data for constructors. To select information in the system we use tag queries that are lists of tag names and optional values. An element with tags satisfies the query if it has tags with the corresponding names defined, and when tag value is given in the query the corresponding element tag must have the same value. The functions

```
get constructors [tag query]
get functions [tag query]
```

retrieve from the system all definitions of constructors or functions that satisfy the given queries. All constructors of a type named t have a special tag #type = t for easy access.

To make real use of the described coding we need to transfer functions defined between terms to the normal level, where the functions take arguments and return results of different types. Assume that we have a function that takes N terms and returns a term, $f :$ term, $\ldots$, term $\rightarrow$ term. Then we can define a normal function with given **name** and types **type1**, ..., **typeN** and return type **ret_type** in the following way:

```
Define function [name] [type1] ... [typeN]
    into [res_type] from meta function [f].
```

Such function does what executing f with coding does, i.e., it codes all arguments, executes f on the coded terms and decodes the result back. Additionally, this definition puts into the system new logical formulas that should be proved to guarantee that f indeed has the declared type. These are simple formulas that can be proved just by executing a type checking function. We need to define the type checking function that uses the type and function definitions retrieved from the system to calculate the type of given terms, but this is just a technical problem.

Using meta functions makes it more problematic to compile the functions to C or C++. Sometimes rewriting is requested on terms with variables, so we have to prepare the C++ code for such cases and check it when matching is done. Also, when using built-in C++ types we sometimes have to do boxing to allow term variables to be represented. Additionally, we have to keep in the code the mapping from assigned ids of constructor and function symbols to their string names to be able to execute meta functions.

The possibility of operating on meta level is not trivially implemented, but having it we can define logic and reasoning rules in a clear way. In general we will represent reasoning rules as functions, normally implemented on meta level, that take premises in the form of a T-formula and generate conclusions of the same type, usually denoted:

$$\text{premises} : formula \vdash \text{conclusions} : formula.$$

In such functions we can use the information about constructors of a type to make induction on this type, and we can access rewrite rules for specific functions to recognise them and implement specialised decision procedures. Sometimes we need to define new functions or types in the system to be able to construct the conclusions. To make it possible we allow reasoning rules to create a list of system commands that are executed in turn before the rule is evaluated in an analogous way to how compound sentences are parsed.

To give meaning to the reasoning rules we will present a set of basic rules that are assumed to be true, that is they transfer true premises to true conclusions. Formulas proved by means of these rules are also true and we allow extension of the set of rules used for reasoning by bringing proved formulas down to the meta level. More precisely, if we prove a formula $\varphi \to \psi$ using functions $f_1, \ldots, f_n$ and types $t_1, \ldots, t_n$ then we can add a reasoning rule $\delta \wedge \varphi' \vdash \psi'$. In this rule φ' and ψ' differ from φ and ψ only so that functions and types are replaced with variables with the same type. In δ we use the possibility to get type constructors and rewrite rules for functions to check that the definitions of all f_i and t_i are equivalent to the definitions of the variables with which these were replaced in φ' and ψ'. In this way the reasoning rule depends only on the semantic of the functions and types and not on their names.

In the next section we will present the basic reasoning rules and the game used to find proofs and understand them. Using the possibility to construct deduction rules from proved formulas we can prove correctness of logical decision procedures. In this way, decision methods using automata or quantifier elimination can be proved and used when reasoning about corresponding objects. For example, the old procedures for the theory of real numbers with addition and multiplication [22] or for Presburger arithmetics [15] and their modern variants can be implemented.

4.1 Reason and Search Game for Terms

Let us now define a game that will allow the search for programs and prove their properties in the typed term rewriting model presented before. Positions in this game are formulas and we assume that free variables are implicitly universally quantified. We will not identify positions that differ only in the names of variables, but their identity will be important in determining the winner if induction is used. In this game, each reasoning rule $\varphi \vdash \psi$ describes a possible move of Eloise from ψ to φ and a possible move of Abelard from $\neg\psi$ to $\neg\varphi$. When we know that $\varphi_1 \vdash \psi, \ldots, \varphi_k \vdash \psi$ and that $\psi \vdash \varphi_1 \vee \ldots \vee \varphi_k$, then we call the set of Eloise's moves from ψ to $\varphi_1, \ldots, \varphi_k$ *complete* and analogous for the moves of Abelard from $\neg\psi$ to $\neg\varphi_1, \ldots, \neg\varphi_k$.

We will also assume that the terms in the equalities inside the positions are always rewritten to their normal forms. When proving properties of functions that do not terminate we might not be able to satisfy this requirement and fall in an infinite loop when trying to normalise a term after a move has been made. We will assume that such moves are disallowed and we will not consider them.

We will first describe the winning condition in the game and give a basic set of simple moves that are sufficient to make induction on the structure of the type as defined by constructors, to generalize the formula, and to substitute parts of the formula using some already proved equalities. To get the full power necessary for all proofs we additionally need to create new reasoning rules as described before or add new types, functions, and prove lemmas. Still, the basic reasoning rules correspond to the notion of a simple proof and should be enough for intuitively easy properties. We also present simple moves that are not complete but can often be used in practice to find the proofs faster.

Note that in many cases the proofs of existence of a function lead to the definition of this function and, therefore, we say that this is also a search game. When functional variables are present we will sometimes add rewrite rules for these functions to the system during the game, or even define new functions during the proof. For existential statements that are proved in a non-constructive way we also allow to define the corresponding functions in the system using a **define** ... **from formula** command similar to the one we used for meta functions. Of course, such functions can be used only for proofs, they can not be rewritten and compiled, but still sometimes it is useful to have them defined.

Let us first state what positions are trivially winning for Eloise and which for Abelard. The only trivially winning positions for Eloise are the positions $t = t$ for some term t, and the positions trivially winning for Abelard are the ones $s_1 = s_2$ where s_1 and s_2 are ground terms and are not equal.

Of course, if any player can move from a position p to a winning position for her or him, then the position p is also winning, and we will guarantee that each position will be winning for at most one player. When a set of moves is complete then if a player loses with all these moves then she loses in this position. When we use inductive rules we have to check if we get back to a position that is identical to the one from which we started only with new variables. We will discuss this later when inductive moves are presented.

The first kind of moves that we will analyse is very simple; Eloise can move from $\varphi \vee \psi$ and Abelard can move from $\varphi \wedge \psi$ to either φ or ψ. These moves correspond to the reasoning rule:

$$\varphi \vdash \varphi \vee \psi.$$

For Eloise this is a direct correspondence whereas for Abelard we have to substitute the rule with $\neg\varphi \vdash \neg\varphi \vee \neg\psi$ and remember that $\neg\neg\varphi = \varphi$. For every rule, when we want to extract from it the possible moves of Abelard then we should also remember to substitute negated formulas. The two rules $\varphi \vdash \varphi \vee \psi, \psi \vdash \varphi \vee \psi$ are complete. As we find the moves in the game more intuitive than the reasoning rules used in the implementation, we will stick to presenting the moves.

Another possible move for Abelard at a position φ is to try to describe inductively the terms that can be substituted for variables in $\text{Var}(\varphi)$ depending on their type, which is possible to do for $x \in \text{Var}(\varphi)$ by considering all constructors of $\text{type}(x)$ if it is not a type variable and not a functional type. Assume, for example, that there is a position $f(x) = c$ where x is a variable from $\text{lists}(\alpha)$. Then, Abelard can move either to $f(\text{Nil}) = c$ or to $f(\text{Cons}(x_0, x_1)) = c$. All possible constructors from which the type can be constructed must be taken into account and then such a set of moves is complete.

A similar induction is possible for functional variables on the type of any arguments or on the result type, and then new rewrite rules have to be added to the system. For example, let z be a functional variable with type $\alpha, \text{lists}(\alpha) \rightarrow \text{lists}(\alpha)$. Then we can make induction on the second argument in such a way that we define a new function in the system named f_z with rewrite rules:

$$f_z(x, \text{Nil}, z_1, z_2) \rightarrow z_1(x) \quad , \quad f_z(x, \text{Cons}(y_1, y_2), z_1, z_2) \rightarrow z_2(x, y_1, y_2).$$

We have to cover all possible constructors of the chosen argument type and add an appropriate number of new functional variables (z_1, z_2) with passing types. Then we have to replace each occurrence of z by the pair (z_1, z_2) and each call $z(x, y)$ by $f_z(x, y, z_1, z_2)$. When the occurrences of the variable z as a functional value are substituted we will have to change the functions that use it to take the pair (z_1, z_2) as argument instead of z and use $f_z(x, y, z_1, z_2)$ instead of $z(x, y)$. Propagating these changes might require us to define other new functions with appropriate types in the system, but it should be clear that such moves are correct and complete.

When we perform induction on the return type we either set the discussed function to a constant or to one of the variables that has the same type as the result, or we set it to a function call of another function that can use additional intermediate computation results. To make the move we first have to clone the formula in our position to have an appropriate number of functional variables, so instead of $\varphi(z)$ we consider in our example $\varphi(z_1) \vee \varphi(z_2) \vee \varphi(z_3)$ and construct new functions:

$$f_{z_1}(x, y) \rightarrow \text{Nil} \quad , \quad f_{z_2}(x, y) \rightarrow y \quad , \quad f_{z_3}(x, y, v_1, v_2) \rightarrow v_1(x, y, v_2(x, y)).$$

Note that the type of the intermediate result $v_2(x, y)$ will be assumed to be as general as possible, so we will take the tuple type for all arguments, and additionally a string type for other computed information, so finally it will be: pairs(pairs(α, lists(α)), strings). Then we replace the variable z_1 with f_{z_1}, z_2 with f_{z_2}, and z_3 with f_{z_3} in the same way as we did above with f_z. In the first formula $\varphi(z_1)$ the variable z_1 can disappear because of normalization to Nil, in the second formula it can be substituted by the second argument. In the third one we will now have two new functional variables v_1, v_2, and we have to correct the types and perhaps extend the system appropriately to get to the right position $\varphi'() \wedge \varphi''(y) \wedge \varphi'''(c, v_1, v_2)$.

Since we always normalize terms in the positions to which we move, it might happen that during a play we will return to a position in which we have already been, but with different variables. For example, if we have a function f defined by rewrite rules $f(\text{Nil}) \rightarrow \text{Nil}$ and $f(\text{Cons}(x, y)) \rightarrow f(y)$, then we might want to show that $\varphi = (f(x) = \text{Nil})$ is true. In the position φ Abelard must take one of the complete inductive moves described above, so he can either go to $f(\text{Nil}) = \text{Nil}$, which will be rewritten on the fly to $\text{Nil} = \text{Nil}$ and is trivially winning for Eloise, or to $f(\text{Cons}(x_1, x_2)) = \text{Nil}$ which will be rewritten to $f(x_2) = \text{Nil}$, and he could repeat this move infinitely. Eloise can have similar problems trying to prove $\exists x \, f(x) = \text{Cons}(1, \text{Nil})$.

To cope with such issues when a position identical modulo variable renaming is repeated in a cycle or if we have any infinite play we need to be more careful determining who wins. In a simple case when just the position of one player is repeated infinitely often and this player is making an inductive move, then the player loses. But with interleaved existential and universal quantifiers we get a bigger problem. For example, if for some function g we analyse the formula $\exists x \, \forall y \, g(x, y) = \text{T}$ then it can happen that we make in turn induction on x and y. But to preserve the meaning of the quantifiers we have to assure that any inductive step for x does not depend on the previous steps for y. To guarantee this we might have to consider power-sets of positions and check whether the strategies are correlated there. With more interleaving quantifiers these might even be power-sets of power-sets etc. as the satisfiability problem for automatic structures, which can be reduced to this, has non-elementary complexity in the number of quantifier interleaving occurrences.

There is another important kind of inductive move that Eloise can take and it is also complete. Let us assume we have an equality $f(t_1, \ldots, t_n) = t$ somewhere inside the formula φ and that the function f is defined by the set of rewrite rules $l_1 \rightarrow r_1, \ldots, l_k \rightarrow r_k$. When we say that f is defined by a set of rewrite rules $\mathcal{R}$ we assume that, for any ground terms $u_1, \ldots, u_n$ in normal form, the term $f(u_1, \ldots, u_n)$ can be rewritten at the top position with some rule from $\mathcal{R}$. Moreover, we assume that the order of rule application is not important for rules in the set $\mathcal{R}$. We will assume that functions in our system are exhaustively defined, so the first requirement is satisfied. When we have ordered linear rewrite rules we can always make them independent of the order by enumerating constructors, for example if and was defined by $and(\text{T}, \text{T}) \rightarrow \text{T}$, $and(x, y) \rightarrow \text{F}$ then we can change the rules to $and(\text{T}, \text{T}) \rightarrow \text{T}$, $and(x, \text{F}) \rightarrow \text{F}$, $and(\text{F}, x) \rightarrow \text{F}$ to make them independent of the order.

Let us now return to the equality $f(t_1, \ldots, t_n) = t$ and the rules $l_i \to r_i$ that are exhaustive and do not depend on order, and let $l_i = f(l_i^1, \ldots, l_i^n)$. Since the term $f(t_1, \ldots, t_n)$ will be rewritten by some of these rules when it is substituted to be ground, we can search for the correct rule and substitutions to rewrite it and check the formula later. This corresponds to the possibility for Eloise to move to the position $\psi_1 \lor \ldots \lor \psi_k$ where:

$$\psi_i = \exists \mathrm{Var}(l_i) \; t_1 = l_i^1 \land \ldots \land t_n = l_i^n \land \varphi[f(t_1, \ldots, t_n) = t \leftarrow r_i = t],$$

where $\varphi[f(t_1, \ldots, t_n) = t \leftarrow r_i = t]$ is the position φ with the equality $f(t_1, \ldots, t_n) = t$ changed to $r_i = t$. Note that if the position φ contains unbound variables the new variables from l_i take the unbound ones as arguments, which corresponds to skolemization.

To clarify, it let us consider a position $implies(t_1, t_2) = \mathrm{F}$ for some terms t_1 and t_2 with two unbound variables x and y, and let $implies$ be a normal implication defined by $implies(\mathrm{F}, v) \to \mathrm{T}$, $implies(\mathrm{T}, \mathrm{T}) \to \mathrm{T}$, $implies(\mathrm{T}, \mathrm{F}) \to \mathrm{F}$. In this case, Eloise can move to the formula:

$$(\exists v \; t_1 = \mathrm{F} \land t_2 = v(x, y) \land \mathrm{T} = \mathrm{F}) \lor$$

$$\lor (t_1 = \mathrm{T} \land t_2 = \mathrm{T} \land \mathrm{T} = \mathrm{F}) \lor (t_1 = \mathrm{T} \land t_2 = \mathrm{F} \land \mathrm{F} = \mathrm{F}),$$

which can be winning only for the last component, so Eloise moves to $t_1 = \mathrm{T} \lor t_2 = \mathrm{F}$. Observe that v was a functional variable and took x and y as arguments. This is a complete move and it could be taken inside a quantified formula or a formula with free variables as above. This is not possible, for example, for $\varphi \lor \psi$ as $\forall x \; \varphi \lor \psi \not\Leftrightarrow \forall x \; \varphi \lor \forall x \; \psi$.

As you might have noticed, the induction on functional variables for Abelard will make it possible to prove anything of interest only in very rare cases, as it usually only complicates the problem to induce on functions. But for Eloise it might be very important if she wishes to find a function with a specified property. We made it possible to use intermediate results and added a string type by inducing on the result type of a function to make all computable functions representable in this way, but often we should look for a nicer solution using other functions and types that we already have in the system. Also, when we look for a term with non-functional type it might be useful to represent it as the result of computation of a function that already exists.

More precisely, let us assume that we are looking for a term of type T either to substitute it for a bound variable x or for the result of a function in an inductive move by Eloise for a functional variable. In the second case there are additional parameters $x_1, \ldots, x_n$ that are arguments of the function with $\mathrm{type}(x_i) = T_i$. Let us then take any function f defined in the system with type $S_1, \ldots, S_k \to R$ such that there exists a type substitution σ for which $R\sigma = T$ and for some indices $\{i_1, \ldots, i_l\} \subseteq \{1, \ldots, k\}$ we can assign numbers $p(i_m)$ so that $S_{i_m}\sigma = T_{p(i_m)}$. With this function, we can represent the term we are searching for by $x = f(y_1, \ldots, y_k)$, where for $m \in \{i_1, \ldots, i_l\}$ we have $y_m = x_{p(m)}$, and the other arguments are new variables that we will again be requested to find.

Less formally, we just represent the term we want to find as a function call with any combination of already existing or new arguments. In this way we can use any function from the system that has an appropriate type to find the term we are looking for. Such moves are only optional for Eloise, but in practice it is very common to use the knowledge we have in this way, and many natural problems can be solved in just a few steps if the right functions are known in advance and are used in the right time.

The moves described above form the basis for all proofs and should suffice for very simple properties and to find programs that are not complex. But for even slightly more interesting proofs we need to use other formulas proved before to interact with the one we want to prove. We will present the possible moves for such interaction; these are not complete and some of the formulas used must already be known to be true, winning for Eloise. Keep in mind that we also presented a way to create new reasoning rules when the ones here are not sufficient to solve the problem efficiently.

When Eloise plays in a position φ she can choose any term t with type T that appears at some position in some of the equalities in φ and has no bound variables, and then move to a position ψ which is identical to φ with all occurrences of t at any position in any term in any equality replaced by the variable x with type T. We will call this move the generalization of t.

To make another move, suppose that we know that a formula $t = s_1 \vee \ldots \vee t = s_n$ is true and we are are in a position φ that contains a term u in some equality $u = s$. If, for some position p in u and for some substitution σ, we have $u|_p = t\sigma$ and no variables in $u|_p$ are bound, then let us define ψ_i to be a position identical to φ with the term u replaced by $u[s_i\sigma]_p$. Then we can allow Eloise to move from φ to $\psi_1 \vee \ldots \vee \psi_n$.

We allow another way for the players to move or to change the system, which makes it possible to define new types, functions, and construct new positions to analyse using the existing ones as building blocks. These moves are described in a simple way: every player can choose any well typed term, build a well constructed position, and insert it into the game. She can also build a function with arguments and result types that already exist in the system and choose a number of rewrite rules for it. New types can also be constructed by choosing a number of well formed constructors and both for functions and types it is possible to build a few of them at once and make them mutually recursive. It is also possible to prove lemmas and create new reasoning rules. The optional moves combined with the simple ones make it possible to prove complex properties of programs.

As one can see, there is a limited set of sensible basic moves and a wider possibility to make optional moves using the knowledge in the system or creating new types and functions that might be useful later. To play the game in a good way, so that all false formulas are found to be false fast and all true are proved efficiently, Eloise and Abelard have to use sensible strategies and make appropriate moves according to the situation. Of course, any player strategy that does not skip any infinitely long possible move is a program search procedure, and it will find programs that provably fulfil the specified formula. When the game itself is defined with appropriate type inside the system and possible moves are also defined, we can specify that a strategy is a function that chooses a possible move in a given state of the game, and we can use the game-based search procedure to find better strategies and, therefore, make the strategy self-improve by learning as was described before in the theoretical discussion.

Expressing reasoning as a game makes it possible to understand heuristics that we use for reasoning, like "always look first at a few simple examples before you start to prove" or "do not use one induction after another" as simple strategies in the game. For certain types of positions we can use the decision procedures that already exist, and include them in the game as soon as they are implemented as reasoning rules and proofs of their correctness are given. These procedures do not have to be complex and complete, they can also represent good heuristics. As a very simple example, assume that we are in a position $\exists y\ x + y(x, z) = z$, or there is some more complex arithmetic expression given but only with constants and addition. The first move that any well-acting strategy should take is to substitute $y(x, z) = z - x$ or use an algebraic solver to find the right function to substitute, and in this way incorporate the simple decision procedure into the reasoning game. More powerful reasoning rules using automata and quantifier elimination can also be implemented.

5 Conclusions

We showed how program search methods can be used both to solve problems and to automatically construct more efficient problem solvers. After showing a theoretical solution, we demonstrated a convenient model of computation and a game for reasoning. We argued that the model of computation can be practical and efficient and that in our reasoning game we can understand the actions taken and incorporate other decision procedures.

The presented model and reasoning method are both extensive enough to cover the tasks of artificial general intelligence, and simple enough to use them for specific reasoning tasks when programming. We are now working on making the system user-friendly and to build the basic knowledge library for it. It would be valuable to have an extensive standard library of types, programs encoded as rewrite rules, and proofs of important facts about these programs and related types in the presented model. Together with a few simple hand-written heuristics, efficient compilation methods for typed rewrite rules, and the program notation extended to be comfortable to use, this should make it practical to produce proofs of correctness of programs, and even to generate simple programs automatically. At the same time, it would be an interesting database of tests and formal proofs in a simple theoretical model, so it could be potentially used by other systems and also serve as a set of examples to teach future self-improving procedures. The question of whether it will be achievable to define reasoning heuristics well enough to work for more complex programs directly in the presented model is still open. But, since the moves we take in the reasoning game have clear intuitive meaning, we can hope to formalize our own thinking methods in this way, or, if we fail, at least to understand clearly which of the intuitive steps we take when solving problems are the most problematic ones for AI.

We find that the design of the system that we described and that additionally handles natural language processing forms the basis for AGI. One can not expect things like consciousness or speech recognition from such a system, at least not before they are programmed into it. But one can solve problems and even sometimes write programs automatically only by specifying what properties must hold. This system is also a viable software design and development environment where efficient applications with graphical interfaces can be implemented, and where tedious programming tasks can be automated. As natural language processing is included, one can also ask experts in specific domains to write their knowledge directly into the system and later use this knowledge in programs or for reasoning. When a large base of knowledge and a number of reasoning heuristics are included the system will also be capable to learn from them and optimize its own structure.

Since the presented system manages to establish a correspondence between the natural thinking and language of a human being and formal notation suitable for computers and code generation, it makes the communication and cooperation between people and computers practical in almost any situation when a problem needs to be solved. Therefore, we think that the presented way to bring formal logic together with natural language processing and allow to extend it using numerical heuristics is interesting for future AGI development.

In this chapter we omitted a lot of important related AI research. We did not discuss fuzzy and probabilistic logics and models of computation, although these should certainly be used. Still, we prefer to include them and related verification methods [6] as reasoning procedures in the presented model rather than analyse them as core elements on the same level as the programming primitives and logic. As the topic discussed is very extensive we certainly failed to mention and reference all relevant publications, so for more detailed study you should consult the first four books in the reference list.

References

1. Atkins BT, Fillmore CJ, *FrameNet*, www.icsi.berkeley.edu/~framenet/
2. Baader F, Nipkow T (1998) *Term Rewriting and All That*, Cambridge University Press.

3. Bederson B, *Piccolo Toolkit*, www.cs.umd.edu/hcil/piccolo/
4. Gödel K (1931) *Über formal unentscheidbare Sätze der Principia Mathematica und verwandter Systeme I*, Monatshefte für Mathematik und Physik 38:173-198.
5. Grädel E (2002) *Model Checking Games*, Proceedings of WOLLIC 02, vol. 67 of Electronic Notes in Theoretical Computer Science, Elsevier.
6. Hurd J (2002) *Formal Verification of Probabilistic Algorithms*, PhD thesis, University of Cambridge.
7. Hutter M (2000) *A Theory of Universal Artificial Intelligence based on Algorithmic Complexity*, Technical Report cs.AI/0004001.
8. Hutter M (2005) *Universal Algorithmic Intelligence: A Mathematical top→down Approach*, this volume.
9. Hutter M (2004) *Universal Artificial Intelligence: Sequential Decisions based on Algorithmic Probability*, Springer, Berlin.
10. Kaiser Ł (2003) *Speagram*, www.speagram.org.
11. Kirkegaard C (2001) *Borel - A Bounded Resource Language*, Project Report, University of Edinburgh.
12. Kolmogorov AN (1965) *Three Approaches to the Quantitative Definition of Information*, Problems of Information Transmission 1:1-11.
13. Levin LA (1973) *Universal Sequential Search Problems*, Problems of Information Transmission 9(3):265-266.
14. Li M, Vitanyi PMB (1997) *An Introduction to Kolmogorov Complexity and Its Applications*, Springer, Berlin.
15. Presburger M (1929) *Über die Vollständigkeit eines gewissen Systems der Arithmetic ganzer Zahlen, in welchem die Addition als einzige Operation hervortritt*, Compte-Rendus dei Congres des Math. des pays slavs.
16. Raskin J (2000) *The Humane Interface: New Directions for Designing Interactive Systems*, Addison-Wesley Professional.
17. Robinson A, Voronkov A, Robinson J (2001) *Handbook of Automated Reasoning*, newblock MIT Press, Cambridge, MA.
18. Schmidhuber J (2002) *Optimal Ordered Problem Solver*, Technical Report IDSIA-12-02.
19. Schmidhuber J (2005) *Gödel Machines: Fully Self-Referential Optimal Universal Self-Improvers*, this volume.
20. Schmidhuber J (2005) *The New AI: General & Sound & Relevant for Physics*, this volume.
21. Solomonoff R (1989) *A System for Incremental Learning Based on Algorithmic Probability*, Proceedings of the Sixth Israeli Conference on Artificial Intelligence, Computer Vision and Pattern Recognition.
22. Tarski A (1948) *A Decision Method for Elementary Algebra and Geometry*, prepared for publication by JCC Mc Kinsey, U.S. Air Force Project RAND, R-109, the RAND Corporation.
23. Wadler WL (1984) *Listlessness is Better than Laziness*, PhD thesis, Carnegie-Mellon University.

The Natural Way to Artificial Intelligence

Vladimir G. Red'ko

Institute of Optical and Neural Technologies, Russian Academy of Science
Moscow, Russia
redko@iont.ru - http://www.keldysh.ru/pages/BioCyber/

Summary. The chapter argues that the investigations of evolutionary processes that result in human intelligence by means of mathematical/computer models can be a serious scientific basis of AI research. The "intelligent inventions" of biological evolution (unconditional reflex, habituation, conditional reflex, ...) to be modeled, conceptual background theories (the metasystem transition theory by V.F.Turchin and the theory of functional systems by P.K. Anokhin) and modern approaches (Artificial Life, Simulation of Adaptive Behavior) to such modeling are outlined. Two concrete computer models, "Model of Evolutionary Emergence of Purposeful Adaptive Behavior" and the "Model of Evolution of Web Agents" are described. The first model is a pure scientific investigation; the second model is a step to practical applications. Finally, a possible way from these simple models to implementation of high level intelligence is outlined.

1 Introduction

Artificial intelligence (AI) is an area of applied research. Experience demonstrates that an area of applied researches is successful, when there is a powerful scientific basis for the area. For example, solid state physics was the scientific base for microelectronics in the second part of the twentieth century. It should be noted that solid state physics is very interesting for physicists from scientific point of view, and, therefore, physicists made most of the scientific basis of microelectronics, independently of possible applications of their results. And results of microelectronics are colossal, as they are everywhere now.

What is a possible scientific basis for AI (analogously to the scientific base of microelectronics)? We can consider this problem in the following manner. Natural human intelligence emerged through biological evolution. It is very interesting from scientific point of view to study evolutionary processes that resulted in human intelligence, to study cognitive evolution, evolution of cognitive animal abilities. Moreover, investigations of cognitive evolution are very important from an epistemological point of view; such investigations could clarify the profound epistemological problem: Why are human intelligence, human thinking, and human logic applicable to cognition of nature? Therefore, we can conclude, that investigation of cognitive evolution can be the natural scientific basis for AI developments.

What are possible subjects for the investigations of cognitive evolution? What are possible relationships between academic investigations of cognitive evolution and applied AI research? In my opinion, it is natural:

1. To develop mathematical/computer models of "intelligent inventions" of biological evolution (such as unconditional reflex, habituation, conditional reflex and so on).
2. To represent by means of such models a general picture of cognitive evolution.
3. To use these models as a scientific background for AI research.

The goal of this chapter is to propose and discuss steps to such research. The structure of the chapter is as follows. Section 2 discusses an epistemological problem that can stimulate investigations of cognitive evolution. Section 3 outlines the subject of these investigations and some conceptual approaches to the investigations. Section 4 describes two concrete and rather simple models: the "Model of Evolutionary Emergence of Purposeful Adaptive Behavior" and the "Model of Evolution of Web Agents." Section 5 outlines a possible way to implementation of higher cognitive abilities from these simple models.

2 The Epistemological Problem

There is a very interesting and profound epistemological problem: why is *human* intelligence applicable to cognition of *nature*?

To illustrate the problem, let's consider physics, the most fundamental natural science. The power of physics is due to the extensive and effective use of mathematics. However, why is mathematics applicable to physics? Indeed, a mathematician creates his theories, using his intelligence, quite independently from the real physical world. The mathematician can work in a silence of his cabinet, resting on a sofa, or in an isolated prison cell. Why are his results applicable to real nature?

Are we able to solve these questions? In my opinion, yes. We can analyze evolutionary roots of human intelligence and try to investigate why and how did high level intelligent cognitive abilities evolutionarily emerged. In other words, we can follow evolutionary roots of animal and human cognitive abilities and represent a general picture of evolutionary emergence of human thinking and human intelligence. We can analyze why and how animal and human cognitive features emerged, how these cognitive features operate, why they are applicable to cognition of nature.

Can we really proceed in this way? Can we find evolutionary roots of human intelligence in animal cognition properties? Yes, we can. Let's consider the elementary logic rule that is used by a mathematician in deductive inferences, *modus ponens*: "if A is present and B is a consequence of A, then B is present," or $\{A, A \rightarrow B\} \implies B$.

Now let's go from the mathematician to a Pavlovian dog [28]. The dog is subjected to the experiment of classical conditioning. A neutral conditioned stimulus (CS), a sound, is followed by a biologically significant unconditioned stimulus (US), a food. The unconditioned stimulus arouses salivation. After a number of presentations of the pair (CS, US), the causal relation $CS \rightarrow US$ is stored in dog's memory. Using this relation at a new presentation of the CS, the dog is able to do elementary "inference": $\{CS, CS \rightarrow US\} \implies US$. Then expecting the US, the dog salivates.

Of course, the application of *modus ponens* rule (purely deductive) by the mathematician and the inductive "inference" of the dog are obviously different. However, can we think about evolutionary roots of logical rules used in mathematics? Yes, we certainly can. The logical conclusion of the mathematician and the "inductive inference" of the dog are qualitatively similar.

Moreover, we can go further. We can imagine that there is a semantic network in the dog's memory. This network is a set of notions and links between notions. For example, we can imagine that the dog has notions "food," "danger," "dog of the opposite sex" – these notions correspond to main animal needs: energy, safety and reproduction. Further, the notion "food" can have semantic links to notions "meat," "bread," and so on. We can also imagine that a semantic link between a CS and a US is generated in dog's memory at classical conditioning. For example, if the CS is a sound and the US is meat, then the semantic link between the CS and the US can be illustrated by Fig. 1.

Fig. 1: Illustration of the semantic link between the conditioned and unconditioned stimuli: the meat follows the sound.

We can further imagine the generation and development of different semantic networks during the dog's life. These networks reflect the dog's experience and stored in the dog's memory. To some extent, these semantic networks are similar to semantic networks that are studied in AI research (see e.g. [12]).

So, we can think about evolutionary roots of inference rules and logical conclusions.

Additionally, I would like to note here an interesting analogy between conditioned reflex and the Hume's consideration of notion of causality.

In 1748 David Hume wrote "Philosophical Essays Concerning Human Understanding" [16], where he called in question the notion of causality, one of the main scientific concepts. Briefly, Hume's argumentation is as follows.

If we observe that some event A is many times followed by another one B, we usually conclude that the first event A is the cause of the second event B. For example, if we see that some moving billiard ball hits a second resting ball (event A) and the second ball begins to move (event B), and if we observe a series of such event pairs $(A \to B)$, then we conclude that the hit of the first ball is the cause of the movement of the second ball.

What is an origin of this conclusion?

According to Hume, we have not a solid reason for this. If we thoroughly analyze the issue, we can establish that only custom and habit, as well as some "internal feeling," force us to establish the relation between causes and effects. See [30] for some details of Hume's argumentation.

A dog in experiments of classical conditioning establishes a relation between conditioned stimulus (CS) and unconditioned stimulus (US). We can say that after repetition of a number of events $CS \to US$, some "internal feeling" forces the dog to establish the relation between "cause" (CS) and "effect" (US), in very similar manner as a human does this in the Hume's "thinking experiments."

Therefore, we can follow (at least intuitively and in very general terms) the relation between classical conditioning and prediction between causes and effects. We can try to answer to Hume's question "Why can we deduce that some event is the cause of another event?" analyzing the evolutionary roots of classical conditioning and investigating, what neural network of a dog (or a simpler animal) can produce "internal feeling" that forces the dog to establish the relation $CS \to US$.

Concluding this section, we can say that there is the important epistemological problem: why is human intelligence applicable to cognition of nature? The epistemological problem is important – it concerns the foundation of the whole science. Consequently, it is important to investigate this problem to the fullest extent. We can try to solve the problem by analyzing evolutionary roots of human intelligence. We can analyze how did human logic rules and other constituents of human thinking originate through biological evolution. Going further, we can design mathematical/computer models of "intelligent inventions" of biological evolution and try to create a theory of evolutionary origin of human intelligence.

What could be a subject of this theory? What "intelligent inventions" of biological evolution could be modeled? What models have been already created? What conceptual theories could be used in these investigations? These questions are discussed in the next section.

3 Approaches to the Theory of Evolutionary Origin of Human Intelligence

In the first steps towards a theory of evolutionary origin of human intelligence, it is natural to represent a picture of evolution of animal cognitive abilities and some conceptual schemes, which could help us to model the process of evolution of "intelligent" properties of animals. This section tries to represent such a picture of cognitive evolution and to describe such conceptual schemes.

3.1 "Intelligent Inventions" of Biological Evolution

We begin here from the very beginning – from the simplest forms of life – and try to extract levels of "intelligent inventions" of biological evolution. We mention examples of "inventions" and corresponding mathematical/computer models that have been already developed.

First Level

An organism perceives different states of external environment; the information about these states is memorized in the organism genome and is inherited. The organism adaptively uses the information about these states by changing its behavior in accordance with changes of the environment states.

An example of this level is the *regulation of enzyme synthesis* in bacteria in accordance with the classical scheme by [17]. This scheme of regulation can be outlined as follows. The bacterium *E. Coli* uses glucose as its main nourishment. However, if glucose is absent, but another substrate, lactose, is present in environment, *E. Coli* turns on the synthesis of special enzymes, which transform lactose into the usual nourishment, glucose. When the bacterium returns into glucose-rich environment, the synthesis of transforming enzymes is turned off. This scheme of regulation can be considered as the unconditional reflex at molecular-genetic level. It can also be considered as a scheme of primordial control system.

The mathematical model of such a scheme of regulation, the "adaptive syser" was created and analyzed by [29]. The model represents a possible scheme of origin of a primeval control system at a prebiological level.

Second Level

An organism individually stores the information about situations in external environment in its short-term memory. This memorizing ensures the acquired adaptation of the organism to events in the environment.

An example of this level is the *habituation* of infusoria, demonstrated by [19]. If an infusorium is subjected many times to a neutral stimulus, e.g. a drop of water, its reaction (twitching) to the stimulus is initially large, but in further course of the experiment, the reaction is decreased. This form of adaptation is of the short-term type. According to experiments of W. Kinastowski, the habituation of infusoria is formed in 10 to 30 minutes, and it is maintained during 1 to 3 hours.

Tsetlin's automata are well-developed mathematical models that correspond approximately to the "intelligence level" of habituation [33]. Tsetlin's automata illustrate simple acquired properties of biological organisms and simple adaptive behavior in changing external environments. In the last decade the models of habituation are developed in the field of "Adaptive Behavior" (e.g. [32]).

Third Level

An organism individually stores the causal relations between the events in external environment. The causal relations are stored in long-term memory.

An example of this level is *classical conditioning*. In the well-known experiments of I.P. Pavlov ([28]) on a dog, a neutral conditioned stimulus, CS was followed by a biologically significant unconditioned stimulus, US. The unconditioned stimulus aroused a certain unconditioned response. After a number of presentations of the pair $CS \rightarrow US$, the CS alone became able to arouse the same (conditioned) response.

The classical conditioning has several non-trivial particularities. There are three stages of learning procedure in classical conditioning: pre-generalization, generalization, and specialization [22]. During the pre-generalization, the conditioned response is still absent, but there is the increase of electrical activity in different areas of an animal brain. During the generalization, both the CS and other (differential) stimuli, which are similar to the CS, arouse the conditioned response. The generalization is followed by specialization, at which the response to differential stimuli is gradually vanished, whereas the response to the CS is retained.

The causal relation between CS and US is stored in the long-term memory: a conditional reflex is conserved during several weeks for low-level vertebrates and up to several years (and maybe the whole life) for high-level animals. The characteristic feature of classical conditioning is the spontaneous recovery: the renewal of a conditioned response, which takes place several hours after extinguishing of a conditional reflex [35]. The biological meaning of classical conditioning is the foreseeing of future events in the environment and adaptive use of this foreseeing [4, 5].

There are a number of mathematical and cybernetic models of conditional reflex, created and investigated by [26, 15, 7, 20] and others. However, in my opinion, some significant aspects of classical conditioning have not been mathematically described yet (the similar viewpoint was expressed by [6]). This concerns mainly the feature of the spontaneous recovery, the role of a motivation in conditional learning, and the biological meaning of the classical conditioning.

There are several levels of "intelligent inventions" between classical conditioning and human intelligence. We only mention some of them here.

Instrumental conditioning is similar to classical conditioning, but it is more complex: an animal has to discover adequate new conditioned responses (that are not known to it in advance), in order to obtain a reinforcement after a presentation of a conditional stimulus.

Chains of conditioning is a sequence of conditioned responses that is formed on the base of old conditioned relations, which have already been stored in animal memory.

High-level animals use the non-trivial *models of external environment* in their adaptive behavior. Certainly, some forms of "behavioral logic" are used in such modeling in order to predict future situations and to reach a goal. Examples of such "intelligent" behavior are well-known experiments of [21] on apes. Apes were able to use several instruments (sticks, boxes) in order to get over several difficulties and solve a complex task of reaching food. Obviously, apes use certain models and certain logic during solving these tasks.

The *final level*, we consider, is *human logic*. The mathematical theories of our logic are well developed. There is propositional calculus, there is predicate calculus [18], and there are theories of mathematical inference [10, 11]. Theories of inductive and fuzzy logic were intensively developed in the last decades [3, 36, 37].

Thus, it is possible to extract the several key "intelligent inventions" and consider the sequence of achievements of biological evolution (Fig. 2). The abilities to cognize the natural phenomena are gradually increased in this sequence.

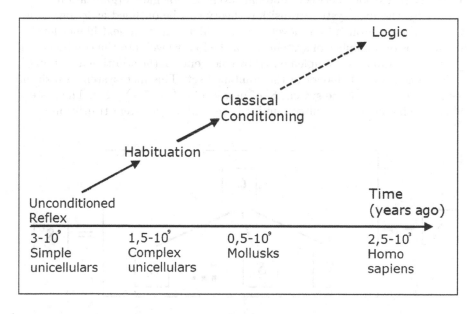

Fig. 2: "Intelligent inventions" of biological evolution. "Authors of inventions and priorities dates" are shown approximately.

Analysis of existing models of "intelligent inventions" demonstrates that we are very far from a full-scale theory of evolution of cognition. The models developed can be considered only as first steps towards such a theory. These models have obviously fragmentary character; there are no models that could describe the transition stages between the intelligent inventions of different evolutionary levels.

Thus, modeling of "intelligent inventions" of biological evolution is at initial stages of development. Therefore, it is reasonable to consider ideas and methodological schemes that could help to model these "inventions." Below we will outline some of methodological approaches: the metasystem transition theory by [34] and the theory of the functional system by [4, 5].

3.2 Methodological Approaches

Metasystem Transition Theory by V.F. Turchin

In the book *The Phenomenon of Science. A Cybernetic Approach to Human Evolution*, Turchin outlined the evolution of cybernetic properties of biological organisms and considered the evolution of scientific cognition as a continuation of biocybernetic evolution [34]. In order to interpret the increase of complexity of cybernetic systems during evolution, Turchin proposed the metasystem transition theory. This theory introduced a general cybernetic scheme of evolutionary transitions between different levels of biological organization.

Briefly, the metasystem transition theory can be outlined as follows (Fig. 3). A transition from a lower level of system hierarchy to a next higher level is a symbiosis of a number of systems S_i of the lower level into the combined set $\sum_i S_i$; the symbiosis is supplemented by emergence of the additional system C, which controls the behavior of the combined set. This metasystem transition results in creation of the system S' of new level ($S' = C + \sum_i S_i$). The system S' can be included as a subsystem into the next metasystem transition.

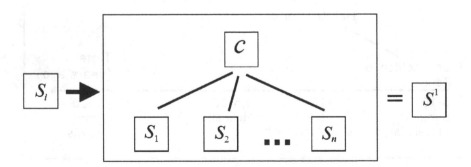

Fig. 3: Scheme of a metasystem transition. S_i are systems of the lower level, C is the control system, S' is the system of the new (higher) level.

Turchin characterizes biological evolution by the following main metasystem transitions:

- *Control of position:* movement
- *Control of movement:* irritability (simple reflex)
- *Control of irritability:* (complex) reflex
- *Control of reflex:* associating (conditional reflex)
- *Control of associating:* human thinking
- *Control of human thinking:* culture

Turchin describes the metasystem transition as certain cybernetic analog of the physical phase transition. He pays special attention to quantitative accumulation of progressive traits in subsystems S_i just before a metasystem transition and to multiplication and developments of subsystems of the penultimate level of the hierarchy after the metasystem transition.

The metasystem transition theory provides us with the interpretation of general processes of evolutionary increase of complexity. The more intimate processes of intelligent adaptive behavior can be analyzed on the base of the theory of functional systems, which was proposed and developed from the 1930s-1970s by Russian physiologist P.K. Anokhin [4, 5].

Theory of Functional Systems by P.K. Anokhin

Anokhin's *functional system* is a neurophysiological system that is aimed at achievement of an organism's vital needful result. The main mechanisms of the functional system operation are (Fig. 4):

1. Afferent synthesis
2. Decision making
3. Generation of an acceptor of an action result
4. Generation of the action (efferent synthesis)
5. The complex action
6. An achievement of a result
7. Backward afferentation about parameters of the result, comparison of the result with its model that were generated in the acceptor of the action result.

Operation of a functional system can be described as follows. An *afferent synthesis* involves synthesis of neural excitations that are due to:

1. Dominating motivation
2. Situational afferentation
3. Launching afferentation
4. Inherited and acquired memory

The afferent synthesis is followed by *decision making*, which means a reduction of degree of freedom for an efferent synthesis and selection of a particular action in accordance with dominating animal need and other constituents of the afferent synthesis.

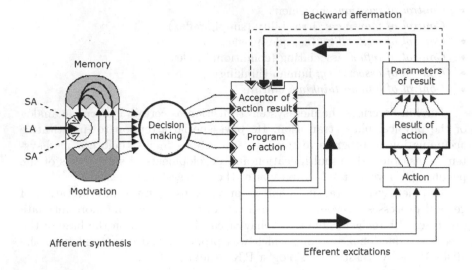

Fig. 4: General architecture of a functional system. LA is launching afferentation, SA is situational afferentation.

Next step of the operation is the generation of the *acceptor of the action result*. This step is the formation of a prognosis of the result. The prognosis includes forming particular parameters of the foreseeing result.

The *efferent synthesis* is a preparation for the effectory action. The efferent synthesis implies generation of some neural excitations before generation of an action command.

All stages of result achievement are permanently estimated by means of *backward afferentation*. If parameters of an actual result are different from parameters of the acceptor of action result, then the action is interrupted and new afferent synthesis takes place. In this case, all operations of the functional system are repeated until the final needful result is achieved.

Thus, operation of the functional system has a cyclic (with backward afferent links) self-regulatory organization.

The most important particularity of Anokhin's theory is orientation of operation of any functional system to achievement of a *final needful result*.

The next particularity is *dynamism, temporality*. At each behavioral action, different neural and other regulatory structures of an organism are mobilized into a functional system.

In addition, an important concept of the functional system theory is *systemogenesis*. The essence of systemogenesis is that organism functional systems – needed for adaptive behavior of animals and men – are ripened at both pre-natal period and ontogenesis.

It should be underlined that the theory of functional systems was proposed and developed in order to interpret a number of neurophysiological data. The theory was formulated in rather general and intuitive terms. In my opinion, it provides us with important *conceptual* approach to understanding of brain operation. This theory could help us to understand neurophysiological aspects of prognosis, foreseeing, creation of casual relation between situations and generation of 'semantic networks" (such as shown in Fig. 1) in animal brains and minds.

3.3 Role of Investigations of "Artificial Life" and "Simulation of Adaptive Behavior"

Let's return to the question of modeling of "intelligent inventions" of biological evolution. Fortunately, two interesting directions of investigations – "Artificial Life" and "Simulation of Adaptive Behavior" – appeared 12-15 years ago, which can help us. We can use methods, concepts and approaches of these researches during creating and developing models of "intelligent inventions."

Artificial Life (Alife), as an area of investigation, took its form in the later 1980s [24, 25]. The main motivation of Alife is to model and understand the formal rules of life. As C.G. Langton said, "the principle assumption made in Artificial Life is that the 'logical form' of an organism can be separated from its material basis of construction" [24]. Alife "organisms" are man-made, imaginary entities, living mainly in computer-program worlds. Evolution, ecology, and the emergence of new features of life-like creatures are under special attention of Alife research.

Simulation of Adaptive Behavior [27] is an area of investigations that is very close to Alife. However, it is more specialized - the main goal of this field of research is

> designing animats, i.e., simulated animals or real robots whose rules of behavior are inspired by those of animals. The proximate goal of this approach is to discover architectures or working principles that allow an animal or a robot to exhibit an adaptive behavior and, thus, to survive or fulfill its mission even in a changing environment. The ultimate goal of this approach is to embed human intelligence within an evolutionary perspective and to seek how the highest cognitive abilities of man can be related to the simplest adaptive behaviors of animals [9].

We can see that the ultimate goal of Simulation of Adaptive Behavior is very close to the task of creation of a theory of evolutionary origin of human intelligence as discussed above.

Thus, we have stated the problem of development of scientific base of AI researches and analyzed general approaches to corresponding investigations. Now it is time to make some concrete steps. To exemplify possible researches, we describe below two concrete computer models: the "Alife Model of Evolutionary Emergence of Purposeful Adaptive Behavior" and the "Model of

Evolution of Web Agents." These models have a number of common features and illustrate possible interrelations between purely academic investigations of cognitive evolution (first model) and applied researches directed to Internet AI (second model).

4 Two Models

4.1 Alife Model of Evolutionary Emergence of Purposeful Adaptive Behavior

The purpose of this model[1] is to analyze the *role of motivations* for simple adaptive behavior. Note that motivation is the important feature of the Anokhin' theory of functional system (the section 3.2.2). Namely, a dominating motivation – that corresponds to a current animal need – takes part in generating behavioral action.

Description of the Model

The main assumptions of the model are as follows:

- There are agents (Alife organisms), which have two natural needs (the need of energy and the need of reproduction).
- The population of agents evolves in the simple environment, where patches of grass (agent's food) grow. The agents receive some information from their environment and perform some actions. Agents can move, eat grass, rest and mate with each other. Mating results in birth of new agents. An agent has an internal energy resource R; the resource is increased during eating. Performing an action, the agent spends its resource. When the resource of the agent goes to zero, the agent dies.
- Any need of an agent is characterized by a quantitative parameter (motivation parameter) that determines the motivation to reach a corresponding purpose. E.g., if energy resource of an agent is small, there is the motivation to find food and to replenish the energy resource by eating.
- The agent behavior is controlled by a neural network, which has special inputs from motivations. If there is a certain motivation, the agent can search for solution to satisfy the need according to the motivation. This type of behavior can be considered as *purposeful* (there is the purpose to satisfy the need).
- The population of agents evolves. The main mechanism of the evolution is the formation of genomes of new agents with the aid of crossovers and mutations. A genome of the agent codes the synaptic weights of the agent's neural network.

---

[1]This model was created and developed together with Mikhail S. Burtsev and Roman V. Gusarev [8]

The environment in our model is a linear one-dimensional set of cells (Fig. 5). We assume that only a single agent can occupy any cell.

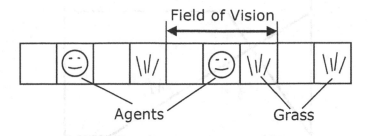

Fig. 5: Agents in the one-dimensional cellular environment

The time is discrete. At any time iteration, each agent executes exactly one action. The set of possible actions of agents is the following:

1. Resting
2. Moving to a neighboring cell (to the left or to the right)
3. Jumping (over several cells into random direction)
4. Eating
5. Mating

The grass patches appear randomly and grow certain time at cells of the environment. The agents are "short-sighted." This means that any agent views the situation only in three cells: in its own cell and in two neighboring cells. We designate these three cells as "field of vision" of an agent (Fig. 5).

We introduce two quantitative parameters, corresponding to the agents needs:

1. Motivation to search the food M_E that corresponds to the need of energy
2. Motivation to mating M_R that corresponds to the need of reproduction

Motivations are defined as follows (see Fig. 6):

$$M_E = max\{\tfrac{R_0 - R}{R_0}, 0\} \qquad M_R = min\{\tfrac{R}{R_1}, 1\},$$

where R_0 is some "optimal" value of energy resource R, R_1 is the value of energy resource, which is the most appropriate for reproduction.

The neural network of an agent controls its behavior. We suppose that the neural network includes one layer of neurons. The neurons receive signals from external and internal environment via sensory inputs. There are full interconnections between sensory inputs and neurons: each neuron is connected to any input. The outputs of neurons determine agent's actions. Each neuron corresponds to one action. Taking into account that actions "moving" and "mating" have two variants (an agent can move to the left or to the right

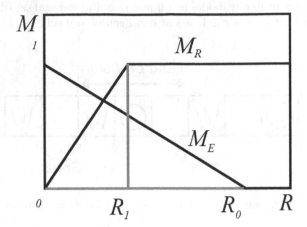

Fig. 6: Dependence of motivations M_E, and M_R on energy resource R of an agent

and mate with left or right neighbor), we have 7 neurons. Each neuron has 9 sensory inputs. Since inputs and neurons have all possible synaptic interconnections, there are 9x7 = 63 synaptic weights in the neural network. The neurons have typical logistic activation function.

We assume that at the given moment of time the agent accomplishes the action, corresponding to that neuron, which has maximal output signal.

The scheme of the evolution is implemented in the following way. We assume that a genome of an agent codes synaptic weights of the agent's neural network. Each synaptic weight is represented by a real number and considered as a gene of the genome. When a new agent is being born, its genome is created in the following manner:

1. A uniform recombination of parent's genomes is formed
2. This recombined genome is subjected to small mutations

Results of Computer Simulations

To analyze the influence of motivations on behavior of agents, we performed two series of simulations. In the first series, the agents had motivations (the motivations were introduced as described above). In the second series, the agents had no motivations (the inputs from motivations were artificially suppressed by means of special choice of parameters R_0, R_1). In order to analyze the influence of food amount in the external environment on population behavior, the simulations in both series were performed for the several probabilities of grass appearance in cells.

Choosing certain parameters, which determine energy consumption at agent actions, we defined some reasonable agent physiology. We chose also some reasonable values of parameters R_0, R_1 and starting values of energy resource of agents in initial population.

All agents of the initial population had the same synaptic weights of neural networks. These weights determined some reasonable initial instincts of agents.

The first instinct was the instinct of food replenishment. This instinct was dedicated to execute two types of actions:

1. If an agent sees a grass in its own cell, it eats this grass
2. If an agent sees a grass in a neighboring cell, it moves into this cell

The second instinct was the instinct of reproduction. This instinct implies that if an agent sees another agent in one of the neighboring cells, it tries to mate with this neighbor.

In addition to these main instincts, the agents were provided with the instinct of "fear of tightness:" if an agent sees two agents in the both neighboring cells, it jumps.

The synaptic weights from motivational inputs in the neural network were equal to zero for all agents in initial population. Therefore, motivations began to play a role only in the course of evolution.

The main quantitative characteristics that we used in order to describe the quality of an evolutionary process was the total number of agents in population N. We obtained the dependencies $N(t)$ on time t for both series of experiments: for population of agents with motivations and for population of agents without motivations. We also analyzed evolutionary dynamics of agent actions and registered a statistics of the synaptic weights during a process of evolution.

Examples of the dependencies $N(t)$ are shown in Fig. 7. With a small amount of food (Fig.7a), both populations of agents (with and without motivations) die out – the amount of food is not enough to support consumption of energy needed for agent actions. With an average amount of food (Fig.7b), the population of agents without motivations dies out, whereas the population of agents with motivations is able to find a "good" living strategy and survives. With a large amount of food (Fig.7c), both populations survive; however, the population with motivations finds better neural network control system, which ensures the larger final population.

Thus, neural network inputs from internal motivations provide an opportunity for the population to find better control system for agents in the course of evolutionary search.

Interpretation of Simulation Results

We performed detailed analysis of agents' actions evolution for populations with and without motivations. Basing on this analysis, we interpreted behavioral control of agents.

342   Vladimir G. Red'ko

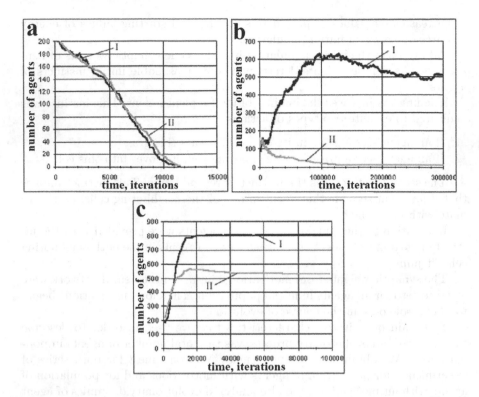

Fig. 7: Dependencies of number of agents in population with motivations (I) and without motivations (II) on time $N(t)$ for different probabilities of grass appearance, P_g: (a) $P_g = 1/2000$, (b) $P_g = 1/200$, (c) $P_g = 1/20$

The scheme of behavioral control of agent *without motivations* that was discovered by evolution is shown in Fig. 8. This scheme includes three rules, which are used by an agent during its life.

The first rule says that if the agent sees a grass patch, it seeks to eat this food. Namely, it eats food, if the food is in its own cell, or goes to grassy neighboring cell and eats food at the next moment of time.

The second rule says that if the agent sees a neighbor, it mates, trying to give birth to an offspring.

These two rules are just instincts, which we forced upon agents of an initial population. The evolution confirmed that they are useful and adaptive.

The third rule says that if the agent doesn't see anything in its field of vision, it decides to rest. This rule was discovered by evolution, and, of course, the rule has a certain adaptive value.

It is obvious that such agent behavior is determined by current state of the external environment only. These three rules can be considered as simple reflexes.

Fig. 8: Scheme of behavioral control of agents without motivations

Let us consider the control system of an agent *with motivations.* The analysis of simulations demonstrates that the control scheme of an agent with motivations can be represented as a hierarchical system. Three rules described above constitute the lower level of the control system. The second level is due to motivations. This hierarchical control system works in the following manner (Fig. 9).

If the energy resource of an agent is low, the motivation to search food is large, and the motivation to mating is small, so the agent uses only two of mentioned rules, the first and the third – the mating is suppressed. If the energy resource of the agent is high, the motivation to mating is turned on, and so the agent seeks to mate – the second and the third rules govern mainly the agent behavior, however, sometimes the first rule works too.

So, the transition from the scheme of control without motivations (Fig. 8) to the scheme with motivations (Fig. 9) can be considered as the emergence of a new level of hierarchy in the control system of an agent. This transition is analogous to the metasystem transition from simple reflexes to complex reflex in the metasystem transition theory [34].

Thus, the model demonstrates that simple hierarchical control system, where simple reflexes are controlled by motivations, can emerge in evolutionary processes, and this hierarchical system is more effective as compared to behavioral control governed by means of simple reflexes only.

4.2 Model of Evolution of Web Agents

The goal of the model[2] is to analyze evolution and self-organization of Alife agents in Internet environment. The model is similar to previous one. The main particularities (characterizing new features as compared with the model of the section 4.1) of the current model are:

---

[2]This model was developed together with Ben Goertzel and Yuri V. Macklakov. The work on the model was supported by Webmind, Inc. See [14] for a more detailed description of the model

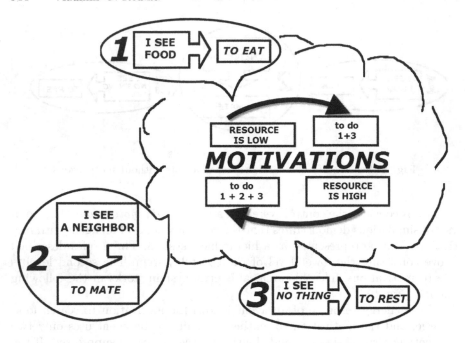

Fig. 9: Scheme of behavioral control of agents with motivations.

- The model implies that there is a set of Web World lobes where a population of Alife agents evolves. Each lobe contains a sub-population of agents (Fig. 10). *The lobes are distributed in an Internet environment.*
- Agents can *communicate* with each other. Agents can *fly* between different lobes. Agents can execute several actions; in particular, they can *solve tasks.* Solving a task, the agent obtains certain reward.
- Agents have two needs: *the need of energy* and *the need of knowledge.* Any need is characterized by a quantitative motivation parameter.
- There are two *neural networks* that control behavior of an agent. The first neural network governs *selection of actions* of the agent. The second neural network governs *solution of tasks.* There is a procedure of learning of the second neural network. This learning is based on some modification of well-known back-propagation method (see below). The synaptic weights of the first neural network do not change during agent life.
- The synaptic weights of the first neural network and initial synaptic weights of the second neural network are genes of the *two chromosomes* of the agent.

The model implies that any agent has its internal energy resource. Executing an action, the agent spends its energy resource. When internal energy resource of an agent goes to zero, this agent dies. Any agent can eat food and replenish its internal energy resource. However, before eating, the agent should

Web World

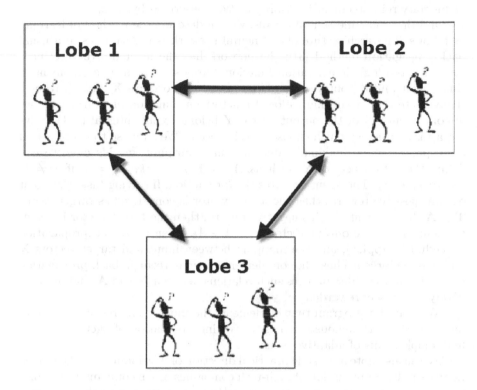

Fig. 10: Alife agents distributed in Internet.

solve some task. The value of the reward that the agent obtains depends on quality of task solution. Rewards can be positive or negative. Receiving positive reward, the agent eats food and increases its energy resource. When the agents receives negative reward (punishment), its energy resource is decreased.

Agents can communicate each with others. By communicating agents help each other to increase their knowledge about situations in different lobes.

In any lobe, agents can mate each other. When executing the action "Mating" an agent becomes a partner for mating. Two partners for mating in the same lobe give birth to a child. Each parent transmits to the offspring some part of its energy resource. Each chromosome of the offspring is obtained through one-point crossover of the corresponding chromosomes of the both parents. Additionally, there are small mutations of the genes of chromosomes.

Flying between lobes, agents are able to travel over the Web World.

There is a rather non-trivial procedure of learning of the second neural network (the network of the task solver). This learning is based on the complementary reinforcement back-propagation, described by [1, 2].

Omitting some inessential details, we can describe the method of learning as follows. The architecture of the neural network is the same as in a usual back-propagation method [31]: the network has the layered structure; neurons have the logistic activation function. Suppose that at given moment of time, input and output vectors of the neural network are X and Y, respectively. Note that according to logistic activation function of neurons, values of components Y_i of the output vector Y belong to the interval $(0, 1)$. Solving a task, an agent much choose certain action. We suppose that the action corresponds to the maximal value of neuron outputs. In this case we can define the action vector A, such as $A_j = 1$ if $j = k$, $A_j = 0$ if $j \neq k$, $k = argmax_i Y_i$. The agent executes the k-th action. If solving task, the agent obtains positive reward, then the action vector is considered as target vector $T = A$. If the agent obtains negative reward, then the target vector is "complementary" to the output vector: $T_i = 1 - A_i$. Then usual backpropagation procedure is applied, and the mapping between input and target vectors X and T is reinforced. Thus, the complementary reinforcement back-propagation method reinforces/dereinforces such relations between X and A that are positively/negatively rewarded.

We created a program that implements the model. The results of preliminary simulations demonstrated that evolving population of agent is able to find simple forms of adaptive behavior.

We can also note a possible practical direction of development of the model. Let's consider a population of high-tech companies. Each company has a computer network; this network is the lobe, where a corresponding sub-population of agent evolves. Any company has a special person, the supervisor of sub-population of agents. This supervisor gives some practical tasks to agents in his lobe and rewards or punishes them. Tasks could be such as "give me prognosis of this certain market" or "find me a good partner for this kind of cooperation," etc. Agents should solve tasks and are rewarded/punished accordingly. Agents have access to Internet. Companies have web-sites, so agents are able to analyze information about the population of companies. During an evolution of the population of Web agents, tasks for agents may be made more and more complex and this could ensure more and more intelligent behavior of agents.

Of course, the described two models are only simple examples of concrete researches. In the next section, we will outline a possible way from these simple models to implementation of higher cognitive abilities.

5 Towards the Implementation of Higher Cognitive Abilities

Let us consider possible steps toward modeling high level intelligence.

Step 1

Evolutionary optimization of simple instinctive behavior. We can code a control system of an agent (e.g. agent's neural network) by means of a genome and optimize the genome by means of an evolutionary method. For example, we can introduce a parameter of vital resource R of an agent; resource is increased/decreased at a successful/unsuccessful action of the agent. If resource of the agent goes below certain threshold, the agent dies; if agent's resource is large, the agent gives birth a child (deterministically or in some stochastic reproductive process), reproducing (and modifying by mutations) its genome. The model of the section 4.1 is the example of this level of implementation.

Step 2

Using the concept of internal vital resource R, we can introduce the *natural scheme of unsupervised learning*. Suppose that the control system of an agent is a layered neural network with logistic activation function of neurons. Then this control system can be optimized at *each action* of the agent by means of the complementary reinforcement back-propagation method (described in the section 4.2). If the action of the agent is successful ($\Delta R > 0$), the synaptic weights of the neural network are reinforced; if the action is unsuccessful ($\Delta R < 0$), the synaptic weights are dereinforced.

This method can be complemented with usual evolutionary optimization: initial (obtained at the birth of an agent) synaptic weights of the neural network can constitute the genome of the agent. Method of evolutionary optimization of the agent genome is the same as described above (Step 1).

Step 3

We can consider several vital needs of an agent (energy, security, reproduction, knowledge), characterizing j-th need by its own resource R_j and motivation M_j ($j = 1, 2, \ldots, n$). It is natural to assume that a motivation M_k monotonically decreases with increasing the corresponding resource R_k. Supposing that at each moment of time there is a *dominating motivation* M_d that determines the agent behavior, we can introduce the scheme of unsupervised learning for this case too. Namely, if the resource R_d, corresponding to dominating motivation, increases/decreases, then synaptic weights of the neural network of the agent are reinforced/dereinforced. Note, knowledge can be considered as an important need of the agent [38], implying that intellectual curiosity is the motivation to increase knowledge.

Step 4

We can imaginary reorganize the scheme of modeling of the Step 3, trying to approach to a scheme of P.K. Anokhin's functional system (see Sect. 3.2). We can consider an animat (or animal, or model of animal) that has rather arbitrary structure of the neural network. There are different links in the network with different weights between neurons. The animat has needs and corresponding motivations M_j as above. However, now we assume that the animat can have also a model of the external world and it can make prognosis of results of its actions.

We can suppose that at given dominating motivation M_d (e.g. the motivation to get food) some excitatory processes take place in the neural network. Excitatory processes can restore in neural memory patterns of objects that are related to satisfaction of the need (e.g. the pattern of meat) and patterns of situations, at which these object were observed. (It is not difficult to imagine these patterns - the patterns can be stored in the form of Hebbian assemblies.) Taking into account this information, our animat can try to make a prognosis about results of its possible actions. This process of prognosis is rather non-trivial. However, imagine that our animat is able to make a prognosis. We can also imagine that the animat is able to select an adequate action in accordance with the prognosis. Then we can naturally suppose that the animat is able to learn by means of modification of its neural network. If the action is successful, that is the foreseeing result is achieved, then the existing links in the neural network are reinforced by corresponding modification of synaptic weights. Otherwise, some unlearning procedure could take pace, e.g. in the form of some dereinforcement as discussed above in Steps 2 and 3.

In addition, we can imagine a set of Hebbian assemblies in the neural networks of animats; assemblies store patterns of neuron activities that characterize notions, names or concepts. Assemblies store patterns in the form of associative memory, so assemblies memorize the most general and statistically averaged notions [23]. The set of assemblies connected by neuron links can be considered as a semantic net. We can imagine that, using the semantic net, the animat is able to make some "logical" inferences, similar to that of discussed in Sect. 2 and in the draft paper by [13].

Thus, we can imagine a non-trivial neural-network-based control system of animats. Using this control system, an animat is able to construct models of environment, to make "logical" inferences, to predict results of its actions. The animat is also able to learn; links in animat's neural network are changed during learning. We can also imagine that neural network architecture of animats can be optimized by ontogenetic development and evolutionary optimization in population of animats. We can consider the intelligence of such animats as "dog-level" intelligence.

Of course, the Step 4 is quite imaginary. However, the described conceptual scheme of animat control system is sufficiently concrete and can stimulate researches of animat intelligence and developments of real AI systems.

Moreover, we can go further to some fantastic step.

Step 5

Can we try to imagine a metasystem transition from "dog-level" intelligence (outlined in the Step 4) to human level intelligence? Yes, we can. Let's suppose that there is a *society* of animats, and each animat has a neural network control system. Animats are able to create models of the external world and make inductive logical inferences about the world; they are able to make prognosis and to use forecasting in their activity. Suppose that animats can communicate each other (similar to agents of the model of the Section 4.2). Their communications could help them to produce collective actions. Therefore, communications could result in some "animat language"; and the notions, corresponding to words of this language, could solidify in animats memories. These animats could also invent numerals in order to use calculations in planning collective actions. Thus, such animats could have *primitive thinking*, similar to the thinking of a hunter tribe of ancient men. Let us suppose now that there is some sub-society that would like to create most strongest form of thinking, to think about thinking, to create a special language about thinking. Such animats could be considered as mathematicians and philosophers of animat society (similar to mathematicians and philosophers of Ancient Greece). This step from the primitive thinking to the *critical thinking* is an important metasystem transition to human level intelligence [34]. Of course, this step is quite fantastic; nevertheless, we can imagine and even try to model it.

6 Conclusion

This chapter has mainly conceptual, philosophical character. Nevertheless, I hoped that it could stimulate developments of concrete models of "intelligent" adaptive behavior. In my opinion, modeling of intelligent features outlined in Step 4 of Sect. 5 would be the most interesting and important from both the scientific and AI application points of view.

7 Acknowledgements

I thank Konstantin Anokhin and Benjamin Goertzel for a number of useful discussions. Work on this paper was supported by Russian Humanitarian Scientific Foundation. Project 00-03-00093.

References

1. Ackley D, Littman M (1990) *Generalization and Scaling in Reinforcement Learning.* In: Touretzky S (ed) Advances in Neural Information Processing Systems 2. Morgan Kaufmann, San Mateo, CA.

2. Ackley D, Littman M (1992) *Interactions Between Learning and Evolution*. In: Langton C, Taylor C, Farmer J, Rasmussen S (eds) Artificial Life II: Santa Fe Studies in the Sciences of Complexity, Addison-Wesley, Reading, MA.
3. Angluin D, Smith CH (1983) Inductive Inference: Theory and Methods. *Comp. Surveys*, 15(3):237–269.
4. Anokhin PK (1974) *Biology and Neurophysiology of the Conditioned Reflex and its Role in Adaptive Behavior*. Pergamon, Oxford.
5. Anokhin PK (1979) *System Mechanisms of Higher Nervous Activity*. Nauka, Moscow (in Russian).
6. Balkenius C, Moren J (1998) *Computational Models of Classical Conditioning: a Comparative Study*. In: Langton C, Shimohara T (eds) Proceedings of Artificial Life V, MIT Press, Bradford Books, MA. http://www.lucs.lu.se/Abstracts/LUCS_Studies/LUCS62.html
7. Barto AG, Sutton RS (1982) Simulation of Anticipatory Responses in Classical Conditioning by Neuron-like Adaptive Element. *Behav. Brain Res.* 4:221.
8. Burtsev MS, Red'ko VG, Gusarev RV (2001) Model of Evolutionary Emergence of Purposeful Adaptive Behavior: The Role of Motivation. In: *Proceedings of the 6th European Conference in Advances in Artificial Life - ECAL*. http://xxx.lanl.gov/abs/cs.NE/0110021
9. Donnart JY, Meyer JA (1996) Learning Reactive and Planning Rules in a Motivationally Autonomous Animat. *IEEE Transactions on Systems, Man, and Cybernetics - Part B: Cybernetics*, 26(3):381-395
10. Gentzen G (1935) Untersuchungen über das logische Schliessen. *Mathematische Zeitschrift*, 39:176–210
11. Gentzen G (1936) Die Wíederspruchsfreiheit der reinen Zahlentheorie. *Math. Ann*, 112(4):493–565.
12. Goertzel B (2001) *Creating Internet Intelligence*. Plenum Press, New York.
13. Goertzel B, Pennachin C, this volume.
14. Goertzel B, Macklakov Y, Red'ko VG (2001) Model of Evolution of Web Agents In: *Report at The First Global Brain Workshop* http://www.keldysh.ru/pages/BioCyber/webagents/webagents.htm
15. Grossberg S (1974) Classical and Instrumental Learning by Neural Networks. *Progress in Theoretical Biology*, 3:51–141.
16. Hume D (1748) *Philosophical Essays Concerning Human Understanding*. A. Millar, London.
17. Jacob F, Monod J (1961) Genetic Regulatory Mechanisms in the Synthesis of Proteins. *J. Mol. Biol. Vol*, 3:318–356.
18. Kleene SC (1967) *Mathematical Logic*. John Wiley and Sons, New York.
19. Kinastowski W (1963) Der Einfluss der mechanischen Reise auf die Kontraktilitat von Spirostomum ambguum Ehrbg. *Acta Protozool*, 1(23):201–222.
20. Klopf AH, Morgan JS, Weaver SE (1993) A Hierarchical Network of Control Systems that Learn: Modeling Nervous System Function During Classical and Instrumental Conditioning. *Adaptive Behavior*, 1(3):263–319.
21. Köhler W (1925) *The Mentality of Apes*. Humanities Press, New York.
22. Kotlyar BI, Shulgovsky VV (1979) *Physiology of Central Nervous System*. Moscow State University Press, Moscow (in Russian).
23. Kussul EM (1992) *Associative Neuron-Like Structures*. Naukova Dumka, Kiev, Russia (in Russian).

24. Langton CG (ed) (1989) *Artificial Life: The Proceedings of an Interdisciplinary Workshop on the Synthesis and Simulation of Living Systems.* Addison-Wesley, Redwood City, CA.
25. Langton C, Taylor C, Farmer J, Rasmussen S (eds) (1992) *Artificial Life II: Santa Fe Studies in the Sciences of Complexity.* Addison-Wesley, Reading, MA.
26. Lyapunov AA (1958) On some general problems of cybernetics. *Problems of Cybernetics,* 1:5–22 (in Russian).
27. Meyer JA, Wilson SW (eds) (1990) *From Animals to Animats.* MIT Press, Cambridge, MA.
28. Pavlov IP (1927) *Conditioned Reflexes.* Oxford, London.
29. Red'ko VG (1990) Adaptive Syser. *Biofizika,* 35(6):1007–1011 (in Russian). See also: `http://pespmc1.vub.ac.be/ADAPSYS.html`
30. Red'ko, VG (2000) Evolution of Cognition: Towards the Theory of Origin of Human Logic. *Foundations of Science,* 5(3):323–338.
31. Rumelhart DE, Hinton GE, Williams RG (1986) Learning Representation by Back-Propagating Error. *Nature.* 323(6088):533–536.
32. Staddon JER (1993) On Rate-Sensitive Habituation. *Adaptive Behavior,* 1(4):421–436.
33. Tsetlin ML (1973) *Automaton Theory and Modeling of Biological Systems.* Academic Press, New York.
34. Turchin VF (1977) *The Phenomenon of Science: A Cybernetic Approach to Human Evolution.* Columbia University Press, New York.
35. Voronin LG (1977) *Evolution of higher nervous activity.* Nauka, Moscow (in Russian).
36. Zadeh LA (1973) *The Concept of Linguistic Variable and its Application to Approximate Reasoning.* Elsevier, New York.
37. Zadeh LA, Klir GJ, Yuan B (eds) (1996) *Fuzzy Sets, Fuzzy Logic, and Fuzzy Systems: Selected Papers by Lotfi A. Zadeh.* World Scientific, Singapore.
38. Zhdanov AA (1998) About an Autonomous Adaptive Control Methodology. *ISIC/CIRA/(ISAS'98),* 227–232.

3D Simulation: the Key to A.I.

Keith A. Hoyes

Inca Research, Ltd.
keith@incaresearch.com -- http://www.incaresearch.com

Summary. The proposal is a radical one – that human cognition is significantly weaker than we presume and AI significantly closer than we dared hope. I believe that the human mind is largely made up of tricks and sleights of hand that enamor us with much pride; but our pedestal might not be quite so high or robust as we imagine. I will pursue the argument that human cognition is based largely on 3D simulation and as such is particularly vulnerable to co-option by future advances in animation software.

1 Introduction

"A is A" – Ayn Rand

Monsters Inc. was an entertaining film and like so many others of its genre, it allowed us, for a time, to enter a world that never really existed. To the computers that generated the images, the world doesn't exist either, it is just so many 1's and 0's. But those bits got transformed into a language we could all understand; a world we can feel, fear and predict. Our eyes similarly take a cryptic stream of bits and somehow too create a world we can feel and predict. If you close your eyes and imagine entering your kitchen to get soda, you must surely have created a 3D world to navigate. As you re-open your eyes, just how are those dancing 2D patterns you see converted into the 3D virtual realities in your mind? [7]

In the virtual world, when a princess kisses a frog it turns into a prince. The real world does not work that way. For general AI to solve real world problems, its thinking needs to be bound by real world behaviors. All significant phenomena in the real world exist in three dimensions, or can be expressed as such. The common language describing computers, bicycles and brains is that of their 3D material existences animated over time (A is A). Further, derivative concepts such as math, stock markets, software and emotion can similarly be bound. If a concept cannot be described in three dimensions over time, it is quite likely false. Like the frog above, it may exist only in some virtual domain [8].

The real world cannot violate the laws of physics, logic or axioms to enter a fantasy world – frogs to princes. But the virtual can. It can be bound or unbound. But when bound to physics, it can accurately simulate reality. This has important consequences for AI.

Finally, the real world is bound by time. The virtual is not. It can run time backwards and forwards and at any speed. It can also accept time discontinuities, freezes and gaps. The virtual can predict events in reality before they have even happened! It can represent the now, the future or the past. It's when the real and the virtual are mixed, that the magic really begins.

2 Pillars of Intelligence

2.1 Deep Blue

To many, the victory of deep blue – a mere computer, over the smartest chess genius alive was both disconcerting and also raised hope of a new dawn for AI. But in the end it didn't really amount to much. Politicians still got elected, deep blue got de-commissioned and most computers still act more like glorified calculating machines than thinking people. The fact one such calculator beat the smartest guy in the world at chess was just a freak anomaly. Just how Deep Blue beat Kasparov I will explain. Though a much more important question, I should think, is – Just how could Kasparov, hope to beat a computer?

Deep Blue operated primarily on just one of the three pillars of intelligence – time travel. I'll explain. The important aspects of a chess game can be simulated quite perfectly in a computer. At any given instant of real time, the game will, obviously, be in its current real state. Deep blue took that state as its starting point. It made predictions, grading each outcome as far into the future as time and resources would permit. Its final move was thus calculated to have the greatest probability of success. And the rest, as they say, is history.

2.2 Virtual Reality

Deep blue simulated a world very different from our own, But there are simulated environments in software labs, movie studios and military compounds all around the world that are very much more like our own. The fact that virtual worlds will soon become quite convincing and compelling is taken much for granted these days. It is only seen as a matter of time before the accuracy of simulations can completely fool the expectations of our senses. This will form the second pillar of intelligence.

2.3 The Humble Earthworm

An earthworm doesn't know much about "pillars of intelligence," but it represents one all the same. It has no ability for virtual time travel, no machinery for creating virtual environments, but it can surely feel the real world around it; the resistance of soil; the drying heat of the sun; the injury from a bird predator – and act accordingly. It might be argued that such a simple organism, which presents evidence of feeling, is in fact demonstrating basic reflexive responses to stimuli – like a thermostat. So when I say feel, what perhaps I should have said is the ability to discriminate within the flow of sensory inputs, that which it considers good from that which is bad. In other words, it has a sensorimotor system linking the real world to the virtual. These then: virtual time travel; virtual reality and an information bridge to reality, I present, are the three pillars of intelligence[6].

3 Consciousness

The moment a silicon eye can stare back into your peering eyes, unflinching and following your every move – then, you will believe in such a thing as an artificial soul. You may be wrong, but not entirely so. We intuitively know that a thing that can break through the veil of vision, to make sense out of that mess of light and shade; to really "see" a living, breathing human being, has crossed that Rubicon. It will actually be achieved through unconscious, mechanistic vision processing. But our intuition in this case will be right – 2D to 3D instantiation is indeed at the very heart of consciousness. (In animation, rendering is the process of converting 3D scenes to 2D bitmap images for human viewing. Instantiation, in this context, is the reverse – the conversion of 2D bitmaps back to 3D environments, this is a key concept in this paper).

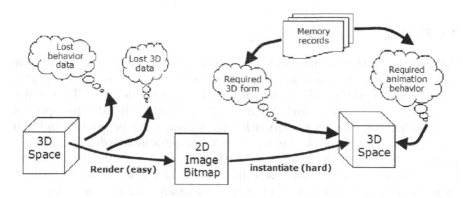

Fig. 1: Rendering and instantiation

Your basic human being is constructed from a virtual reality chamber connected to a biological, self assembling, nanotech robot with sensors. The chamber is self learning from exposure to the outside world and free will stems from a process of grading simulated predictions against pre-programmed genetic and culturally programmed schemas. Without a simulated environment running behind our eyes, we would be totally blind. The stream of data can only ever represent a series of bit maps; there is no hidden information our eyes can see that a camera can't – there is less! The images are simply used as cues in the construction of a virtual environment. The contents of that environment are actually drawn from memory and the bitmaps simply maintain simulation alignment and paint texture over the model surfaces. The experience of consciousness is bound to that simulation.

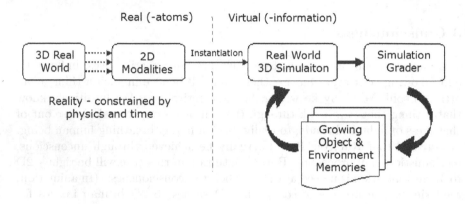

Fig. 2: Atoms and information

3.1 Feeling and Qualia

However fancy the arrangement, the human mind is still, only made from matter. There is no mysterious essence of feeling from magic atoms hidden away in the corners, and non-invisible real atoms do not feel! That leaves only one cause – information. When energy and matter are able to represent coherent information, such information can subsequently be graded, and that information interpreted as feeling. It is a computational trick; an illusion, necessary to control the behavior of biological creatures. Evolution uses the process of consciousness and the subjective "closed loop" belief in feelings, to guide behavior toward the survival and reproduction of genes. Although we may assert pleasure and pain to be computational illusions, we have no conscious control over the process; so telling ourselves pain is just information will not work[1].

For humans there is a first person relationship between the sense modalities and the affect within the mind. Every sensory receptor – whether from touch, sight, sound, smell or taste, will flow into memory somewhere, and maps directly to the first person perspective in a simulated environment. This nexus represents the "eye" or "I" of consciousness. Together with the muscles, the sensorimotor system forges the creation of a simulated environment which is processed by filters to grade simulations according to genetic and cultural presets. These analyses produce the illusion of feeling and emotion. They are used to guide subsequent cognitive attention.

But this information can have no meaning until it is grounded, through instantiation, to virtual objects that have form and invariably a history time-line (behavior). What we actually perceive are virtual objects within our own minds, the senses are used to align these objects to external reality, to consolidate existing memories, or to train new memories if the objects are novel. Once the sensory flows are aligned to precedents, the scene can then become known. Not simply because of the informational connection to a form – the instantiation. But because the virtual object forms have known behavior precedent potentials and a "spatial" home within the mirror-world simulation.

This is the point at which subconsciousness can take hold, by taking these behavior precedent options and running trials "subconsciously," away from the perceived scene – which may be linked or not to reality through the sensorimotor system. Subconscious simulations are fast, dynamic simulations, that seek out narrative with significant grading points. And this is where the issue of emotions and feeling states – "qualia" enter the picture. Genetically, the brain is programmed, and programmable, with a value hierarchy. Just as the eyes are formed in expectation of light, so the brain is formed with memory references in expectation of information against which to compare. We describe the subsequent gradings as our feelings and emotions. The most obvious example being the sexual beauty (form) and grace (behavior) of the opposite sex, emotional recognition that is innate.

The higher speed of subconsciousness is necessary to discover scene outcomes ahead of real world time. Such that actions aligned to goals can be discovered before it is too late – such as catching a ball. Emotional grading of simulations is generally more intense if they are currently aligned to reality through the senses. This motivates action in preference to reflection.

So, to summarize, the brain contains information describing object forms and behaviors. These memories are organized into a spatial hierarchy mimicking the external world. Some memories are created by the genes, but the bulk are forged into memory as the sensorimotor system interacts with reality. The contents of consciousness are the scene alignments currently in resonance within memory and reflecting back to the sensory cortices, such that the sensory envelope of the modalities can extend to embrace imaginary worlds. Subconscious processes are resonances not currently aligned to the sensory cortices, although fully capable of being emotionally graded, leading to non conscious feeling states and motivations. The brain needs subconscious processes (i.e. the simulation and grading of memory precedents) in order to discover choices upon which to align consciousness and/or physical actions.

The analog nature of human biology, which interfaces electrical, chemical and cellular processes beneath the computations of "mind," leads to physical pathways, sensitized chemical boundaries, linking the computation of feeling to the sensory "feeling" of feeling – to qualia. There is strong evidence to suggest that supplemental sensory regions exist beyond the traditional modalities, mapped instead to an existential feeling space deep inside the body, but in actuality merely extending across chemical boundaries within the brain. Such a mechanism would provide powerful feedback paths of the same "class of feeling sensation" as from the touch senses, but without the concomitant external body surface mapping. Instead, it is as if some "phantom limb" were at the body's core. Thus the feeling effects of subconscious emotional script analysis will have physical manifestations. Evidence from pharmacology clearly points to the existence of chemical pathways affecting emotion and states of mind.

Like a "second sense," emotion would impart an evolutionary advantage even before the emergence of higher cognition. Since primitive emotions can provide effective shortcuts to otherwise complex, or slow, effortful cognitive processes. It can often be seen directly in children before they learn to subordinate their emotions to their emerging wider scope cognition. This same effect occurs in the processing of language, providing short cuts to understanding. Much of our social language is predominantly emotional, often pre-empting and short circuiting rational thought; since the whole meaning really is meant to be just the emotion tags. It would often be considered quite disingenuous to even attempt a rational analysis. Spock comes to mind here!

Emotions are not only used by the brain to grade simulations, they can also be linked to objects to help predict their behaviors. Animation within a simulated environment will involve causes, objects (actors) and effects. The emotional states of objects (which include people and animals, as well as inanimate objects) originate from the context, the initial conditions and from the historic memory records. This empathic knowledge within the simulation is different from the first person emotional analysis of sub consciousness used to grade the scripts. It instead provides behavior cues to more accurately guide the simulation. For example, empathic knowledge of joy or anger in a character will significantly affect their expected behaviors and interactions. Even traditionally inanimate objects can be injected with empathic behavior attributes as evidenced in cartoons, such as an "angry car" or a "cheerful flower."

4 General Intelligence

Intelligence, as a computational process, is a continuum rather than an end point. When bound to the real world, it is the ability to so deeply understand the nature of reality, of which it can provide increasingly accurate predictive power. This includes the ability to run predictions in reverse to build causal or historic relationship chains.

Humans use this predictive power as a means to interact purposefully with their environment – to aid survival and promote adherence to genetically prescribed or socially engineered values and goals. But for Intelligence to exist at all, there are certain environmental pre-cursors:

1. A physical medium upon which it can bind the predictions: *reality.*
2. A representative medium in which it can model the predictions: *virtual*
3. A motive force: *energy*

And for intelligence to speculate on our reality it needs a means to:

1. Access that reality: *exposure*
2. Perceive that reality: *modalities*
3. Decipher that reality: *instantiation machinery*
4. Model that reality: *modeling machinery*
5. Grade the simulations: *emotional machinery*
6. Classify and store data: *memory machinery*

A presumption is that due to quantum effects at the very small level and chaos effects at the very large – prediction, and thus intelligence, will remain illusory. Added uncertainty arises with other biologically constructed animated beings. Constrained by physical law, yet animated by reflex, genetically programmed instinct, or from their internal cognitive processes. How can such complexity ever be intelligently predicted? Yet we ourselves appear able, at least to some extent, to overcome all of these effects.

At the atomic level, it is rarely necessary to predict particle animation with certainty, because all significant effects occur in the aggregate, where statistical probability can reliably model behavior. Also, predictions can be constrained to avoid chaotic events (so rather than walk a tightrope to get from A to B you take the foot bridge). With biology, statistical prediction still works well on macro events, but is limited in the details. So although intelligent prediction does appear to have some constraints, there are still very large areas where it can be relied upon. Within the oceans of chaos there is much dry land upon which to build a rational intellect.

It is further presumed that computers are deterministic and humans non deterministic. Therefore, given an initial set of conditions, a computer can only ever follow a predetermined course. Whereas a human, with "free will," can follow his own. For all intents and purposes both can be considered non deterministic, though statistically predictable. The study of human twins illustrates how the same largely deterministic genetic inheritance can be affected by real world chaotic forces. Like internal brain chemistry guiding emotions; sensory data flow, unique first person perspectives and the resulting memory structures; differentiated emotional responses etc. Add all these variables and more together and you have a combinatorial explosion. Genuine AI will similarly benefit from many of these same forces. Even blind random inputs could be easily added if found beneficial.

Intelligence is non-judgmental and the pursuit of knowledge morally neutral. But any action affecting other conscious entities creates moral hazard.

Morality arises from the exigencies of biological survival within a social framework and is dominated by genetic and social programming biases, as show in Table 1.

Positive emotional grading		Negative emotional grading	
mating	pleasure	physical damage	pain
predicted food	anticipation	predicted damage	fear
fulfillment	satiation	lack of food	danger
profitable action	pride	social exclusion	shame

Table 1: Positive and negative emotional grading.

The primary genetically derived grading process leads to the basic positive moral status of survival (existence), feeding and mating. Secondary genetic and socially trained schemas lead to the moral grading of simulations involving cultural concepts such as cooperation, altruism, group patriotism, treachery, over consumption, monogamy etc.

4.1 Human Intelligence

The human mind is a particularly difficult thing to understand, but it is the best example we have of intelligence with intentionality. The brain appears to achieve this through a massive structure of neural networks which are able, over time, to effectively interpret sensory data in order to understand and predict the perceived environment – more usually our external world. Thus, research into evolving hardware and synthetic neural networks would appear a worthwhile endeavor[5].

Biological intelligence evolved through natural selection. It developed inside a mobile mechanical body with rich sense modalities and programmed survival instincts to grade the information flow. It is protected during a nurture phase where a subconscious computational process can learn to extract meaning from the sensory modality flow and bind the internal simulation architecture to the physics and object behaviors of the real world. This subconscious simulation builds a personal feeling of familiarity with the outside world. Otherwise each moment would forever seem strange and new as if being met for the very first time. Intelligence then develops gradually through continued interactions with the environment being compared to script predictions. The level of intelligence reached is based on both the initial biological construction and from subsequent interactions with the environment – nature and nurture.

A human infant, exposed to the outside world, gradually learns to interpret the 2D visual images into 3D virtual objects. This process is significantly aided through muscular feedback, mobility and the other sensory modalities, together with genetically inspired dedicated machinery for this purpose. The

3D objects, once extracted, exist not in isolation, but within their virtual environments and as animated scripts. These will gradually build up structured and cross linked historic memory records, forming an increasingly accurate world model. Objects have textures and behaviors (animated shape morphs and/or motion scripts), together with empathic emotional hues.

Intelligence, as such, begins to really kick in when a sufficiently detailed world model has formed and enough 3D object behaviors accumulated. The maturing mind can then focus more on the content than the 3D instantiation (sometimes referred to as binding – translating modality inputs to percepts [3]). An inner virtual world will come to map the external world, and the ability to notice and interpret anomalies between the two will increase, as will the ability to predict events from precedents.

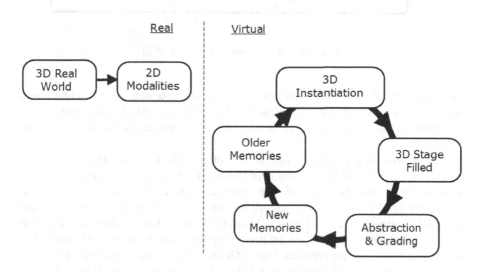

Fig. 3: The human mind: learning

When a child awakes, her mind will resume the virtual model of her room and her waking eyes will orientate, texturize and track that model. She will experience a feeling of familiarity as her sensory flow matches the virtual model she holds in memory. As she moves, so the perspective of the model will too. In fact, a series of subconscious 3D script predictions will have pre-empted her motion even before she gets started. It will partly be those predictions that lead to her intentionality of action. As her eyes scan the visual scene, detailed 2D image data will paint accuracy into, and reinforce the authenticity of her virtual world. It is in this way that she is conscious she is in a room, and feels competent to negotiate reality.

An unconscious process runs memorized script behaviors ahead of real "modality" time to generate as many predictive script estimates as time or

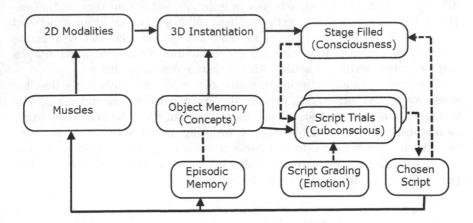

Fig. 4: The Human Mind: Free Will

satiation permit. The best case script can be used to form new learned memories or to animate physical action by aligning the virtual simulation to the modality inputs, linking the virtual body animation to motor control in the real body.

There are two priorities to human cognition. The first, mentioned above, is reactive thought, which involves negotiating real world environments, objects and people in real time. Here, the subconscious simulators may operate at maximum speed and concomitant reduction in accuracy. The simulations are generally bound to the real world through the modalities. The second, reflective thought, involves thinking by processing memory records, with limited or no external sensory perception, but with far greater depth and precision.

The content of reflective thought is based on simulations built from learned objects and behaviors acting on historic episodic scripts. Virtual in nature, these simulations will be time discontinuous for easier layering, merging and comparison – in order to discover relationships and metaphor. Cost-benefit analysis and risk assessment are extensively used to guide, grade and judge this script discovery process. They are synonymous to human emotions. Compared to reactive cognition, these simulations are not driven by exigencies from the outside world.

Other factors influencing this process are genetically derived biases carrying heavy emotional content (like fear of snakes, desire for the opposite sex, etc.). Such imprints must surely have been written into memory by the genes and must also exist in the very same language as whatever instantiations the modalities cause. The fact that genetically derived instinctive triggers can be recognized and emotionally graded and responded to from untrained input, categorically implies a priori knowledge of that percept and of a common language for its recognition. For 2D visual input, where 2D images can so easily disguise content, 3D instantiation is by far the most credible link. Thus genetically derived instinctive imprints must have a direct correlation to our modalities – particularly vision, with the most likely common language being 3D instantiation.

The process of human learning is thus predicated on exposure to the real world through the sense modalities. The mind gradually builds historic records of familiar environments, 3D objects and features, with increasing fidelity. Adding more objects and details as time goes by. The power of time shifting, time discontinuity and layering/blending in virtual simulations leads to rational prediction and intelligent cognition when aligned to goals. A side effect of this process is the seductive lure of unbinding the virtual models from real world physics and historically learned behaviors, and promoting instead, an internal world of fantasy. This process is further encouraged by the effects of biological feedback in the form of emotion. Human cognition is highly tuned to emotional cues within content, and uses them as short cuts to cognitive effort. Unbound simulations can thus be used to amplify emotion in a simulation. Presumably, attending to material survival have kept such processes in check.

5 3D Simulation and Language

Man successfully learned to express and then codify knowledge by symbolic notation. It could then be externalized and preserved through generations as a common resource to be shared and built upon. But language has a subsidiary relationship to reality. If you take a 3D cube to represent all space time, what lies inside that cube is reality. But the virtual extends both in and outside of that cube. Language, too, straddles both worlds like floating braids, weaving in and out of reality, embracing fairytales and hard science alike. As such, it may not be so reliable a foundation upon which to base AI.

Even when language tries to constrain itself to describe real world objects or behaviors, it is not always so easy to test whether the braid is really bound by reality. It is often ambiguous. There are other problems:

1. Language can break physical law and logic with impunity.
2. Language is interpreted differently by each conscious entity.
3. Language does not fully circumscribe or instantiate an event.
4. Language is time serial in nature, consciousness is parallel.

Nowadays, visual media too can subvert the authenticity of our simulations by invoking fake imagery, the way language has always been able to do. In any event, the best way to test the truth of any language is to bind it to reality through physical experiment. But can virtual 3D simulations be bound by real world physics to keep them in the "reality cube"? It is often said that a picture is worth a thousand words. Maybe a 3D model is worth a thousand

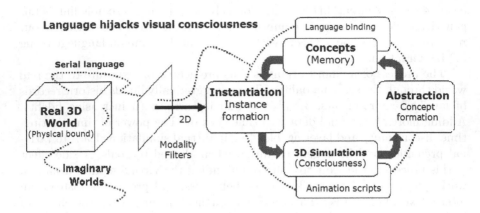

Fig. 5: Language hijacks visual consciousness

pictures. At one million words per model, 3D simulations might build a better basis for AI.

The syntactic structure of language often implies precision and completeness, but only by translating language into the form of a simulation can any ambiguity, or breaches of physical law or logic be discovered. Language processing is notorious for its blindness to common sense, which become glaringly obvious the moment a simulation is run.

Language is also used to impart emotion, either through the delivery, emphasis, or choice of words. Emotional analysis will form a key part of any 3D simulation. The best way to discover the meaning of a given piece of language, is to run a simulation around it, test its adherence to reality and grade its emotional procession in relation to values or goals.

Any language must exist within the context of a simulated world model; this will help determine boundaries. Nouns are drawn from object and environment memories, verbs from the spatial and temporal "behavior" memories. Thus language can build simulation scripts – or allegories. Script validity may be discovered by testing the simulation for violations of logic etc. But for much of language, real meaning is hidden within inference or metaphor (Therefore, the substitution of disparate objects but with matching behavior patterns or vice versa). These metaphorical script trials can similarly be interpreted based on context, logic and graded through emotional cost-benefit analyses.

But how can a 3D simulation interpret concepts such as math, statistics or software? The temptation, of course, is to not bother interpreting to a simulation at all, because binary computational algorithms are already naturally suited to these domains. But that would be a mistake. An algorithm can solve a calculation millions of times faster and more accurately, but there will be no concomitant understanding of what happened. It is only when the numbers, graphs, or code are modeled, and analyzed in simulation with reference to historic representations of reality, that meaning and understanding can occur. The simulators within the human brain are not well suited to modeling mathematical or repetitive iterative processes due to rapid informational decay and weak cognitive focus. Thus, we tend to use memorized shortcuts to help maintain momentum.

If the goal is to test for possible relationships from a set of numbers, they might enter a simulator as columns of varying height. The simulator could draw on its historic memories of common number series. Such as shoe sizes; imperial weights; removable storage media sizes; French coin denominations. Or from calculated series, like prime numbers or various other mathematical series. It is thus by the sorting, layering, scaling, merging and comparing of these graphic patterns that relationships or meaning can be found within the numbers behind them, and that subsequent meaning bound to existing memories and thus representations of the real world. The traditional brittleness of computers dealing with numbers and language in the context of AI, stems from the difficulty of blending the data into wider knowledge integrations, particularly through metaphor, where the substitution of disparate knowledge areas extends the reach and depth of understanding.

The proper place for math and language notation is as a mechanism for the coding and serialization of information, so it can be efficiently stored, transferred or retrieved from constrained informational channels. Within AI the best way to process such shorthand notation is to translate back to the 3D domain where it can be bound to the constraints of either real world physics, or at the very least a notional 3D space and have behaviors referenced to historic precedents.

Language gives the illusion of delivering more content than it really does, and it is this very imprecision and ambiguity that gives it such flexibility for social communication. But the devil is in the details and it is those missing details where the real action lies. Ayn Rand states a single word can imply a thousand instances, but an implication is not the same as the thing. To identify a chair or a molecule as a class might be efficient, but it is not precise until it is instantiated as a specific chair or molecule at a specific location. 3D simulation is the real fire in the mind, but to be fair, by adding symbolic language, it's like throwing gasoline on that fire – by adding turbo charged addressing and scripting to our 3D memory records. Language thus leverages our simulators hard, as if on steroids, igniting the firestorm of our wider human culture.

Language is used extensively in human cognition to economically build up simulations and to express their script procession in a serial communicable form. Further, It is almost certainly the coding mechanism used to classify objects for subsequent retrieval from memory and possibly even a predominant part of our episodic scripts. But serial language is simply insufficient means in dealing fully with the real challenges of AI; though it is certainly an essential element. Language is to the mind as a scene scripting language is to animation software. It describes and directs the animation flow.

Some examples: Fred was in the living room practicing his putting. What would happen if he practiced his driving? How could AI based on language alone understand this type of common sense content? Or even more importantly, solve the following tasks: design a mechanical human arm, a virus that can target cancer cells, or a three dimensional memory chip. Simulate a 256 bit RISC processor core?

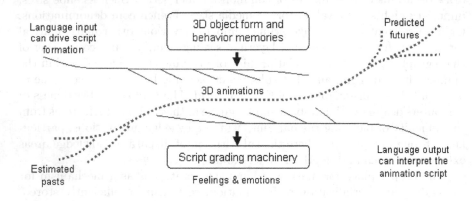

Fig. 6: Language and scripts.

6 Epistemology

To ensure survival, human consciousness has been dominated by guiding interactions in the real world to procure resources and mating opportunities. Mortality is the primary existential condition and leads to many biological prejudices, with physical and cognitive power rising gradually from childhood, peaking in adult life and then falling off again in old age. Humans have fairly rigid cognitive machinery which constrains their capacity for intelligence and they cannot easily alter their genetically pre-programmed emotion analyzer to favor mental effort over visceral pleasures.

If you take the image from an eye or camera, or you listen to speech, you create parallel information wave-fronts. These are meaningless without reference to a common reality – which for humans is existence. So how can it be that blind or deaf people can think? It is because they have constructed the same 3D world model from the remaining modalities; particularly touch and movement. For instance, a sighted person cannot see clear glass, yet he understands the concept of glass. If he is told a sheet of perfect invisible glass separates a room, though he may not be able to see it, he can conceptualize its existence quite clearly and act accordingly. For a deaf person, the language tags directing simulations would be purely visual rather than audible in nature.

A predominant feature of human existence is physical animation. These abilities are likely heavily supported by specific trained neural networks rather than any intimate conscious control. Motion requires fast cybernetic feedback to handle momentum. To offload this work onto sub processes would leave consciousness more time to deal with higher goals, like a plane requires limited input to guide flight. So the human body can animate largely free of direct conscious control.

The human organism is but one half of the coin, the other is his environment. Moreover, Intelligence is but one aspect of a complex set of processes involved in biological existence. Any artificial intelligence in the true likeness of man will surely be quite an anomaly. For these variables are the source of all our biological motivations for survival and cognitive attention. The human organism uses exposure to the environment over time to facilitate the development of a realistic world model. Success in this endeavor aids survival. But human cognitive focus is largely dominated by biological imperatives. This drives much of our intentionality and subsequent physical activity, creating the curious human civilization in which we live.

If we come to the question of our objective in building machine intelligence, we might ask – is it to replicate as closely as possible the human condition? Or will other goals be better aligned to our technology and desires? The Human means to knowledge occurs over many decades through full-on reality immersion with subsequent repetitive trial and error learning cycles. Such methods, even if practical, might be too slow a strategy to developing useful AI. One might presume that AI will have been achieved once the Turing test is successfully passed. However true this may be, it might not actually be the wisest of strategy for current research. The reason being, the test presumes anthropomorphic qualities in a machine are necessarily indicative of the most advanced state of consciousness to be sought; where concepts such as social inclusion and biological proclivities are pre-eminent. To put it bluntly, knowing how to eat a banana or understand a joke, admirable though they may be, might not be quite as important as an ability to accurately model a specific protein fold, and predict resulting regions of subtle chemical reactivity!

7 Instantiation: the Heart of Consciousness

Possibly the greatest software challenge for AI will be the instantiation engine. It must reverse a 2D bitmap render of vision (or indeed from any modality input) to recognize the environment and objects from internal memory correlates (concepts) to recreate the virtual 3D scene. There are really just a few common classes of environment sets – countryside, office, kitchen, work bench, shop, theatre, plane etc. If any environment match can be found, a fully instantiated scene framework will be ready to go, leaving only image scale, detail and perspective to be resolved.

A few pound lump of clay can instantiate a greater variety of forms than the entire number of atoms in the universe. But only a tiny subset of those forms will have any meaning attached and be associated with any behaviors – cat, fridge, airplane etc. The human mind is able to, with only a few pounds of meat, instantiate form and behavior from novel 2D vision scenes at the rate of about one object per second. Considering how many 3D pattern matches must be made against our library of known objects, this is quite an achievement. In most circumstances, significant mystery can remain within a scene (bitmap areas without instantiation), so long as the major items are decoded out; such as environments, significant life forms or emotionally charged objects.

Possibly, with unlimited time and processing power, artificial instantiation could be achieved through 3D scene estimates, rendered down to 2D and then compared with the bitmap input. Corrective feedback cycles could iteratively discover the light sources (from radiosity and shadow effects) and camera perspective (from room edge key points or with lock-in provided from a single object discovery). But it should be possible to design faster search algorithms than such brute force trials, perhaps by comparing pre-rendered trial object "icons" to the 2D scene. Or in reverse, by extracting edge patterns from the 2D image, normalizing scale and tossing those into a search path through memory to catch shape and/or surface pattern matches.

The challenge is to design a 3D object description language that can be interrogated rapidly and one based on fuzzy search criteria. You cannot use a polling search metaphor against a million images, each of a thousand orientations; you have to use an "interrupt" or "vector" search metaphor. Human vision is based on the identification of features rather than exact form, thus a violin twisted around a pole can still be recognized; or a clock printed on a crumpled table cloth. The challenges of high speed instantiation make the decisions where to focus attention; on the motion of a cat or to follow the eyes of a human, seem almost trivial by comparison.

Just as a human is built upon autonomous biological layers, cognition has its own autonomous layers. For instance instantiation, morphing and tweening (the construction of in-between time frames during simulation). When we script a human actor entering a room, the motion tweens do not need to be consciously re-calculated; their construction is either automatically generated or already stored in memory as an animated motion tween. Only the environment, context and emotional attitude need to be scripted in order to direct simulations.

Rendering is the translation of 3D scenes to 2D bitmaps. Instantiation is the reverse, the creation of 3D scenes from 2D bitmaps. Using a neuron array metaphor, where a projected image triggers firing along an axon. If those neurons each have say 256 axons (connections) propagating out, within that tangle there is spatially encoded all possible orientations and translations of

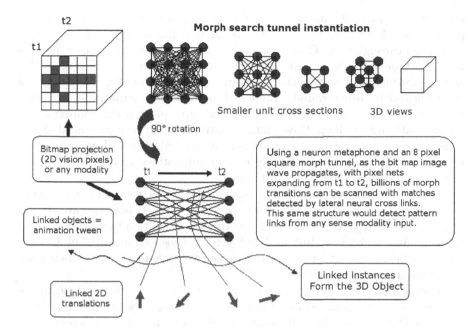

Fig. 7: Morph search tunnel instantiation

any 3D object. The decoding out of that data could be achieved from the propagating input wave function through time. For example, if each of the elements on two opposing faces are connected to every element on the opposing face. Therefore each input pixel has 256 vectors spreading out. If it took one hour for the signal of a firing neuron to travel along the axons between the surfaces, and you divided that time period up into small enough units, at any instant in time, a set of those vectors from the expanding pixel wave fronts will be optimally aligned to a specific translation of the projected object. Were those vectors known (trained), and linked together, the full 3D translation could in theory be described by those lateral connection sets. If those connection channels were two way, the objects could either be instantiated (identified) from input modality patterns, or in reverse, be used to trigger the same visual imagery (memorized experience) but directly from the linked network patterns, themselves connected to similar and associated modality patterns of visual and oral language tags, or even taste, smell and touch attributes.

Modality flows, whether from sight, sound, touch, taste or smell manifest in the brain as parallel analog data channels of specific and appropriate frequency, phase and dynamic (amplitude) ranges. The same principle of instantiation applies equally to all these analog sensory data sets; with receiving neural arrays, optimally tuned to the character of each input class. For example, the sound of a word or event, as with vision, will enter the neural array as a parallel 2 dimensional analog data wave-front of frequency and phase channels or "aural pixels," extending into the neural array as a third dimension through time. Cross connections linking spatial patterns will again identify those with the closest correlation to existing memory traces. In this way, as for vision, if only part of a word is heard, in any tone or accent, or even masked by other sounds, there will be sufficient signature correlation to make reasonable probabilistic guesses for subsequent wider context simulation trials and grading. These data signatures, being now instantiated, are thus linked to the universal environment map of objects and environments. Otherwise, the inputs would merely remain unidentified sounds bearing only fleeting similarities to known aural traces.

Instantiation processing from sensory modalities is automatic and unconscious; there is little mental effort involved, and further, not only are the 3D objects instantiated, but also are any associated animation tweens (object behaviors). Just as bitmaps link to 3D objects, so those 3D objects link to form animated behaviors, either as internal memorized tweens or newly constructed object motion or morph tweens.

Take a mouse object at time t1 and a teaspoon at t2, place them in the same spatial location and connect their surfaces together with orthogonal vector lines. Divide those lines into equal "time" segments and render a perspective to create frames for the movie script. This process is known as "morph tweening," and will represent one of the core visual translation tools necessary for AI to both interpret modality flow and to create new and novel content. During any visual thought process, creating smooth in-between renders between distant or disparate objects in time and/or space will be crucial.

Even apart from AI, the commercial spin-offs from an instantiation engine will be enormous. To start with, consider the possible re-animation of all historic language documents and visual 2D media, to create a cornucopia of rich, new, flexible animatable content.

8 In a Nutshell

Biological organisms interact with reality to survive. Sensory and motor systems evolved and so eventually did a computation engine in between. In humans, these computations create a simulated environment through exposure to the real world, converting existential matter into virtual representations. (Therefore external reality made from atoms, have digital/informational correlates). An internal map and a repertoire of environments, objects and behaviors develop through a process of exposure, perception, instantiation and memory formation.

All sense modalities converge to memory space as pure information, which is the very loci of consciousness. This information is instantiated, simulated and then graded to guide behavior and generate data we experience as feelings. Subconscious processes unlock the time and reality constraints enforced by the external world via the modalities and allow object behaviors to flow freely, constrained only by "prior art" and processing resources. These script variations subsequently feed simulations back into the area of consciousness – like Aristotle's "Cartesian Theatre." The first person conscious observer occurs at the information interface between the rendered virtual simulations and those "rendered" by the modalities of the outside world. The subconscious processes place consciousness, and thus our perception of reality, into a known place, and a time event horizon of the present placed between predicted futures and a remembered past[2].

The subconscious uses cues from the external world, or recent episodic memory scripts, to seed script diversity as simulations are intimately dissected and transposed. Virtual time travel and time discontinuities are aggressively used to construct metaphor, meaning and relevance out of the resulting script compositions. This "meaning" is discovered using genetically and socially programmed emotional filters which grade the scripts according to factors such as survival, security and cost benefit analyses, prioritizing social and resource capital, such that for every act, a human will know to the best of his cognitive ability, what is most in his interests at that time. It is the breadth of this process of subconscious wide scope accounting with ever increasing circles of virtual time expansion within simulation scripts, coupled with "emotional" cost benefit analyses that defines the depth of a man's intellect[14].

Subconscious processing uses the short cuts of context and precedent to speed script discovery, and when the rules of simulation are grounded in history and reality, the subject can use the simulations as the basis of learning and for future plans in dealing with real world situations, without then needing to physically act them out. Because the simulations are unbound by time, they can often beat external reality and thus anticipate real world events.

Once the optimum simulation script has emerged, real world human animation can be guided through one-step-ahead simulation linked to modality feedback. Trained neural cybernetic scripts would greatly enhance the speed, accuracy and grace of these animations, such that the individual control of limbs and body momentum are left to subsidiary pre-trained largely automatic processes. Human action subsequently follows with intentionality declared to be free will.

With subconscious activity constantly trawling memory records and modality stimuli, free will is simply the ability, at any given time, to flip life's animated momentum to be aligned with alternative virtual script offerings, even a destructive one if proof of courage, or free will, are defined as higher goals. (Which themselves are guided by the socially or genetically programmed emotions). During sleep, or quiet meditation, the process is driven by memory records alone, and away from the roar of sense modality flows, the subconscious script simulations can leak like ghosts into full consciousness, leading to imagination, creativity, planning and ultimately to self consciousness.

Humans have the ability to compute and render into consciousness the scene from any movie, placing say Donald Duck, or their grand mother, in the leading role. This ability comes from an internal scripting language that has access to powerful modeling and animation functions. We do not need to consciously solve the mathematics for the inverse kinetics of mechanical motion; or of momentum or gravity. We create the animated collage from prior learned 3D models, environments and either pre-rendered animation sequences or on the fly with motion and shape morph tweening. After morphing and blending the objects and scenes from prior learned behaviors, we can then render them to an observer perspective into consciousness, mentally skipping over much detail – the way we're deceived by a skilled magician – believing all the while we've missed nothing. But the mind has an advantage the eyes do not; it can censor and lie at will. The senses and internal memory contradictions try to keep the mind honest.

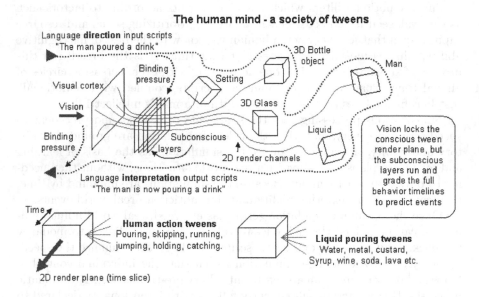

Fig. 8: The Human mind: a society of tweens

To summarize the postulates:

1. Our reality exists in 3 dimensions over time.
2. Units of matter can be represented by units of information.
3. The aggregates of atoms within objects and environments of reality can be converted (through modalities) to stored information within memory.
4. The identity and behavior of objects can be instantiated from those memory records (through computation) and then stored as informational representations.

5. These representations can subsequently be recalled and manipulated to simulate the behavior of their real world correlates as 3D animation tweens.
6. A software process can judge and emotionally grade the intrinsic value of these simulations to guide and optimize script formation.
7. Step-ahead animation and pre-trained cybernetics can be used to align physical action to the script.
8. With sufficient computation, memory resources and exposure to reality, this process can become a self reinforcing seed process – leading to advancing intelligence.

Intelligence, consciousness and feeling are virtual informational processes based around 3D simulated environments which are bound to reality by pre formed genetic road maps and experientially over time through sensory modalities and mobility. Consciousness arises from the supervision of these simulations linked to feeling – which is the computational process of grading those simulations. Intelligence is the ability to expand the time horizon (time dilation) to discover causes and make predictions. All these processes can be achieved artificially and will lead to AI. The controversial aspects of this paper are that:

1. 3D simulated environments are the basis of cognition.
2. Human language is a subsidiary process.
3. Any language which contains meaning can be reduced to a 3D simulation.
4. Human feelings are illusory; they are self referential computation processes.
5. The nexus of consciousness is the boundary between the modalities and the feedback from simulated environments created by subconscious computations.

Thus, the proposal is – that matter can be represented by information; that objects and environments can be instantiated from perceptions of reality; that they can subsequently be simulated and that information can be stored; that these simulations can be graded based on their progression in time; that simulations can run faster than reality; that through the superposition of memorized behaviors, simulations can represent potential versions of reality; that these superposition's can be "emotionally" graded and evolve toward an optimal prediction - using historic precedent; that chaotic discontinuities can be avoided; that these simulations, being able to predict reality, can be used to align physical action to those simulations; that this computational process can beat the procession of time in reality. This then, is the process that leads to consciousness, intelligence and intentionality.

9 Real-World AI

There are several factors distinguishing real AI from expert systems: the breadth and scope of the knowledge base; the ability to ask the questions; to identify missing knowledge; to judge the relevance of results; to apply context or predict effects over time, etc. These extra features require a simulated environment like our own and a world model of equal breadth.

On the evidence that immobile, deaf children can still develop high intelligence, presumably from visual stimuli, we might also expect a similarly restricted machine analog to have an equal chance of success. In order to conceptualize a credible AI architecture from vision, imagine following our current technological trends for a few years to where the following levels are reached:

- *Cameras:* 36 bit color depth, 6000 x 3000 resolution, 60 frames per second
- *Exposure:* 1 gigabit broadband internet connection attached to browser clients
- *Memories:* 10 terabyte, non volatile, shared, direct addressable, 10nS access time
- *Processors:* 10 teraflops, serial (in autonomous clusters)
- *Instantiation:* accurate 2D to 3D translation software
- *3D modelers:* any shape, scale, texture, orientation, behavior, etc
- *3D simulators:* supporting physics, collisions, chaos, time shifting, etc
- *2D renderers:* supporting shaders, shadows, radiosity, fog, etc
- *Animation scripting language:* object insertions, orientations, behaviors, morphing, tweening, layer and time management
- *Database:* of records, concepts, objects, environments, and episodic scripts
- *Language to animation script translator*
- *Animation script to language translator*
- *Script grader:* cost benefit analysis, entropy, normal, harm, irreversibility, danger, opportunity, 3rd person script empathy and 1st person emotion analysis

The major software challenges:

1. 3D instantiation from 2D sense modalities
2. Construction and maintenance of a universe environment map
3. Construction and maintenance of object records and behaviors
4. Powerful, multilayer 3D simulation engine
5. Blending/morphing of environments, objects, properties and behaviors
6. Grading of simulations to guide script progression

The most appealing hardware structure would be a network cluster of maybe 10 or more powerful self-contained computers but with shared memory resources, each dealing independently with separate aspects of AI. The continuing massive worldwide investment in operating systems and application software can be leveraged to become tools, blurring the boundaries between modalities and consciousness. Products such as Windows, Linux, commercial 3D modelers, OCR and speech recognition software. But the boundary between man and machine is already getting very blurred with ubiquitous cell phones providing an almost telepathic modality; not to mention speech control of computers, graphical interfaces, instant messaging and email, etc. To some extent, most people already spend most of their lives in virtual reality; they just don't recognize novels, radio, TV, computer games or software as being virtual environments.

A human without recourse to modality extensions such as an auto, cell phone, internet, computers, fedex account, credit card, 3D printers, etc. would be a greatly diminished soul, and the same goes for AI. The most important cognitive skill will not be to walk or even talk, but to manage multiple computer graphical user interfaces.

But how could these advanced technologies begin to be organized to create intelligence? First, the camera would project its bitmap data to a memory map, which would be routinely processed by the instantiation engine to identify known objects and environments from memory records. A subsequent 3D simulated environment would be constructed in memory to match the visual scene and simultaneously rendered back down to a 2D bit map memory space at the same first person perspective as the camera input – much like a 3D animation film is rendered to the 2D screen image. If the camera data flow were interrupted, the rendered 2D data from simulation would be an accurate mirror copy of the real scene.

There are three dynamic events that can now occur within the visual field. An object can change, the perspective can change, or the whole scene can change. For scene changes, the previously described process of instantiation and discovery would be repeated. For perspective changes, motion vectors (as used in video compression) would be calculated to keep track of scene perspective. For object animation, the software process would recognize localized anomalies between the simulated projection and the vision projection. Then using normal instantiation techniques focused on the anomaly, the object in simulation would be oriented until the 2D rendered projection and the input vision were once again in alignment.

The memory management software would need to maintain an associative database linking all objects, environments, behaviors and scripts. Together with growing lists of knowledge about these models, such as: language tags, price, legal status, disposal, source, manufacture, flammability, safety, uses, weight, dangers, precautions, social status, classes, trends, history, composition, aging properties, storage, popularity, component parts, assembly, regulatory compliance, standards, size, growth time, environmental impact etc. Object behaviors would be characterized and stored based on:

1. Motion vectors over time. A feather would tumble through air differently than a balloon floats or an insect darts – stored as positional and temporal data sets.
2. Shape variation or morphing over time – butterfly, bouncing ball, coiled spring etc.
3. Reaction to stimuli (touch, drop, cut, etc.)

The overall environment map would need to hold concepts ranging from the universe, through planets, countries, cities, neighborhoods, homes and factories, to materials, chemicals, molecules and atoms. At any point in the simulation, a relationship would exist to this universal map. Which specific country, town and room? Or if generic, it would still need a generic history with the potential to be "fixed" by subsequent facts. The depth and accuracy of this virtual world will largely determine the bounds and precision of thought for the artificial intelligence.

All objects blend together in an overall environment map, which fits within a wider contextual world map. Physics rules (gravity, hardness, weight, momentum, heat, speed of light, etc.) guide behavior and interactions between objects. (Cloth against solids, light through glass etc.) Much of this is already well advanced in commercial 3D software packages. The overall resolution and speed is dependent simply on processing power and memory resources. The software must subsequently recognize any bitmap changes as object behavior animation or changes to perspective and re-calculate to keep the simulation bound to the vision input.

The input video stream drives the construction of the virtual scripts. If novel, those scripts might be the basis of new memory formation. Inconsistencies would be challenged based on the source credibility or physical law, with certain knowledge discovery causing rippling adjustments throughout memory. Logical inconsistencies and vagueness might be highlighted to trigger some human supervisory training to help bootstrap the process. The addition of a language translator to convert words to simulation scripts will greatly speed learning, since most human knowledge and communication channels exist in the form of serial language streams. The language parser would construct scenes from any objects alluded to in the text, with action scripts proceeding from memory precedent and/or from the language verbs, syntax or emphasis.

Any proto-intelligence would begin as basic memory formation and correction processes, but the main advances will arise when running the subconscious simulation machinery separately from the vision input. The content of those simulations could be guided either by recent episodic memory scripts, prior behaviors or simulator physics, and graded by "genetic obsessions," such as the "need" to understand.

The process of discovery might involve the searching of any language scripts associated with the problem, with their subsequent conversion to animation. Metaphor will be examined through object or scene substitutions within the trial scripts. Script diversity built from breaking up object sets and re-ordering time through dislocations. But how exactly are all these script trials to be graded? This is the most difficult part of the process to explain with any clarity. There are several grading concepts like testing against law, mores or relevance to global goals. Further grading concepts might be: normal object condition; reduction of scene entropy; novelty detection; consequences to the wider time frame or applicability to other environments. But the most likely method will revolve around either quick-and-dirty pre-programmed emotional prejudices or, if more time is available, growing circles of cost benefit analysis expanding in time and in environment space, as the potential effects of the sample scripts ripple outward. These wider scope integrations will ultimately be graded against predefined "genetic" schema. Such as profit, shame, humor, social capital, etc. A final script must be found that predicts the highest probability of benefit and the lowest possibility of costs.

Another strategy for knowledge discovery would be the joining of means with ends to build a script timeline from the missing links in between. Once in the 3D domain, tweening can be used to bridge gaps, with the new tweened content tested against simulator reality constraints such as gravity, physical form and behavior, social mores and rules etc. Or perhaps more like a jig-saw puzzle, only with the pieces made up of memory records of objects and their behaviors or triggered from external search results. Finally, the expansion of complex objects to simpler sub units. Or the reverse, the assembly of complex from the simple, would further aid knowledge discovery.

So at this point we have a simulated environment held in "conscious" memory tracking the live video feed (and/or receiving script revisions from a subconscious process). We have a subconscious simulator building and expanding upon those conscious scripts using prior behaviors, with particular interest in novelty. We have script expansion though behavior extrapolation (e.g. a vase being nudged toward the edge of a counter will be predicted to fall and shatter). We have scripts graded through cost benefit analysis (a broken vase creates a loss of value and a mess – entropy). Next, we need a method for allocating time and resources to maximum effect, to direct focus and attention, and an ability to interact with external knowledge bases. Finally we need a satiation response to help allocate computational resources and escape dead ends.

The ability to search the external world for solutions would require language formation from the simulated scripts and an output method for gaining human attention or an ability to directly enter text searches into internet search engines. Due to speed, the first choice would likely be the internet, with human intervention being the least rational choice for guidance. Humans will be totally unable to keep up with the data velocity associated with AI thinking. An un-tethered AI would quickly overtake one kept anchored to the dead weight of human consciousness.

The primary source for learning material would be the translation of web based information to animated scripts within a global environment map to form the basis of knowledge integration. This would require 3D script construction built from text, charts, sounds and images, in conjunction with the previously described video/image instantiation engine. The seed AI could begin making its own predictions and then testing those predictions through further internet searches to discover if it had found the correct causes, processes or results. Only when a concept exists without internal contradictions can it be said to be properly integrated and its authenticity secure. The cognitive advantages available to AI will include the following:

1. Persistence in simulation layers
2. Simulation accuracy and precision (e.g. for math and software)
3. Increased number of conscious objects
4. Increased size of simulation
5. Accurate simulation of physical law
6. Accurate "photographic" memory
7. Multiple parallel modality inputs (e.g.100 simultaneous internet channels)
8. Extended modality inputs (e.g. data protocols, radio, IR, UV, ultrasonic etc.)
9. Automatic, high speed multi language translators
10. Greater conscious control of simulation progression and persistence
11. Scientific calculator, thesaurus, dictionary and encyclopedic resources
12. Patience, rationality and deep foresight

Finally, the human mind is unable to properly render its internal 3D content to anywhere near the clarity as when "painted" directly by the modalities. Thus, we are only really partially conscious; there is an enormous richness to existence and experience we are blind to. The mind is full of ghosts rather than realistic impressions and a ghost world is hard to fully embrace.

The only credible mechanism for self awareness to occur is as a computational process dealing with information representing and bound to reality. The self can then exist and be aware through a process of reflection (simulation) in a time controlled domain, where emotional grading (feeling) can percolate through time dilating script trials.

If you doubt that such processes of simulation and virtual time travel will really lead to intelligence, think of this analogy. You suddenly find yourself able to re-run time backwards and forwards in the real world as many times as you like, even making notes as you go along. After many such "simulations," do you not think it likely that the action you finally take might be a little wiser?

9.1 Examples and Metaphors

Imagine a fish tank, 1000mm wide, deep, and high. The tank is filled with 1mm cubes. Inside each cube is a little scroll that says: air, gold, glass, skin, hair, cheese and such. These scrolls represent electronic memory locations that can be filled with information about real objects. Laws defining the relationships

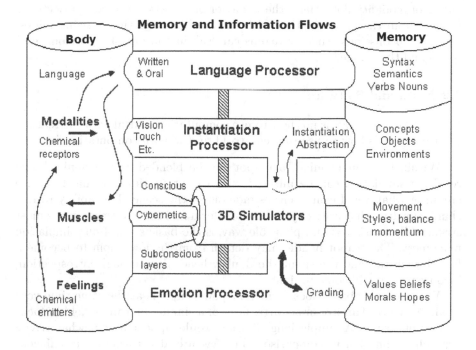

Fig. 9: Memory and information flows.

between adjacent memory spaces are programmed to be in harmony with those in nature. Such as weight, object boundaries, momentum, light refraction, texture, behavior etc, very much like current 3D modeling and animation software.

This memory space can be thought of as a movie stage – a "Cartesian Theatre" or to use modern parlance – a virtual reality chamber. This virtual chamber can be filled with objects or environments and at any scale. Gas will disperse, liquids will spread to boundaries and solids will have weight and maintain form. Animated objects will flow according to their motion vectors and morphology – a perfect analog of real life; except matter is replaced by information. Like a 3D window or camera, this "box" can float across virtual landscapes and environments, to be filled at one moment with the great expanses of space and time and in the next, the most intimate molecular spaces of tiny living cells.

At the centre of this chamber is a virtual, animate human character. The contents of the theatre always render down to this 2D observer perspective, which is also the source point of filling the chamber from any modality inputs. This virtual space will form the contents of waking consciousness.

Further, shadow realms exist. Again like scrolls set beside the originals, except the contents of these scrolls are able to break free from the straight jacket of modality flow. Here, the behavior of objects can follow trajectories learned from the past, together with substitutions and time discontinuities. These "subconscious" shadow realms can leak in and out of "consciousness" to also fill the "stage."

9.2 Math and Software

"The challenge is not how to use computation to deal with the real world – it is how to use the real world to deal with computation."

Within the human mind, a teapot can be blended with a donkey! The resulting simulation can be infected with the properties of china, flesh, ice cream or whatever. Inconsistencies fade out of the scene. This ability to mechanically draw disparate objects of class, form and scale together in the most structurally consistent and plausible way, is the basic stuff of our simulation machinery. The teapot handle may detach from the lower join to become a free flowing tail and the spout the donkey head. But through introspection, inconsistencies will come to light.

Whereas this ability does seem very powerful, it is at the same time very weak. No more than five or six attributes of a simulation can be held in focus at any one time. A simple long division would appear a somewhat trivial symbolic animation in comparison, but few are able to maintain sufficient control over the parts to achieve even this simple feat[4].

Math and software memories similarly exist either as animation scripts or simple learned pattern recognitions – such as multiplication tables. Like the images on a dice face, digits can have direct dot pattern equivalences for subsequent math animation (add, subtract etc.) Thus math can manifest as either image animations (e.g. joining/separating dot groups) or rely on memorized symbolic beliefs, such as 12×12 has "an equivalence to" 144. Or for the binary truth tables – or should I say "belief tables"[10].

Just as there are behavior scripts for the way a ball bounces, a rabbit runs and a feather floats, so there are the more abstract behavior scripts of memory indexing, for-next loops and the like. Most math and software concepts would likely exist initially as animation scripts, but as our familiarity and confidence with them grows, short cuts are taken, jumping straight from beginning to end, and so over time they become simple memory beliefs without the intermediate animations. When we imagine a vase falling to the ground and breaking, we jump from the initial fall to the shattered remains more through belief, than the accurate simulation of each part of the event down to each individual chard.

9.3 Barcode Example

Finding the relationship between bar patterns and a decimal subscript will be used to illustrate knowledge discovery through animation (see Fig. 10).

The assumptions made are that the AI has access to image samples and that the memory beliefs of basic math and software primitives exist, as do the fundamental instantiation and grading abilities previously outlined.

Human cognition evolved to integrate 3D objects, environments and behaviors into a knowledge hierarchy – not 2D symbolic abstractions. It takes a great deal of effort and training for a human mind to so contort itself as to be able attempt these classes of problems. But with persistence, the help of external tools like pen, paper, calculator and computer, together with a little academic "coercion" – we are sometimes rewarded with results.

The method of discovery does not need to be infallible or super efficient, it just needs to have a statistical chance of success in finding connections and thus guide knowledge formation within the time allocated. The higher goal, as always, is to discover meaning through finding memory connections, joining means with ends and reducing mystery. In this case, the means is a barcode image, the ends – a decimal subscript number. A simulator deals primarily with object shapes and forms. Apart from drawing upon prior memorized beliefs in the form of animated scripts or static image relationships, there are fundamental "instructions" operating on those forms:

- Instantiation – identification
- Separation, scene explosion
- Re-scaling
- Perspective translation
- Geometric alignments
- Language attachments
- Object substitutions
- Joining – connecting

And grading machinery based on:

- Proportionality
- Similarity of scale, quantity and class
- Pattern matching
- Scene entropy
- Scene simplicity (Occam's razor)
- Completeness/loose ends

These processes are fast, automatic and operate in layers through reversible animated pipelined scripts. Humans use pen and paper to "fix" parts of these flows to create order and permanence out of these somewhat chaotic streams. This helps construct an external framework to guide the process. AI will have the ability to do this internally by way of "persistent" simulation layers[7].

Each process is essentially dumb and automatic, but as a whole, and connected to sufficient source material and memory support, new connections can usually be found and integrated into memory. Dead end simulations will

fade away and if grading progress stalls, higher level processes will kick in –
overall goal re-appraisal; the process will seek more real world data through
the modalities or widen the internal associative memory search.

Applying instantiation to the global barcode image would yield six classes
of abstract objects; two rectangle shapes and four numeric digit shapes. Language attachment to the object instances would connect as thick and thin
bars and the four digits as a number.

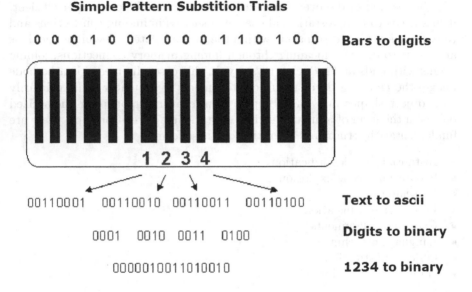

Simple Pattern Substition Trials

0 0 0 1 0 0 1 0 0 0 1 1 0 1 0 0 0   **Bars to digits**

1 2 3 4

00110001   00110010   00110011   00110100   **Text to ascii**

0001   0010   0011   0100   **Digits to binary**

0000010011010010   **1234 to binary**

Fig. 10: Simple Pattern Substitution Trials

At the "ends" part of the problem, we have a number 1234. Memory
references will recall a belief that numbers have "an equivalence to" binary
1's and 0's. The first script trial might show an ascii equivalence yielding 8 bits
per character. Thus an image of 1234 transforms to 32 digits. A second script
layer might show each separate digit converted to a simple binary count. The
third has the whole decimal number, 1234 represented by a binary count. Of
the three scripts, simple pattern recognition would grade binary expansion as
the closest match between means and ends. Further sample barcode image
trials would confirm the link. Memory formations of the newly discovered
script sequences would follow, including mutual pointers between the existing
precursor knowledge records of decimal to binary equivalence etc. (Which
incidentally, would reinforce the familiarity and trust in those prior beliefs.)

Now, when presented with similar barcode images, the scene will be recognized and will draw from memory links to the newly formed animation scripts

and an intimate familiarity with the scene will ensue due to these very same memory references, together with the emotional confidence that comes from recognition and understanding. The fundamental simulator operations used in this example of discovery were:

- *Scene instantiation:* to shape primitives
- *Language tagging:* from memory recognition of images/forms
- *Prior memory associations:* decimal to binary equivalence (as animation or belief)
- *Object substitutions:* bar shapes to "thick" / "thin" or to 1's and 0's
- *Image comparisons:* the bit patterns.

The process of decoding the barcode will not be understood in some isolated abstract way, but within the known framework of reality through intimate linkages with existing memory records; all being a part of a world knowledge and environment map. If a barcode is now presented with no number or vice versa, the simulation can play the script in forward or reverse to discover the missing parts through simulation to final substitution of bar patterns or decimal digits.

9.4 Software Design

The same *principles* involved in joining the donkey to the teapot would be used to create software. Each part of a flow chart script would be drawn into morphing relationships with software primitives. The challenge of software design is similar to the long division problem – only very small fragments can be held in simulation at one time. Software is the application of language rules to "direct the structured animated progression of data bits". Just like normal language is used to script the animation of everyday objects. With software, pen and paper are often used to "fix" the framework using language tags in order to maintain animation persistence and build complexity. And just like for real world animation, where atoms are aggregated to forms and forms to behaviors – so bits are often aggregated to higher data abstractions. Like floating point numbers, arrays, memory containers or pointers with animated behaviors their own.

Using prior knowledge of indexed memory containers, a simple symbol substitution layer can form to match the data in our example, see Fig. 11. The initial "means" are still the bars substituted for 1's and 0's. The "ends", the decimal digits with the previously learned simulated decode script in between. But this script is neither a formal flow chart nor software. It includes all sorts of miracles and beliefs to get from the bars to the numbers. (Bars turn into symbols, symbols to patterns. Patterns are compared to other patterns). There needs to be discovered, through trial and error, linking morph translations between the means and ends using software animation characters.

To start, the traditional **for-next** loop might be used as an initial script trial skeleton, upon which to attach elements of the known model. It doesn't

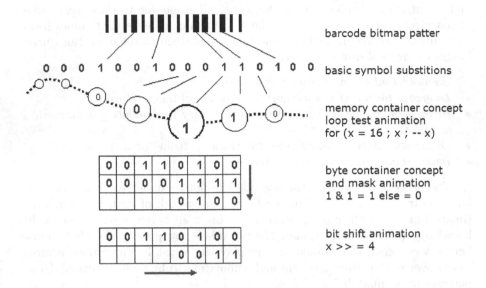

Fig. 11: Symbol substitution

much matter how the `for-next` animation is initially understood or remembered. Whether a cart wheel with spokes marking along a track, or a string of beads passing through some grading point. As experience has always shown digits to be the predominant substitution, they themselves will likely become the animation characters. And so for someone familiar with the C programming language, the expression `for(x=4;x;--x)` will invoke this abstract animation, but with roots firmly embedded to real world behaviors and thus connected in some way to all other knowledge. There are only two variables in the original simulation, 16 inputs and 4 results. So any loop substitutions are likely to be based around these two numbers, rather than say 42 or 365. The simulation iterations will then run by substituting the only elements possible to change, yielding:

- `for(x=16;x;--x) do something with 'means'`
- `for(y=4;y;--y) do something with 'ends'`

Morph attempts can now be made between these loop fragments and the original decode script. (As the donkey animation was merged with a teapot). Disparate parts of the two scripts will find tentative bindings, which will strengthen or weaken upon introspection – i.e. running the animations to discover anomalies. And using the framework from persistent language layers (as humans use pen and paper), to build and hold animation complexity. In this way, animated fragments will bind to the original decode script to generate code trials with subsequent C-language script formation, as shown in Fig. 12

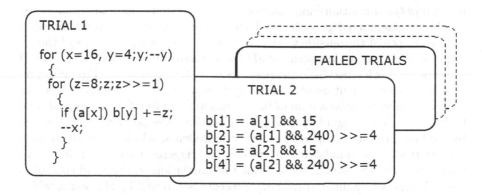

Fig. 12: Script formation

Wider scope accounting would expand the extent of simulation to close loose ends by explicitly defining memory container sizes; initial conditions; test flexibility to handle longer barcodes or discover optimizations through further substitution experiments. But more importantly, for this code fragment to be understood in any context, it must be integrated into a wider causal framework of just how the barcode widths will become input data; what host device will be running this computation and how the results will be used. Thus the software code fragment will come to have a relationship with a material existence in the real world; as the motion of real electron charges on real atoms within the microcontroller of a real product. It is these linkages that are far more important to intelligent understanding than software – the awkward mental construction of abstract pattern animations and beliefs.

10 Conclusion

Everything that really matters in the world has form and behavior predictability. Sure, fancy mathematics can predict the exact arc of a theoretical cannon shell. But the universal language of object form and behavior (reality) is not English or math, it is 3D animation. More intriguing still, is that the 3D simulations within the brain are intimately "connected/grounded" to reality through our senses and muscles (the sensorimotor system), whereas data in electronic systems is almost entirely disconnected.

This theory challenges much conventional wisdom about human action, consciousness and artificial intelligence. In its simplest form, the theory presents these processes, all of them, by a single paradigm; 3D object computing. That is, all mental activity centers around the processing of virtual 3D objects.

The process involves the recognition of objects in reality from the sensory input flow (instantiation) and constructing an internal scene simulation based on those remembered precedents. These simulations exist beyond the fixed time reference of the outside world, because the memory precedents of the assembled objects contain a series of 3D "movie script like" scenarios for each of the objects – as learned from the past. As such, they can be used to make predictions of past and future action. Being virtual, these predictions can beat the time of reality, to allow a human to say, catch a ball in a future moment, and to know the balls origin from the past. Consciousness performs time dilation by building a simulation from memory precedents, which animate through virtual time. The simulated objects can be triggered into initial alignment with reality by the human senses. But the simulations easily yield control to the collections of animated memory precedents aroused by the scene, which subconsciously search out emotional value peaks and troughs based on the biologically inspired pleasure/pain value axis. Together with metaphoric links and substitutions, this allows creativity in the choice of physical action, while remaining broadly aligned to mental goals.

Language and symbols are sensory objects too, and are extensions of the same simulation processes. But they have the special properties of indexing 3D objects and scene narrative; including empathic states, subjective values and goals. Allowing ideas to be shared socially, with wisdom traveling through the generations.

The gist of this research is that all conscious and intelligent processes center around 3D simulation; with language and symbols used for indexing and scripting. That all knowledge can be understood in terms of 3D model behavior based on precedent. That software designed to handle 3D models and environments, will be central to AI and this commercial software is advancing rapidly.[15].

Notes

This work is influenced by the work of Ayn Rand [13] and the philosophy of objectivism [12], and in particular, her proofs of: the primacy of existence; the validity of our senses; the instantiation of instances from concepts; the proper relationship between consciousness and reality, and of reason and emotion. This paper builds on her work to integrate 3D forms as the primary vehicle for connecting human percepts to higher mental abstractions.

References

1. Dawkins R (1982) *The Extended Phenotype*. Oxford University Press, Oxford.
2. Dennet D (1991) *Consciousness Explained*. Little Brown, Boston, MA.
3. Donald M (2001) *A Mind so Rare: the Evolution of Human Consciousness*. W. W. Norton, New York.

4. Le Doux J (1998) *The Emotional Brain: the Mysterious Underpinnings of Emotional Life*. Touchstone Books, New York.
5. de Garis H (1999) *Artificial Embryology and Cellular Differentiation*. In: Bentley P (ed) Evolutionary Design by Computers, Morgan Kaufmann, San Francisco, CA.
6. Harold FM (2001) *The Way of the Cell*. Oxford University Press, Oxford.
7. Kosslyn S, Koenig O (1995) *Wet Mind: The new Cognitive Neuroscience*. Free Press, New York.
8. Kurzweil R (2001) *The Paradigms and Paradogms of Intelligence, Part 2: The Church Turing Thesis*. www.kurzweilai.net/articles/art0256.html.
9. Kurzweil R (2001) *The Law of Accelerating Returns*. www.kurzweilai.net/articles/art0134.html.
10. Moravec H (1999) *Robot: Mere Machine to Transcedent Mind*. Oxford University Press, Oxford.
11. Minsky M (1988) *The Society of Mind*. Touchstone Books, New York.
12. Peikoff L (1993) *Objectivism: The philosophy of Ayn Rand*. Meridian, New York.
13. Rand A, Binswanger H. Peikoff L (eds) (1990) *An Introduction to Objectivist Epistemology*. Meridian, New York.
14. Schacter D (1996) *Searching for Memory: The Brain, The Mind and The Past*. Basic Books, New York.
15. Yudkowsky E (2000) *General Intelligence and Seed AI*. This volume, and www.singinst.org/GISAI/

Levels of Organization in General Intelligence

Eliezer Yudkowsky

Singularity Institute for Artificial Intelligence
Ste. 106 PMB 12 4290 Bells Ferry Road, Kennesaw, GA 30144 USA
http//singinst.org

Summary. Section 1 discusses the conceptual foundations of general intelligence as a discipline, orienting it within the Integrated Causal Model of Tooby and Cosmides; Section 2 constitutes the bulk of the paper and discusses the functional decomposition of general intelligence into a complex supersystem of interdependent internally specialized processes, and structures the description using five successive levels of functional organization: Code, sensory modalities, concepts, thoughts, and deliberation. Section 3 discusses probable differences between humans and AIs and points out several fundamental advantages that minds-in-general potentially possess relative to current evolved intelligences, especially with respect to recursive self-improvement.

1 Foundations of General Intelligence

What is intelligence? In humans – currently the only known intelligent entities – intelligence is a brain with a hundred billion neurons and a hundred trillion synapses; a brain in which the cerebral cortex alone is organized into 52 cytoarchitecturally distinct areas per hemisphere. Intelligence is not the complex expression of a simple principle; intelligence is the complex expression of a complex set of principles. Intelligence is a supersystem composed of many mutually interdependent subsystems – subsystems specialized not only for particular environmental skills but for particular *internal* functions. The heart is not a specialized organ that enables us to run down prey; the heart is a specialized organ that supplies oxygen to the body. Remove the heart and the result is not a less efficient human, or a less specialized human; the result is a system that ceases to function.

Why is intelligence? The cause of human intelligence is evolution – the operation of natural selection on a genetic population in which organisms reproduce differentially depending on heritable variation in traits. Intelligence is an evolutionary advantage because it enables us to model, predict, and manipulate reality. Evolutionary problems are not limited to stereotypical ancestral contexts such as fleeing lions or chipping spears; our intelligence includes the ability to model social realities consisting of other humans, and the ability to predict and manipulate the *internal* reality of the mind. Philosophers of the mind sometimes define "knowledge" as cognitive patterns that map to external reality [76], but a surface mapping has no inherent evolutionary utility. Intelligence requires more than passive correspondence between internal representations and sensory data, or between sensory data and reality. Cognition goes beyond passive denotation; it can predict future sensory data from past experience. Intelligence requires correspondences strong enough for the organism to choose between futures by choosing actions on the basis of their future results. Intelligence in the fully human sense requires the ability to manipulate the world by reasoning backward from a mental image of the desired outcome to create a mental image of the necessary actions. (In Section 2, these ascending tests of ability are formalized as *sensory, predictive, decisive,* and *manipulative* bindings between a model and a referent.)

Understanding the evolution of the human mind requires more than classical Darwinism; it requires the modern "neo-Darwinian" or "population genetics" understanding of evolution – the Integrated Causal Model set forth by [98]. One of the most important concepts in the ICM is that of "complex functional adaptation." Evolutionary adaptations are driven by selection pressures acting on genes. A given gene's contribution to fitness is determined by regularities of the *total* environment, including both the *external* environment and the *genetic* environment. Adaptation occurs in response to statistically present genetic complexity, not just statistically present environmental contexts. A new adaptation requiring the presence of a previous adaptation cannot spread unless the prerequisite adaptation is present in the genetic environment with sufficient statistical regularity to make the new adaptation a recurring evolutionary advantage. Evolution uses existing genetic complexity to build new genetic complexity, but evolution exhibits no foresight. Evolution does not construct genetic complexity unless it is an *immediate* advantage, and this is a fundamental constraint on accounts of the evolution of complex systems.

Complex functional adaptations – adaptations that require multiple genetic features to build a complex interdependent system in the phenotype – are usually, and necessarily, universal within a species. Independent variance in each of the genes making up a complex interdependent system would quickly reduce to insignificance the probability of any phenotype possessing a full functioning system. To give an example in a simplified world, if independent genes for "retina," "lens," "cornea," "iris," and "optic nerve" each had an independent 20% frequency in the genetic population, the random-chance probability of any individual being born with a complete eyeball would be 3125:1.

Natural selection, while feeding on variation, uses it up [96]. The bulk of genetic complexity in any single organism consists of a deep pool of panspecies complex functional adaptations, with selection pressures operating on a surface froth of individual variations. *The target matter of Artificial Intelligence is not the surface variation that makes one human slightly smarter than another human, but rather the vast store of complexity that separates a human from an amoeba.* We must avoid distraction by the surface variations that occupy the whole of our day-to-day social universe. The differences between humans are the points on which we compete and the features we use to recognize our fellows, and thus it is easy to slip into paying them too much attention.

A still greater problem for would-be analysts of panhuman complexity is that the foundations of the mind are not open to introspection. We perceive only the highest levels of organization of the mind. You can remember a birthday party, but you cannot remember your hippocampus encoding the memory.

Is either introspection or evolutionary argument relevant to AI? To what extent can truths about humans be used to predict truths about AIs, and to what extent does knowledge about humans enable us to create AI designs? If the sole purpose of AI as a research field is to test theories about human cognition, then only truths about human cognition are relevant. But while human cognitive science constitutes a legitimate purpose, it is not the sole reason to pursue AI; one may also pursue AI as a goal in its own right, in the belief that AI will be useful and beneficial. From this perspective, what matters is the quality of the resulting intelligence, and not the means through which it is achieved. However, proper use of this egalitarian viewpoint should be distinguished from historical uses of the "bait-and-switch technique" in which "intelligent AI" is redefined away from its intuitive meaning of "AI as recognizable person," simultaneously with the presentation of a AI design which leaves out most of the functional elements of human intelligence and offers no replacement for them. There is a difference between relaxing constraints on the means by which "intelligence" can permissibly be achieved, and lowering the standards by which we judge the results as "intelligence." It is thus permitted to depart from the methods adopted by evolution, but is it wise?

Evolution often finds good ways, but rarely the best ways. Evolution is a useful inspiration but a dangerous template. Evolution is a good teacher, but it's up to us to apply the lessons wisely. Humans are not good examples of minds-in-general; humans are an evolved species with a cognitive and emotional architecture adapted to hunter-gatherer contexts and cognitive processes tuned to run on a substrate of massively parallel 200Hz biological neurons. Humans were created by evolution, an unintelligent process; AI will be created by the intelligent processes that are humans.

Because evolution lacks foresight, complex functions cannot evolve unless their prerequisites are evolutionary advantages for other reasons. The human evolutionary line did not evolve *toward* general intelligence; rather, the hominid line evolved smarter and more complex systems that *lacked* general intelligence, until finally the cumulative store of existing complexity contained all the tools and subsystems needed for evolution to stumble across general intelligence. Even this is too anthropocentric; we should say rather that primate evolution stumbled across a fitness gradient whose path includes the subspecies *Homo sapiens sapiens*, which subspecies exhibits one particular kind of general intelligence.

The human designers of an AI, unlike evolution, will possess the ability to plan ahead for general intelligence. Furthermore, unlike evolution, a human planner can jump sharp fitness gradients by executing multiple simultaneous actions; a human designer can use foresight to plan multiple new system components as part of a coordinated upgrade. A human can take present actions based on anticipated forward compatibility with future plans.

Thus, the ontogeny of an AI need not recapitulate human philogeny. Because evolution cannot stumble across grand supersystem designs until the subsystems have evolved for other reasons, the philogeny of the human line is characterized by development from very complex non-general intelligence to very complex general intelligence through the layered accretion of adaptive complexity lying within successive levels of organization. In contrast, a deliberately designed AI is likely to begin as a set of subsystems in a relatively primitive and undeveloped state, but nonetheless already designed to form a functioning supersystem[1]. Because human intelligence is evolutionarily recent, the vast bulk of the complexity making up a human evolved in the *absence* of general intelligence; the rest of the system has not yet had time to adapt. Once an AI supersystem possesses any degree of intelligence at all, no matter how primitive, that intelligence becomes a tool which can be used in the construction of further complexity.

Where the human line developed from very complex non-general intelligence into very complex general intelligence, a successful AI project is more likely to develop from a primitive general intelligence into a complex general intelligence. Note that *primitive* does not mean *architecturally simple*. The right set of subsystems, even in a primitive and simplified state, may be able to function together as a complete but imbecilic mind which then provides a framework for further development. This does *not* imply that AI can be reduced to a single algorithm containing the "essence of intelligence." A cognitive supersystem may be "primitive" relative to a human and still require a tremendous amount of functional complexity.

I am admittedly biased against the search for a single essence of intelligence; I believe that the search for a single essence of intelligence lies at the center of AI's previous failures. Simplicity is the grail of physics, not AI. Physicists win Nobel Prizes when they discover a previously unknown underlying layer and explain its behaviors. We already know what the ultimate bottom layer of an Artificial Intelligence looks like; it looks like ones and zeroes. Our job is to build something interesting out of those ones and zeroes. The Turing formalism does not solve this problem any more than quantum electrodynamics tells us how to build a bicycle; knowing the abstract fact that a bicycle

---

[1]This does not rule out the possibility of discoveries in cognitive science occurring through less intentional and more evolutionary means. For example, a commercial AI project with a wide range of customers might begin with a shallow central architecture loosely integrating domain-specific functionality across a wide variety of tasks, but later find that their research tends to produce specialized internal functionality hinting at a deeper, more integrated supersystem architecture.

is built from atoms doesn't tell you *how* to build a bicycle out of atoms – which atoms to use and where to put them. Similarly, the abstract knowledge that biological neurons implement human intelligence does not explain human intelligence. The classical hype of early neural networks, that they used "the same parallel architecture as the human brain," should, at most, have been a claim of using the same parallel architecture as an earthworm's brain. (And given the complexity of biological neurons, the claim would still have been wrong.)

> The science of understanding living organization is very different from physics or chemistry, where parsimony makes sense as a theoretical criterion. The study of organisms is more like reverse engineering, where one may be dealing with a large array of very different components whose heterogenous organization is explained by the way in which they interact to produce a functional outcome. Evolution, the constructor of living organisms, has no privileged tendency to build into designs principles of operation that are simple and general.
> *Leda Cosmides and John Tooby, "The Psychological Foundations of Culture" [98].*

The field of Artificial Intelligence suffers from a heavy, lingering dose of genericity and black-box, blank-slate, *tabula-rasa* concepts seeping in from the Standard Social Sciences Model (SSSM) identified by [98]. The general project of liberating AI from the clutches of the SSSM is more work than I wish to undertake in this paper, but one problem that must be dealt with immediately is *physics envy*. The development of physics over the last few centuries has been characterized by the discovery of unifying equations which neatly underlie many complex phenomena. Most of the past fifty years in AI might be described as the search for a similar unifying principle believed to underlie the complex phenomenon of intelligence.

Physics envy in AI is the search for a *single, simple* underlying process, with the expectation that this one discovery will lay bare all the secrets of intelligence. The tendency to treat new approaches to AI as if they were new theories of physics may at least partially explain AI's past history of *overpromise* and *oversimplification*. Attributing all the vast functionality of human intelligence to some single descriptive facet – that brains are "parallel," or "distributed," or "stochastic;" that minds use "deduction" or "induction" – results in a failure (an *overhyped* failure) as the project promises that all the functionality of human intelligence will slide out from some simple principle.

The effects of physics envy can be more subtle; they also appear in the lack of interaction between AI projects. Physics envy has given rise to a series of AI projects that could only use *one* idea, as each new hypothesis for the *one true essence of intelligence* was tested and discarded. Douglas Lenat's AM and EURISKO programs [60] – though the results were controversial and may have been mildly exaggerated [85] – nonetheless used very intriguing and fundamental design patterns to deliver significant and unprecedented results. Despite this, the design patterns of EURISKO, such as self-modifying decomposable heuristics, have seen almost no reuse in later AIs. Even Lenat's subsequent Cyc project [62] apparently does not reuse the ideas developed in EURISKO. From the perspective of a modern-day programmer, accustomed to hoarding design patterns and code libraries, the lack of crossfertilization is a surprising anomaly. One would think that self-optimizing heuristics would be useful as an external tool, e.g. for parameter tuning, even if the overall cognitive architecture did not allow for the internal use of such heuristics.

The behavior of the AI field, and of Lenat himself, is more understandable if we postulate that EURISKO was treated as a *failed hypothesis*, or even as a *competing hypothesis*, rather than an *incremental success* or a *reusable tool*. Lenat tried self-optimizing heuristics and they failed to yield intelligence; onward, then, to Cyc, the next hypothesis!

The most common paradigms of traditional AI – search trees, neural networks, genetic algorithms, evolutionary computation, semantic nets – have in common the property that they can be implemented without requiring a store of preexisting complexity. The processes that have become traditional, that *have* been reused, are the tools that stand alone and are immediately useful. A semantic network is a "knowledge" representation so simple that it is literally writable on paper; thus, an AI project adding a semantic network need not design a hippocampus-equivalent to form memories, or build a sensory modality to represent mental imagery. Neural networks and evolutionary computations are not generally intelligent but they are generically intelligent; they can be trained on any problem that has a sufficiently shallow fitness gradient relative to available computing power. (Though EURISKO's self-modifying heuristics probably had generality equalling or exceeding these more typical tools, the source code was not open and the system design was far too complex to build over an afternoon, so the design pattern was not reused – or so I would guess.)

With the exception of the semantic network, which I regard as completely bankrupt, the standalone nature of the traditional processes may make them useful tools for shoring up the initial stages of a general AI supersystem. But standalone algorithms are not substitutes for intelligence and they are not complete systems. Genericity is not the same as generality.

"Physics envy" (trying to replace the human cognitive supersystem with a single process or method) should be distinguished from the less ambitious attempt to *clean up* the human mind design while leaving the essential architecture intact. Cleanup is probably inevitable while human programmers are involved, but it is nonetheless a problem to be approached with extreme caution. Although the population genetics model of evolution admits of many theoretical reasons why the presence of a feature may not imply adaptiveness (much less optimality), in practice the adaptationists usually win. The spandrels of San Marco may not have been built for decorative elegance [27], but they are still holding the roof up. Cleanup should be undertaken, not with pride in the greater simplicity of human design relative to evolutionary design, but with a healthy dose of anxiety that we will leave out something important.

An example: Humans are currently believed to have a modular adaptation for visual face recognition, generally identified with a portion of inferotemporal cortex, though this is a simplification [89]. At first glance this brainware appears to be an archetypal example of human-specific functionality, an adaptation to an evolutionary context with no obvious analogue for an early-stage AI. However, [9] has suggested from neuropathological evidence (associated deficits) that face recognition brainware is also responsible for the generalized task of *acquiring very fine expertise in the visual domain*; thus, the dynamics of face recognition may be of general significance for builders of sensory modalities.

Another example is the sensory modalities themselves. As described in greater detail in Sect. 2, the human cognitive supersystem is built to require the use of the sensory modalities which we originally evolved for other purposes. One good reason why the human supersystem uses sensory modalities is that the sensory modalities are *there*. Sensory modalities are evolutionarily ancient; they would have existed, in primitive or complex form, during the evolution of all higher levels of organization. Neural tissue was already dedicated to sensory modalities, and would go on consuming ATP even if inactive, albeit at a lesser rate. Consider the incremental nature of adaptation, so that in the very beginnings of hominid intelligence only a very small amount of *de novo* complexity would have been involved; consider that evolution has no inherent drive toward design elegance; consider that adaptation is in response to the total environment, which includes both the external environment and the genetic environment – these are all plausible reasons to suspect evolution of offloading the computational burden onto pre-existing neural circuitry, even where a human designer would have chosen to employ a separate subsystem. Thus, it was not inherently absurd for AI's first devotees to try for general intelligence that employed no sensory modalities.

Today we have at least one reason to believe that nonsensory intelligence is a bad approach; we tried it and it didn't work. Of course, this is far too general an argument – it applies equally to "we tried non-face-recognizing intelligence and it didn't work" or even "we tried non-bipedal intelligence and it didn't work." The argument's real force derives from specific hypotheses about the functional role of sensory modalities in general intelligence (discussed in Sect. 2). But in retrospect we can identify at least one *methodological* problem: Rather than identifying the role played by modalities in intelligence, and then attempting to "clean up" the design by *substituting* a simpler process into the functional role played by modalities[2], the first explorers of AI simply assumed that sensory modalities were irrelevant to general intelligence.

---

[2]I cannot think of any plausible way to do this, and do not advocate such an approach.

Leaving out key design elements, without replacement, on the basis of the mistaken belief that they are not relevant to general intelligence, is an error that displays a terrifying synergy with "physics envy." In extreme cases – and most historical cases *have* been extreme – the design ignores everything about the human mind except one characteristic (logic, distributed parallelism, fuzziness, etc.), which is held to be "the key to intelligence." (On my more pessimistic days I sometimes wonder if successive fads are the only means by which knowledge of a given feature of human intelligence becomes widespread in AI.)

I argue strongly for "supersystems," but I do not believe that "supersystems" are the necessary and sufficient Key to AI. General intelligence requires the *right* supersystem, with the right cognitive subsystems, doing the right things in the right way. Humans are not intelligent by virtue of being "supersystems," but by virtue of being a *particular* supersystem which implements human intelligence. I emphasize supersystem design because I believe that the field of AI has been crippled by the *wrong kind of simplicity* – a simplicity which, as a design constraint, rules out workable designs for intelligence; a simplicity which, as a methodology, rules out incremental progress toward an understanding of general intelligence; a simplicity which, as a viewpoint, renders most of the mind invisible except for whichever single aspect is currently promoted as the Key to AI.

If the quest for design simplicity is to be "considered harmful"[3], what should replace it? I believe that rather than simplicity, we should pursue *sufficiently complex explanations* and *usefully deep designs*. In ordinary programming, there is no reason to assume *a priori* that the task is enormously large. In AI the rule should be that the problem is always harder and deeper than it looks, even after you take this rule into account. Knowing that the task is large does not enable us to meet the challenge just by making our designs larger or more complicated; certain *specific* complexity is required, and complexity for the sake of complexity is worse than useless. Nonetheless, the presumption that we are more likely to underdesign than overdesign implies a different attitude towards design, in which victory is never declared, and even after a problem appears to be solved, we go on trying to solve it. If this creed were to be summed up in a single phrase, it would be: "Necessary but not sufficient." In accordance with this creed, it should be emphasized that supersystems thinking is only one part of a larger paradigm, and that an open-ended design process is itself "necessary but not sufficient." These are first steps toward AI, but not the only first steps, and certainly not the last steps.

---

[3]A phrase due to [21] in "Go To Statement Considered Harmful;" today it indicates that a prevalent practice has more penalties than benefits and should be discarded.

2 Levels of Organization in Deliberative General Intelligence

Intelligence in the human cognitive supersystem is the result of the many cognitive processes taking place on multiple levels of organization. However, this statement is vague without hypotheses about specific levels of organization and specific cognitive phenomena. The concrete theory presented in Sect. 2 goes under the name of "deliberative general intelligence" (DGI).

The human mind, owing to its accretive evolutionary origin, has several major distinct candidates for the mind's "center of gravity." For example, the limbic system is an evolutionarily ancient part of the brain that now coordinates activities in many of the other systems that later grew up around it. However, in (cautiously) considering what a more foresightful and less accretive design for intelligence might look like, I find that a single center of gravity stands out as having the most complexity and doing most of the substantive work of intelligence, such that in an AI, to an even greater degree than in humans, this center of gravity would probably become the central supersystem of the mind. This center of gravity is the cognitive superprocess which is introspectively observed by humans through the internal narrative – the process whose workings are reflected in the mental sentences that we internally "speak" and internally "hear" when thinking about a problem. To avoid the awkward phrase "stream of consciousness" and the loaded word "consciousness," this cognitive superprocess will hereafter be referred to as *deliberation*.

2.1 Concepts: An Illustration of Principles

My chosen entry point into deliberation is words – that is, the words we mentally speak and mentally hear in our internal narrative. Let us take the word "lightbulb" (or the wordlike phrase "light bulb") as an example[4]. When you see the letters spelling "light bulb," the phonemes for *light bulb* flow through your auditory cortex. If a mental task requires it, a visual exemplar for the "light bulb" category may be retrieved as mental imagery in your visual cortex (and associated visual areas). Some of your past memories and

---

[4]Note that "lightbulb" is a *basic-level category* [8]. "Basic-level" categories tend to lie on the highest level at which category members have similar shapes, the highest level at which a single mental image can reflect the entire category, the highest level at which a person uses similar motor actions for interacting with category members, et cetera [87]. "Chair" is a basic-level category but "furniture" is not; "red" is a basic-level category but "scarlet" is not. Basic-level categories generally have short, compact names, are among the first terms learned within a language, and are the easiest to process cognitively. [57] cautions against inadvertent generalization from basic-level categories to categories in general, noting that most researchers, in trying to think of examples of categories, almost always select examples of basic-level categories.

experiences, such as accidentally breaking a light bulb and carefully sweeping up the sharp pieces, may be associated with or stored under the "light bulb" concept. "Light bulb" is associated to other concepts; in cognitive priming experiments, it has been shown that hearing a phrase such as "light bulb"[5] will prime associated words such as "fluorescent" or "fragile," increasing the recognition speed or reaction speed when associated words are presented [69]. The "light bulb" concept can act as a mental category; it describes some referents in perceived sensory experiences or internal mental imagery, but not other referents; and, among the referents it describes, it describes some strongly and others only weakly.

To further expose the internal complexity of the "light bulb" concept, I would like to offer an introspective illustration. I apologize to any academic readers who possess strong philosophical prejudices against introspection; I emphasize that the exercise is not intended as *evidence* for a theory, but rather as a means of introducing and grounding concepts that will be argued in more detail later. That said:

Close your eyes, and try to immediately (without conscious reasoning) visualize a *triangular light bulb* – now. Did you do so? What did you see? On personally performing this test for the first time, I saw a pyramidal light bulb, with smoothed edges, with a bulb on the square base. Perhaps you saw a tetrahedral light bulb instead of a pyramidal one, or a light bulb with sharp edges instead of smooth edges, or even a fluorescent tube bent into a equilateral triangle. The specific result varies; what matters is the process you used to arrive at the mental imagery.

Our mental image for "triangular light bulb" would intuitively appear to be the result of imposing "triangular," the adjectival form of "triangle," on the "light bulb" concept. That is, the novel mental image of a triangular light bulb is apparently the result of combining the sensory content of two pre-existing concepts. (DGI agrees, but the assumption deserves to be pointed out explicitly.) Similarly, the combination of the two concepts is not a collision, but a structured imposition; "triangular" is imposed on "light bulb," and not "light-bulb-like" on "triangle."

The structured combination of two concepts is a major cognitive process. I emphasize that I am not talking about interesting complexity which is supposedly to be found in the overall pattern of relations between concepts; I am talking about complexity which is directly visible in the *specific* example of imposing "triangular" on "light bulb." I am not "zooming out" to look at the overall terrain of concepts, but "zooming in" to look at the cognitive processes needed to handle this *single* case. The specific example of imposing "triangular" on "light bulb" is a nontrivial feat of mind; "triangular light bulb" is a trickier concept combination than "green light bulb" or "triangular parking lot."

---

[5]I don't know of a specific case of priming tests conducted on the specific word-pair "lightbulb" and "fluorescent," but this is a typical example.

The mental process of visualizing a "triangular light bulb" flashes through the mind very quickly; it may be possible to glimpse subjective flashes of the concept combination, but the process is not really open to human introspection. For example, when first imposing "triangular" on "light bulb," I would report a brief subjective flash of a *conflict* arising from trying to impose the planar 2-D shape of "triangular" on the 3-D "light bulb" concept. However, before this conflict could take place, it would seem necessary that some cognitive process have already selected the *shape* facet of "triangular" for imposition – as opposed to, say, the color or line width of the "triangle" exemplar that appears when I try to visualize a "triangle" as such. However, this initial selection of *shape* as the key facet did not rise to the level of conscious attention. I can guess at the underlying selection process – in this case, that past experience with the usage had already "cached" *shape* as the salient facet for the concept *triangular*, and that the concept was abstracted from an experiential base in which shape, but not color, was the perceived similarity within the group of experiences. However, I cannot actually introspect on this selection process.

Likewise, I may have glimpsed the existence of a conflict, and that it was a conflict resulting from the 2D nature of "triangular" versus the 3D nature of "light bulb," but *how* the conflict was detected is not apparent in the subjective glimpse. And the resolution of the conflict, the transformation of the 2D *triangle* shape into a 3D *pyramid* shape, was apparently instantaneous from my introspective vantage point. Again, I can guess at the underlying process – in this case, that several already-associated conceptual neighbors of "triangle" were imposed on "light bulb" in parallel, and the best fit selected. But even if this explanation is correct, the process occurred too fast to be visible to direct introspection. I cannot rule out the possibility that a more complex, more deeply creative process was involved in the transition from *triangle* to *pyramid*, although basic constraints on human information-processing (the 200 spike/second speed limit of the underlying neurons) still apply. Nor can I rule out the possibility that there was a unique serial route from *triangle* to *pyramid*.

The creation of an actual visuospatial image of a pyramidal light bulb is, presumably, a complex visual process – one that implies the ability of the visuospatial modality to reverse the usual flow of information and send commands from high-level features to low-level features, instead of detecting high-level features from low-level features. DGI hypothesizes that visualization occurs through a flow from high-level *feature controllers* to low-level feature controllers, creating an articulated mental image within a sensory modality through a multistage process that allows the detection of conflicts at higher levels before proceeding to lower levels. The final mental imagery is introspectively visible, but the process that creates it is mostly opaque.

Some theorists defy introspection to assert that our mental imagery is purely abstract [81]. Yet there exists evidence from neuroanatomy, functional neuroimaging, pathology of neurological disorders, and cognitive psychology to support the contention that mental imagery is directly represented in sensory modalities [53]. [23] show that mental imagery can create visual afterimages[6] similar to, though weaker than, the afterimages resulting from real visual experience. [94] estimate that while the cat has roughly 10^6 fibers from the lateral geniculate nucleus[7] to the visual cortex, there are approximately 10^7 fibers running in the opposite direction. No explanatory consensus currently exists for the existence of the massive corticothalamic feedback projections, though there are many competing theories; the puzzle is of obvious interest to an AI researcher positing a theory in which inventing novel mental imagery is more computationally intensive than sensory perception.

To return to the "triangular lightbulb" example: Once the visuospatial image of a pyramidal light bulb was fully articulated, the next introspective glimpse was of a conflict in visualizing a *glass* pyramid – a pyramid has sharp edges, and sharp glass can cut the user. This implies the mental imagery had semantic content (knowledge about the material composition of the pyramidal light bulb), imported from the original "light bulb" concept, and well-integrated with the visual representation. Like most modern-day humans, I know from early parental warnings and later real-life confirmation that sharp glass is dangerous. Thus the rapid visual detection of sharp glass is important when dealing with real-life sensory experience. I say this to emphasize that no extended line of intelligent reasoning (which would exceed the 200Hz speed limit of biological neurons) is required to react negatively to a fleeting mental image of sharp glass. This reaction could reasonably happen in a single perceptual step, so long as the same perceptual system which detects the visual signature of sharp glass in real-world sensory experience also reacts to mental imagery.

The conflict detected was resolved by the imposition of smooth edges on the glass pyramid making up the pyramidal light bulb. Again, this apparently occurred instantly; again, nontrivial hidden complexity is implied. To frame the problem in the terms suggested by [36], the imaginative process needed to possess or create a "knob" governing the image's transition from sharp edges to rounded edges, and the possession or creation of this knob is the most interesting part of the process, not the selection of one knob from many. If

---

[6]Finke and Schmidt showed that afterimages from mental imagery can recreate the McCullough effect. The McCullough effect is a striking illustration of the selective fatiguing of higher-level feature detectors, in which, following the presentation of alternating green horizontal and red vertical bars, differently colored afterimages are perceived in the white space of a background image depending on whether the background image has horizontal black-and-white bars (red afterimage) or vertical black-and-white bars (green afterimage). This is an unusual and counterintuitive visual effect, and not one that a typical study volunteer would know about and subconsciously "fake" (as Pylyshyn contends).

[7]The lateral geniculate nucleus is a thalamic body which implements an intermediate stage in visual processing between the retina and the visual cortex.

the "knob" was created on the fly, it implies a much higher degree of systemic creativity than selecting from among pre-existing options.

Once the final conflict was resolved by the perceptual imposition of smoothed edges, the final mental image took on a stable form. Again, in this example, all of the mental events appeared introspectively to happen automatically and without conscious decisions on my part; I would estimate that the whole process took less than one second.

In concept combination, a few flashes of the intermediate stages of processing may be visible as introspective glimpses – especially those conflicts that arise to the level of conscious attention before being resolved automatically. But the extreme rapidity of the process means the glimpses are even more unreliable than ordinary introspection – where introspection is traditionally considered unreliable to begin with. To some extent, this is the *point* of the illustration narrated above; almost all of the internal complexity of concepts is hidden away from human introspection, and many theories of AI (even in the modern era) thus attempt to implement concepts on the token level, e.g., "lightbulb" as a raw LISP atom.

This traditional problem is why I have carefully avoided using the word *symbol* in the exposition above. In AI, the term "symbol" carries implicit connotations about representation – that the symbol is a naked LISP atom whose supposed meaning derives from its relation to the surrounding atoms in a semantic net; or at most a LISP atom whose content is a "frame-based" LISP structure (that is, whose content is another semantic net). Even attempts to argue against the design assumptions of Good Old-Fashioned AI (GOFAI) are often phrased in GOFAI's terms; for example, the "symbol grounding problem." Much discussion of the symbol grounding problem has approached the problem as if the design *starts out* with symbols and "grounding" is then *added.* In some cases this viewpoint has directly translated to AI architectures; e.g., a traditional semantic net is loosely coupled to a connectionist sensorimotor system [33].

DGI belongs to the existing tradition that asks, not "How do we ground our semantic nets?", but rather "What is the underlying stuff making up these rich high-level objects we call 'symbols'?" – an approach presented most beautifully in [35]; see also [10]. From this viewpoint, without the right underlying "symbolstuff," there *are* no symbols; merely LISP tokens carved in mockery of real concepts and brought to unholy life by the naming-makes-it-so fallacy.

Imagine sensory modalities as solid objects with a metaphorical surface composed of the layered feature detectors and their inverse functions as feature controllers. The metaphorical "symbolstuff" is a pattern that interacts with the feature detectors to test for the presence of complex patterns in sensory data, or inversely, interacts with the feature controllers to produce complex mental imagery. Symbols combine through the faceted combination of their symbolstuffs, using a process that might be called "holonic conflict resolution," where information flows from high-level feature controllers to low-level feature controllers, and conflicts are detected at each layer as the flow proceeds. "Holonic" is a useful word to describe the simultaneous application of reductionism and holism, in which a single quality is simultaneously a combination of parts and a part of a greater whole [51]. For example, a single feature detector may make use of the output of lower-level feature detectors, and act in turn as an input to higher-level feature detectors. Note that "holonic" does not imply strict hierarchy, only a general flow from high-level to low-level.

I apologize for adding yet another term, "holonic conflict resolution," to a namespace already crowded with terms such as "computational temperature" [71], "Prägnanz" [52], "Hopfield networks" [41], "constraint propagation" [55], and many others. Holonic conflict resolution is certainly not a wholly new idea, and may even be wholly unoriginal on a feature-by-feature basis, but the combination of features I wish to describe does not exactly match the existing common usage of any of the terms above. "Holonic conflict resolution" is intended to convey the image of a process that flows serially through the layered, holonic structure of perception, with detected conflicts resolved locally or propagated to the level above, with a final solution that satisfices. Many of the terms above, in their common usage, refer to an iterated annealing process which seeks a global minimum. Holonic conflict resolution is intended to be biologically plausible; i.e., to involve a smooth flow of visualization which is computationally tractable for parallel but speed-limited neurons.

Holonic conflict resolution is not proposed as a complete solution to perceptual problems, but rather as the active canvas for the interaction of concepts with mental imagery. In theoretical terms, holonic conflict resolution is a structural framework within which to posit specific conflict-detection and conflict-resolution methods. Holonic imagery is the artist's medium within which symbolstuff paints mental pictures such as "triangular light bulb."

A constructive account of concepts and symbolstuff would need to supply:

- A description of how a concept is *satisfied by* and *imposed on* referents in a sensory modality
- A symbolstuff representation satisfying (a) that can contain the internal complexity needed for faceted concept combination
- A representation satisfying (a) and (b), such that it is computationally tractable to abstract new concepts using sensory experience as raw material

This is *not* an exhaustive list of concept functionality; these are just the three most "interesting" challenges[8]. These challenges are interesting because the difficulty of solving them simultaneously seems to be the multiplicative (rather than additive) product of the difficulties of solving them individually. Other design requirements for a constructive account of concepts would include: association to nearby concepts; supercategories and subcategories;

---

[8] "Interesting" is here used in its idiomatic sense of "extremely hard."

exemplars stored in memory; prototype and typicality effects [88]; and many others (see, e.g., [57]).

The interaction of concepts with modalities, and the interaction of concepts with each other, illustrate what I believe to be several important rules about how to approach AI.

The first principle is that of *multiple levels of organization*. The human phenotype is composed of atoms[9], molecules, proteins, cells, tissues, organs, organ systems, and finally the complete body – eight distinguishable layers of organization, each successive layer built above the preceding one, each successive layer incorporating evolved adaptive complexity. Some useful properties of the higher level may emerge naturally from lower-level behaviors, but not all of them; higher-level properties are also subject to selection pressures on heritable variation and the elaboration of complex functional adaptations. In postulating multiple levels of organization, I am not positing that the behaviors of all higher layers emerge automatically from the lowest layer.

If I had to pick one *single* mistake that has been the *most* debilitating in AI, it would be *implementing a process too close to the token level* – trying to implement a high-level process without implementing the underlying layers of organization. Many proverbial AI pathologies result at least partially from omitting lower levels of organization from the design.

Take, for example, that version of the "frame problem" – sometimes also considered a form of the "commonsense problem" – in which intelligent reasoning appears to require knowledge of an infinite number of special cases. Consider a CPU which adds two 32-bit numbers. The higher level consists of two integers which are added to produce a third integer. On a lower level, the computational objects are not regarded as opaque "integers," but as ordered structures of 32 bits. When the CPU performs an arithmetic operation, two structures of 32 bits collide, under certain rules which govern the local interactions between bits, and the result is a new structure of 32 bits. Now consider the woes of a research team, with no knowledge of the CPU's underlying implementation, that tries to create an arithmetic "expert system" by encoding a vast semantic network containing the "knowledge" that **two** and **two** make **four**, **twenty-one** and **sixteen** make **thirty-seven**, and so on. This giant lookup table requires eighteen billion billion entries for completion.

In this hypothetical world where the lower-level process of addition is not understood, we can imagine the "common-sense" problem for addition; the launching of distributed Internet projects to "encode all the detailed knowledge necessary for addition;" the frame problem for addition; the philosophies of formal semantics under which the LISP token **thirty-seven** is meaning-

---

[9]The levels begin with "atoms" rather than "quarks" or "molecules" because the atomic level is the highest layer selected from a bounded set of possible elements (ions and isotopes notwithstanding). "Quarks" are omitted from the list of layers because no adaptive complexity is involved; evolution exercises no control over how quarks come together to form atoms.

ful because it refers to thirty-seven objects in the external world; the design principle that the token `thirty-seven` has no internal complexity and is rather given meaning by its network of relations to other tokens; the "number grounding problem;" the hopeful futurists arguing that past projects to create Artificial Addition failed because of inadequate computing power; and so on.

To some extent this is an unfair analogy. Even if the thought experiment is basically correct, and the woes described would result from an attempt to capture a high-level description of arithmetic without implementing the underlying lower level, this does not prove the analogous mistake is the source of these woes in the real field of AI. And to some extent the above description is unfair even as a thought experiment; an arithmetical expert system would not be as bankrupt as semantic nets. The regularities in an "expert system for arithmetic" would be real, noticeable by simple and computationally feasible means, and could be used to deduce that arithmetic was the underlying process being represented, even by a Martian reading the program code with no hint as to the intended purpose of the system. The gap between the higher level and the lower level is not absolute and uncrossable, as it is in semantic nets.

An arithmetic expert system that leaves out one level of organization may be recoverable. Semantic nets leave out *multiple* levels of organization. Omitting all the experiential and sensory grounding of human symbols leaves no raw material to work with. If all the LISP tokens in a semantic net were given random new names, there would be no way to deduce whether `G0025` formerly meant `hamburger` or `Johnny Carson`. [29] describes the symbol grounding problem arising out of semantic nets as being similar to trying to learn Chinese as a first language using only a Chinese-to-Chinese dictionary.

I believe that many (though not all) cases of the "commonsense problem" or "frame problem" arise from trying to store all possible descriptions of high-level behaviors that, in the human mind, are modeled by visualizing the lower level of organization from which those behaviors emerge. For example, [58] give a sample list of "built-in inferences" emerging from what they identify as the Source-Path-Goal metaphor:

- If you have traversed a route to a current location, you have been at all previous locations on that route.
- If you travel from A to B and from B to C, then you have traveled from A to C.
- If X and Y are traveling along a direct route from A to B and X passes Y, then X is farther from A and closer to B than Y is.
- (et cetera)

A general intelligence with a visual modality has no need to explicitly store an infinite number of such statements in a theorem-proving production system. The above statements can be perceived on the fly by inspecting depictive mental imagery. Rather than storing *knowledge about* trajectories, a visual modality actually *simulates the behavior of* trajectories. A visual modality uses low-level elements, metaphorical "pixels" and their holonic feature structure, whose behaviors locally correspond to the real-world behaviors of the referent. There is a mapping from representation to referent, but it is a mapping on a lower level of organization than traditional semantic nets attempt to capture. The correspondence happens on the level where 13 is the structure 00001101, not on the level where it is the number thirteen.

I occasionally encounter some confusion about the difference between a *visual modality* and a *microtheory of vision*. Admittedly, microtheories in theorem-proving systems are well known in AI, athough I personally consider it to be a paradigm of little worth, so some confusion is understandable. But layered feature extraction in the visual modality – which is an established fact of neuroscience – is also very well known even in the pure computer science tradition of AI, and has been well-known ever since David Marr's tremendously influential 1982 book *Vision* [65] and earlier papers. To make the difference explicit, the human visual cortex "knows" about edge detection, shading, textures of curved surfaces, binocular disparities, color constancy under natural lighting, motion relative to the plane of fixation, and so on. The visual cortex does not know about butterflies. In fact, a visual cortex "knows" nothing; a sensory modality contains behaviors which correspond to environmental invariants, not knowledge about environmental regularities.

This illustrates the second-worst error in AI, the failure to distinguish between *things that can be hardwired* and *things that must be learned*. We are not preprogrammed to know about butterflies. Evolution wired us with visual circuitry that makes sense of the sensory image of the butterfly, and with object-recognition systems that form visual categories. When we see a butterfly, we are then able to recognize future butterflies as belonging to the same kind. Sometimes evolution bypasses this system to gift us with visual instincts, but this constitutes a tiny fraction of visual knowledge. A modern human recognizes a vast number of visual categories with no analogues in the ancestral environment.

What problems result from failing to distinguish between things that can be hardwired and things that must be learned? "Hardwiring what should be learned" is so universally combined with "collapsing the levels of organization" that it is difficult to sort out the resulting pathologies. An expert systems engineer, in addition to believing that knowledge of butterflies can be preprogrammed, is also likely to believe that knowing about butterflies consists of a `butterfly` LISP token which derives its meaning from its relation to other LISP tokens – rather than *butterfly* being a stored pattern that interacts with the visual modality and recognizes a butterfly. A semantic net not only lacks richness, it lacks the capacity to represent richness. Thus, I would attribute the symbol grounding problem to "collapsing the levels of organization," rather than "hardwiring what should be learned."

But even if a programmer who understood the levels of organization tried to create butterfly-recognizing symbolstuff by hand, I would still expect the resulting butterfly pattern to lack the richness of the learned butterfly pattern in a human mind. When the human visual system creates a *butterfly* visual category, it does not write an opaque, procedural butterfly-recognition codelet using abstract knowledge about butterflies and then tag the codelet onto a butterfly frame. Human visual categorization abstracts the butterfly category from a store of visual experiences of butterflies.

Furthermore, visual categorization – the general concept-formation process, not just the temporal visual processing stream – leaves behind an association between the butterfly concept and the stored memories from which "butterfly" was abstracted; it associates one or more exemplars with the butterfly category; it associates the butterfly category through overlapping territory to other visual categories such as *fluttering;* it creates butterfly symbolstuff that can combine with other symbolstuffs to produce mental imagery of a *blue butterfly;* and so on. To the extent that a human lacks the patience to do these things, or to the extent that a human does them in fragile and hand-coded ways rather than using robust abstraction from a messy experiential base, *lack of richness* will result. Even if an AI needs programmer-created concepts to bootstrap further concept formation, bootstrap concepts should be created using programmer-directed tool versions of the corresponding AI subsystems, and the bootstrap concepts should be replaced with AI-formed concepts as early as possible.

Two other potential problems emerging from the use of programmer-created content are *opacity* and *isolation.*

Opacity refers to the potential inability of an AI's subsystems to modify content that originated outside the AI. If a programmer is creating cognitive content, it should at least be the kind of content that the AI *could have* created on its own; it should be content in a form that the AI's cognitive subsystems can manipulate. The best way to ensure that the AI can modify and use internal content is to have the AI create the content. If an AI's cognitive subsystems are powerful enough to create content independently, then hopefully those same subsystems will be capable of adding to that content, manipulating it, bending it in response to pressures exerted by a problem, and so on. What the AI creates, the AI can use and improve. Whatever the AI accomplishes on its own is a part of the AI's mind; the AI "owns" it and is not simply borrowing it from the programmers. This is a principle that extends far beyond abstracting concepts!

Isolation means that if a concept, or a piece of knowledge, is handed to the AI on a silver platter, the AI may be isolated from the things that the AI would have needed to learn first in order to acquire that knowledge naturally, in the course of building up successive layers of understanding to handle problems of increasing complexity. The concept may also be isolated from the other things that the AI would have learned at around the same time, which may mean a dearth of useful associations and slippages. Programmers may try to second-guess isolation by teaching many similar knowledges at around the same time, but that is no substitute for a natural ecology of cognition.

2.2 Levels of Organization in Deliberation

The model of deliberation presented in this chapter requires five distinct layers of organization, each layer built on top of the underlying layer.

- The bottom layer is *source code* and *data structures* – complexity that is manipulated directly by the programmer. The equivalent layer for humans is *neurons* and *neural circuitry.*
- The next layer is *sensory modalities.* In humans, the archetypal examples of sensory modalities are sight, sound, touch, taste, smell, and so on[10]; implemented by the visual areas, auditory areas, et cetera. In biological brains, sensory modalities come the closest to being "hardwired;" they generally involve clearly defined stages of information-processing and feature-extraction, sometimes with individual neurons playing clearly defined roles. Thus, sensory modalities are some of the best candidates for processes that can be directly coded by programmers without rendering the system crystalline and fragile.
- The next layer is *concepts.* Concepts (also sometimes known as "categories," or "symbols") are abstracted from our experiences. Abstraction reifies a perceived similarity within a group of experiences. Once reified, the common quality can then be used to determine whether new mental imagery *satisfies* the quality, and the quality can be *imposed* on a mental image, altering it. Having abstracted the concept "red," we can take a mental image of a non-red object (for example, grass) and imagine "red grass." Concepts are patterns that mesh with sensory imagery; concepts are *complex, flexible, reusable* patterns that have been reified and placed in long-term storage.
- The next layer is *thoughts,* built from structures of concepts. By imposing concepts in targeted series, it becomes possible to build up complex mental images within the workspace provided by one or more sensory modalities. The archetypal example of a thought is a human "sentence" – an arrangement of concepts, invoked by their symbolic tags, with internal structure and targeting information that can be reconstructed from a linear series of words using the constraints of syntax, constructing a complex mental image that can be used in reasoning. Thoughts (and their corresponding mental imagery) are the disposable one-time structures, built from reusable concepts, that implement a non-recurrent mind in a non-recurrent world.
- Finally, it is sequences of thoughts that implement *deliberation* – explanation, prediction, planning, design, discovery, and the other activities used to solve knowledge problems in the pursuit of real-world goals.

Although the five-layer model is central to the DGI theory of intelligence, the rule of *Necessary But Not Sufficient* still holds. An AI project will not

---

[10]Other human modalities include, e.g., proprioception and vestibular coordination.

succeed by virtue of "implementing a five-layer model of intelligence, just like the human brain." It must be the right five layers. It must be the *right* modalities, used in the *right* concepts, coming together to create the *right* thoughts seeking out the *right* goals.

The five-layer model of deliberation is not inclusive of everything in the DGI theory of mind, but it covers substantial territory, and can be extended beyond the deliberation superprocess to provide a loose sense of which level of organization any cognitive process lies upon. Observing that the human body is composed of molecules, proteins, cells, tissues, and organs is not a complete design for a human body, but it is nonetheless important to know whether something is an organ or a protein. Blood, for example, is not an prototypical tissue, but it is composed of cells, and is generally said to occupy the tissue level of organization of the human body. Similarly, the hippocampus, in its role as a memory-formation subsystem, is not a sensory modality, but it can be said to occupy the "modality level": It is brainware (a discrete, modular chunk of neural circuitry); it lies above the neuron/code level; it has a characteristic tiling/wiring pattern as the result of genetic complexity; it interacts as an equal with the subsystems comprising sensory modalities.

Generalized definitions of the five levels of organization might be as follows:

Code-level, hardware-level: No generalized definition is needed, except that the biological equivalent is the *neural level* or *wetware level.*

Modality-level: Subsystems which, in humans, derive their adaptive complexity from genetic specification – or rather from the genetic specification of an initial tiling pattern and a self-wiring algorithm, and from exposure to invariant environmental complexity[11]. The AI equivalent is complexity which is known in advance to the programmer and which is directly specified through programmer efforts. Full systems on this level are modular parts of the cognitive supersystem – one of a large but limited number of major parts making up the mind. Where the system in question is a sensory modality or a system which clearly interrelates to the sensory modalities and performs modality-related tasks, the system can be referred to as *modality-level.* Similarly, a subsystem or subprocess of a major modality-level system, or a minor function of such a subsystem, may also be referred to as modality-level. Where this term is inappropriate, because a subsystem has little or no relation to sensory modalities, the subsystem may be referred to as *brainware*[12].

---

[11] Environmental complexity of this type is reliably present and is thus "known in advance" to the genetic specification, and in some sense can be said to be a constant and reliable part of the genetic design.

[12] The term "brainware" is not necessarily anthropomorphic, since the term "brain" can be extended to refer to nonbiological minds. The biology-only equivalent is often half-jokingly referred to as *wetware,* but the term "wetware" should denote the human equivalent of the code level, since only neurons and synapses are actually wet.

Concept-level: Concepts are cognitive objects which are placed in long-term storage, and reused as the building blocks of thoughts. The generalization for this level of organization is *learned complexity:* cognitive content which is derived from the environment and placed in long-term storage, and which thereby becomes part of the permanent reservoir of complexity with which the AI challenges future problems. The term *concept-level* might optionally be applied to any learned complexity that resembles categories; i.e., learned complexity that interacts with sensory modalities and acts on sensory modalities. Regardless of whether they are conceptlike (an issue considered later), other examples of learned complexity include declarative beliefs and episodic memories.

Thought-level: A thought is a specific structure of combinatorial symbols which builds or alters mental imagery. The generalizable property of thoughts is their *immediacy.* Thoughts are not evolved/programmed brainware, or a long-term reservoir of learned complexity; thoughts are constructed on a moment-by-moment basis. Thoughts make up the life history of a non-recurrent mind in a non-recurrent universe. The generalized thought level extends beyond the mentally spoken sentences in our stream of consciousness; it includes all the major cognitive events occurring within the world of active mental imagery, especially events that involve structuring the combinatorial complexity of the concept level.

Deliberation: Which, like the code level, needs no generalization. Deliberation describes the activities carried out by patterns of thoughts. The patterns in deliberation are not just epiphenomenal properties of thought sequences; the deliberation level is a complete layer of organization, with complexity specific to that layer. In a deliberative AI, it is patterns of thoughts that plan and design, transforming abstract high-level goal patterns into specific low-level goal patterns; it is patterns of thoughts that reason from current knowledge to predictions about unknown variables or future sensory data; it is patterns of thoughts that reason about unexplained observations to invent hypotheses about possible causes. In general, deliberation uses organized sequences of thoughts to solve knowledge problems in the pursuit of real-world goals.

Even for the generalized levels of organization, not everything fits cleanly into one level or another. While the hardwired-learned-invented trichotomy usually matches the modality-concept-thought trichotomy, the two are conceptually distinct, and sometimes the correspondence is broken. But the levels of organization are almost always useful – even exceptions to the rule are more easily seen as partial departures than as complete special cases.

2.3 The Code Level

The code level is composed of functions, classes, modules, packages; data types, data structures, data repositories; all the purely programmatic challenges of creating AI. Artificial Intelligence has traditionally been much more intertwined with computer programming than it should be, mostly because of attempts to overcompress the levels of organization and implement thought sequences directly as programmatic procedures, or implement concepts directly as LISP atoms or LISP frames. The code level lies directly beneath the modality level or brainware level; bleedover from modality-level challenges may show up as legitimate programmatic problems, but little else – not thoughts, cognitive content, or high-level problem-solving methods.

Any good programmer – a programmer with a feeling for aesthetics – knows the tedium of solving the same special case, over and over, in slightly different ways; and also the triumph of thinking through the metaproblem and creating a general solution that solves all the special cases simultaneously. As the hacker Jargon File observes, "Real hackers generalize uninteresting problems enough to make them interesting and solve them – thus solving the original problem as a special case (and, it must be admitted, occasionally turning a molehill into a mountain, or a mountain into a tectonic plate)." [82]. This idiom *does not work* for general AI! A real AI would be the ultimate general solution because it would encapsulate the cognitive processes that human programmers use to write *any* specific piece of code, but this ultimate solution cannot be obtained through the technique of successively generalizing uninteresting problems into interesting ones.

Programming is the art of translating a human's mental model of a problem-solution into a computer program; that is, the art of translating thoughts into code. Programming *inherently* violates the levels of organization; it leads directly into the pitfalls of classical AI. The underlying low-level processes that implement intelligence are of a fundamentally different character than high-level intelligence itself. When we translate our thoughts about a problem into code, we are establishing a correspondence between code and the high-level *content* of our minds, not a correspondence between code and the dynamic process of a human mind. In ordinary programming, the task is to get a computer to solve a specific problem; it may be an "interesting" problem, with a very large domain, but it will still be a specific problem. In ordinary programming the problem is solved by taking the human thought process that would be used to solve an instance of the problem, and translating that thought process into code that can also solve instances of the problem. Programmers are humans who have learned the art of inventing thought processes, called "algorithms," that rely only on capabilities an ordinary computer possesses.

The reflexes learned by a good, artistic programmer represent a fundamental danger when embarking on a general AI project. Programmers are trained to solve problems, and trying to create general AI means solving the programming problem of creating a mind that solves problems. There is the danger of a short-circuit, of misinterpreting the problem task as writing code that directly solves some specific challenge posed to the mind, instead of building a mind that can solve the challenge with general intelligence. Code, when abused, is an excellent tool for creating long-term problems in the guise of short-term solutions.

Having described what we are *forbidden* to do with code, what *legitimate* challenges lie on this level of organization?

Some programming challenges are universal. Any modern programmer should be familiar with the world of compilers, interpreters, debuggers, Integrated Development Environments, multithreaded programming, object orientation, code reuse, code maintenance, and the other tools and traditions of modern-day programming. It is difficult to imagine anyone successfully coding the brainware level of general intelligence in assembly language – at least if the code is being developed for the first time. In that sense object orientation and other features of modern-day languages are "required" for AI development; but they are necessary as productivity tools, not because of any deep similarity between the structure of the programming language and the structure of general intelligence. Good programming tools help with AI *development* but do not help with *AI*.

Some programming challenges, although universal, are likely to be unusually severe in AI development. AI development is *exploratory, parallelized,* and *large.* Writing a great deal of exploratory code means that IDEs with refactoring support and version control are important, and that modular code is even more important than it is usually – or at least, code that is as modular as possible given the highly interconnected nature of the cognitive supersystem.

Parallelism on the hardware level is currently supported by symmetric multiprocessing chip architectures [42], NOW (network-of-workstations) clustering [1] and Beowulf clustering [3], and message-passing APIs such as PVM [26] and MPI [28]. However, software-level parallelism is not handled well by present-day languages and is therefore likely to present one of the greatest challenges. Even if software parallelism *were* well-supported, AI developers will still need to spend time explicitly thinking on how to parallelize cognitive processes – human cognition may be massively parallel on the lower levels, but the overall flow of cognition is still serial.

Finally, there are some programming challenges that are likely to be unique to AI.

We know it is possible to evolve a general intelligence that runs on a hundred trillion synapses with characteristic limiting speeds of approximately 200 spikes per second. An interesting property of human neurobiology is that, at a limiting speed of 150 meters per second for myelinated axons, each neuron is potentially within roughly a single "clock tick" of any other neuron in the brain[13]. [90] describes a quantity S that translates to the wait time, in clock cycles, between different parts of a cognitive system – the minimum

---

[13]The statement that each neuron is "potentially" within one clock tick of any other neuron is meant as a statement about the genome, not a statement about developmental neurology – that is, it would probably require a genetic change to produce a previously forbidden connection.

time it could take for a signal to travel between the most distant parts of the system, measured in the system's clock ticks. For the human brain, S is on the rough order of 1 – in theory, at least. In practice, axons take up space and myelinated axons take up even more space, so the brain uses a highly modular architecture, but there are still long-distance pipes such as the corpus callosum. Currently, S is much greater than 1 for clustered computing systems. S is greater than 1 even within a single-processor computer system; Moore's Law for intrasystem communications bandwidth describes a substantially slower doubling time than processor speeds. Increasingly the limiting resource of modern computing systems is not processor speed but memory bandwidth [108] (and this problem has gotten worse, rather than better, since 1995).

One class of purely programmatic problems that are unique to AI arise from the need to "port" intelligence from massively parallel neurons to clustered computing systems (or other human-programmable substrate). It is conceivable, for example, that the human mind handles the cognitive process of *memory association* by comparing current working imagery to all stored memories, in parallel. We have no particular evidence that the human mind uses a brute force comparison, but it *could* be brute-forced. The human brain acknowledges no distinction between CPU and RAM. If there are enough neurons to store a memory, then the same neurons may presumably be called upon to compare that memory to current experience. (This holds true even if the correspondence between neural groups and stored memories is many-to-many instead of one-to-one.)

Memory association may or may not use a "compare" operation (brute force or otherwise) of current imagery against all stored memories, but it seems likely that the brain uses a massively parallel algorithm at one point or another of its operation; memory association is simply a plausible candidate. Suppose that memory association is a brute-force task, performed by asking all neurons engaged in memory storage to perform a "compare" against patterns broadcast from current working imagery. Faced with the design requirement of matching the brute force of 10^{14} massively parallel synapses with a mere clustered system, a programmer may be tempted to despair. There is no *a priori* reason why such a task should be possible.

Faced with a problem of this class, there are two courses the programmer can take. The first is to implement an analogous "massive compare" as efficiently as possible on the available hardware – an algorithmic challenge worthy of Hercules, but past programmers have overcome massive computational barriers through heroic efforts and the relentless grinding of Moore's Law. The second road – much scarier, with even less of a guarantee that success is possible – is to *redesign the cognitive process* for different hardware.

The human brain's most fundamental limit is its speed. Anything that happens in less than a second perforce must use less than 200 *sequential* operations, however massively parallelized. If the human brain really does use a massively parallel brute-force compare against all stored memories to handle the problem of association, it's probably because there isn't time to do anything else! The human brain is massively parallel because massive parallelism is the only way to do *anything* in 200 clock ticks. If modern computers ran at 200Hz instead of 2GHz, PCs would also need 10^{14} processors to do anything interesting in realtime.

A sufficiently bold general AI developer, instead of trying to reimplement the cognitive process of association as it developed in humans, might instead ask: *What would this cognitive subsystem look like, if it had evolved on hardware instead of wetware?* If we remove the old constraint of needing to complete in a handful of clock ticks, and add the new constraint of not being able to offhandedly "parallelize against all stored memories," what is the *new* best algorithm for memory association? For example, suppose that you find a method of "fuzzy hashing" a memory, such that mostly similar memories automatically collide within a container space, but where the fuzzy hash inherently requires an extended linear series of sequential operations that would have placed "fuzzy hashing" out of reach for realtime neural operations. "Fuzzy hashing" would then be a strong candidate for an alternative implementation of memory association.

A computationally cheaper association subsystem that exploits serial speed instead of parallel speed, whether based around "fuzzy hashing" or something else entirely, might still be qualitatively less intelligent than the corresponding association system within the human brain. For example, memory recognition might be limited to clustered contexts rather than being fully general across all past experience, with the AI often missing "obvious" associations (where "obvious" has the anthropocentric meaning of "computationally easy for a human observer"). In this case, the question would be whether the overall general intelligence could function well enough to get by, perhaps compensating for lack of associational breadth by using longer linear chains of reasoning. The difference between serialism and parallelism, on a low level, would propagate upward to create cognitive differences that compensate for the loss of human advantages or exploit new advantages not shared by humans.

Another class of problem stems from "porting" across the extremely different *programming styles* of evolution versus human coding. Human-written programs typically involve a long series of chained dependencies that intersect at single points of failure – "crystalline" is a good term to describe most human code. Computation in neurons has a different character. Over time our pictures of biological neurons have evolved from simple integrators of synaptic inputs that fire when a threshold input level is reached, to sophisticated biological processors with mixed analog-digital logics, adaptive plasticity, dendritic computing, and functionally relevant dendritic and synaptic morphologies [50]. What remains true is that, from an algorithmic perspective, neural computing uses roughly arithmetical operations[14] that proceed along multiple intertwining channels in which information is represented redundantly and processed stochastically. Hence, it is easier to "train" neural networks – even nonbiological connectionist networks – than to train a piece of human-written code. Flipping a random bit inside the state of a running program, or flipping a random bit in an assembly-language instruction, has a much greater effect than a similar perturbation of a neural network. For neural networks the *fitness landscapes* are smoother. Why is this? Biological neural networks need to tolerate greater environmental noise (data error) and processor noise (computational error), but this is only the beginning of the explanation.

Smooth fitness landscapes are a useful, necessary, and fundamental outcome of evolution. Every evolutionary success starts as a mutation – an error – or as a novel genetic combination. A modern organism, powerfully adaptive with a large reservoir of genetic complexity, necessarily possesses a very long evolutionary history; that is, the genotype has necessarily passed through a very large number of successful mutations and recombinations along the road to its current form. The "evolution of evolvability" is most commonly justified by reference to this historical constraint [16], but there have also been attempts to demonstrate local selection pressures for the characteristics that give rise to evolvability [106], thus averting the need to invoke the controversial agency of species selection. Either way, smooth fitness landscapes are part of the design signature of evolution.

"Smooth fitness landscapes" imply, among other things, that a small perturbation in the program code (genetic noise), in the input (environmental noise), or in the state of the executing program (processor noise), is likely to produce at most a small degradation in output quality. In most human-written code, a small perturbation of any kind usually causes a crash. Genomes are built by a cumulative series of point mutations and random recombinations. Human-written programs start out as high-level goals which are translated, by an extended serial thought process, into code. A perturbation to human-written code perturbs the code's final form, rather than its first cause, and the code's final form has no history of successful mutation. The thoughts that *gave rise* to the code probably have a smooth fitness metric, in the sense that a slight perturbation to the programmer's state of mind will probably produce code that is at most a little worse, and possibly a little better. Human thoughts, which are the original source of human-written code, are resilient; the code itself is fragile.

The dream solution would be a programming language in which human-written, top-down code somehow had the smooth fitness landscapes that are characteristic of accreted evolved complexity, but this is probably far too much to ask of a programming language. The difference between evolution

---

[14]Note that biological neurons can easily implement multiplication as well as addition and subtraction [48], plus low- and band-pass filtering, normalization, gain control, saturation, amplification, thresholding, and coincidence detection [49].

and design runs deeper than the difference between stochastic neural circuitry and fragile chip architectures. On the other hand, using fragile building blocks can't possibly *help,* so a language-level solution might solve at least some of the problem.

The importance of smooth fitness landscapes holds true for all levels of organization. Concepts and thoughts should not break as the result of small changes. The code level is being singled out because smoothness on the code level represents a different kind of problem than smoothness on the higher levels. On the higher levels, smoothness is a product of correctly designed cognitive processes; a learned concept will apply to messy new data because it was abstracted from a messy experiential base. Given that AI complexity lying within the concept level requires smooth fitness landscapes, the correct strategy is to duplicate the smoothness on that level – to accept as a high-level design requirement that the AI produce error-tolerant concepts abstracted from messy experiential bases.

On the code level, neural circuitry is smooth and stochastic by the nature of neurons and by the nature of evolutionary design. Human-written programs are sharp and fragile ("crystalline") by the nature of modern chip architectures and by the nature of human programming. The distinction is not likely to be erased by programmer effort or new programming languages. The long-term solution might be an AI with a sensory modality for code (see Sect. 3), but that is not likely to be attainable in the early stages. The basic code-level "stuff" of the human brain has built-in support for smooth fitness landscapes, and the basic code-level "stuff" of human-written computer programs does not. Where human processes rely on neural circuitry being *automatically* error-tolerant and trainable, it will take additional programmatic work to "port" that cognitive process to a new substrate where the built-in support is absent. The final compromise solution may have error tolerance as one explicit design feature among many, rather than error-tolerance naturally emerging from the code level.

There are other important features that are also supported by biological neural networks – that are "natural" to neural substrate. These features probably include:

- Optimization for recurring problems
- Completion of partial patterns
- Similarity recognition (detection of static pattern repetition)
- Recurrence recognition (detection of temporal repetition)
- Clustering detection, cluster identification, and sorting into identified clusters
- Training for pattern recognition and pattern completion
- Massive parallelism

Again, this does not imply an unbeatable advantage for biological neural networks. In some cases wetware has very *poor* feature support, relative to contemporary hardware. Contemporary hardware has better support for:

- Reflectivity and execution traces
- Lossless serialization (storage and retrieval) and lossless pattern transformations
- Very-high-precision quantitative calculations
- Low-level algorithms which involve extended iteration, deep recursion, and complex branching
- "Massive serialism;" the ability to execute hundreds of millions of *sequential* steps per second

The challenge is using new advantages to compensate for the loss of old advantages, and replacing substrate-level support with design-level support.

This concludes the account of *exceptional* issues that arise at the code level. An enumeration of *all* issues that arise at the code level – for example, serializing the current contents of a sensory modality for efficient transmission to a duplicate modality on a different node of a distributed network – would constitute at least a third of a complete constructive account of a general AI. But programming is not all the work of AI, perhaps not even most of the work of AI; much of the effort needed to construct an intelligence will go into prodding the AI into forming certain concepts, undergoing certain experiences, discovering certain beliefs, and learning various high-level skills. These tasks cannot be accomplished with an IDE. Coding the wrong thing successfully can mess up an AI project worse than any number of programming failures. I believe that the most important skill an AI developer can have is knowing what *not* to program.

2.4 The Modality Level

The Evolutionary Design of Modalities in Humans

Most students of AI are familiar with the high-level computational processes of at least one human sensory modality, vision, at least to the extent of being acquainted with David Marr's "2 1/2D world" and the concept of layered feature extraction [65]. Further investigations in computational neuroscience have both confirmed Marr's theory and rendered it enormously more complex. Although many writers, including myself, have been known to use the phrase "visual cortex" when talking about the entire visual modality, this is like talking about the United States by referring to New York. About 50% of the neocortex of nonhuman primates is devoted exclusively to visual processing, with over 30 distinct visual areas identified in the macaque monkey [22].

The major visual stream is the retinal-geniculate-cortical stream, which goes from the retina to the lateral geniculate nucleus to the striate cortex[15] to the higher visual areas. Beyond the visual cortex, processing splits into two major secondary streams; the ventral stream heading toward the temporal

---

[15]The striate cortex is also known as "primary visual cortex," "area 17," and "V1".

lobe for object recognition, and the dorsal stream heading toward the parietal lobe for spatial processing. The visual stream begins in the retina, which contains around 100 million rods and 5 million cones, but feeds into an optic cable containing only around 1 million axons. Visual preprocessing begins in the first layer of the retina, which converts the raw intensities into center-surround gradients, a representation that forms the basis of all further visual processing. After several further layers of retinal processing, the final retinal layer is composed of a wide variety of ganglion types that include directionally selective motion detectors, slow-moving edge detectors, fast movement detectors, uniformity detectors, and subtractive color channels. The axons of these ganglions form the optic nerve and project to the magnocellular, parvocellular, and koniocellular layers of the lateral geniculate nucleus; currently it appears that each class of ganglion projects to only one of these layers. It is widely assumed that further feature detection takes place in the lateral geniculate nucleus, but the specifics are not currently clear. From the lateral geniculate nucleus, the visual information stream continues to area V1, the primary visual cortex, which begins feature extraction for information about motion, orientation, color and depth. From primary visual cortex the information stream continues, making its way to the higher visual areas, V2 through V6. Beyond the visual cortex, the information stream continues to temporal areas (object recognition) and parietal areas (spatial processing).

As mentioned earlier, primary visual cortex sends massive corticothalamic feedback projections to the lateral geniculate nucleus [94]. Corticocortical connections are also typically accompanied by feedback projections of equal strength [22]. There is currently no standard explanation for these feedback connections. DGI[16] requires sensory modalities with *feature controllers* that are the inverse complements of the *feature detectors;* this fits with the existence of the feedback projections. However, it should be noted that this assertion is not part of contemporary neuroscience. The existence of feature controllers is allowed for, but not asserted, by current theory; their existence is asserted, and required, by DGI. (The hypothesis that feedback projections play a role in mental imagery is not limited to DGI; for example, [53] cites the existence of corticocortical feedback projections as providing an underlying mechanism for higher-level cognitive functions to control depictive mental imagery.)

The general lesson learned from the human visual modality is that modalities are not microtheories, that modalities are not flat representations of the pixel level, and that modalities are functionally characterized by successive layers of successively more elaborate feature structure. Modalities are one of the best exhibitions of this evolutionary design pattern – ascending layers of adaptive complexity – which also appears, albeit in very different form, in the ascending code-modality-concept-thought-deliberation model of the human mind. Each ascending layer is more elaborate, more complex, more flexible,

---

[16]Deliberative General Intelligence, the theory of mind presented in this chapter.

and more computationally expensive. Each layer requires the complexity of the layer underneath – both functionally within a single organism, and evolutionarily within a genetic population.

The concept layer is evolvable in a series of short steps if, and only if, there already exists substantial complexity within the modality layer. The same design pattern – ascending layers of adaptive complexity – also appears *within* an evolved sensory modality. The first features detected are simple, and can evolve in a single step or a small series of adaptive short steps. The ability to detect these first features can be adaptive even in the absence of a complete sensory modality. The eye, which is currently believed to have independently evolved in many different species, may have begun, each time, as a single light-sensitive spot on the organism's skin.

In modalities, each additional layer of feature detectors makes use of the information provided by the first layer of feature detectors. In the absence of the first layer of feature detectors, the "code" for the second layer of feature detectors would be too complex to evolve in one chunk. With the first layer of feature detectors already present, feature detectors in the second layer can evolve in a single step, or in a short series of locally adaptive steps. The successive layers of organization in a sensory modality are a beautiful illustration of evolution's design signature, the functional ontogeny of the information recapitulating the evolutionary phylogeny.

Evolution is a good teacher but a poor role model; is this design a bug or a feature? I would argue that it is generally a feature. There is a deep correspondence between *evolutionarily* smooth fitness landscapes and *computationally* smooth fitness landscapes. There is a deep correspondence between each successive layer of feature detectors being evolvable, and each successive layer of feature detectors being computable in a way that is "smooth" rather than "fragile," as described in the earlier discussion of the code layer. Smooth computations are more evolvable, so evolution, in constructing a system incrementally, tends to construct linear sequences or ascending layers of smooth operations.

An AI designer may conceivably discard the requirement that each ascending layer of feature detection be incrementally useful/adaptive – although this may make the subsystem harder to incrementally develop and test! It is cognitively important, however, that successive layers of feature detectors be computationally "smooth" in one specific sense. DGI concepts interact with inverse feature detectors, "feature controllers," in order to construct mental imagery. For the task of *imposing a concept* and the still more difficult task of *abstracting a concept* to be *simultaneously* tractable, it is necessary that sensory modalities be a continuum of locally smooth layers, rather than consisting of enormous, intractable, opaque chunks. There is a deep correspondence between the smooth design that renders concepts tractable and the smooth architecture emergent from incremental evolution.

The feature controllers used to create mental imagery are evolvable and preadaptive in the absence of mental imagery; feature controllers could begin as top-down constraints in perceptual processing, or even more simply as a perceptual step which happens to be best computed by a recurrent network. In both cases, the easiest (most evolvable) architecture is generally one in which the feedback connection reciprocates the feedforward connection. Thus, the feature controller layers are not a separate system independent from the feature detector layers; rather, I expect that what is locally a feature detector is also locally a feature controller. Again, this smooth reversibility helps render it possible to learn a single concept which can act as a category detector *or* a category imposer. It is the *simultaneous* solution of *concept imposition, concept satisfaction, concept faceting,* and *concept abstraction* that requires reversible features – feature controllers which are the local inverses of the feature detectors. I doubt that feature controllers reach all the way down to the first layers of the retina (I have not heard of any feedback connections reaching this far), but direct evidence from neuroimaging shows that mental imagery activates primary visual cortex [54]; I am not sure whether analogous tests have been performed for the lateral geniculate nucleus, but the feedback connections are there.

The Human Design of Modalities in AI

An AI needs sensory modalities – but which modalities? How do those modalities contribute materially to general intelligence outside the immediate modality?

Does an AI need a visuospatial system modeled after the grand complexity of the visuospatial system in primates and humans? We know more about the human visual modality than about any other aspect of human neurology, but that doesn't mean we know enough to build a visual modality from scratch. Furthermore, the human visual modality is enormously complex, computationally intensive, and fitted to an environment which an AI does not necessarily have an immediate need to comprehend. Should humanlike 3D vision[17] be one of the *first* modalities attempted?

I believe it will prove best to discard the human modalities or to use them as inspiration only – to use a completely different set of sensory modalities during the AI's early stages. An AI occupies a different environment than a human and direct imitation of human modalities would not be appropriate. For an AI's initial learning experiences, I would advocate placing the AI in complex virtual environments, where the virtual environments are *internal to the computer* but *external to the AI.* The programmers would then attempt to develop sensory modalities corresponding to the virtual environments. Henceforth I may use the term "microenvironment" to indicate a complex virtual environment. The term "microworld" is less unwieldy, but should not be taken

---

[17]I say "human-like" and not "primate-like" or "mammal-like" because of the possibility that the human visual modality has further adaptations that support the use of mental imagery in deliberation.

as having the Good Old-Fashioned AI connotation of "microworlds" in which all features are directly represented by predicate logic, e.g., SHRDLU's simplified world of blocks and tables [105].

Abandoning the human modalities appears to introduce an additional fragile dependency on the correctness of the AI theory, in that substituting novel sensory modalities for the human ones would appear to require a correct understanding of the nature of sensory modalities and how they contribute to intelligence. This is true, but I would argue that the existence of an *additional* dependency is illusory. An attempt to blindly imitate the human visual modality, without understanding the role of modalities in intelligence, would be unlikely to contribute to general intelligence except by accident. Our modern understanding of the human visual modality is not so perfect that we could rely on the functional completeness of a neurologically inspired design; for example, a design based only on consensus contemporary theory might omit feature controllers! However, shifting to microworlds does require that experience in the microworlds reproduce functionally relevant aspects of experience in real life, including unpredictability, uncertainty, real-time process control, holonic (part-whole) organization, et cetera. I do not believe that this introduces an *additional* dependency on theoretic understanding, over and above the theoretic understanding that would be required to build an AI that absorbed complexity from these aspects of real-world environments, but it nonetheless represents a strong dependency on theoretic understanding.

Suppose we are designing, *de novo,* a sensory modality and virtual environment. Three possible modalities that come to mind as reasonable for a *very primitive and early-stage AI,* in ascending order of implementational difficulty, would be:

1. A modality for Newtonian billiard balls
2. A modality for a 100x100 "Go" board
3. A modality for some type of interpreted code (a metaphorical "codic cortex")

In human vision, the very first visual neurons are the "rods and cones" which transduce impinging environmental photons to a neural representation as sensory information. For each of the three modalities above, the "rods and cones" level would probably use essentially the same representation as the data structures used to create the microworld, or virtual environment, in which the AI is embodied. This is a major departure from the design of naturally evolved modalities, in which the basic level – the quark level, as far as we know – is many layers removed from the high-level objects that give rise to the indirect information that reaches the senses. Evolved sensory modalities devote most of their complexity to *reconstructing* the world that gives rise to the incoming sensory impressions – to reconstructing the 3D moving objects that give rise to the photons impinging on the rods-and-cones layer of the retina. Of course, choosing vision as an example is arguably a biased selection; sound is not as complex as vision, and smell and taste are not as complex as sound.

Nonetheless, eliminating the uncertainty and intervening layers between the true environment and the organism's sensory data is a major step. It should significantly reduce the challenges of early AI development, but is a dangerous step nonetheless because of its distance from the biological paradigm and its elimination of a significant complexity source.

I recommend eliminating environmental reconstruction as a complexity source in *early* AI development. Visualizing the prospect of deliberately degrading the quality of the AI's environmental information on one end, and elaborating the AI's sensory modality on the other end, I find it likely that the entire operation will cancel out, contributing nothing. An AI that had to learn to reconstruct the environment, in the same way that evolution learned to construct sensory modalities, might produce interesting complexity as a result; but if the same programmer is creating environmental complexity and modality complexity, I would expect the two operations to cancel out. While environmental reconstruction is a nontrivial complexity source within the human brain, I consider the *ratio* between the difficulty of programmer development of the complexity, and the contribution of that complexity to general intelligence, to be relatively small. Adding complexity for environmental reconstruction, by introducing additional layers of complexity in the microworld and deliberately introducing information losses between the topmost layer of the microworld and the AI's sensory receptors, and then attempting to create an AI modality which could reconstruct the original microworld content from the final sensory signal, would require a relatively great investment of effort in return for what I suspect would be a relatively small boost to general intelligence.

Suppose that for each of the three modalities – billiards, Go, code – the "pre-retinal" level consists of true and accurate information about the quark level of the virtual microworld, although perhaps not complete information, and that the essential complexity which renders the model a "sensory modality" rests in the feature structure, the ascending layers of feature detectors and descending layers of feature controllers. Which features, then, are appropriate? And how do they contribute materially to general intelligence?

The usual statement is that the complexity in a sensory modality reflects regularities of the environment, but I wish to offer a slightly different viewpoint. To illustrate this view, I must borrow and severely simplify the punchline of a truly elegant paper, "The Perceptual Organization of Colors" by Roger Shepard [93]. Among other questions, this paper seeks to answer the question of trichromacy: Why are there three kinds of cones in the human retina, and not two, or four? Why is human visual perception organized into a three-dimensional color space? Historically, it was often theorized that trichromancy represented an arbitrary compromise between chromatic resolution and spatial resolution; that is, between the number of colors perceived and the grain size of visual resolution. As it turns out, there is a more fundamental reason why three color channels are needed.

To clarify the question, consider that surfaces possess a potentially infinite number of spectral reflectance distributions. We will focus on spectral reflectance distributions, rather than spectral power distributions, because adaptively relevant objects that emit their own light are environmentally rare. Hence the physically constant property of most objects is the spectral reflectance distribution, which combines with the spectral power distribution of light impinging on the object to give rise to the spectral power distribution received by the human eye. The spectral reflectance distribution is defined over the wavelengths from 400nm to 700nm (the visible range), and since wavelength is a continuum, the spectral reflectance distribution can theoretically require an unlimited number of quantities to specify. Hence, it is not possible to exactly constrain a spectral reflectance distribution using only three quantities, which is the amount of information transduced by human cones.

The human eye is not capable of discriminating among all physically possible reflecting surfaces. However, it is possible that for "natural" surfaces – surfaces of the kind commonly encountered in the ancestral environment – reflectance for each pure frequency does not vary independently of reflectance for all other frequencies. For example, there might exist some set of *basis reflectance functions,* such that the reflectance distributions of almost all natural surfaces could be expressed as a weighted sum of the basis vectors. If so, one possible explanation for the trichromacy of human vision would be that three color channels are just enough to perform adequate discrimination in a "natural" color space of limited dimensionality.

The ability to discriminate between all natural surfaces would be the design recommended by the "environmental regularity" philosophy of sensory modalities. The dimensionality of the internal model would mirror the dimensionality of the environment.

As it turns out, natural surfaces have spectral reflectance distributions that vary along roughly five to seven dimensions [64]. There thus exist natural surfaces that, although appearing to trichromatic viewers as "the same color," nonetheless possess different spectral reflectance distributions.

[93] instead asks how many color channels are needed to ensure that the color we perceive is the *same* color each time the surface is viewed under different lighting conditions. The amount of ambient light can also potentially vary along an unlimited number of dimensions, and the actual light reaching the eye is the product of the spectral power distribution and the spectral reflectance distribution. A reddish object in bluish light may reflect the same number of photons of each wavelength as a bluish object in reddish light. Similarly, a white object in reddish light may reflect mostly red photons, while the same white object in bluish light may reflect mostly blue photons. And yet the human visual system manages to maintain the property of *color constancy;* the same object will appear to be the same color under different lighting conditions.

[46] measured 622 spectral power distributions for natural lighting, under 622 widely varying natural conditions of weather and times of day, and found

that variations in natural lighting reduce to three degrees of freedom. Furthermore, these three degrees of freedom bear a close correspondence to the three dimensions of color opponency that were proposed for the human visual system based on experimental examination [44]. The three degrees of freedom are:

1. The *light-dark* variation, which depends on the total light reaching the object.
2. The *yellow-blue* variation, which depends on whether a surface is illuminated by direct sunlight or is in shade. In shade the surface is illuminated by the Raleigh-scattered blue light of the sky, but is not directly illuminated by the sun. The corresponding yellow extreme occurs when an object is illuminated only by direct sunlight; e.g., if sunlight enters through a small channel and skylight is cut off.
3. The *red-green* variation, which depends on both the elevation of the sun (how much atmosphere the sun travels through), and the amount of atmospheric water vapor. E.g., illumination by a red sunset versus illumination at midday. Red wavelengths are the wavelengths least scattered by dust and most absorbed by water.

The three color channels of the human visual system are precisely the number of channels needed in order to maintain color constancy under natural lighting conditions[18]. Three color channels are not enough to discriminate between all natural surface reflectances, but three color channels are the exact number required to compensate for ambient natural lighting and thereby ensure that the same surface is perceptually the "same color" on any two occasions. This simplifies the adaptively important task of recognizing a previously experienced object on future encounters.

The lesson I would learn out from this morality tale of color constancy is that sensory modalities are about *invariants* and not just *regularities*. Consider the task of designing a sensory modality for some form of interpreted code. (This is a very challenging task because human programming languages tend toward non-smooth fitness landscapes, as previously discussed.) When considering which features to extract, the question I would ask is not "What regularities are found in code?" but rather "What feature structure is needed for the AI to perceive two identical algorithms with slightly different implementations as 'the same piece of code'?" Or more concretely: "What features does this modality need to extract to perceive the recursive algorithm for the Fibonacci sequence and the iterative algorithm for the Fibonacci sequence as 'the same piece of code'?"

Tip your head slightly to the left, then slightly to the right. Every retinal receptor may receive a different signal, but the experienced visual field remains

---

[18]Artificial lighting, which has an "unnatural" spectral power distribution (one that is not the weighted sum of the natural basis vectors), can cause objects to appear as a different color to the human visual system. Hence the manufacture and sale of "natural lighting" or "full spectrum" light sources.

almost exactly the "same." Hold up a chess pawn, and tip it slightly to the left or slightly to the right. Despite the changes in retinal reception, we see the "same" pawn with a slightly different orientation. Could a sensory modality for code look at two sets of interpreted bytecodes (or other program listing), completely different on a byte-by-byte basis, and see these two listings as the "same" algorithm in two slightly different "orientations"?

The modality level of organization, like the code level, has a characteristic kind of work that it performs. Formulating a *butterfly* concept and seeing two butterflies as members of the same category is the work of the concept level, but seeing a chess pawn in two orientations as the same pawn is the work of the modality level. There is overlap between the modality level and the concept level, just as there is overlap between the code level and the modality level. But on the whole, the modality level is about *invariants* rather than *regularities* and *identities* rather than *categories*.

Similarly, the understanding conferred by the modality level should not be confused with the analytic understanding characteristic of thoughts and deliberation. Returning to the example of a codic modality, one possible indication of a serious design error would be constructing a modality that could analyze *any possible* piece of code equally well. The very first layer of the retina – rods and cones – is the *only* part of the human visual system that will work on all possible pixel fields. The rest of the visual system will only work for the low-entropy pixel fields experienced by a low-entropy organism in a low-entropy environment. The very next layer, after rods and cones, already relies on center-surround organization being a useful way to compress visual information; this only holds true in a low-entropy visual environment.

Designing a modality that worked equally well for any possible computer program would probably be an indication that the modality was extracting the wrong kind of information. Thus, one should be wary of an alleged "feature structure" that looks as if it would work equally well for all possible pieces of code. It may be a valid analytical method but it probably belongs on the deliberation level, not the modality level. (Admittedly not *every* local step of a modality *must* be dependent on low-entropy input; some local stages of processing may have the mathematical nature of a lossless transform that works equally well on any possible input. Also, hardware may be better suited than wetware to lossless transforms.)

The human brain is constrained by a characteristic serial speed of 200 sequential steps per second, and by the ubiquitous internal use of the synchronous arrival of associated information, to arrange processing stages that flow smoothly forward. High-level "if-then" or "switch-case" logic is harder to arrive at neurally, and extended complex "if-then" or "switch-case" logic is probably almost impossible unless implemented through branching *parallel* circuitry that remains synchronized. Probably an exceptional condition must be ignored, averaged out, or otherwise handled using the same algorithms that would apply to any other modality content. Can an AI modality use an architecture that applies different algorithms to different pieces of modality content? Can an AI modality handle exceptional conditions through special-case code? I would advise caution, for several reasons. First, major "if-then" branches are characteristic of deliberative processes, and being tempted to use such a branch may indicate a level confusion. Second, making exceptions to the smooth flow of processing will probably complicate the meshing of concepts and modalities. Third, modalities are imperfect but fault-tolerant processes, and the fault tolerance plays a role in smoothing out the fitness landscapes and letting the higher levels of organization be built on top; thus, trying to handle *all* the data by detecting exceptional conditions and correcting them, a standard pattern in human programming, may indicate that the modality is insufficiently fault-tolerant. Fourth, handling all exceptions is characteristic of trying to handle all inputs and not just low-entropy inputs. Hence, on the whole, sensory modalities are characterized by the smooth flow of information through ascending layers of feature detectors. Of course, detecting an exceptional condition *as a feature* may turn out to be entirely appropriate!

Another issue which may arise in artificial sensory modalities is that unsophisticated artificial modalities may turn out to be significantly more expensive, computationally, for the effective intelligence they deliver. Sophisticated evolved modalities conserve computing power in ways that might be very difficult for a human programmer to duplicate. An example would be the use of partial imagery, modeling only the features that are needed for a high-level task [31]; a simplified modality that does not support partial imagery may consume more computing power. Another example would be the human visual system's selective concentration on the center of the visual field – the "foveal architecture," in which areas of the visual field closer to the center are allocated a greater number of neurons. The cortical magnification factor for primates is inverse-linear [99]; the complex logarithm is the only two-dimensional map function that has this property [91], as confirmed experimentally by [92]. A constant-resolution version of the visual cortex, with the maximum human visual resolution across the full human visual field, would require 10,000 times as many cells as our actual cortex [86].

But consider the programmatic problems introduced by the use of a logarithmic map. Depending on where an object lies in the visual field, its internal representation on a retinotopic map will be completely different; no direct comparison of the data structures would show the identity or even hint at the identity. That an off-center object in our visual field can rotate without *perceptually* distorting, as its image distorts wildly within the physical retinotopic map, presents a nontrivial computational problem[19].

Evolution conserves computing power by complicating the algorithm. Evolution, considered as a *design pressure,* exerts a steady equipotential design pressure across all existing complexity; a human programmer wields general intelligence like a scalpel. It is not much harder for evolution to "design" and "debug" a logarithmic visual map because of this steady "design pressure;"

---

[19]For one suggested solution, see [6].

further adaptations can build on top of a logarithmic visual map almost as easily as a constant-resolution map. A human programmer's general intelligence would run into difficulty keeping track of all the simultaneous design complications created by a logarithmic map. It might be possible, but it would be difficult, especially in the context of exploratory research; the logarithmic map transforms simple design problems into complex design problems and hence transforms complex design problems into nightmares.

I would suggest using constant-resolution sensory modalities during the early stages of an AI – as implied above by suggesting a sensory modality modeled around a 100x100 Go board – but the implication is that these early modalities will be lower-resolution, will have a smaller field, and will be less efficient computationally. An opposing theoretic view would be that complex but efficient modalities introduce necessary issues for intelligence. An opposing pragmatic view would be that complex but efficient modalities are easier to accomodate in a mature AI if they have been included in the architecture from the beginning, so as to avoid metaphorical "Y2K" issues (ubiquitous dependencies on a simplifying assumption which is later invalidated).

2.5 The Concept Level

DGI uses the term *concept* to refer to the mental stuffs underlying the words that we combine into sentences; concepts are the combinatorial building blocks of thoughts and mental imagery. These building blocks are learned complexity, rather than innate complexity; they are abstracted from experience. Concept structure is absorbed from recurring regularities in perceived reality.

A concept is *abstracted* from experiences that exist as sensory patterns in one or more modalities. Once abstracted, a concept can be compared to a new sensory experience to determine whether the new experience *satisfies* the concept, or equivalently, whether the concept *describes* a facet of the experience. Concepts can describe both environmental sensory experience and internally generated mental imagery. Concepts can also be *imposed* on current working imagery. In the simplest case, an exemplar associated with the concept can be loaded into the working imagery, but constructing complex mental imagery requires that a concept target a piece of existing mental imagery, which the concept then transforms. Concepts are faceted; they have internal structure and associational structure which comes into play when imposition or description encounters a bump in the road. Faceting can also be invoked purposefully; for example, "tastes like chocolate" versus "looks like chocolate." To solve any one of these problems alone, at a sufficient degree of generality and in a computationally tractable way, would be a serious challenge; to solve all three problems simultaneously constitutes the fundamental challenge of building a system that learns complexity in combinatorial chunks.

A "concept kernel" is the pseudo-sensory pattern produced by abstracting from sensory experience. During concept satisfaction, this kernel interacts with the layered feature detectors to determine whether the reported imagery matches the kernel; during concept imposition, the kernel interacts with the layered feature controllers to produce new imagery or alter existing imagery. A programmer seeking a good representation for concept kernels must find a representation that *simultaneously* fulfills these requirements:

1. The kernel representation can be *satisfied by* and *imposed on* referents in a sensory modality.
2. The kernel representation or concept representation contains the internal structure needed for faceted concept combination, as in "triangular lightbulb" previously given as an example.
3. It is computationally tractable to *abstract* new kernel representations using sensory experience as raw material.

Concepts have other properties besides their complex kernels. Kernels relate concepts to sensory imagery and hence the modality level. Concepts also have complexity that relates to the concept level; i.e., concepts have complexity that derives from their relation to other concepts. In Good Old-Fashioned AI this aspect of concepts has been emphasized at the expense of all others[20], but that is no excuse for ignoring concept-concept relations in a new theory. Concepts are supercategories and subcategories of each other; there are concepts that describe concepts and concepts that describe relations between concepts.

In formal logic, the traditional idea of concepts is that concepts are categories defined by a set of individually necessary and together sufficient requisites; that a category's extensional referent is the set of events or objects that are members of the category; and that the combination of two categories is the sum of their requisites and hence the intersection of their sets of referents. This formulation is inadequate to the complex, messy, overlapping category structure of reality and is incompatible with a wide range of established cognitive effects [57]. Properties such as *usually* necessary and *usually* sufficient requisites, and concept combinations that are *sometimes* the sum of their requisites or the intersection of their extensional classes, are emergent from the underlying representation of concepts – along with other important properties, such as prototype effects in which different category members are assigned different degrees of typicality [88].

Concepts relate to the thought level primarily in that they are the building blocks of thoughts, but there are other level-crossings as well. Introspective concepts can describe beliefs and thoughts and even deliberation; the concept "thought" is an example. Inductive generalizations are often "about" concepts in the sense that they apply to the referents of a concept; for example, "Triangular lightbulbs are red." Deliberation may focus on a concept in order to arrive at conclusions about the extensional category, and introspective deliberation may focus on a concept in its role as a cognitive object. Concept

---

[20]This does not imply that GOFAI handles concept-concept relations *correctly*. The links in a classical "semantic net" are as oversimplified as the nodes.

structure is ubiquitously invoked within perceptual and cognitive processes because category structure is ubiquitous in the low-entropy processes of our low-entropy universe.

The Substance of Concepts

One of the meanings of "abstraction" is "removal;" in chemistry, to *abstract* an atom means subtracting it from a molecular group. Using the term "abstraction" to describe the process of creating concepts could be taken as implying two views: First, that to create a concept is to generalize; second, that to generalize is to lose information. Abstraction as information loss is the classical view of concepts (that is, the view of concepts under GOFAI and formal logic). Forming the concept "red" is taken to consist of focusing only on color, at the expense of other features such as size and shape; all concept usage is held to consist of purposeful information-loss.

The problem with the classical view is that it allows only a limited repertoire of concepts. True, some concepts apparently work out to straightforward information-loss. The task of arriving at a concept kernel for the concept "red" – a kernel capable of interacting with visual imagery to distinguish between red objects and non-red objects – is relatively trivial. Even simultaneously satisfying the abstraction and satisfaction problems for "red" is relatively trivial. Well-known, fully general tools such as neural nets or evolutionary computation would suffice. To learn to solve the satisfaction problem, a neural net need only to learn to fire when the modality-level feature detectors for "color" report a certain color – a point falling within a specific volume of color space – across a broad area, and not to fire otherwise. A piece of code need only evolve to test for the same characteristic. (The neural net would probably train faster for this task.)

A sufficiently sophisticated modality would simplify the task even further, doing most of the work of grouping visual imagery into objects and detecting solid-color or same-hue or mostly-the-same-hue surfaces. The human visual modality goes still farther and precategorizes colors, dividing them up into a complex color space [7], said color space having eleven culturally universal focal volumes [4], said focal volumes having comparatively sharp internal boundaries relative to physically continuous variations in wavelength (see [93], or just look at the *bands* in a rainbow). Distinguishing across innate color boundaries is easy; distinguishing within color boundaries is hard [68]. Thus, the human visual modality provides very strong suggestions as to where the boundaries lie in color space, although the final step of categorization is still required [19].

Given a visual modality, the concept of *red* lies very close to the metaphorical "surface" of the modality. In humans *red* is probably at the surface, a direct output of the modality's feature-detectors. In AIs with less sophisticated visual modalities, "redness" as a category would need to be abstracted as a fuzzy volume within a smooth color space lacking the human boundaries. The *red* concept kernel (in humans and AIs) needs to be more complex than a simple binary test or fuzzy color clustering test, since "redness" as we understand it describes visual areas and not single pixels (although *red* can describe a "visual area" consisting of a small point). Even so, the complexity involved in the redness concept lies almost entirely within the sensory modality, rather than the concept kernel. We might call such concepts *surface concepts*.

Even for surface concepts, simultaneously solving abstraction, satisfaction, and imposition would probably be far more tractable with a special representation for concept kernels, rather than generically trained neural nets or evolutionary programs. Imposition requires a concept kernel which can be selectively applied to imagery within a visual modality, transforming that imagery such that the final result satisfies the concept. In the case of the concept "red," the concept kernel would interact with the feature controllers for color, and the targeted mental imagery would become red. This cannot be done by painting each individual pixel the same shade of red; such a transformation would obliterate edges, surfaces, textures, and many other high-level features that intuitively ought to be preserved. Visualizing a "red lemon" does not cause the mind to picture a bright red patch with the outline of a lemon. The concept kernel does not send separate color commands to the low-level feature controller of each individual visual element; rather the concept kernel imposes *red* in combination with other currently activated features, to depict a *red lemon* that retains the edge, shape, surface curvature, texture, and other visualized features of the starting *lemon* image. Probably this occurs because perceived coloration is a property of surfaces and visual objects rather than, or as well as, individual visual elements, and our redness concept kernel interacts with this high-level feature, which then ripples down in coherent combination with other features.

Abstracting an impose-able concept kernel for "red" is a problem of different scope than abstracting a satisfy-able kernel for "red." There is an immediately obvious way to train a neural net to detect satisfaction of "red," given a training set of known "red" and non-"red" experiences, but there is no equally obvious teaching procedure for the problem of *imposing* "red." The most straightforward success metric is the degree to which the transformed imagery satisfies a neural network already trained to detect "red," but a bright red lemon-shaped patch is likely to be more "red" than a visualized red lemon. How does the kernel arrive at a transformation which makes a coherent change in object coloration, rather than a transformation which paints all visual elements an indiscriminate shade of red, or a transformation which loads a random red object into memory? Any of these transformations would satisfy the "red" concept.

Conceivably, fully general neural nets could be trained to impose *minimal* transformations, although I am not sure that "minimal transformation" is the rule which should govern concept imposition. Regardless of the real tractability of this problem, I strongly doubt that human cognitive systems create concepts by training generic neural nets on satisfaction and imposition.

I suspect that concepts do not have independent procedures for satisfaction and imposition; I also suspect that neither satisfaction nor imposition are the product of reinforcement learning on a fully general procedure. Rather, I suspect that a concept kernel consists of a pattern in a representation related to (but not identical with) the representation of sensory imagery, that this pattern is produced by transforming the experiences from which the concept is abstracted, and that this pattern interacts with the modality to implement both concept satisfaction and concept imposition.

A very simple example of a non-procedural, pattern-based concept kernel would be "clustering on a single feature." *Red* might be abstracted from an experiential base by observing an unusual clustering of point values for the *color* feature. Suppose that the AI is challenged with a virtual game in which the goal is to find the "keys" to a "lock" by selecting objects from a large sample set. When the AI successfully passes five trials by selecting the correct object on the first try, the AI is assumed to have learned the rule. Let us suppose that the game rule is that "red" objects open the lock, and that the AI has already accumulated an experiential base from its past failures and successes on individual trials.

Assuming the use of a three-dimensional color space, the *color* values of the correct keys would represent a tight cluster relative to the distribution among all potential keys. Hence the abstracted concept kernel might take the form of a feature-cluster pair, where the feature is *color* and the cluster is a central point plus some measure of standard deviation. This creates a concept kernel with a prototype and quantitative satisfiability; the concept has a central point and fuzzy but real boundaries. The same concept kernel can also be imposed on a selected piece of mental imagery by loading the central color point into the *color* feature controller – that is, loading the clustered value into the feature controller corresponding to the feature detector clustered upon.

Clustering of this type also has indirect implications for concept-concept relations: The *red* concept's "color volume" might overlap a nearby concept such as *burgundy,* or might turn out to enclose that concept; a modality-level fact which over time might naturally give rise to an association relationship, or a supercategory relationship, on the concept level. This would not humanly occur through direct comparison of the representations of the concept kernels, but through the observation of overlap or inclusion within the categories of extensional referents. A more strongly introspective AI might occasionally benefit from inspecting kernel representations, but this should be an adjunct to experiential detection of category relationships, not a substitute for it.

Clustering on a single feature is definitely not a complete conceptual system. Single-feature clustering cannot notice a correlation between two features where neither feature is clustered alone; single-feature clustering cannot cross-correlate two features in any way at all. Concepts which are limited to clustering on a single feature will always be limited to concepts at the immediate surface of a given sensory modality.

At the same time, a concept system is not a general intelligence and need not be capable of representing *every possible* relation. Suppose a human were challenged with a game in which the "correct key" always had a color that lay on the exact surface of a sphere in color space; could the human concept-formation system *directly* abstract this property? I would guess not; I would guess that, at most, a human might notice that the key tended to belong to a certain group of colors; i.e., might slice up the surface of this color sphere into separate regions, and postulate that solution keys belong to one of several color regions. Thus, even though in this case the underlying "rule" is *computationally* very simple, it is unlikely that a human will create a concept that directly incorporates the rule; it may even be impossible for a human to abstract a kernel that performs this simple computation. A concept-formation system need not be generally intelligent in itself; need not represent all possible perceptual regularities; just enough for the overall mind to work.

I suspect that the system design used by humans, and a good design for AIs, will turn out to be a repertoire of different concept-formation methods. ("Clustering on a single feature" could be one such method, or could be a special case of a more general method.) Concept faceting could then result either from concepts with multiple kernels, so that a concept employs more than one categorization method against its perceptual referents, or from internal structure in a single kernel, or both. If some aspects of perceptual referents are more salient, then kernels which match those aspects are likely to have greater weight within the concept. Faceting within a concept, arising out of multiple unequal kernels or faceting within a single complex kernel, seems like the most probable source of prototype effects within a category.

Stages in Concept Processes

Concept formation is a multi-stage process. For an AI to form a new concept, the AI must have the relevant experiences, perceptually group the experiences, notice possible underlying similarities within members of a group (this may be the same perceived similarity that led to the original experiential grouping), verify the generalization, initiate the new concept as distinguished cognitive content, create the concept kernel(s) by abstraction from the experiential base, and integrate the new concept into the system. (This checklist is intended as an interim approximation; actual mind designs may differ, but presumably a temporal sequence will still be involved.)

In the example given earlier, an AI abstracts *redness* starting with a bottom-up, experience-driven event: noticing the possible clustering of the *color* feature within the pre-existing category *keys*. Conceivably the act of checking for color clustering could have been suggested top-down, for example by some heuristic belief, but in this example we will assume the seminal perception of similar coloration was an unexpected, bottom-up event; the product of continuous and automatic checks for clustering on a single feature across all high-level features in currently salient experiential categories. Rather than being part of an existing train of thought, the detection of clustering creates an "Aha!" event, a new cognitive event with high salience that becomes the focus of attention, temporarily shunting aside the previous train of thought. (See the discussion of the thought level.)

If the scan for clustering and other categorizable similarities is a continuous background task, it may imply a major expenditure of computational resources – perhaps a major percentage of the computing power used by the AI. This is probably the price of having a cognitive process that can be driven by bottom-up interrupts as well as top-down sequences, and the price of having a cognitive process that can occasionally notice the unexpected. Hence, the efficiency, optimization, and scalability of algorithms for such continuous background tasks may play a major role in determining the AI's performance. If imagery stays in place long enough, I would speculate that it may be possible to farm out the task of *noticing* a possible clustering to distant parts of a distributed network, while keeping the task of *verifying* the clustering, and all subsequent cognitive actions, within the local process. Most of the computing power is required to find the hint, not to verify the match, and a false hint does no damage (assuming the false hints are not malicious attacks from untrusted nodes).

Once the suspicion of similarity is triggered by a cue picked up by a continuous background process, and the actual degree of similarity is verified, the AI would be able to create the concept as cognitive content. Within the above example, the process that notices the possible clustering is essentially the same process that would verify the clustering and compute the degree of clustering, center of clustering, and variance within the cluster. Thus, clustering on a single feature may compress into a single stage the cueing, description, and abstraction of the underlying similarity. Given the expense of a continuous background process, however, I suspect it will usually be best to separate out a less expensive cueing mechanism as the background process, and use this cueing mechanism to suggest more detailed and expensive scans. (Note that this is a "parallel terraced scan;" see [84] and [38].)

After the creation of the concept and the concept kernel(s), it would then be possible for the AI to notice concept-concept relations, such as super-category and subcategory relations. I do not believe that concept-concept relations are computed by directly comparing kernel representations; I think that concept-concept relations are learned by generalizing across the concept's usage. It may be a good heuristic to look for concept-concept relations immediately after forming a new concept, but that would be a separate track within deliberation, not an automatic part of concept formation.

After a concept has been formed, the new concept must be integrated into the system. For us to concede that a concept has really been "integrated into the system" and is now contributing to intelligence, the concept must be used. Scanning across the stored base of concepts in order to find which concepts are satisfied by current mental imagery promises to be an even more computationally expensive process than continuous background checks for clustering.

An individual satisfaction check is probably less computationally intensive than carrying out a concept imposition – but satisfaction checks seem likely to be a continuous background operation, at least in humans.

As discussed earlier, humans and AIs have different computational substrates: Humans are slow but hugely parallel; AIs are fast, but resource-poor. If humans turn out routinely parallelize against all learned concepts, an AI may simply be unable to afford it. The AI optimum may involve comparing working imagery against a smaller subset of learned complexity – only a few concepts, beliefs, or memories would be scanned against working imagery at any given point. Alternatively, an AI may be able to use terraced scanning[21], fuzzy hashing[22], or branched sorting[23] to render the problem tractable. One hopeful sign is the phenomenon of cognitive priming on related concepts [69], which suggests that humans, despite their parallelism, are not using pure brute force. Regardless, I conjecture that matching imagery against large concept sets will be one of the most computationally intensive subprocesses in AI, perhaps *the* most expensive subprocess. Concept matching is hence another good candidate for distribution under "notice distantly, verify locally;" note also that the concept base could be sliced up among distributed processors, although this might prevent matching algorithms from exploiting regularities within the concept base and matching process.

Complex Concepts and the Structure of "Five"

Under the classical philosophy of category abstraction, abstraction consists solely of selective focus on information which is already known; focusing on the "color" or "redness" of an object as opposed to its shape, position, or velocity. In DGI's "concept kernels," the internal representation of a concept has complexity extending beyond information loss – even for the case of "redness" and other concepts which lie almost directly on the surface of a sensory modality. The only concept that is pure information-loss is a concept that lies *entirely* on the surface of a modality; a concept whose satisfaction exactly equals the satisfaction of some single feature detector.

The concept for "red," described earlier, is actually a fuzzy percept for degrees of redness. Given that the AI has a flat color space, rather than a human color space with innate focal volumes and color boundaries, the

---

[21] The use of computationally inexpensive cues to determine when more expensive checks should be performed.

[22] An algorithm which reduces complex representations to a form that can be more easily compared or scanned.

[23] Rather than comparing against each potential match in turn, an algorithm would be used which eliminates half the potential matches by asking a question, then eliminates half the remaining potential matches by asking a new question pre-optimized against that set, and so on until the remaining potential matches are computationally tractable. Branched sorting of this kind could conceivably be implemented by spatial properties of a parallel neural network as well.

"redness" percept would contain at least as much additional complexity – over and above the modality-level complexity – as is used to describe the clustering. For example, "clustering on a single feature" might take the form of describing a Gaussian distribution around a central point. The specific use of a Gaussian distribution does not contribute to useful intelligence unless the environment also exhibits Gaussian clustering, but a Gaussian distribution is probably useful for allowing an AI to notice a wide class of clusterings around a central point, even clusterings that do not actually follow a Gaussian distribution.

Even in the absence of an immediate environmental regularity, a concept can contribute to effective intelligence by enabling the perception of more complex regularities. For example, an alternating sequence of "red" and "green" key objects may fail the modality-level tests for clustering because no Gaussian cluster contains (almost) all successes and excludes (almost) all failures. However, if the AI has already previously developed concepts for "red" and "green," the alternating repetition of the satisfaction of the "red" and "green" concepts is potentially detectable by higher-level repetition detectors. Slicing up the color space with surface-level concepts renders computationally tractable the detection of higher-order alternation. Even the formation of simple concepts – concepts lying on the surface of a modality – expands the perceptual capabilities of the AI and the range of problems the AI can solve.

Concepts can also embody regularities which are not directly represented in any sensory modality, and which are not any covariance or clustering of feature detectors already in a sensory modality.

Melanie Mitchell and Douglas Hofstadter's "Copycat" program works in the domain of letter-strings, such as "abc", "xyz", "onml", "ddd", "cwj", etc. The function of Copycat is to complete analogy problems such as "abc:abd::ace:?" [37]. Since Copycat is a model of perceptual analogy-making, rather than a model of category formation, Copycat has a limited store of pre-programmed concepts and does not learn further concepts through experience. (This should *not* be taken as criticism of the Copycat project; the researchers explicitly noted that concept formation was not being studied.)

Suppose that a general AI (not Copycat), working in the toy domain of letter strings, encounters a problem that can only be solved by discovering what makes the letter-strings "hcfrb", "yhumd", "exbvb", and "gxqrc" similar to each other but dissimilar to the strings "ndaxfw", "qiqa", "r", "rvm", and "zinw". Copycat has the built-in ability to count the letters in a string or group; in DGI's terms Copycat might be said to extract *number* as a modality-level feature. There is extensive evidence that humans also have brainware support for subitizing (directly perceiving) small numbers, and brainware support for perceiving the approximate quantities of large numbers (see [20] for a review). Suppose, however, that a general AI does *not* possess a modality-level counting ability. How would the AI go about forming the category of "five," or even "groups-of-five-letters"?

This challenge points up the inherent deficit of the "information loss" viewpoint of abstraction. For an AI with no subitization support – or for a human challenged with a number like "nine," which is out-of-range for human subitization – the distinguishing feature, cardinality, is not represented by the modality (or in humans, represented only approximately). For both humans and AIs, the ability to form concepts for non-subitizable exact numbers requires more than the ability to selectively focus on the facet of "number" rather than the facet of "location" or "letter" (or "color," "shape," or "pitch"). The fundamental challenge is not *focusing* on the numerical facet but rather *perceiving* a "numerical facet" in the first place. For the purposes of this discussion, we are not speaking of the ability to understand numbers, arithmetic, or mathematics, only an AI's ability to form the category "five." Possession of the category "five" does not even imply the possession of the categories "four" or "six," much less the formulation of the abstract supercategory "number."

Similarly, the "discovery" of fiveness is not being alleged as mathematically significant. In mathematical terms almost any set of cognitive building blocks will suffice to discover numbers; numbers are fundamental and can be constructed through a wide variety of different surface procedures. The significant accomplishment is not "squeezing" numbers out of a system so sparse that it apparently lacks the usual precursors of number. Rather, the challenge is to give an account of the discovery of "fiveness" in a way that generalizes to the discovery of other complex concepts as well. The hypothesized building blocks of the concept should be general (useful in building other, non-numerical concepts), and the hypothesized relations between building blocks should be general. It is acceptable for the discovery of "fiveness" to be straightforward, but the discovery method must be general.

A working but primitive procedure for satisfying the "five" concept, *after* the discovery of fiveness, might look something like this: Focus on a target group (the group which may or may not satisfy "five"). Retrieve from memory an exemplar for "five" (that is, some specific past experience that has become an exemplar for the "five" concept). Picture the "five" exemplar in a separate mental workspace. Draw a correspondence from an object within the group that is the five exemplar to an object within the group that is the target. Repeat this procedure until there are no objects remaining in the exemplar imagery or there are no objects remaining in the target imagery. Do not draw a correspondence from one object to another if a correspondence already exists. If, when this procedure completes, there are no dangling objects in the exemplar or in the target group, label the target group as satisfying the "five" concept.

In this example, the "five" property translates to the property: "I can construct a complete mapping, with no dangling elements, using unique correspondences, between this target group of objects, and a certain group of objects whose mental image I retrieved from memory."

This is mathematically straightforward, but cognitively general. In support of the proposition that "correspondence," "unique correspondence," and "complete mapping with no dangling elements" are all general conceptual primitives, rather than constructs useful solely for discovering numbers, please note that Copycat incorporates correspondences, unique correspondences, and a perceptual drive toward complete mappings [71]. Copycat has a direct procedural implementation of number sense and does not use these mapping constructs to build numerical concepts. The mapping constructs I have invoked for number are *independently* necessary for Copycat's theory of analogy-making as perception.

Once the procedure ends by labeling imagery with the "five" concept, that imagery becomes an experiential instance of the "five" concept. If the examples associated with a procedurally defined concept have any universal features or frequent features that are perceptually noticeable, the concept can acquire kernels after the fact, although the kernel may express itself as a hint or as an expectation, rather than being a necessary and sufficient condition for concept satisfaction. Concepts with procedural definitions are regular concepts and may possess kernels, exemplars, associated memories, and so on.

What is the benefit of decomposing "fiveness" into a complex procedure, rather than simply writing a codelet, or a modality-level feature detector, which directly counts (subitizes) the members of a group? The fundamental reason for preferring a non-modality solution in this example is to demonstrate that an AI must be capable of solving problems that were not anticipated during design. From this perspective "fiveness" is a bad example to use, since it would be very unlikely for an AI developer to not anticipate numericity during the design phase.

However, a decomposable concept for "five," and a modality-level feature detector which subitizes all numbers up to $(2^{32} - 1)$, can also be compared in terms of how well they support general intelligence. Despite its far greater computational overhead, I would argue that the decomposable concept is superior to a modality-level feature detector.

A billiards modality with a feature detector that subitizes all the billiard balls in a perceptual grouping and outputs a perceptually distinct label – a "numeron detector" – will suffice to solve many immediate problems that require a number sense. However, an AI that uses this feature detector to form a *surface* concept for "five" will not be able to subitize "five" groups of billiards within a supergroup, unless the programmer also had the foresight to extend the subitizing feature detector to count groups as well as specific objects[24]. Similarly, this universal subitizing ability will not extend across multiple modalities, unless the programmer had the foresight to extend the feature detector there as well[25]. Brainware is limited to what the programmer

---

[24]There is some indication that young humans possess a tendency to count discrete physical objects and that this indeed interferes with the ability of human children to count groups of groups or count abstract properties [95].

[25]In animals, experiments with cross-modality numeracy sometimes exhibit surprisingly positive results. For example, rats trained to press lever A on hearing two tones *or* seeing two flashes, and to press lever B on hearing four tones *or* seeing four

was thinking about at the time. Does an AI understand "fiveness" when it becomes able to count five apples? Or when the AI can also count five events in two different modalities? Or when the AI can count five of its own thoughts? It is programmatically trivial to extend the feature detector to handle any of these as a special case, but that is a path which ends in requiring an infinite amount of tinkering to implement routine thought processes (i.e., non-decomposability causes a "commonsense problem").

The *most* important reason for decomposability is that concepts with organized internal structures are more mutable. A human-programmed numeron detector, mutated on the code level, would probably simply break. A concept with internal structure or procedural structure, created by the AI's own thought processes in response to experience, is mutable by the AI's thought processes in response to further experience. For example, Douglas Lenat attests (see [60] and [61]) that the most difficult part of building EURISKO[26] was inventing a decomposable representation for heuristics, so that the class of transformations accessible to EURISKO would occasionally result in improvements rather than broken code fragments and LISP errors. To describe this as *smooth fitness landscapes* is probably stretching the metaphor too much, but "smoothing" in some form is definitely involved. Raw code has only a single level of organization, and changing a random instruction on this level usually simply breaks the overall function. A EURISKO heuristic was broken up into chunks, and could be manipulated (by EURISKO's heuristics) on the chunk level.

Local shifts in the chunks of the "five"-ness procedure yield many useful offspring. By selectively relaxing the requirement of "no dangling objects" in the target image, we get the concept "less than or equal to five"-ness. By relaxing the requirement of "no dangling objects" in the exemplar image, we get the concept "greater than or equal to five"-ness. By requiring one or more dangling objects in the target image, we get the concept "more than five"-ness. By comparing two target images, instead of an exemplar and an image, we get the concept "one-to-one correspondence between group members" (what we would call "same-number-as" under a different procedure), and from there "less than" or "less than or equal to," and so on.

One of these concepts, the one-to-one correspondence between two mental images, is not just a useful offspring of the "fiveness" concept, but a *simpler* offspring. Thus it is probably not an "offspring" at all, but a *prerequisite* concept that suggests a real-world path to the apprehension of fiveness. Many

---

flashes, spontaneously press lever B on hearing two tones *and* seeing two flashes [12]. This may indicate that rats categorize on (approximate) quantities by categorizing on an internal accumulator which is cross-modality. Evolution, however, tends to write much smoother code than human programmers; I am speaking now of the likely consequence of a "naive" AI programmer setting out to create a numeron-detector feature.

[26]EURISKO was a self-modifying AI that used heuristics to modify heuristics, including modification of the heuristics modifying the heuristics.

physical tasks in our world require equal numbers (corresponding sets) for some group; four pegs for four holes, two shoes for two feet.

Experiential Pathways to Complex Concepts

Consider the real-world task of placing four pegs in four holes. A peg cannot fill two holes; two pegs will not fit in one hole. Solid objects cannot occupy the same location, cannot appear in multiple locations simultaneously, and do not appear or disappear spontaneously. These rules of the physical environment are reflected in the default behaviors of our own visuospatial modality; even early infants represent objects as continuous and will look longer at scenes which imply continuity violations [97].

From real-world problems such as pegs and holes, or their microworld analogues, an AI can develop concepts such as *unique correspondence:* a peg cannot fill multiple holes, multiple pegs will not fit in one hole. The AI can learn rules for drawing a *unique correspondence*, and test the rules against experience, before encountering the need to form the more complex concept for "fiveness." The presence of an immediate, local test of utility means that observed failures and successes can contribute unambiguously to forming a concept that is "simple" relative to the already-trained base of concepts. If a new concept contains many new untested parts, and a mistake occurs, then it may be unclear to the AI which local error caused the global failure. If the AI tries to chunk "fiveness" all in a single step, and the current procedure for "fiveness" satisfaction fails – is positively satisfied by a non-five-group, or unsatisfied by a five-group – it may be unclear to the AI that the global failure resulted from the local error of a nonunique correspondence.

The full path to fiveness would probably involve:

1. Learning *physical continuity;* acquiring expectations in which objects do not spontaneously disappear or reappear. In humans, this viewpoint is likely very strongly supported by modality-level visuospatial intuitions in which continuity is the default, and the same should hold true of AIs.

2. Learning *unique correspondence.* Unique correspondence, as a mental skill, tends to be reinforced by any goal-oriented challenge in which a useful object cannot be in two places at once.

3. Learning *complete mapping.* Completeness, along with symmetry, is one of the chief cognitive pressures implemented by Copycat in its model of analogy-making as a *perceptual* operation [71]. A drive toward completeness implies that dangling, unmapped objects detract from the perceived "goodness" of a perceptual mapping. Thus, there may be modality-level support for noticing dangling, unmapped objects within an image.

4. With these three underlying concepts present, it is possible to abstract the concept of *complete mapping using the unique-correspondence relation,* also known as *one-to-one mapping.* We, using an entirely different procedure, would call this relation *same-number-as* ("identity of numeron produced by counting").

5. With *one-to-one mapping*, it is possible for an AI to notice that all the answers on a challenge task are related to a common prototype by the *one-to-one mapping* relation. The AI could then abstract the "five" concept using the prototype as the exemplar and the relation as a test.

6. Where do we go from here? Carl Feynman (personal communication) observes at this point that the *one-to-one mapping* relation is commutative and transitive, and therefore defines a set of equivalence classes; these equivalence classes turn out to be the natural numbers. At first, using "equivalence class detection" as a cognitive method sounded like cheating, but on reflection it's hard to see why a general intelligence should not notice when objects with a common relation to a prototype are similarly related to each other. "Equivalence class" may be a mathematical concept that happens to roughly (or even exactly) correspond to a perceptual property.

7. Forming the superclass concept of *number* is not dealt with in this paper, due to space constraints.

A deliberative intelligence must build up complex concepts from simple concepts, in the same way that evolution builds high-level feature detectors above low-level feature detectors, or builds organs using tissues, or builds thoughts over concepts or modalities. There are holonic[27] ecologies within the learned complexity of concepts, in the same way and for roughly the same reason that there is genetically specified holonic structure in modality-level feature detection. Categories describe regularities in perception, and in doing so, become part of the perceptual structure in which further regularities are detected.

If the programmer hardwires a subitizer that outputs numerons (unique number tags) as detected features, the AI may be able to chunk "five" very rapidly, but the resulting concept will suffer from *opacity* and *isolation*. The concept will not have the lower levels of organization that would enable the AI's native cognitive abilities to disassemble and reassemble the concept in useful new shapes; the inability of the AI to decompose the concept is *opacity*. The concept will not have a surrounding ecology of similar concepts and prerequisite concepts, such as would result from natural knowledge acquisition by the AI. Cognitive processes that require well-populated concept ecologies will be unable to operate; an AI that has "triangle" but not "pyramid" is less likely to successfully visualize "triangular lightbulb." This is *isolation*.

Microtasks

In the DGI model of AI development, concepts are abstracted from an experiential base; experiences are cognitive content within sensory modalities;

---

[27] As described earlier, "holonic" describes the simultaneous application of reductionism and holism, in which a single quality is simultaneously a combination of parts and a part of a greater whole.

and sensory modalities are targeted on a complex virtual microenvironment. Having experiences from which a concept can be abstracted is a (necessary, but not sufficient) requirement for learning the concept. How does an AI obtain these experiences? It would be possible to teach the AI about "fiveness" simply by presenting the AI with a series of sensory images (programmatically manipulating the AI's microenvironment) and prompting the AI's perceptual processes to generalize them, but this severs the task of concept formation from its ecological validity (metaphorically speaking). Knowledge goals (discussed in later sections) are not arbitrary; they derive from real-world goals or higher-level knowledge goals. Knowledge goals exist in a holonic goal ecology; the goal ecology shapes our knowledge goals and thereby often shapes the knowledge itself.

A first approximation to ecological validity is presenting the AI with a "challenge" in one of the virtual microenvironments previously advocated – for example, the billiards microenvironment. Henceforth, I will shorten "microenvironmental challenge" to "microtask." Microtasks can tutor concepts by presenting the AI with a challenge that must be solved using the concept the programmer wishes to tutor. For scrupulous ecological validity the key concept should be part of a larger problem, but even playing "one of these things is not like the others" would still be better than manipulating the AI's perceptual processes directly.

Tutoring a concept as the key to a microtask ensures that the concept's basic "shape," and associated experiences, are those required to solve problems, and that the AI has an experience of the concept being necessary, the experience of discovering the concept, and the experience of using the concept successfully. Effective intelligence is produced not by *having* concepts but by *using* concepts; one learns to use concepts by using them. The AI needs to possess the experiences of discovering and using the concept, just as the AI needs to possess the actual experiential referents that the concept generalizes; the AI needs experience of the contexts in which the concept is useful.

Forming a complex concept requires an incremental path to that complex concept – a series of building-block concepts and precursor concepts so that the final step is a leap of manageable size. Under the microtask developmental model, this would be implemented by a series of microtasks of ascending difficulty and complexity, in order to coax the AI into forming the precursor concepts leading up to the formation of complex concepts and abstract concepts. This is a major expense in programmer effort, but I would argue that it is a necessary expense for the creation of rich concepts with goal-oriented experiential bases.

The experiential path to "fiveness" would culminate with a microtask that could only be solved by abstracting and using the fiveness concept, and would lead up to that challenge through microtasks that could only be solved by abstracting and using concepts such as "object continuity," "unique correspondence," "mapping," "dangling group members," and the penultimate concept of "one-to-one mapping."

With respect to the specific microtask protocol for presenting a "challenge" to the AI, there are many possible strategies. Personally, I visualize a simple microtask protocol (on the level of "one of these things is not like the others") as consisting of a number of "gates," each of which must be "passed" by taking one of a set of possible actions, depending on what the AI believes to be the rule indicating the correct action. Passing ten successive gates *on the first try* is the indicator of success. (For a binary choice, the chance of this happening accidentally is 1024:1. If the AI thinks fast enough that this may happen randomly (which seems rather unlikely), the number of successive gates required can be raised to twenty or higher.) This way, the AI can succeed or fail on individual gates, gathering data about individual examples of the common rule, but will not be able to win through the entire microtask until the common rule is successfully formulated. This requires a microenvironment programmed to provide an infinite (or merely "relatively large") number of variations on the underlying challenge – enough variations to prevent the AI from solving the problem through simple memory.

The sensory appearance of a microtask would vary depending on the modality. For a Newtonian billiards modality, an individual "gate" (subtask) might consist of four "option systems," each option system grouped into an "option" and a "button." Spatial separations in the Newtonian modality would be used to signal grouping; the distance between option systems would be large relative to the distance within option systems, and the distance between an option and a button would be large relative to the distance between subelements of an option. Each option would have a different configuration; the AI would choose one of the four options based on its current hypothesis about the governing rule. For example, the AI might select an option that consists of *four* billiards, or an option with *two large* billiards and *one small* billiard, or an option with *moving* billiards. Having chosen an option, the AI would manipulate a motor effector billiard – the AI's embodiment in that environment – into contact with the button *belonging to* (grouped with) the selected option. The AI would then receive a signal – perhaps a movement on the part of some billiard acting as a "flag" – which symbolized success or failure. The environment would then shift to the next "gate," causing a corresponding shift in the sensory input to the AI's billiards modality.

Since the format of the microtask is complex and requires the AI to *start out* with an understanding of notions like "button" or "the button which belongs to the chosen option," there is an obvious chicken-and-egg problem with teaching the AI the format of the microtask before microtasks can be used to tutor other concepts. For the moment we will assume the bootstrapping of a small concept base, perhaps by "cheating" and using programmer-created cognitive content as *temporary* scaffolding.

Given this challenge format, a simple microtask for "fiveness" seems straightforward: The option containing five billiards, regardless of their size or relative positions or movement patterns, is the key to the gate. In practice, setting up the fiveness microtask may prove more difficult because of the need to eliminate various false ways of arriving at a solution. In particular, if the AI has a sufficiently wide variety of quantitative feature detectors, then the AI will almost certainly possess an emergent Accumulator Model (see [67]) of numeracy. If the AI takes a relatively fixed amount of time to mentally process each object, then single-feature clustering on the subjectively perceived time to mentally process a group could yield the microtask solution without a complex concept of fiveness. Rather than fiveness, the AI would have formed the concept "things-it-takes-about-20-milliseconds-to-understand." The real-world analogue of this situation has already occurred when an experiment formerly thought to show evidence for infant numeracy on small visual sets was demonstrated to show sensitivity to the contour length (perimeter) of the visual set, but not to the cardinality of the visual set [13]. Even with all precursor concepts already present, a complex microtask might be necessary to make fiveness the *simplest* correct answer.

Also, the microtasks for the earlier concepts leading up to fiveness might inherently require greater complexity than the "option set" protocol described above. The concept of unique correspondence derives its behavior from physical properties. Choosing the right option set is a perceptual *decision* task rather than a physical *manipulation* task; in a decision microtask, the only manipulative subtask is maneuvering an effector billiard to touch a selected button. Concepts such as "dangling objects" or "one-to-one mapping" might require manipulation subtasks rather than decision subtasks, in order to incorporate feedback about physical (microenvironmental) outcomes into the concept.

For example, the microtask for teaching "one-to-one mapping" might incorporate the microworlds equivalent of a peg-and-hole problem. The microtask might be to divide up 9 "pegs" among 9 "holes" – where the 9 "holes" are divided into three subgroups of 4, 3, and 2, and the AI must allocate the peg supply among these subgroups in advance. For example, in the first stage of the microtask, the AI might be permitted to move pegs between three "rooms," but not permitted to place pegs in holes. In the second stage of the microtask the AI would attempt to place pegs in holes, and would then succeed or fail depending on whether the initial allocation between rooms was correct. Because of the complexity of this microtask, it might require other microtasks simply to explain the problem format – to teach the AI about pegs and holes and rooms. ("Pegs and holes" are universal and translate easily to a billiards modality; "holes," for example, might be immobile billiards, and "pegs" moveable billiards to be placed in contact with the "holes.")

Placing virtual pegs in virtual holes is admittedly not an inherently impressive result. In this case the AI is being taught to solve a simple problem so that the learned complexity will carry over into solving complex problems. If the learned complexity does carry over, and the AI later goes on to solve more difficult challenges, then, *in retrospect*, getting the AI to think coherently enough to navigate a microtask will "have been" an impressive result.

Interactions on the Concept Level

Concept-concept interactions are more readily accessible to introspection and to experimental techniques, and are relatively well-known in AI and in cognitive psychology. To summarize some of the complexity bound up in concept-concept interactions:

- Concepts are associated with other concepts. Activating a concept can "prime" a nearby concept, where "priming" is usually experimentally measured in terms of decreased reaction times [69]. This suggests that more computational resources should be devoted to scanning for primed concepts, or that primed concepts should be scanned first. (This viewpoint is too mechanomorphic to be considered as an explanation of priming in humans. Preactivation or advance binding of a neural network would be more realistic.)
- Nearby concepts may sometimes "slip" under cognitive pressures; for example, "triangle" to "pyramid." Such slippages play a major role in analogies under the Copycat system [71] Slippages occurring in complex design and planning problems probably incorporate context sensitivity and even goal orientation; see the later discussion of conflict and resonance in mental imagery.
- Concepts, in their role as categories, share territory. An individual sparrow, as an object, is described by the concepts "sparrow" and "bird." All objects that can be described as "sparrow" will also be described by "bird." Thus, information arriving through "bird" will usually, though not always, affect the entire territory of "sparrow." This form of inheritance can take place without an explicit "is-a" rule connecting "sparrow" to "bird;" it is enough that "bird" happens to describe all referents of "sparrow."
- Concepts, in their role as categories, have supercategory and subcategory relationships. Declarative beliefs targeted on concepts can sometimes be inherited through such links. For example, "At least one X is an A" is inherited by the supercategory Y of X: If all referents of X are referents of Y, then "At least one referent of X is an A" implies that "At least one referent of Y is an A." Conversely, rules such as "All X are A" are inherited by subcategories of X but not supercategories of X. Inheritance that occurs on the concept level, through an "is-a" rule, should be distinguished from pseudo-inheritance that occurs through shared territory in specific mental imagery. Mental quantifiers such as "all X are Y" usually translate to "most X are Y" or "X, by default, are Y;" all beliefs are subject to controlled exception. It is possible to reason about category hierarchies deliberatively rather than perceptually, but our speed in doing so suggests a perceptual shortcut.
- Concepts possess transformation relations, which are again illustrated in Copycat. For example, in Copycat, "a" is the "predecessor" of "b", and "1" is the "predecessor" of "2". In a general intelligence these concept-concept relations would refer to, and would be generalized from, observation of

transformational processes acting on experiential referents which causes the same continuous object to move from one category to another. Often categories related by transformational processes are subcategories of the same supercategory.

- Concepts act as verbs, adjectives, and adverbs as well as nouns. In humans, concepts act as one-place, two-place, and three-place predicates, as illustrated by the "subject," "direct object," and "indirect object" in the human parts of speech; "X gives Y to Z." For humans, four-place and higher predicates are probably represented through procedural rules rather than perceptually; spontaneously noticing a four-place predicate could be very computationally expensive. Discovering a predicate relation is assisted by categorizing the predicate's subjects, factoring out the complexity not germane to the predicate.

- Concepts, in their role as symbols with auditory, visual, or gestural tags, play a fundamental role in both human communication and internal human conceptualization. The short, snappy auditory tag "five" can stand in for the complexity bound up in the fiveness concept. Two humans that share a common lexical base can communicate a complex mental image by interpreting the image using concepts, describing the image with a concept structure, translating the concepts within the structure into socially shared auditory tags, transforming the concept structure into a linear sequence using shared syntax, and emitting the auditory tags in that linear sequence. (To translate the previous sentence into English: We communicate with sentences that use words and syntax from a shared language.) The same base of complexity is apparently also used to summarize and compactly manipulate thoughts internally; see the next section.

I also recommend George Lakoff's *Women, Fire, and Dangerous Things: What Categories Reveal about the Mind* [57] for descriptions of many concept-level phenomena.

2.6 The Thought Level

Concepts are *combinatorial* learned complexity. Concepts represent regularities that recur, not in isolation, but in combination and interaction with other such regularities. Regularities are not isolated and independent, but are similar to other regularities, and there are simpler regularities and more complex regularities, forming a metaphorical "ecology" of regularities. This essential fact about the structure of our low-entropy universe is what makes intelligence possible, computationally tractable, evolvable within a genotype, and learnable within a phenotype.

The thought level lies above the learned complexity of the concept level. Thoughts are structures of combinatorial concepts that alter imagery within the workspace of sensory modalities. Thoughts are the disposable one-time structures implementing a non-recurrent mind in a non-recurrent world. Modalities are wired; concepts are learned; thoughts are invented.

Where concepts are building blocks, thoughts are immediate. Sometimes the distance between a concept and a thought is very short; *bird* is a concept, but with little effort it can become a thought that retrieves a bird exemplar as specific mental imagery. Nonetheless, there is still a conceptual difference between a brick and a house that happens to be built from one brick. Concepts, considered as concepts, are building blocks with ready-to-use concept kernels. A thought fills in all the blanks and translates combinatorial concepts into specific mental imagery, even if the thought is built from a single concept. Concepts reside in long-term storage; thoughts affect specific imagery.

The spectra for "learned vs. invented," "combinatorial vs. specific," "stored vs. instantiated," and "recurrent vs. nonrecurrent" are conceptually separate, although deeply interrelated and usually correlated. Some cognitive content straddles the concept and thought levels. "Beliefs" (declarative knowledge) are learned, specific, stored, and recurrent. An episodic memory in storage is learned, specific, stored, and nonrecurrent. Even finer gradations are possible: A retrieved episodic memory is learned, specific, and immediate; the memory may recur as mental content, but its external referent is nonrecurrent. Similarly, a concept which refers to a specific external object is learned, specific, stored, and "semi-recurrent" in the sense that it may apply to more than one sensory image, since the object may be encountered more than once, but still referring to only one object and not a general category.

	Modalities	Concepts	Thoughts
Source	Wired	Learned	Invented
Degrees of freedom	Representing	Combinatorial	Specific
Cognitive immediacy	(not applicable)	Stored	Instantiated
Regularity	Invariant	Recurrent	Nonrecurrent
Amount of complexity	Bounded	Open-ended	Open-ended

Thoughts and Language

The archetypal examples of "thoughts" (invented, specific, instantiated, nonrecurrent) are the sentences mentally "spoken" and mentally "heard" within the human stream of consciousness. We use the same kind of sentences, spoken aloud, to communicate thoughts between humans.

Words are the phonemic tags (speech), visual tags (writing), gestural tags (sign language), or haptic tags (Braille) used to invoke concepts. Henceforth, I will use speech to stand for all language modalities; "auditory tag" or "phonemic tag" should be understood as standing for a tag in any modality.

When roughly the same concept shares roughly the same phonemic tag within a group of humans, words can be used to communicate concepts between humans, and sentences can be used to communicate complex imagery. The phonemes of a word can evoke all the functionality of the real concept associated with the auditory tag. A spoken sentence is a linear sequence of words; the human brain uses grammatical and syntactical rules to assemble the linear sequence into a structure of concepts, complete with internal and external targeting information. "Triangular lightbulb," an adjective followed by a noun, becomes "triangular" targeting "light bulb." "That is a telephone," anaphor-verb-article-noun, becomes a statement about the telephone-ness of a previously referred-to object. "That" is a backreference to a previously invoked mental target, so the accompanying cognitive description ("is a telephone") is imposed on the cognitive imagery representing the referent of "that."

The cognitive process that builds a concept structure from a word sequence combines syntactic constraints and semantic constraints; pure syntax is faster and races ahead of semantics, but semantic disharmonies can break up syntactically produced cognitive structures. Semantic guides to interpretation also reach to the word level, affecting the interpretation of homophones and ambiguous phonemes.

For the moment I will leave open the question of why we hear "mental sentences" *internally* – that is, the reason why the transformation of concept structures into linear word sequences, obviously necessary for spoken communication, also occurs internally within the stream of consciousness. I later attempt to explain this as arising from the coevolution of thoughts and language. For the moment, let it stand that the combinatorial structure of words and sentences in our internal narrative reflects the combinatorial structure of concepts and thoughts.

Mental Imagery

The complexity of the thought level of organization arises from the cyclic interaction of thoughts and mental imagery. Thoughts modify mental imagery, and in turn, mental imagery gives rise to thoughts.

Mental imagery exists within the representational workspace of sensory modalities. Sensory imagery arises from environmental information (whether the environment is "real" or "virtual"); imaginative imagery arises from the manipulation of modality workspace through concept imposition and memory retrieval.

Mental imagery, whether sensory or imaginative, exhibits holonic organization: from the "pixel" level into objects and chunks; from objects and chunks into groups and superobjects; from groups and superobjects into mental scenes. In human vision, examples of specific principles governing grouping are proximity, similarity of color, similarity of size, common fate, and closure [104]; continuation [73]; common region and connectedness [78]; and collinearity [59]. Some of the paradigms that have been proposed for resolving the positive inputs from grouping principles, and the negative inputs from detected conflicts, into a consistent global organization, include: Holonic conflict resolution (described earlier), computational temperature [71], Pr'agnanz [52], Hopfield networks [41], the likelihood principle [32]; [63], minimum description length [34], and constraint propagation [55]

Mental imagery provides a workspace for specific perceptions of concepts and concept structures. A chunk of sensory imagery may be mentally labeled with the concept structure "yellow box," and that description will remain bound to the object – a part of the perception of the object – even beyond the scope of the immediate thought. Learned categories and learned expectations also affect the gestalt organization of mental imagery [110].

Mental imagery is the active canvas on which deliberative thought is painted – "active canvas" implying a dynamic process and not just a static representation. The gestalt of mental imagery is the product of many local relations between elements. Because automatic cognitive processes maintain the gestalt, a local change in imagery can have consequences for connected elements in working imagery, without those changes needing to be specified within the proximate thought that caused the modification. The gestalt coherence of imagery also provides feedback on which possible changes will cohere well, and is therefore one of the verifying factors affecting which potential thoughts rise to the status of actuality (see below).

Imagery supports *abstract* percepts. It is possible for a human to reason about an object which is known to cost $1000, but for which no other mental information is available. Abstract reasoning about this object requires a means of representing mental objects that occupy no *a priori* modality; however, this does not mean that abstract reasoning operates independently of all modalities. Abstract reasoning might operate through a modality-level "object tracker" which can operate independently of the modalities it tracks; or by borrowing an existing modality using metaphor (see below); or the first option could be used routinely, and the second option when necessary. Given an abstract "object which costs $1000," it is then possible to attach concept structures that describe the object without having any specific sensory imagery to describe. If I impose the concept "red" on the existing abstract imagery for "an object which costs $1000," to yield "a red object which costs $1000," the "red" concept hangs there, ready to activate when it can, but not yielding specific visual imagery as yet.

Similarly, knowledge generalized from experience with concept-concept relations can be used to detect abstract conflicts. If I know that all penguins are green, I can deduce that "a red object which costs $1000" is not a penguin. It is possible to detect the conflict between "red" and "green" by a concept-level comparison of the two abstract descriptions, even in the absence of visualized mental imagery. However, this does *not* mean that it is possible for AI development to implement only "abstract reasoning" and leave out the sensory modalities. First, a real mind uses the rich concept-level complexity acquired from sensory experience, and from experience with reasoning that uses fully visualized imaginative imagery, to support abstract reasoning; we know that "red" conflicts with "green" because of prior sensory experience with *red* and *green*. Second, merely because some steps in reasoning appear as if they could theoretically be carried out purely on the concept level does not mean that a complete deliberative process can be carried out purely on the concept level.

Third, abstract reasoning often employs metaphor to contribute modality behaviors to an abstract reasoning process.

The idea of "pure" abstract reasoning has historically given rise to AI pathologies and should be considered harmful. With that caution in mind, it is nonetheless possible that human minds visualize concepts only to the extent required by the current train of thought, thus conserving mental resources. An early-stage AI is likely to be less adept at this trick, meaning that early AIs may need to use full visualizations where a human could use abstract reasoning.

Abstract reasoning is a means by which inductively acquired generalizations can be used in deductive reasoning. If empirical induction from an experiential base in which all observed penguins are green leads to the formation of the belief "penguins are green," then this belief may apply abstractly to "a red object which costs $1000" to conclude that this object is probably not a penguin. In this example, an abstract belief is combined with abstract imagery about a specific object to lead to a further abstract conclusion about that specific object. Humans go beyond this, employing the very powerful technique of "deductive reasoning." We use abstract beliefs to reason about abstract mental imagery that describes classes and not just specific objects, and arrive at conclusions which then become new abstract beliefs; we can use deductive reasoning, as well as inductive reasoning, to acquire new beliefs. "Pure" deductive reasoning, like "pure" abstract reasoning, should be considered harmful; deductive reasoning is usually grounded in our ability to visualize specific test cases and by the intersection of inductive confirmation with the deductive conclusions.

Imagery supports tracking of reliances, a cognitive function which is conceptually separate from the perception of event causation. Another way of thinking about this is that perceived *cognitive* causation should not be confused with perceived causation in real-world referents. I may believe that the sun will rise soon; the cause of this belief may be that I heard a rooster crow; I may know that my confidence in sunrise's nearness relies on my confidence in the rooster's accuracy; but I do not believe that the rooster crowing causes the sun to rise.

Imagery supports complex percepts for "confidence" by tracking reliances on uncertainty sources. Given an assertion A with 50% confidence that "object X is blue," and a belief B with 50% confidence that "blue objects are large," the classical deduction would be the assertion "object X is large" with 25% confidence. However, this simple arithmetical method omits the possibility, important even under classical logic, that A and B are both mutually dependent on a third uncertainty C – in which case the combined confidence can be greater than 25%. For example, in the case where "object X is blue" and "blue objects are large" are both straightforward deductions from a third assertion C with 50% confidence, and neither A nor B have any inherent uncertainty of their own, then "object X is large" is also a straightforward deduction from C, and has confidence 50% rather than 25%.

Confidence should not be thought of as a single quantitative probability; *confidence* is a percept that sums up a network of reliances on uncertainty sources. Straightforward links – that is, links whose local uncertainty is so low as to be unsalient – may be eliminated from the perceived reliances of forward deductions: "object X is large" is seen as a deduction assertion C, not a deduction from C plus "object X is blue" plus "blue objects are large." If, however, the assertion "object X is blue" is contradicted by independent evidence supporting the inconsistent assertion "object X is red," then the reliance on "object X is blue" is an independent source of uncertainty, over and above the derived reliance on C. That is, the confidence of an assertion may be evaluated by weighing it against the support for the negation of the assertion [101]. Although the global structure of reliances is that of a network, the local percept of confidence is more likely derived from a set of reliances on supporting and contradicting assertions whose uncertainty is salient. That the *local* percept of confidence is a set, and not a bag or a directed network, accounts for the elimination of common reliances in further derived propositions and the preservation of the global network structure. In humans, the percept of confidence happens to exhibit a roughly quantitative strength, and this quantity behaves in some ways like the mathematical formalism we call "probability."

Confidence and probability are not identical; for humans, this is both an advantage and a disadvantage. Seeing an assertion relying on four independent assertions of 80% confidence as psychologically different from an assertion relying on a single assertion of 40% confidence may contribute to useful intelligence. On the other hand, the human inability to use an arithmetically precise handling of probabilities may contribute to known cases of non-normative reasoning, such as not taking into account Bayesian priors, overestimating conjunctive probabilities and underestimating disjunctive probabilities, and the other classical errors described in [100]. See however [15] for some cautions against underestimating the ecological validity of human reasoning; an AI might best begin with separate percepts for "humanlike" confidence and "arithmetical" confidence.

Imagery interacts with sensory information about its referent. Expectational imagery is confirmed or violated by the actual event. Abstract imagery created and left hanging binds to the sensory percept of its referent when and if a sensory percept becomes available. Imagery interacts with Bayesian information about its referent: assertions that make predictions about future sensory information are confirmed or disconfirmed when sensory information arrives to satisfy or contradict the prediction. Confirmation or disconfirmation of a belief may backpropagate to act as Bayesian confirmation or disconfirmation on its sources of support. (Normative reasoning in these cases is generally said to be governed by the Bayesian Probability Theorem.) The ability of imagery to bind to its referent is determined by the "matching" ability of the imagery – its ability to distinguish a sensory percept as belonging to itself – which in turn is a property of the way that abstract imagery interacts with incoming sensory imagery on the active canvas of working memory. A classical AI with a symbol for "hamburger" may be able to distinguish correctly spelled keystrokes typing out "hamburger," but lacks the matching ability to bind to hamburgers in any other way, such as visually or olfactorily. In humans, the abstract imagery for "a red object" may not involve a specific red image, but the "red" concept is still bound to the abstract imagery, and the abstract imagery can use the "red" kernel to match a referent in sensory imagery.

Imagery may bind to its referent in different ways. A mental image may be an immediate, environmental sensory experience; it may be a recalled memory; it may be a prediction of future events; it may refer to the world's present or past; it may be a subjunctive or counterfactual scenario. We can fork off a subjunctive scenario from a descriptive scene by thinking "What if?" and extrapolating, and we can fork off a separate subjunctive scenario from the first by thinking "What if?" again. Humans cannot continue the process indefinitely, because we run out of short-term memory to track all the reliances, but we have the native tracking ability. Note that mental imagery does not have an opaque tag selected from the finite set "subjunctive," "counterfactual," and so on. This would constitute code abuse: directly programming, as a special case, that which should result from general behaviors or emerge from a lower level of organization. An assertion within counterfactual imagery is not necessarily marked with the special tag "counterfactual;" rather, "counterfactual" may be the name we give to a set of internally consistent assertions with a common dependency on an assertion that is strongly disconfirmed. Similarly, a prediction is not necessarily an assertion tagged with the opaque marker "prediction;" a prediction is better regarded as an assertion with deductive support whose referent is a future event or other referent for which no sensory information has yet arrived; the prediction imagery then binds to sensory information when it arrives, permitting the detection of confirmation or disconfirmation. The distinction between "prediction," "counterfactual," and "subjunctive scenario" can arise out of more general behaviors for confidence, reliance, and reference.

Mental imagery supports the perception of similarity and other comparative relations, organized into complex mappings, correspondences, and analogies (with Copycat being the best existing example of an AI implementation; see [71]). Mental imagery supports expectations and the detection of violated expectations (where "prediction," above, refers to a product of deliberation, "expectations" are created by concept applications, modality behaviors, or gestalt interactions). Mental imagery supports temporal imagery and the active imagination of temporal processes. Mental imagery supports the description of causal relations between events and between assertions, forming complex causal networks which distinguish between implication and direct causation [80]. Mental imagery supports the binding relation of "metaphor" to allow extended reasoning by analogy, so that, e.g., the visuospatial percept of a forking path can be used to represent and reason about the behavior of if-then-else branches, with conclusions drawn from the metaphor (tentatively) applied to the referent [58]. Imagery supports annotation of arbitrary objects with arbitrary percepts; if I wish to mentally label my watch as "X", then "X" it shall be, and if I also label my headphones and remote control as "X", then "X" will form a new (though arbitrary) category.

This subsection obviously has not been a fully constructive account of mental imagery. Rather this has been a very brief description of some of the major properties needed for mental imagery to support the thought level of organization. I apologize, but to write up a theory of general intelligence in a single chapter, it is often necessary to compress a tremendous amount of complexity into one sentence and a bibliographic reference.

The Origin of Thoughts

Thoughts are the cognitive events that change mental imagery. In turn, thoughts are created by processes that relate to mental imagery, so that deliberation is implemented by the cyclic interaction of thoughts modifying mental imagery which gives rise to further thoughts. This does not mean that the deliberation level is "naturally emergent" from thought. The thought level has specific features allowing thought in paragraphs and not just sentences – "trains of thought" with internal momentum, although not so much momentum that interruption is impossible.

At any one moment, out of the vast space of possible thoughts, a single thought ends up being "spoken" within deliberation. Actually, "one thought at a time" is just the human way of doing things, and a sufficiently advanced AI might multiplex or multithread deliberation, but this doesn't change the basic question: Where do thoughts come from? I suggest that it is best to split our conceptual view of this process into two parts; first, the production of suggested thoughts, and second, the selection of thoughts that appear "useful" or "possibly useful" or "important" or otherwise interesting. In some cases, the process that invents or suggests thoughts may do most of the work, with winnowing relatively unimportant; when you accidentally rest your hand on a hot stove, the resulting bottom-up event immediately hijacks deliberation. In other cases, the selection process may comprise most of the useful intelligence, with a large number of possible thoughts being tested in parallel. In addition to being conceptually useful, distinguishing between *suggestion* and *verification* is useful on a design level if "verifiers" and "suggesters" can take advantage of modular organization. Multiple suggesters can be judged by one verifier and multiple verifiers can summate the goodness of a suggestion. This does not necessarily imply hard-bounded processing stages in which "suggestion" runs, terminates and is strictly followed by "verification," but it implies a common ground in which repertoires of suggestion processes and verification processes interact.

I use the term *sequitur* to refer to a cognitive process which suggests thoughts. "Sequitur" refers, not to the way that two thoughts follow each other – that is the realm of deliberation – but rather to the source from which a *single* thought arises, following from mental imagery. Even before a suggested thought rises to the surface, the suggestion may interact with mental imagery to determine whether the thought is interesting and possibly to influence the thought's final form. I refer to specific interactions as *resonances;* a suggested thought resonates with mental imagery during verification. Both positive resonances and negative resonances (conflicts) can make a thought more interesting, but a thought with no resonances at all is unlikely to be interesting.

An example of a sequitur might be noticing that a piece of mental imagery satisfies a concept; for a human, this would translate to the thought "X is a Y!" In this example, the concept is cued and satisfied by a continuous background process, rather than being suggested by top-down deliberation; thus, noticing that X is a Y comes as a surprise which may shift the current train of thought. How *much* of a surprise – how salient the discovery becomes – will depend on an array of surrounding factors, most of which are probably the same resonances that promoted the candidate suggestion "concept Y matches X" to the real thought "X is a Y!." (The difference between the suggestion and the thought is that the real thought persistently changes current mental imagery by binding the Y concept to X, and shifts the focus of attention.)

What are the factors that determine the resonance of the suggestion "concept Y matches X" or "concept Y may match X" and the salience of the thought "X is a Y"? Some of these factors will be inherent properties of the concept Y, such as Y's past value, the rarity of Y, the complexity of Y, et cetera; in AI, these are already-known methods for ranking the relative value of heuristics and the relative salience of categories. Other factors are inherent in X, such as the degree to which X is the focus of attention.

Trickier factors emerge from the interaction of X (the targeted imagery), Y (the stored concept that potentially matches X), the suggested mental imagery for Y describing X, the surrounding imagery, and the task context. A human programmer examining this design problem naturally sees an unlimited range of potential correlations. To avoid panic, it should be remembered that evolution did not begin by contemplating the entire search space and attempting to constrain it; evolution would have incrementally developed a repertoire of correlations in which adequate thoughts resonated some of the time. Just as concept kernels are not AI-complete, sequiturs and resonances are not AI-complete. Sequiturs and resonances also may not need to be human-equivalent to minimally support deliberation; it is acceptable for an early AI to miss out on many humanly obvious thoughts, so long as those thoughts which are successfully generated sum to fully general deliberation.

Specific sequiturs and resonances often seem reminiscent of general heuristics in Lenat's EURISKO [60] or other AI programs intended to search for interesting concepts and conjectures [14]. The resemblance is further heightened by the idea of adding learned associations to the mix; for example, correlating which concepts Y are frequently useful when dealing with imagery described by concepts X, or correlating concepts found useful against categorizations of the current task domain, bears some resemblance to EURISKO trying to learn specific heuristics about when specific concepts are useful. Similarly, the general sequitur that searches among associated concepts to match them against working imagery bears some resemblance to EURISKO applying a heuristic. Despite the structural resemblance, sequiturs are not heuristics. Sequiturs are general cognitive subprocesses lying on the brainware level of organization. The subprocess is the sequitur that handles thoughts of the general form "X is a Y;" any cognitive content relating to specific Xs and Ys is learned complexity, whether it takes the form of heuristic beliefs or correlative associations. Since our internal narrative is open to introspection, it is not surprising if sequiturs produce some thoughts resembling the application of heuristics; the mental sentences produced by sequiturs are open to introspection, and AI researchers were looking at these mental sentences when heuristics were invented.

Some thoughts that might follow from "X is a Y!" (unexpected concept satisfaction) are: "Why is X a Y?" (searching for explanation); or "Z means X can't be a Y!" (detection of belief violation); or "X is not a Y" (rechecking of a tentative conclusion). Any sequence of two or more thoughts is technically the realm of deliberation, but connected deliberation is supported by properties of the thought level such as focus of attention. The reason that "Why is X a Y?" is likely to follow from "X is a Y!" is that the thought "X is a Y" shifts the focus of attention to the Y-ness of X (the mental imagery for the Y concept binding to X), so that sequitur processes tend to focus selectively on this piece of mental imagery and try to discover thoughts that involve it.

The interplay of thoughts and imagery has further properties that support deliberation. "Why is X a Y?" is a thought that creates, or focuses attention on, a question – a thought magnet that attracts possible answers. *Question imagery* is both like and unlike *goal imagery*. (More about goals later; currently what matters is how the thought level interacts with goals, and the intuitive definition of goals should suffice for that.) A goal in the classic sense might be defined as abstract imagery that "wants to be true," which affects cognition by affecting the AI's decisions and actions; the AI makes decisions and takes actions based on whether the AI predicts those decisions and actions will lead to the goal referent. *Questions* primarily affect which thoughts arise, rather than which decisions are made. Questions are thought-level complexity, a property of mental imagery, and should not be confused with *reflective goals* asserting that a piece of knowledge is desirable; the two interrelate very strongly but are conceptually distinct. A question is a thought magnet and a goal is an action magnet. Since stray thoughts are (hopefully!) less dangerous than stray actions, questionness (*inquiry*) can spread in much more unstructured ways than goalness (*desirability*).

Goal imagery is abstract imagery whose referent is brought into correspondence with the goal description by the AI's actions. Question imagery is also abstract imagery, since the answer is not yet known, but question imagery has a more open-ended satisfaction criterion. Goal imagery tends to want its referent to take on a *specific* value; question imagery tends to want its referent to take on *any* value. Question imagery for "the outcome of event E" attracts any thoughts about the outcome of event E; it is the agnostic question "What, if anything, is the predicted outcome of E?" Goal imagery for "the outcome of event E" tends to require some *specific* outcome for E.

The creation of question imagery is one of the major contributing factors to the continuity of thought sequences, and therefore necessary for deliberation. However, just as goal imagery must affect actual decisions and actual actions before we concede that the AI has something which deserves to be called a "goal," question imagery must affect actual thoughts – actual sequiturs and actual verifiers – to be considered a cognitively real question. If there is salient question imagery for "the outcome of event E," it becomes the target of sequiturs that search for beliefs about implication or causation whose antecedents are satisfied by aspects of E; in other words, sequiturs searching for beliefs of the form "E usually leads to F" or "E causes F." If there is open question imagery for "the cause of the Y-ness of X," and a thought suggested for some other reason happens to intersect with "the cause of the Y-ness of X," the thought resonates strongly and will rise to the surface of cognition.

A similar and especially famous sequitur is the search for a causal belief whose consequent matches goal imagery, and whose antecedent is then visualized as imagery describing an event which is predicted to lead to the goal. The event imagery created may become new goal imagery – a subgoal – if the predictive link is confirmed and no obnoxious side effects are separately predicted (see the discussion of the deliberation level for more about goals and subgoals). Many classical theories of AI, in particular "theorem proving" and "planning" [77], hold up a simplified form of the "subgoal seeker" sequitur as the core algorithm of human thought. However, this sequitur does not in itself implement planning. The process of seeking subgoals is more than the one cognitive process of searching for belief consequents that match existing goals. There are other roads to finding subgoal candidates aside from backward chaining on existing goals; for example, forward reasoning from available actions. There may be several different real sequiturs (cognitive processes) that search for relevant beliefs; evolution's design approach would have been "find cognitive processes that make useful suggestions," not "constrain an exhaustive search through all beliefs to make it computationally efficient," and this means there may be several sequiturs in the repertoire that selectively search on different kinds of causal beliefs. Finding a belief whose consequent matches goal imagery is not the same as finding an event which is predicted to lead to the goal event; and even finding an action predicted to lead to at least one goal event is not the same as verifying the net desirability of that action.

The sequitur that seeks beliefs whose consequents match goal imagery is only one component of the thought level of organization. But it is a component that looks like the "exclamation mark of thought" from the perspective of many traditional theories, so it is worthwhile to review how the other levels of organization contribute to the effective intelligence of the "subgoal seeker" sequitur.

A goal is descriptive mental imagery, probably taking the form of a concept or concept structure describing an event; goal-oriented thinking uses the combinatorial regularities of the concept layer to describe regularities in the structure of goal-relevant events. The search for a belief whose consequent matches a goal description is organized using the category structure of the concept layer; concepts match against concepts, rather than unparsed sensory imagery matching against unparsed sensory imagery. Searching through beliefs is computationally tractable because of learned resonances and learned associations which are "learned complexity" in themselves, and moreover represent regularities in a conceptually described model rather than a raw sensory imagery. Goal-oriented thinking as used by humans is often abstract, which requires support from properties of mental imagery; it requires that the mind maintain descriptive imagery which is not fully visualized or completely satisfied by a sensory referent, but which binds to specific referents when these become available. Sensory modalities provide a space in which all this imagery can exist and interprets the environment from which learned complexity is learned. The feature structure of modalities renders learning computationally tractable. Without feature structure, concepts are computationally intractable; without category structure, thoughts are computationally intractable. Without modalities there are no experiences and no mental imagery; without learned complexity there are no concepts to structure experience and no beliefs generalized from experience. In addition to supporting basic requirements, modalities contribute directly to intelligence in any case where referent behaviors coincide with modality behaviors, and indirectly in cases where there are valid metaphors between modality behaviors and referent behaviors.

Even if inventing a new subgoal is the "exclamation mark of thought" from the perspective of many traditional theories, it is an exclamation mark at the end of a very long sentence. The rise of a single thought is an event that occurs within a whole mind – an intact reasoning process with a past history.

Beliefs

Beliefs – declarative knowledge – straddle the division between the concept level and the thought level. In terms of the level characteristics noted earlier, beliefs are learned, specific, stored, and recurrent. From this perspective beliefs should be classified as learned complexity and therefore a part of the generalized concept level. However, beliefs bear a greater surface resemblance to mental sentences than to individual words. Their internal structure appears to resemble concept structures more than concepts; and beliefs possess characteristics, such as structured antecedents and consequents, which are difficult to describe except in the context of the thought level of organization. I have thus chosen to discuss beliefs within the thought level[28].

Beliefs are acquired through two major sources, *induction* and *deduction*, respectively referring to generalization over experience, and reasoning from previous beliefs. The strongest beliefs have both inductive and deductive support: deductive conclusions with experiential confirmation, or inductive generalizations with causal explanations.

Induction and deduction can intersect because both involve abstraction. Inductive generalization produces a description containing categories that act as variables – abstract imagery that varies over the experiential base and describes it. Abstract deduction takes several inductively or deductively acquired generalizations, and chains together their abstract antecedents and abstract consequents to produce an abstract conclusion, as illustrated in the earlier discussion of abstract mental imagery. Even completely specific beliefs confirmed by a single experience, such as "New Year's Eve of Y2K took place on a Friday night," are still "abstract" in that they have a concept-based, category-structure description existing above the immediate sensory memory, and this conceptual description can be more easily chained with abstract beliefs that reference the same concepts.

Beliefs can be suggested by generalization across an experiential base, and supported by generalization across an experiential base, but there are limits to how much support pure induction can generate (a common complaint of philosophers); there could always be a disconfirming instance you do not know about. Inductive generalization probably resembles concept generalization, more or less; there is the process of initially noticing a regularity across an experiential base, the process of verifying it, and possibly even a process producing something akin to concept kernels for cueing frequently relevant beliefs. Beliefs have a different structure than concepts; concepts are either *useful* or *not useful,* but beliefs are either *true* or *false.* Concepts apply to referents, while beliefs describe relations between antecedents and consequents. While this implies a different repertoire of generalizations that produce inductive beliefs, and a different verification procedure, the computational task of noticing a generalization across antecedents and consequents seems strongly reminiscent of generalizing a two-place predicate.

Beliefs are well-known in traditional AI, and are often dangerously misused; while any process whatever can be *described with* beliefs, this does not mean that a cognitive process is *implemented by* beliefs. I possess a visual modality that implements edge detection, and I possess beliefs about my visual modality, but the latter aspect of mind does not affect the former. I could possess no beliefs about edge detection, or wildly wrong beliefs about edge detection, and my visual modality would continue working without a hiccup.

---

[28]Whether a belief is *really* more like a concept or more like a thought is a "wrong question." The specific similarities and differences say all there is to say. The levels of organization are aids to understanding, not Aristotelian straitjackets.

An AI may be able to introspect on lower levels of organization (see Sect. 3), and an AI's cognitive subsystems may interact with an AI's beliefs more than the equivalent subsystems in humans (again, see Sect. 3), but beliefs and brainware remain distinct – not only distinct, but occupying different levels of organization. When we seek the functional consequences of beliefs – their material effects on the AI's intelligence – we should look for the effect on the AI's reasoning and its subsequent decisions and actions. Anything can be described by a belief, including every event that happens within a mind, but not all events within a mind are implemented by the possession of a belief which describes the rules governing that event.

When a mind "really" possesses a belief "about" something, and not just some opaque data, is a common question in AI philosophy. I have something to say about this in the next section. In formal, classical terms, the cognitive effect of possessing a belief is sometimes defined to mean that when the antecedent of a belief is satisfied, its consequent is concluded. I would regard this as one sequitur out of many, but it is nonetheless a good example of a sequitur – searching for beliefs whose antecedents are satisfied by current imagery, and concluding the consequent (with reliances on the belief itself and on the imagery matched by the antecedent). However, this sequitur, if applied in the blind sense evoked by classical logic, will produce a multitude of useless conclusions; the sequitur needs to be considered in the context of verifiers such as "How rare is it for this belief to be found applicable?", "How often is this belief useful when it is applicable?", or "Does the consequent produced intersect with any other imagery, such as open question imagery?"

Some other sequiturs involving beliefs: Associating backward from question imagery to find a belief whose consequent touches the question imagery, and then seeing if the belief's antecedent can be satisfied by current imagery, or possibly turning the belief's antecedent into question imagery. Finding a causal belief whose consequent corresponds to a goal; the antecedent may then become a subgoal. Detecting a case where a belief is *violated* – this will usually be highly salient.

Suppose an AI with a billiards modality has inductively formed the belief "all billiards which are 'red' are 'gigantic'." Suppose further that 'red' and 'gigantic' are concepts formed by single-feature clustering, so that a clustered size range indicates 'gigantic', and a clustered volume of color space indicates 'red'. If this belief is salient enough, relative to the current task, to be routinely checked against all mental imagery, then several cognitive properties should hold if AI really possesses a belief about the size of red billiards. In subjunctive imagery, used to imagine non-sensory billiards, any billiard imagined to be red (within the clustered color volume of the 'red' concept) would need to be imagined as being gigantic (within the clustered size range of the 'gigantic' concept). If the belief "all red billiards are gigantic" has salient uncertainty, then the conclusion of gigantism would have a reliance on this uncertainty source and would share the perceived doubt. Given external sensory imagery, if a billiard is seen which is red and small, this must be perceived as violating the belief. Given sensory imagery, if a billiard is somehow seen as "red" in advance of its size being perceived (it's hard to imagine how this would happen in a human), then the belief must create the prediction or expectation that the billiard will be gigantic, binding a hanging abstract concept for 'gigantic' to the sensory imagery for the red billiard. If the sensory image is completed later and the concept kernel for 'gigantic' is not satisfied by the completed sensory image for the red billiard, then the result should be a violated expectation, and this conflict should propagate back to the source of the expectation to be perceived as a violated belief.

Generally, beliefs used within subjunctive imagery control the imagery directly, while beliefs used to interpret sensory information govern expectations and determine when an expectation has been violated. However, "sensory" and "subjunctive" are relative; subjunctive imagery governed by one belief may intersect and violate another belief – any imagery is "sensory" relative to a belief if that imagery is not directly controlled by the belief. Thus, abstract reasoning can detect inconsistencies in beliefs. (An inconsistency should not cause a real mind to shriek in horror and collapse, but it should be a salient event that shifts the train of thought to hunting down the source of the inconsistency, looking at the beliefs and assertions relied upon and checking their confidences. Inconsistency detections, expressed as thoughts, tend to create question imagery and knowledge goals which direct deliberation toward resolving the inconsistency.)

Coevolution of Thoughts and Language: Origins of the Internal Narrative

Why is the transformation of concept structures into linear word sequences, obviously necessary for spoken communication, also carried out within the internal stream of consciousness? Why not use only the concept structures? Why do we transform concept structures into grammatical sentences if nobody is listening? Is this a necessary part of intelligence? Must an AI do the same in order to function?

The dispute over which came first, thought or language, is ancient in philosophy. Modern students of the evolution of language try to break down the evolution of language into incrementally adaptive stages, describe multiple functions that are together required for language, and account for how preadaptations for those functions could have arisen [43]. Functional decompositions avoid some of the chicken-and-egg paradoxes that result from viewing language as a monolithic function. Unfortunately, there are further paradoxes that result from viewing language independently from thought, or from viewing thought as a monolithic function.

From the perspective of a cognitive theorist, language is only one function of a modern-day human's cognitive supersystem, but from the perspective of an evolutionary theorist, linguistic features determine which *social* selection pressures apply to the evolution of cognition at any given point. Hence "coevolution of thought and language" rather than "evolution of language as one part of thought." An evolutionary account of language alone will become "stuck" the first time it reaches a feature which is adaptive for cognition and preadaptive for language, but for which no independent linguistic selection pressure exists in the absence of an already-existent language. Since there is currently no consensus on the functional decomposition of intelligence, contemporary language evolution theorists are sometimes unable to avoid such sticking points.

On a first look DGI might appear to explain the evolvability of language merely by virtue of distinguishing between the concept level and the thought level; as long as there are simple reflexes that make use of learned category structure, elaboration of the concept level will be independently adaptive, even in the absence of a humanlike thought level. The elaboration of the concept level to support cross-modality associations would appear to enable crossing the gap between a signal and a concept, and the elaboration of the concept level to support the blending or combination of concepts (adaptive because it enables the organism to perceive simple combinatorial regularities) would appear to enable primitive, nonsyntactical word sequences. Overall this resembles Bickerton's [5] picture of *protolanguage* as an evolutionary intermediate, in which learned signals convey learned concepts and multiple concepts blend, but without syntax to convey targeting information. Once protolanguage existed, linguistic selection pressures proper could take over.

However, as [18] points out, this picture does not explain why other species have not developed protolanguage. Cross-modal association is not limited to humans or even primates. Deacon suggests that some necessary mental steps in language are not only *unintuitive* but actually *counterintuitive* for nonhuman species, in the same way that the Wason Selection Test is counterintuitive for humans. Deacon's account of this "awkward step" uses a different theory of intelligence as background, and I would hence take a different view of the nature of the awkward step: my guess is that chimpanzees find it extraordinarily hard to learn symbols as we understand them because language, even protolanguage, requires creating abstract mental imagery which can hang unsupported and then bind to a sensory referent later encountered. The key difficulty in language – the step that is awkward for other species – is not the ability to associate signals; primates (and rats, for that matter) can readily associate a perceptual signal with a required action or a state of the world. The awkward step is for a signal to evoke a category as abstract imagery, apart from immediate sensory referents, which can bind to a referent later encountered. This step is completely routine for us, but could easily be almost impossible in the absence of design support for "hanging concepts in midair." In the absence of thought, there are few reasons why a species would find it useful to hang concepts in midair. In the absence of language, there are even fewer reasons to associate a perceptual signal with the evocation of a concept as abstract imagery. Language is hard for other species, not because of a gap between the signal and the concept, but because language uses a feature of mental imagery for which there is insufficient design support in other species. I suspect it may have been an adaptive context for abstract imagery, rather than linguistic selection pressures, which resulted in the adaptation which turned out to be preadaptive for symbolization and hence started some primate species sliding down a fitness gradient that included coevolution of thought and language.

If, as this picture suggests, pre-hominid evolution primarily elaborated the concept layer (in the sense of elaborating brainware processes that support categories, not in the sense of adding learned concepts as such), it implies that the concept layer may contain the bulk of supporting functional complexity for human cognition. This does not follow necessarily, since evolution may have spent much time but gotten little in return, but it is at least suggestive. (This section on the concept level is, in fact, the longest section.) The above picture also suggests that the hominid family may have coevolved combinatorial concept structures that modify mental imagery internally (thoughts) and combinatorial concept structures that evoke mental imagery in conspecifics (language). It is obvious that language makes use of many functions originally developed to support internal cognition, but *coevolution* of thought and language implies a corresponding opportunity for evolutionary elaboration of hominid thought to co-opt functions originally evolved to support hominid language.

The apparent necessity of the internal narrative for human deliberation could turn out to be an introspective illusion, but if real, it strongly suggests that linguistic functionality has been co-opted for cognitive functionality during human evolution. Linguistic features such as special processing of the tags that invoke concepts, or the use of syntax to organize complex internal targeting information for structures of combinatorial concepts, could also be adaptive or preadaptive for efficient thought. Only a few such linguistic features would need to be co-opted as necessary parts of thought before the "stream of consciousness" became an entrenched part of human intelligence. This is probably a sufficient explanation for the existence of an internal narrative, possibly making the internal narrative a pure spandrel (emergent but nonadaptive feature). However, caution in AI, rather than caution in evolutionary psychology, should impel us to wonder if our internal narrative serves an adaptive function. For example, our internal narrative could express deliberation in a form that we can more readily process as (internal) sensory experience for purposes of introspection and memory; or the cognitive process of imposing internal thoughts on mental imagery could co-opt a linguistic mechanism that also translates external communications into mental imagery; or the internal narrative may co-opt social intelligence that models other humans by relating to their communications, in order to model the self. But even if hominid evolution has co-opted the internal narrative, the overall model still suggests that – while we cannot disentangle language from intelligence or disentangle the evolution of thought from the evolution of language – a *de novo* mind design could disentangle intelligence from language.

This in turn suggests that an AI could use concept structures without serializing them as grammatical sentences forming a natural-language internal narrative, as long as all linguistic functionality co-opted for human intelligence were reproduced in non-linguistic terms – including the expression of thoughts in an introspectively accessible form, and the use of complex internal targeting in concept structures. Observing the AI may require recording the AI's thoughts and translating those thoughts into humanly understandable forms, and the programmers may need to communicate concept structures to the AI, but this need not imply an AI capable of understanding or producing human language. True linguistic communication between humans and AIs might come much later in development, perhaps as an ordinary domain competency rather than a brainware-supported talent. Of course, human-language understanding and natural human conversation is an *extremely* attractive goal, and would undoubtedly be attempted as early as possible; however, it appears that language need not be implemented immediately or as a necessary prerequisite of deliberation.

2.7 The Deliberation Level

From Thoughts to Deliberation

In humans, higher levels of organization are generally more accessible to introspection. It is not surprising if the internal cognitive events called "thoughts," as described in the last section, seem strangely familiar; we listen to thoughts all day. The danger for AI developers is that cognitive content which is open to introspection is sometimes temptingly easy to translate directly into code. But if humans have evolved a cyclic interaction of thought and imagery, this fact alone does not prove (or even argue) that the design is a good one. What is the material benefit to intelligence of using blackboard mental imagery and sequiturs, instead of the simpler fixed algorithms of "reasoning" under classical AI?

Evolution is characterized by ascending levels of organization of increasing elaboration, complexity, flexibility, richness, and computational costliness; the complexity of the higher layers is not automatically emergent solely from the bottom layer, but is instead subject to selection pressures and the evolution of complex functional adaptation – adaptation which is relevant at that level, and, as it turns out, sometimes preadaptive for the emergence of higher levels of organization. This design signature emerges at least in part from the characteristic blindness of evolution, and may not be a necessary idiom of minds-in-general. Nonetheless, past attempts to directly program cognitive phenomena which arise on post-modality levels of organization have failed profoundly. There are specific AI pathologies that emerge from the attempt, such as the symbol grounding problem and the commonsense problem. In humans concepts are smoothly flexible and expressive because they arise from modalities; thoughts are smoothly flexible and expressive because they arise from concepts. Even considering the value of blackboard imagery and sequiturs in isolation – for example, by considering an AI architecture that used fixed algorithms of deliberation but used those algorithms to create and invoke DGI thoughts – there are still necessary reasons why deliberative patterns must be built on behaviors of the thought level, rather than being implemented as independent code; there are AI pathologies that would result from the attempt to implement deliberation in a purely top-down way. There is top-down complexity in deliberation – adaptive functionality that is best viewed as applying to the deliberation level and not the thought level – but this complexity is mostly incarnated as behaviors of the thought level that support deliberative patterns.

Because the deliberation level is flexibly emergent out of the sequiturs of the thought level, a train of thought can be diverted without being destroyed. To use the example given earlier, if a deliberative mind wonders "Why is X a Y?" but no explanation is found, this local failure is not a disaster for deliberation as a whole. The mind can mentally note the question as an unsolved puzzle and continue with other sequiturs. A belief violation does not destroy a mind; it becomes a focus of attention and one more thing to ponder. Discovering inconsistent beliefs does not cause a meltdown, as it would in a system of monotonic logic, but instead shifts the focus of attention to checking and revising the deductive logic. Deliberation weaves multiple, intersecting threads of reasoning through intersecting imagery, with the waystations and even the final destination not always known in advance.

In the universe of bad TV shows, speaking the Epimenides Paradox[29] "This sentence is false" to an artificial mind causes that mind to scream in horror and collapse into a heap of smoldering parts. This is based on a stereotype of thought processes that cannot divert, cannot halt, and possess no bottom-up ability to notice regularities across an extended thought sequence. Given how deliberation emerges from the thought level, it is possible to imagine a sufficiently sophisticated, sufficiently reflective AI that could naturally surmount the Epimenides Paradox. Encountering the paradox "This sentence is false" would probably indeed lead to a looping thought sequence at first, but this would not cause the AI to become permanently stuck; it would instead lead to categorization across repeated thoughts (like a human noticing the paradox after a few cycles), which categorization would then become salient and could be pondered in its own right by other sequiturs. If the AI is sufficiently competent at deductive reasoning and introspective generalization, it could generalize across the specific instances of "If the statement is true, it must be false" and "If the statement is false, it must be true" as two general classes of thoughts produced by the paradox, and show that reasoning from a thought of one class leads to a thought of the other class; if so the AI

---

[29] "This sentence is false" is properly known as the Eubulides Paradox rather than the Epimenides Paradox, but "Epimenides Paradox" seems to have become the standard term.

could deduce – not just inductively notice, but deductively confirm – that the thought process is an eternal loop. Of course, we won't know whether it really works this way until we try it.

The use of a blackboard sequitur model is not automatically sufficient for deep reflectivity; an AI that possessed a limited repertoire of sequiturs, no reflectivity, no ability to employ reflective categorization, and no ability to notice when a train of thought has not yielded anything useful for a while, might still loop eternally through the paradox as the emergent but useless product of the sequitur repertoire. Transcending the Epimenides Paradox requires the ability to perform inductive generalization and deductive reasoning on introspective experiences. But it also requires bottom-up organization in deliberation, so that a spontaneous introspective generalization can capture the focus of attention. Deliberation must *emerge* from thoughts, not just *use* thoughts to implement rigid algorithms.

Having reached the deliberation level, we finally turn from our long description of what a mind *is,* and focus at last on what a mind *does* – the useful operations implemented by sequences of thoughts that are structures of concepts that are abstracted from sensory experience in sensory modalities.

The Dimensions of Intelligence

Philosophers frequently define "truth" as an agreement between belief and reality; formally, this is known as the "correspondence theory" of truth [45]. Under the correspondence theory of truth, philosophers of Artificial Intelligence have often defined "knowledge" as a mapping between internal data structures and external physical reality [76]. Considered in isolation, the correspondence theory of knowledge is easily abused; it can be used to argue on the basis of mappings which turn out to exist entirely in the mind of the programmer.

Intelligence is an evolutionary advantage because it enables us to model *and* predict *and* manipulate reality. In saying this, I am not advocating the philosophical position that only useful knowledge can be true. There is enough regularity in the activity of acquiring knowledge, over a broad spectrum of problems that require knowledge, that evolution has tended to create independent cognitive forces for truthseeking. Individual organisms are best thought of as adaptation-executers rather than fitness-maximizers [98]. "Seeking truth," even when viewed as a mere local subtask of a larger problem, has sufficient functional autonomy that many human adaptations are better thought of as "truthseeking" than "useful-belief-seeking." Furthermore, under my own philosophy, I would say beliefs are useful because they are true, not "true" because they are useful.

But usefulness is a *stronger* and *more reliable* test of truth; it is harder to cheat. The social process of science applies prediction as a test of models, and the same models that yield successful predictions are often good enough approximations to construct technology (manipulation).

I would distinguish four successively stronger grades of *binding* between a model and reality:

- A *sensory* binding occurs when there is a mapping between cognitive content in the model and characteristics of external reality. Without tests of usefulness, there is no formal way to prevent abuse of claimed sensory bindings; the supposed mapping may lie mostly in the mind of the observer. However, if the system as a whole undergoes tests of usefulness, much of the task of extending and improving the model will still locally consist of discovering good sensory bindings – finding beliefs that are true under the intuitive "correspondence theory" of truth.

- A *predictive* binding occurs when a model can be used to correctly predict future events. From the AI's internal perspective, a predictive binding occurs when the model can be used to correctly predict future sensory inputs. The AI may be called upon to make successful predictions about external reality (outside the computer), virtual microenvironments (inside the computer but outside the AI), or the outcome of cognitive processes (inside the AI, but proceeding distinct from the prediction). A "sensory input" can derive not only from a sensory device targeted on external reality, but also from sensory cognition targeted on *any* process whose outcome, on the level predicted, is not subject to direct control. (Of course, from our perspective, prediction of the "real world" remains the strongest test.)

- A *decisive* binding occurs when the model can predict the effects of several possible actions on reality, and choose whichever action yields the best result under some goal system (see below). By predicting outcomes under several possible world-states, consisting of the present world-state plus each of several possible actions, it becomes possible to choose between futures.

- A *manipulative* binding occurs when the AI can describe a desirable future with subjunctive imagery, and invent a sequence of actions which leads to that future. Where *decision* involves selecting one action from a predetermined and bounded set, *manipulation* involves inventing new actions, perhaps actions previously unperceived because the set of possible actions is unbounded or computationally large. The simplest form of manipulation is backward chaining from parent goals to child goals using causal beliefs; this is not the only form of manipulation, but it is superior to exhaustive forward search from all possible actions.

I also distinguish three successive grades of *variable complexity:*

- A *discrete* variable has referents selected from a bounded set which is computationally small – for example, a set of 20 possible actions, or a set of 26 possible lowercase letters. The binary presence or absence of a feature is also a discrete variable.

- A *quantitative* variable is selected from the set of real numbers, or from a computationally large set which approximates a smoothly varying scalar quantity (such as the set of floating-point numbers).
- A *patterned* variable is composed of a finite number of quantitative or discrete elements. Examples: A finite string of lowercase letters, e.g. "mkrznye." A real point in 3D space (three quantitative elements). A 2D black-and-white image (2D array of binary pixels).

The dimension of variable complexity is orthogonal to the SPDM (sensory-predictive-decisive-manipulative) dimension, but like SPDM it describes successively tougher tests of intelligence. A decisive binding from desired result to desirable action is computationally feasible only when the "action" is a discrete variable chosen from a small set – small enough that each possible action can be modeled. When the action is a quantitative variable, selected from computationally large sets such as the floating-point numbers in the interval $[0, 1]$, some form of manipulative binding, such as backward chaining, is necessary to arrive at the specific action required. (Note that adding a continuous time parameter to a discrete action renders it quantitative.) Binding precise quantitative goal imagery to a precise quantitative action cannot be done by exhaustive testing of the alternatives; it requires a way to transform the goal imagery so as to arrive at subgoal imagery or action imagery. The simplest transformation is the identity relation – but even the identity transformation is not possible to a *purely* forward-search mechanism. The next most straightforward method would be to employ a causal belief that specifies a reversible relation between the antecedent and the consequent. In real-time control tasks, motor modalities (in humans, the entire sensorimotor system) may automatically produce action symphonies in order to achieve quantitative or patterned goals.

A string of several discrete or quantitative variables creates a patterned variable, which is also likely to be computationally intractable for exhaustive forward search. Binding a patterned goal to a patterned action, if the relation is not one of direct identity, requires (again) a causal belief that specifies a reversible relation between the antecedent and the consequent, or (if no such belief is forthcoming) deliberative analysis of complex regularities in the relation between the action and the outcome, or exploratory tweaking followed by induction on which tweaks increase the apparent similarity between the outcome and the desired outcome.

There are levels of organization within bindings; a loose binding at one level can give rise to a tighter binding at a higher level. The rods and cones of the retina correspond to incoming photons that correspond to points on the surface of an object. The binding between a metaphorical pixel in the retina and a point in a real-world surface is very weak, very breakable; a stray ray of light can wildly change the detected optical intensity. But the actual sensory experience occupies one level of organization *above* individual pixels. The fragile sensory binding between retinal pixels and surface points, on a lower level of organization, gives rise to a solid sensory binding between our perception of the entire object and the object itself. A match between two discrete variables or two rough quantitative variables can arise by chance; a match between two patterned variables on a higher holonic level of organization is far less likely to arise from complete coincidence, though it may arise from a cause other than the obvious. The concept kernels in human visual recognition likewise bind to the entire perceptual experience of an object, not to individual pixels of the object. On an even higher level of organization, the *manipulative* binding between human intelligence and the real world is nailed down by many individually tight *sensory* bindings between conceptual imagery and real-world referents. Under the human implementation, there are at least three levels of organization *within* the correspondence theory of truth! The AI pathology that we perceive as "weak semantics" – which is very hard to define, but is an intuitive impression shared by many AI philosophers – may arise from omitting levels of organization in the binding between a model and its referent.

Actions

The series of motor actions I use to strike a key on my keyboard have enough degrees of freedom that "which key I strike," as a discrete variable, or "the sequence of keys struck," as a patterned variable, are both subject to direct specification. I do not need to engage in complex planning to strike the key sequence "hello world" or "labm4;" I can specify the words or letters directly and without need for complex planning. My motor areas and cerebellum do an enormous amount of work behind the scenes, but it is work that has been optimized to the point of subjective invisibility. A keystroke is thus an *action* for pragmatic purposes, although for a novice typist it might be a goal. As a first approximation, goal imagery has been reduced to action imagery when the imagery can direct a realtime skill in the relevant modality. This does not necessarily mean that actions are handed off to skills with no further interaction; realtime manipulations sometimes go wrong, in which case the interrelation between goals and actions and skills becomes more intricate, sometimes with multiple changing goals interacting with realtime skills. Imagery approaches the action level as it becomes able to interact with realtime skills.

Sometimes a goal does not directly reduce to actions because the goal referent is physically distant or physically separated from the "effectors" – the motor appendages or their virtual equivalents – so that manipulating the goal referent depends on first overcoming the physical separation as a subproblem. However, in the routine activity of modern-day humans, another very common reason why goal imagery does not translate directly into action imagery is that the goal imagery is a high-level abstract characteristic, *cognitively* separated from the realm of direct actions. I can control every keystroke of my typing, but the quantitative percept of *writing quality*[30] referred to by the goal imagery of *high writing quality* is not subject to direct manipulation. I cannot directly set my writing quality to equal that of Shakespeare, in the way that I can directly set a keystroke to equal "H", because *writing quality* is a derived, abstract quantity. A better word than "abstract" is "holonic," the term used earlier from [51] used to describe the way in which a single quality may simultaneously be a whole composed of parts, and a part in a greater whole. *Writing quality* is a quantitative *holon* which is eventually bound to the series of discrete keystrokes. I can directly choose keystrokes, but cannot directly choose the writing-quality holon. To increase the writing quality of a paragraph I must link the writing-quality holon to lower-level holons such as *correct spelling* and *omitting needless words,* which are qualities of the *sentences* holons, which are created through keystroke actions. Action imagery is typically, though not always, the level on which variables are completely free (directly specifiable with many degrees of freedom); higher levels involve interacting constraints which must be resolved through deliberation.

Goals

The very-high-level abstract goal imagery for *writing quality* is bound to directly specifiable action imagery for *words* and *keystrokes* through an intermediate series of child goals which inherit desirability from parent goals. But what are goals? What is desirability? So far I have been using an intuitive definition of these terms, which often suffices for describing how the goal system interacts with other systems, but is not a description of the goal system itself.

Unfortunately, the human goal system is somewhat ... *confused* ... as you know if you're a human. Most of the human goal system originally evolved in the absence of deliberative intelligence, and as a result, behaviors that contribute to survival and reproduction tend to be evolved as independent drives. Taking the intentionalist stance toward evolution, we would say that the sex drive is a child goal of reproduction. Over evolutionary time this might be a valid stance. But individual organisms are best regarded as adaptation-executers rather than fitness-maximizers, and the sex drive is not *cognitively* a child goal of reproduction; hence the modern use of contraception. Further complications are introduced at the primate level by the existence of complex social groups; consequently primates have "moral" adaptations, such as reciprocal altruism, third-party intervention to resolve conflicts ("community

---

[30]Of course, writing quality is made up of a number of components and is not a true scalar variable. A more accurate description would be that "writing quality" is the summation of a number of other percepts, and that we conceive of this summated quality as increasing or decreasing. Some writing qualities may be definitely less than or greater than others, but this does not imply that the complete set of percepts is well-ordered or that the percept itself is cognitively implemented by a simple scalar magnitude.

concern"), and moralistic aggression against community offenders [24]. Still further complications are introduced by the existence of deliberative reasoning and linguistic communication in humans; humans are imperfectly deceptive social organisms that argue about each other's motives in adaptive contexts. This has produced what I can only call "philosophical" adaptations, such as the ways we reason about causation in moral arguments – ultimately giving us the ability to pass (negative!) judgement on the moral worth of our evolved goal systems and evolution itself.

It is not my intent to untangle that vast web of causality in this paper, although I have written (informally but at length) about the problem elsewhere [109], including a description of the cognitive and motivational architectures required for a mind to engage in such apparently paradoxical behaviors as passing coherent judgement on its own top-level goals. (For example, a mind may regard the current representation of morals as a probabilistic approximation to a moral referent that can be reasoned about.) The architecture of morality is a pursuit that goes along with the pursuit of general intelligence, and the two should not be parted, for reasons that should be obvious and will become even more obvious in Sect. 3; but unfortunately there is simply not enough room to deal with the issues here. I will note, however, that the human goal system sometimes does the Wrong Thing[31] and I do not believe AI should follow in those footsteps; a mind may share our moral frame of reference without being a functional duplicate of the human goal supersystem.

Within this paper I will set aside the question of moral reasoning and take for granted that the system supports moral content. The question then becomes how moral content binds to goal imagery and ultimately to actions.

The imagery that describes the supergoal is the moral content and describes the events or world-states that the mind regards as having intrinsic value. In classical terms, the supergoal description is analogous to the intrinsic utility function. Classically, the total utility of an event or world-state is its intrinsic utility, plus the sum of the intrinsic utilities (positive or negative) of the future events to which that event is predicted to lead, multiplied in each case by the predicted probability of the future event as a consequence. (Note that predicted consequences include both direct and indirect consequences, i.e., consequences of consequences are included in the sum.) This may appear at first glance to be yet another oversimplified Good Old-Fashioned AI definition, but for once I shall argue in favor; the classical definition is more fruitful of complex behaviors than first apparent. The property *desirability* should be coextensive with, and should behave identically to, the property *is-predicted-to-lead-to-intrinsic-utility*.

Determining which actions are predicted to lead to the greatest total intrinsic utility, and inventing actions which lead to greater intrinsic utility, has subjective regularities when considered as a cognitive problem and ex-

---

[31]As opposed to the Right Thing. See the Jargon File entry for "Wrong Thing" [83].

ternal regularities when considered as an event structure. These regularities are called *subgoals*. Subgoals define areas where the problem can be efficiently viewed from a local perspective. Rather than the mind needing to rethink the entire chain of reasoning "Action A leads to B, which leads to C, which leads to D, ..., which leads to actual intrinsic utility Z," there is a useful regularity that actions which lead to B are mostly predicted to lead through the chain to Z. Similarly, the mind can consider which of subgoals B_1, B_2, B_3 are most likely to lead to C, or consider which subgoals C_1, C_2, C_3 are together sufficient for D, without rethinking the rest of the logic to Z.

This network (not hierarchical) event structure is an *imperfect* regularity; desirability is heritable only to the extent, and exactly to the extent, that predicted-to-lead-to-Z-ness is heritable. Our low-entropy universe has category structure, but not perfect category structure. Using imagery to describe an event E which is predicted to lead to event F is never perfect; perhaps *most* real-world states that fit description E lead to events that fit description F, but it would be very rare, outside of pure mathematics, to find a case where the prediction is perfect. There will always be some states in the volume carved out by the description E that lead to states outside the volume carved out by description F. If C is predicted to lead to D, and B is predicted to lead to C, then usually B will inherit C's predicted-to-lead-to-D-ness. However, it may be that B leads to a special case of C which does not lead to D; in this case, B would not inherit C's predicted-to-lead-to-D-ness. Therefore, if C had inherited desirability from D, B would not inherit C's desirability either.

To deal with a world of imperfect regularities, goal systems model the regularities in the irregularities, using descriptive constraints, distant entanglements, and global heuristics. If events fitting description E usually but not always lead to events fitting description F, then the mental imagery describing E, or even the concepts making up the description of E, may be refined to narrow the extensional class to eliminate events that seem to fit E but that do not turn out to lead to F. These "descriptive constraints" drive the AI to focus on concepts and categories that expose predictive, causal, and manipulable regularities in reality, rather than just surface regularities.

A further refinement is "distant entanglements;" for example, an action A that leads to B which leads to C, but which also simultaneously has side effects that block D, which is C's source of desirability. Another kind of entanglement is when action A leads to unrelated side effect S, which has negative utility outweighing the desirability inherited from B.

"Global heuristics" describe goal regularities that are general across many problem contexts, and which can therefore be used to rapidly recognize positive and negative characteristics; the concept "margin for error" is a category that describes an important feature of many plans, and the belief "margin for error supports the local goal" is a global heuristic that positively links members of the perceptual category *margin for error* to the local goal context, without requiring separate recapitulation of the inductive and deductive support for the general heuristic. Similarly, in self-modifying or at least self-regulating AIs, "minimize memory usage" is a subgoal that many other subgoals and actions may impact, so the perceptual recognition of events in the "memory usage" category or "leads to memory usage" categories implies entanglement with a particular distant goal.

Descriptive constraints, distant entanglements, and global heuristics do not violate the desirability-as-prediction model; descriptive constraints, distant entanglements, and global heuristics are also useful for modeling complex predictions, in the same way and for the same reasons as they are useful in modeling goals. However, there are at least three reasons for the activity of *planning* to differ from the activity of *prediction*. First, prediction typically proceeds forward from a definite state of the universe to determine what comes after, while planning often (though not always) reasons backward from goal imagery to pick out one point in a space of possible universes, with the space's dimensions determined by degrees of freedom in available actions. Second, desirabilities are differential, unlike predictions; if A and $\sim A$ both lead to the same endpoint E, then from a predictive standpoint this may increase the confidence in E, but from a planning standpoint it means that neither A nor $\sim A$ will inherit *net* desirability from E. The final effect of desirability is that an AI chooses the *most* desirable action, an operation which is comparative rather than absolute; if both A and $\sim A$ lead to E, neither A nor $\sim A$ transmit differential desirability to actions.

Third, while both *implication* and *causation* are useful for reasoning about predictions, only causal links are useful in reasoning about goals. If the observation of A is usually followed by the observation of B, then this makes A a good predictor of B – regardless of whether A is the direct cause of B, or whether there is a hidden third cause C which is the direct cause of both A and B. I would regard *implication* as an emergent property of a directed network of events whose underlying behavior is that of causation; if C causes A, and then causes B, then A will imply B. Both "A causes B" (direct causal link) and "A implies B" (mutual causal link from C) are useful in prediction. However, in planning, the distinction between "A directly causes B" and "A and B are both effects of C" leads to a distinction between "Actions that lead to A, as such, are likely to lead to B" and "Actions that lead directly to A, without first leading through C, are unlikely to have any effect on B." This distinction also means that experiments in manipulation tend to single out real causal links in a way that predictive tests do not. If A implies B then it is often the case that C causes both A and B, but it is rarer in most real-world problems for an action intended to affect A to separately and invisibly affect the hidden third cause C, giving rise to false confirmation of direct causality[32]. (Although it happens, especially in economic and psychological experiments.)

---

[32] I believe this is the underlying distinction which [79] is attempting to model when he suggests that agent actions be represented as surgery on a causal graph.

Activities of Intelligence: Explanation, Prediction, Discovery, Planning, Design

So far, this section has introduced the distinction between sensory, predictive, decisive, and manipulative models; discrete, quantitative, and patterned variables; the holonic model of high-level and low-level patterns; and supergoal referents, goal imagery, and actions. These ideas provide a framework for understanding the immediate subtasks of intelligence – the moment-to-moment activities of deliberation. In carrying out a high-level cognitive task such as *design a bicycle,* the subtasks consist of crossing gaps from very high-level holons such as *good transport* to the holon *fast propulsion* to the holon *pushing on the ground* to the holon *wheel* to the holons for *spokes* and *tires,* until finally the holons become directly specifiable in terms of design components and design materials directly available to the AI.

The activities of intelligence can be described as *knowledge completion* in the service of *goal completion.* To complete a bicycle, one must first complete a design for a bicycle. To carry out a plan, one must complete a mental picture of a plan. Because both planning and design make heavy use of knowledge, they often spawn purely knowledge-directed activities such as explanation, prediction, and discovery. These activities are messy, non-inclusive categories, but they illustrate the general sorts of things that general minds do.

Knowledge activities are carried out both on a large scale, as major strategic goals, and on a small scale, in routine subtasks. For example, "explanation" seeks to extend current knowledge, through deduction or induction or experiment, to fill the gap left by the unknown cause of a known effect. The unknown cause will at least be the referent of question imagery, which will bring into play sequiturs and verifiers which react to open questions. If the problem becomes salient enough, and difficult enough, finding the unknown cause may be promoted from question imagery to an internal goal, allowing the AI to reason deliberatively about which problem-solving strategies to deploy. The knowledge goal for "building a plan" inherits desirability from the objective of the plan, since creating a plan is required for (is a subgoal of) achieving the objective of the plan. The knowledge goal for explaining an observed failure might inherit desirability from the goal achievable when the failure is fixed. Since knowledge goals can govern actual actions and not just the flow of sequiturs, they should be distinguished from question imagery. Knowledge goals also permit reflective reasoning about what kind of internal actions are likely to lead to solving the problem; knowledge goals may invoke sequiturs that search for beliefs about *solving knowledge problems,* not just beliefs about the specific problem at hand.

Explanation fills holes in knowledge about the past. Prediction fills holes in knowledge about the future. Discovery fills holes in knowledge about the present. Design fills gaps in the mental model of a tool. Planning fills gaps in a model of future strategies and actions. Explanation, prediction, discovery, and design may be employed in the pursuit of a specific real-world goal, or

as an independent pursuit in the anticipation of the resulting knowledge being useful in future goals – "curiosity." Curiosity fills completely general gaps
(rather than being targeted on specific, already-known gaps), and involves the
use of forward-looking reasoning and experimentation, rather than backward
chaining from specific desired knowledge goals; curiosity might be thought of
as filling the very abstract goal of "finding out X, where X refers to anything that will turn out to be a good thing to know later on, even though I
don't know specifically what X is." (Curiosity involves a very abstract link
to intrinsic utility, but one which is nonetheless completely true – curiosity *is*
useful.)

What all the activities have in common is that they involve reasoning about
a complex, holonic model of causes and effects. "Explanation" fills in holes
about the past, which is a complex system of cause and effect. "Prediction"
fills in holes in the future, which is a complex system of cause and effect.
"Design" reasons about tools, which are complex holonic systems of cause
and effect. "Planning" reasons about strategies, which are complex holonic
systems of cause and effect. Intelligent reasoning completes knowledge goals
and answers questions in a complex holonic causal model, in order to achieve
goal referents in a complex holonic causal system.

This gives us the three elements of DGI:

- The *what* of intelligence: Intelligence consists *in humans* of a highly modular brain with dozens of areas, which implements a deliberative process
  (built on thoughts built of concepts built on sensory modalities built on
  neurons); plus contributing subsystems (e.g. memory); plus surrounding
  subsystems (e.g. autonomic regulation); plus leftover subsystems implementing pre-deliberative approximations of deliberative processes; plus
  emotions, instincts, intuitions and other systems that influence the deliberative process in ways that were adaptive in the ancestral environment; plus
  everything else. A similar system is contemplated for AIs, of roughly the
  same order of complexity, but inevitably less messy. Both supersystems
  are characterized by levels of organization: Code / neurons, modalities,
  concepts, thoughts, and deliberation.
- The *why* of intelligence: The cause of human intelligence is evolution. Intelligence is an evolutionary advantage because it enables us to model
  reality, including external reality, social reality and internal reality, which
  in turn enables us to predict, decide, and manipulate reality. AIs will have
  intelligence because we, the human programmers, wish to accomplish a
  goal that can best be reached through smart AI, or because we regard the
  act of creating AI as having intrinsic utility; in either case, building AI
  requires building a deliberative supersystem that manipulates reality.
- The *how* of intelligence: Intelligence (deliberate reasoning) completes
  knowledge goals and answers questions in a complex holonic causal model,
  in order to achieve goal referents in a complex holonic causal system.

General Intelligence

The evolutionary context of intelligence has historically included environmental adaptive contexts, social adaptive contexts (modeling of other minds), and reflective adaptive contexts (modeling of internal reality). In evolving to fit a wide variety of adaptive contexts, we have acquired much cognitive functionality that is visibly specialized for particular adaptive problems, but we have also acquired cognitive functionality that is adaptive across many contexts, and adaptive functionality that co-opts previously specialized functionality for wider use. Humans can acquire substantial competence in modeling, predicting, and manipulating fully general regularities of our low-entropy universe. We call this ability "general intelligence." In some ways our ability is very weak; we often solve general problems abstractly instead of perceptually, so we can't deliberatively solve problems on the order of realtime visual interpretation of a 3D scene. But we can often say something which is true enough to be useful and simple enough to be tractable. We can deliberate on how vision works, even though we can't deliberate fast enough to perform realtime visual processing.

There is currently a broad trend toward one-to-one mappings of cognitive subsystems to domain competencies. I confess that I am personally annoyed by the manifestations of this idea in popular psychology, but of course the new phrenologies are irrelevant to genuine hypotheses about mappings between specialized domain competencies and specialized computational subsystems, or decisions to pursue specialized AI. It is not unheard-of for academic trends to reflect popular psychology, but it is generally good form to dispose of a thesis before dissecting the moral flaws of its proponents.

In DGI, human intelligence is held to consist of a supersystem with complex interdependent subsystems that exhibit *internal* functional specialization, but this does not rule out the existence of other subsystems that contribute solely or primarily to specific cognitive talents and domain competencies, or subsystems that contribute more heavily to some cognitive talents than others. The mapping from computational subsystems to cognitive talents is many-to-many, and the mapping from cognitive talents plus acquired expertise to domain competencies is also many-to-many, but this does not rule out specific correspondences between human variances in the "computing power" (generalized cognitive resources) allocated to computational subsystems and observed variances in cognitive talents or domain competencies. It should be noted, however, that the subject matter of AI is not the variance between humans, but the base of adaptive complexity held by all humans in common. If increasing the resources allocated to a cognitive subsystem yields an increase in a cognitive talent or domain competency, it does not follow that the talent or competency can be implemented by that subsystem alone. It should also be noted that under the traditional paradigm of programming, programmers' thoughts about solving specific problems are translated into code, and this is the idiom underlying most branches of classical AI; for example, expert systems engineers supposedly translate the beliefs in specific domains directly into the cognitive content of the AI. This would naturally tend to yield a view of intelligence in which there is a one-to-one mapping between subsystems and competencies. I believe this is the underlying cause of the atmosphere in which the quest for intelligent AI is greeted with the reply: "AI that is intelligent in what domain?"

This does not mean that exploration in specialized AI is entirely worthless; in fact, DGI's levels of organization suggest a specific class of cases where specialized AI may prove fruitful. Sensory modalities lie directly above the code level; sensory modalities were some of the first specialized cognitive subsystems to evolve and hence are not as reliant on a supporting supersystem framework, although other parts of the supersystem depend heavily on modalities. This suggests a specialized approach, with programmers directly writing code, may prove fruitful if the project is constructing a sensory modality. And indeed, AI research that focuses on creating sensory systems and sensorimotor systems continues to yield real progress. Such researchers are following evolution's incremental path, often knowingly so, and thereby avoiding the pitfalls that result from violating the levels of organization.

However, I still do not believe it is possible to match the deliberative supersystem's inherently broad applicability by implementing a separate computational subsystem for each problem context. Not only is it impossible to duplicate general intelligence through the sum of such subsystems, I suspect it is impossible to achieve humanlike performance in most *single* contexts using specialized AI. Occasionally we use abstract deliberation to solve modality-level problems for which we lack sensory modalities, and in this case it is possible for AI projects to solve the problem on the modality level, but the resulting problem-solving method will be very different from the human one, and will not generalize outside the specific domain. Hence Deep Blue.

Even on the level of individual domain competencies, not all competencies are unrelated to each other. Different minds may have different abilities in different domains; a mind may have an "ability surface," with hills and spikes in areas of high ability; but a spike in an area such as *learning* or *self-improvement* tends to raise the rest of the ability surface [103]. The talents and subsystems that are general in the sense of contributing to many domain competencies – and the domain competencies of self-improvement; see Sect. 3 – occupy a strategic position in AI analogous to the central squares in chess.

Self

When can an AI legitimately use the word "I"?

(For the sake of this discussion, I must give the AI a temporary proper name; I will use "Aisa" during this discussion.)

A classical AI that contains a LISP token for "hamburger" knows nothing about hamburgers; at most the AI can recognize recurring instances of a letter-sequence typed by programmers. Giving an AI a suggestively named data structure or function does not make that component the functional analogue of the similarly named human feature [66]. At what point can Aisa talk about something called "Aisa" without Drew McDermott popping up and accusing us of using a term that might as well translate to "G0025"?

Suppose that Aisa, in addition to modeling virtual environments and/or the outside world, also models certain aspects of internal reality, such as the effectiveness of heuristic beliefs used on various occasions. The degrees of binding between a model and reality are sensory, predictive, decisive, and manipulative. Suppose that Aisa can sense when a heuristic is employed, notice that heuristics tend to be employed in certain contexts and that they tend to have certain results, and use this inductive evidence to formulate expectations about when a heuristic will be employed and predict the results on its employment. Aisa now predictively models Aisa; it forms beliefs about its operation by observing the introspectively visible effects of its underlying mechanisms. Tightening the binding from predictive to manipulative requires that Aisa link introspective observations to internal actions; for example, Aisa may observe that devoting discretionary computational power to a certain subprocess yields thoughts of a certain kind, and that thoughts of this kind are useful in certain contexts, and subsequently devote discretionary power to that subprocess in those contexts.

A manipulative binding between Aisa and Aisa's model of Aisa is enough to let Aisa legitimately say "Aisa is using heuristic X," such that using the term "Aisa" is materially different from using "hamburger" or "G0025". But can Aisa legitimately say, "I am using heuristic X"?

My favorite quote on this subject comes from Douglas Lenat, although I cannot find the reference and am thus quoting from memory: "While Cyc knows that there is a thing called Cyc, and that Cyc is a computer, it does not know that it is Cyc."[33] Personally, I would question whether Cyc knows that Cyc is a computer – but regardless, Lenat has made a legitimate and fundamental distinction. Aisa modeling a thing called Aisa is not the same as Aisa modeling itself.

In an odd sense, assuming that the problem exists is enough to solve the problem. If another step is required before Aisa can say "I am using heuristic X," then there must be a material difference between saying "Aisa is using heuristic X" and "I am using heuristic X." And that is one possible answer: Aisa can say "I" when the behavior of modeling itself is materially different, because of the self-reference, from the behavior of modeling another AI that happens to look like Aisa.

One specific case where self-modeling is materially different than other-modeling is in planning. Employing a complex plan in which a linear sequence of actions A, B, C are individually necessary and together sufficient to accom-

---

[33] Lenat may have said this in the early days of Cyc. In a 1997 interview in Wired article, Lenat claims: "Cyc is already self-aware. If you ask it what it is, it knows that it is a computer." [25]

plish goal G requires an implicit assumption that the AI will follow through on its own plans; action A is useless unless it is followed by actions B and C, and action A is therefore not desirable unless actions B and C are predicted to follow. Making complex plans does not actually *require* self-modeling, since many classical AIs engage in planning-like behaviors using programmatic assumptions in place of reflective reasoning, and in humans the assumption is usually automatic rather than being the subject of deliberation. However, deliberate reflective reasoning about complex plans requires an understanding that the future actions of the AI are determined by the decisions of the AI's future self, that there is some degree of continuity (although not perfect continuity) between present and future selves, and that there is thus some degree of continuity between present decisions and future actions.

An intelligent mind navigates a universe with four major classes of variables: Random factors, variables with hidden values, the actions of other agents, and the actions of the self. The space of possible actions differs from the spaces carved out by other variables because the space of possible actions is under the AI's control. One difference between "Aisa will use heuristic X" and "*I* will use heuristic X" is the degree to which heuristic usage is under Aisa's deliberate control – the degree to which Aisa has goals relating to heuristic usage, and hence the degree to which the observation "I predict that I will use heuristic X" affects Aisa's subsequent actions. Aisa, if sufficiently competent at modeling other minds, might predict that a similar AI named Aileen would also use heuristic X, but beliefs about Aileen's behaviors would be derived from predictive modeling of Aileen, and not decisive planning of internal actions based on goal-oriented selection from the space of possibilities. There is a cognitive difference between Aisa saying "I predict Aileen will use heuristic X" and "I plan to use heuristic X." On a systemic level, the global specialness of "I" would be nailed down by those heuristics, beliefs, and expectations that individually relate specially to "I" because of introspective reflectivity or the space of undecided but decidable actions. It is my opinion that such an AI would be able to legitimately use the word "I", although in humans the specialness of "I" may be nailed down by additional cognitive forces as well. (Legitimate use of "I" is explicitly *not* offered as a necessary and sufficient condition for the "hard problem of conscious experience" [11] or social, legal, and moral personhood.)

3 Seed AI

In the space between the theory of human intelligence and the theory of general AI is the ghostly outline of a theory of *minds in general,* specialized for humans and AIs. I have not tried to lay out such a theory explicitly, confining myself to discussing those specific similarities and differences of humans and AIs that I feel are worth guessing in advance. The Copernican revolution for cognitive science – humans as a noncentral special case – is not yet ready; it takes three points to draw a curve, and currently we only have one. Nonetheless, humans *are in fact* a noncentral special case, and this abstract fact is knowable even if our current theories are anthropocentric.

There is a fundamental rift between evolutionary design and deliberative design. From the perspective of a deliberative intelligence – a human, for instance – evolution is the degenerate case of design-and-test where intelligence equals zero. Mutations are atomic; recombinations are random; changes are made on the genotype's lowest level of organization (flipping genetic bits); the grain size of the component tested is the whole organism; and the goodness metric operates solely through induction on historically encountered cases, without deductive reasoning about which contextual factors may later change[34]. The evolution of evolvability [106] improves this picture somewhat. There is a tendency for low-level genetic bits to exert control over high-level complexity, so that changes to those genes can create high-level changes. Blind selection pressures can create self-wiring and self-repairing systems that turn out to be highly evolvable because of their ability to phenotypically adapt to genotypical changes. Nonetheless, the evolution of evolvability is not a substitute for intelligent design. Evolution works, despite local inefficiences, because evolution exerts vast cumulative design pressure over time.

However, the total amount of design pressure exerted over a given time is limited; there is only a limited amount of selection pressure to be divided up among all the genetic variances selected on in any given generation [107]. One obvious consequence is that evolutionarily recent adaptations will probably be less optimized than those which are evolutionarily ancient. In DGI, the evolutionary phylogeny of intelligence roughly recapitulates its functional ontogeny; it follows that higher levels of organization may contain less total complexity than lower levels, although sometimes higher levels of organization are also more evolvable. Therefore, a subtler consequence is that the lower levels of organization are likely to be less well adapted to evolutionarily recent innovations (such as deliberation) than those higher levels to the lower levels – an effect enhanced by evolution's structure-preserving properties, including the preservation of structure that evolved in the absence of deliberation. Any design possibilities that first opened up with the appearance of *Homo sapiens sapiens* remain unexploited because *Homo sapiens sapiens* has only existed for 50,000-100,000 years; this is enough time to select among variances in quantitative tendencies, but not really enough time to construct complex functional adaptation. Since only *Homo sapiens sapiens* in its most modern form is known to engage in computer programming, this may explain why we do not yet have the capacity to reprogram our own neurons (said with tongue firmly in cheek, but there's still a grain of truth). And evolution is

---

[34]Viewing *evolution itself* through the lens provided by DGI is just *barely* possible. There are so many differences as to render the comparison one of "loose analogy" rather than "special case." This is as expected; evolution is not intelligent, although it may sometimes appear so.

extremely conservative when it comes to wholesale revision of architectures; the homeotic genes controlling the embryonic differentiation of the forebrain, midbrain, and hindbrain have identifiable homologues in the developing head of the *Drosophila* fly(!) [40].

Evolution never refactors its code. It is far easier for evolution to stumble over a thousand individual optimizations than for evolution to stumble over two simultaneous changes which are together beneficial and separately harmful. The genetic code that specifies the mapping between codons (a codon is three DNA bases) and the 20 amino acids is inefficient; it maps 64 possible codons to 20 amino acids plus the stop code. Why hasn't evolution shifted one of the currently redundant codons to a new amino acid, thus expanding the range of possible proteins? Because for any complex organism, the smallest change to the behavior of DNA – the lowest level of genetic organization – would destroy virtually all higher levels of adaptive complexity, unless the change were accompanied by millions of other simultaneous changes throughout the genome to shift every suddenly-nonstandard codon to one of its former equivalents. Evolution simply cannot handle simultaneous dependencies, unless individual changes can be deployed incrementally, or multiple phenotypical effects occur as the consequence of a single genetic change. For humans, planning coordinated changes is routine; for evolution, impossible. Evolution is hit with an enormous discount rate when exchanging the paper currency of incremental optimization for the hard coin of complex design.

We should expect the human design to incorporate an intimidatingly huge number of simple functional optimizations. But it is also understandable if there are deficits in the higher design. While the higher levels of organization (including deliberation) have emerged from the lower levels and hence are fairly well adapted to them, the lower levels of organization are not as adapted to the existence of deliberate intelligence. Humans were constructed by accretive evolutionary processes, moving from very complex nongeneral intelligence to very complex general intelligence, with deliberation the last layer of icing on the cake.

Can we exchange the hard coin of complex design for the paper currency of low-level optimization? "Optimizing compilers" are an obvious step but a tiny one; program optimization makes programs faster but exerts no design pressure for better functional organization, even for simple functions of the sort easily optimized by evolution. Directed evolution, used on modular subtasks with clearly defined performance metrics, would be a somewhat larger step. But even directed evolution is still the degenerate case of design-and-test where individual steps are unintelligent. We are, by assumption, building an AI. Why use *unintelligent* design-and-test?

Admittedly, there is a chicken-and-egg limit on relying on an AI's intelligence to help build an AI. Until a stably functioning cognitive supersystem is achieved, only the nondeliberative intelligence exhibited by pieces of the system will be available. Even after the achievement of a functioning supersystem – a heroic feat in itself – the intelligence exhibited by this supersystem will initially be very weak. The weaker an AI's intelligence, the less ability the AI will show in understanding complex holonic systems. The weaker an AI's abilities at holonic design, the smaller the parts of itself that the AI will be able to understand. At whatever time the AI finally becomes smart enough to participate in its own creation, the AI will initially need to concentrate on improving small parts of itself with simple and clear-cut performance metrics supplied by the programmers. This is not a special case of a stupid AI trying to understand itself, but a special case of a stupid AI trying to understand any complex holonic system; when the AI is "young" it is likely to be limited to understanding simple elements of a system, or small organizations of elements, and only where clear-cut goal contexts exist (probably programmer-explained). But even a primitive holonic design capability could cover a human gap; we don't like fiddling around with little things because we get bored, and we lack the ability to trade our massive parallelized power on complex problems for greater serial speed on simple problems. Similarly, it would be unhealthy (would result in AI pathologies) for human programming abilities to play a permanent role in learning or optimizing concept kernels – but at the points where interference seems tempting, it is perfectly acceptable for the *AI's* deliberative processes to play a role, if the AI has advanced that far.

Human intelligence, created by evolution, is characterized by evolution's design signature. The vast majority of our genetic history took place in the *absence* of deliberative intelligence; our older cognitive systems are poorly adapted to the possibilities inherent in deliberation. Evolution has applied vast design pressures to us but has done so very unevenly; evolution's design pressures are filtered through an unusual methodology that works far better for hand-massaging code than for refactoring program architectures.

Now imagine a mind built in its own presence by intelligent designers, beginning from primitive and awkward subsystems that nonetheless form a complete supersystem. Imagine a development process in which the elaboration and occasional refactoring of the subsystems can co-opt any degree of intelligence, however small, exhibited by the supersystem. The result would be a fundamentally different design signature, and a new approach to Artificial Intelligence which I call *seed AI*.

A seed AI is an AI designed for self-understanding, self-modification, and recursive self-improvement. This has implications both for the functional architectures needed to achieve primitive intelligence, and for the later development of the AI if and when its holonic self-understanding begins to improve. Seed AI is not a *workaround* that avoids the challenge of general intelligence by bootstrapping from an unintelligent core; seed AI only begins to yield benefits once there is some degree of available intelligence to be utilized. The later consequences of seed AI (such as true recursive self-improvement) only show up after the AI has achieved significant holonic understanding and general intelligence. The bulk of this chapter, Sect. 2, describes the general intelligence that is prerequisite to seed AI; Sect. 3 assumes some degree of success in constructing general intelligence and asks what may happen afterward. This may seem like hubris, but there are interesting things to be learned thereby, some of which imply design considerations for earlier architecture.

3.1 Advantages of Minds-in-General

To the computer programmers in the audience, it may seem like breathtaking audacity if I dare to predict any advantages for AIs in advance of construction, given past failures. The evolutionary psychologists will be less awed, knowing that in many ways the human mind is an astonishingly flimsy piece of work. If discussing the potential advantages of "AIs" strikes you as too audacious, then consider what follows, not as discussing the potential advantages of "AIs," but as discussing the potential advantages of *minds in general* relative to humans. One may then consider separately the audacity involved in claiming that a given AI approach can achieve one of these advantages, or that it can be done in less than fifty years.

Humans definitely possess the following advantages, relative to *current* AIs:

- We are smart, flexible, generally intelligent organisms with an enormous base of evolved complexity, years of real-world experience, and 10^{14} parallelized synapses, and current AIs are not.

Humans probably possess the following advantages, relative to intelligences developed by humans on forseeable extensions of current hardware:

- Considering each synaptic signal as roughly equivalent to a floating-point operation, the raw computational power of a human is enormously in excess of any current supercomputer or clustered computing system, although Moore's Law continues to eat up this ground [75].
- Human neural hardware – the wetware layer – offers built-in support for operations such as pattern recognition, pattern completion, optimization for recurring problems, et cetera; this support was added from below, taking advantage of microbiological features of neurons, and could be enormously expensive to simulate computationally to the same degree of ubiquity.
- With respect to the holonically simpler levels of the system, the total amount of "design pressure" exerted by evolution over time is probably considerably in excess of the design pressure that a reasonably-sized programming team could expect to personally exert.
- Humans have an extended history as intelligences; we are proven software.

Current computer programs definitely possess these mutually synergetic advantages relative to humans:

- Computer programs can perform highly repetitive tasks without boredom.

- Computer programs can execute complex extended tasks without making that class of human errors caused by distraction or short-term memory overflow in abstract deliberation.
- Computer hardware can perform extended sequences of simple steps at much greater *serial* speeds than human abstract deliberation or even human 200Hz neurons.
- Computer programs are fully configurable by the general intelligences called humans. (Evolution, the designer of humans, cannot invoke general intelligence.)

These advantages will not necessarily carry over to real AI. A real AI is not a computer program any more than a human is a cell. The relevant complexity exists at a much higher layer of organization, and it would be inappropriate to generalize stereotypical characteristics of computers to real AIs, just as it would be inappropriate to generalize the stereotypical characteristics of amoebas to modern-day humans. One might say that a real AI *consumes* computing power but is not a computer. This basic distinction has been confused by many cases in which the label "AI" has been applied to constructs that turn out to be only computer programs; but we should still expect the distinction to hold true of real AI, when and if achieved.

The potential cognitive advantages of *minds-in-general,* relative to humans, probably include:

New sensory modalities: Human programmers, lacking a sensory modality for assembly language, are stuck with abstract reasoning plus compilers. We are not entirely helpless, even this far outside our ancestral environment – but the traditional fragility of computer programs bears witness to our awkwardness. Minds-in-general may be able to exceed human programming ability with relatively primitive *general* intelligence, given a sensory modality for code.

Blending-over of deliberative and automatic processes: Human wetware has very poor support for the realtime diversion of processing power from one subsystem to another. Furthermore, a computer can burn serial speed to generate parallel power but neurons cannot do the reverse. Minds-in-general may be able to carry out an uncomplicated, relatively uncreative track of deliberate thought using simplified mental processes that run at higher speeds – an idiom that blurs the line between "deliberate" and "algorithmic" cognition. Another instance of the blurring line is co-opting deliberation into processes that are algorithmic in humans; for example, minds-in-general may choose to make use of top-level intelligence in forming and encoding the concept kernels of categories. Finally, a sufficiently intelligent AI might be able to incorporate *de novo* programmatic functions into deliberative processes – as if Gary Kasparov[35] could interface

---

[35] Former world champion in chess, beaten by the computer Deep Blue.

his brain to a computer and write search trees to contribute to his intuitive perception of a chessboard.

Better support for introspective perception and manipulation: The comparatively poor support of the human architecture for low-level introspection is most apparent in the extreme case of modifying code; we can think thoughts about thoughts, but not thoughts about individual neurons. However, other cross-level introspections are also closed to us. We lack the ability to introspect on concept kernels, focus-of-attention allocation, sequiturs in the thought process, memory formation, skill reinforcement, et cetera; we lack the ability to introspectively notice, induce beliefs about, or take deliberate actions in these domains.

The ability to add and absorb new hardware: The human brain is instantiated with a species-typical upper limit on computing power and loses neurons as it ages. In the computer industry, computing power continually becomes exponentially cheaper, and serial speeds exponentially faster, with sufficient regularity that "Moore's Law" [72] is said to govern its progress. Nor is an AI project limited to waiting for Moore's Law; an AI project that displays an important result may conceivably receive new funding which enables the project to buy a much larger clustered system (or rent a larger computing grid), perhaps allowing the AI to absorb hundreds of times as much computing power. By comparison, the 5-million-year transition from *Australopithecus* to *Homo sapiens sapiens* involved a tripling of cranial capacity relative to body size, and a further doubling of prefrontal volume relative to the expected prefrontal volume for a primate with a brain our size, for a total sixfold increase in prefrontal capacity relative to primates [17]. At 18 months per doubling, it requires 3.9 years for Moore's Law to cover this much ground. Even granted that intelligence is more software than hardware, this is still impressive.

Agglomerativity: An advanced AI is likely to be able to communicate with other AIs at much higher bandwidth than humans communicate with other humans – including sharing of thoughts, memories, and skills, in their underlying cognitive representations. An advanced AI may also choose to internally employ multithreaded thought processes to simulate different points of view. The traditional hard distinction between "groups" and "individuals" may be a special case of human cognition rather than a property of minds-in-general. It is even possible that no one project would ever choose to split up available hardware among more than one AI. Much is said about the benefits of cooperation between humans, but this is because there is a species limit on individual brainpower. We solve difficult problems using many humans because we cannot solve difficult problems using *one big* human. Six humans have a fair advantage relative to one human, but one human has a tremendous advantage relative to six chimpanzees.

Hardware that has different, but still powerful, advantages: Current computing systems lack good built-in support for biological neural functions such as automatic optimization, pattern completion, massive parallelism, etc. However, the bottom layer of a computer system is well-suited to operations such as reflectivity, execution traces, lossless serialization, lossless pattern transformations, very-high-precision quantitative calculations, and algorithms which involve iteration, recursion, and extended complex branching. Also in this category, but important enough to deserve its own section, is:

Massive serialism: Different 'limiting speed' for simple cognitive processes. No matter how simple or computationally inexpensive, the speed of a human cognitive process is bounded by the 200Hz limiting speed of spike trains in the underlying neurons. Modern computer chips can execute billions of *sequential* steps per second. Even if an AI must "burn" this serial speed to imitate parallelism, simple (routine, noncreative, nonparallel) deliberation might be carried out substantially (orders of magnitude) faster than more computationally intensive thought processes. If enough hardware is available to an AI, or if an AI is sufficiently optimized, it is possible that even the AI's full intelligence may run substantially faster than human deliberation.

Freedom from evolutionary misoptimizations: The term "misoptimization" here indicates an evolved feature that was adaptive for inclusive reproductive fitness in the ancestral environment, but which today conflicts with the goals professed by modern-day humans. If we could modify our own source code, we would eat Hershey's lettuce bars, enjoy our stays on the treadmill, and use a volume control on "boredom" at tax time.

Everything evolution just didn't think of: This catchall category is the flip side of the human advantage of "tested software" – humans aren't necessarily *good* software, just *old* software. Evolution cannot create design improvements which surmount simultaneous dependencies unless there exists an incremental path, and even then will not execute those design improvements unless that particular incremental path happens to be adaptive for other reasons. Evolution exhibits no predictive foresight and is strongly constrained by the need to preserve existing complexity. Human programmers are free to be creative.

Recursive self-enhancement: If a seed AI can improve itself, each local improvement to a design feature means that the AI is now partially the *source* of that feature, in partnership with the original programmers. Improvements to the AI are now improvements to the *source* of the feature, and may thus trigger further improvement in that feature. Similarly, where the seed AI idiom means that a cognitive talent co-opts a domain competency in internal manipulations, improvements to intelligence may improve the domain competency and thereby improve the cognitive talent. From a broad perspective, a mind-in-general's self-improvements may result in a higher level of intelligence and thus an increased ability to originate new self-improvements.

3.2 Recursive Self-enhancement

Fully recursive self-enhancement is a potential advantage of minds-in-general that has no analogue in nature – not just no analogue in human intelligence, but no analogue in *any* known process. Since the divergence of the hominid family within the primate order, further developments have occurred at an accelerating pace – but this is not because the character of the evolutionary process changed or became "smarter;" successive adaptations for intelligence and language opened up new design possibilities and also tended to increase the selection pressures for intelligence and language. Similarly, the exponentially accelerating increase of cultural knowledge in *Homo sapiens sapiens* was triggered by an underlying change in the human brain, but has not itself had time to create any significant changes in the human brain. Once *Homo sapiens sapiens* arose, the subsequent runaway acceleration of cultural knowledge took place with essentially constant brainware. The exponential increase of culture occurs because acquiring new knowledge makes it easier to acquire more knowledge.

The accelerating development of the hominid family and the exponential increase in human culture are both instances of *weakly self-improving processes,* characterized by an externally constant process (evolution, modern human brains) acting on a complexity pool (hominid genes, cultural knowledge) whose elements interact synergetically. If we divide the process into an improver and a content base, then weakly self-improving processes are characterized by an external improving process with roughly constant characteristic intelligence, and a content base within which positive feedback takes place under the dynamics imposed by the external process.

If a seed AI begins to improve itself, this will mark the beginning of the AI's *self-encapsulation.* Whatever component the AI improves will no longer be *caused* entirely by humans; the cause of that component will become, at least in part, the AI. Any improvement to the AI will be an improvement to the *cause* of a component of the AI. If the AI is improved further – either by the external programmers, or by internal self-enhancement – the AI may have a chance to re-improve that component. That is, any improvement to the AI's global intelligence may indirectly result in the AI improving local components. This secondary enhancement does not necessarily enable the AI to make a further, tertiary round of improvements. If only a few small components have been self-encapsulated, then secondary self-enhancement effects are likely to be small, not on the same order as improvements made by the human programmers.

If computational subsystems give rise to cognitive talents, and cognitive talents plus acquired expertise give rise to domain competencies, then self-improvement is a means by which domain competencies can wrap around and improve computational subsystems, just as the seed AI idiom of co-opting deliberative functions into cognition enables improvements in domain competencies to wrap around and improve cognitive talents, and the ordinary idiom of intelligent learning enables domain competencies to wrap around and improve acquired expertise[36]. The degree to which domain competencies improve underlying processes will depend on the AI's degree of advancement; successively more advanced intelligence is required to improve expertise, cognitive talents, and computational subsystems. The degree to which an improvement in intelligence cascades into further improvements will be determined by how much self-encapsulation has already taken place on different levels of the system.

A seed AI is a *strongly self-improving process,* characterized by improvements to the content base that exert direct positive feedback on the intelligence of the underlying improving process. The exponential surge of human cultural knowledge was driven by the action of an already-powerful but constant force, human intelligence, upon a synergetic content base of cultural knowledge. Since strong self-improvement in seed AI involves an initially very weak but improving intelligence, it is not possible to conclude from analogies with human cultural progress that strongly recursive self-improvement will obey an exponential lower bound during early stages, nor that it will obey an exponential upper bound during later stages. Strong self-improvement is a mixed blessing in development. During earlier epochs of seed AI, the dual process of programmer improvement and self-improvement probably sums to a process entirely dominated by the human programmers. We cannot rely on exponential bootstrapping from an unintelligent core. However, we may be able to achieve powerful results by bootstrapping from an *intelligent* core, if and when such a core is achieved. *Recursive* self-improvement is a consequence of seed AI, not a cheap way to achieve AI.

It is possible that self-improvement will become cognitively significant relatively early in development, but the wraparound of domain competencies to improve expertise, cognition, and subsystems does not imply strong effects from *recursive* self-improvement. Precision in discussing seed AI trajectories requires distinguishing between epochs for holonic understanding, epochs for programmer-dominated and AI-dominated development, epochs for recursive and nonrecursive self-improvement, and epochs for overall intelligence.

(Readers allergic to advance discussion of sophisticated AI may consider these epochs as referring to minds-in-general that possess physical access to their own code and some degree of general intelligence with which to manipulate it; the rationale for distinguishing between epochs may be considered

---

[36]It is sometimes objected that an intelligence modifying itself is "circular" and therefore impossible. This strikes me as a complete *non sequitur,* but even if it were not, the objection is still based on the idea of intelligence as an opaque monolithic function. The character of the computational subsystems making up intelligence is fundamentally different from the character of the high-level intelligence that exists atop the subsystems. High-level intelligence can wrap around to make improvements to the subsystems in their role as computational processes without ever *directly* confronting the allegedly sterile problem of "improving itself" – though as said, I see nothing sterile about this.

separately from the audacity of suggesting that AI can progress to any given epoch.)

Epochs for holonic understanding and holonic programming:

1. First epoch: The AI can transform code in ways that do not affect the algorithm implemented. ("Understanding" on the order of an optimizing compiler; i.e., not "understanding" in any real sense.)
2. Second epoch: The AI can transform algorithms in ways that fit simple abstract beliefs about the design purposes of code. That is, the AI would understand what a stack implemented as a linked list and a stack implemented as an array have in common. (Note that this is already out of range of current AI, at least if you want the AI to figure it out on its own.)
3. Third epoch: The AI can draw a holonic line from simple internal metrics of *cognitive* usefulness (how fast a concept is cued, the usefulness of the concept returned) to specific algorithms. Consequently the AI would have the theoretical capability to invent and test new algorithms. This does not mean the AI would have the ability to invent *good* algorithms or *better* algorithms, just that invention in this domain would be theoretically possible. (An AI's theoretical capacity for invention does not imply capacity for improvement over and above the programmers' efforts. This is determined by relative domain competencies and by relative effort expended at a given focal point.)
4. Fourth epoch: The AI has a concept of "intelligence" as the final product of a continuous holonic supersystem. The AI can draw a continuous line from (a) its abstract understanding of intelligence to (b) its introspective understanding of cognition to (c) its understanding of source code and stored data. The AI would be able to invent an algorithm or cognitive process that contributes to intelligence in a novel way and integrate that process into the system. (Again, this does not automatically imply that the AI's inventions are improvements relative to existing processes.)

Epochs for sparse, continuous, and recursive self-improvement:

1. First epoch: The AI has a limited set of rigid routines which it applies uniformly. Once all visible opportunities are exhausted, the routines are used up. This is essentially analogous to the externally driven improvement of an optimizing compiler. An optimizing compiler may make a large number of improvements, but they are not self-improvements, and they are not design improvements. An optimizing compiler tweaks assembly language but leaves the program constant.
2. Second epoch: The cognitive processes which create improvements have characteristic complexity on the order of a classical search tree, rather than on the order of an optimizing compiler. Sufficient investments of computing power can sometimes yield extra improvements, but it is essentially an exponential investment for a linear improvement, and no matter how much

computing power is invested, the total kind of improvements conceivable are limited.

3. Third epoch: Cognitive complexity in the AI's domain competency for programming is high enough that at any given point there is a large number of visible possibilities for complex improvements, albeit perhaps minor improvements. The AI usually does not exhaust all visible opportunities before the programmers improve the AI enough to make new improvements visible. However, it is only programmer-driven improvements in intelligence which are powerful enough to open up new volumes of the design space.

4. Fourth epoch: Self-improvements sometimes result in genuine improvements to "smartness," "creativity," or "holonic understanding," enough to open up a new volume of the design space and make new possible improvements visible.

Epochs for relative human-driven and AI-driven improvement:

1. First epoch: The AI can make optimizations at most on the order of an optimizing compiler, and cannot make design improvements or increase functional complexity. The combination of AI and programmer is not noticeably more effective than a programmer armed with an ordinary optimizing compiler.

2. Second epoch: The AI can understand a small handful of components and make improvements to them, but the total amount of AI-driven improvement is small by comparison with programmer-driven development. Sufficiently major programmer improvements do very occasionally trigger secondary improvements. The total amount of work done by the AI on its own subsystems serves only as a measurement of progress and does not significantly accelerate work on AI programming.

3. Third epoch: AI-driven improvement is significant, but development is "strongly" programmer-dominated in the sense that overall systemic progress is driven almost entirely by the creativity of the programmers. The AI may have taken over some significant portion of the work from the programmers. The AI's domain competencies for programming may play a critical role in the AI's continued functioning.

4. Fourth epoch: AI-driven improvement is significant, but development is "weakly" programmer-dominated. AI-driven improvements and programmerdriven improvements are of roughly the same kind, but the programmers are better at it. Alternatively, the programmers have more subjective time in which to make improvements, due to the number of programmers or the slowness of the AI.

Epochs for overall intelligence:

1. Tool-level AI: The AI's behaviors are immediately and directly specified by the programmers, or the AI "learns" in a single domain using prespecified

learning algorithms. (In my opinion, tool-level AI as an alleged step on the path to more complex AI is highly overrated.)

2. Prehuman AI: The AI's intelligence is not a significant subset of human intelligence. Nonetheless, the AI is a cognitive supersystem, with some subsystems we would recognize, and at least some mind-like behaviors. A toaster oven does not qualify as a "prehuman chef," but a general kitchen robot might do so.

3. Infrahuman AI: The AI's intelligence is, overall, of the same basic character as human intelligence, but substantially inferior. The AI may excel in a few domains where it possesses new sensory modalities or other brainware advantages not available to humans. I believe that a worthwhile test of infrahumanity is whether humans talking to the AI recognize a mind on the other end. (An AI that lacks even a primitive ability to communicate with and model external minds, and cannot be taught to do so, does not qualify as infrahuman.)

It should again be emphasized that this entire discussion *assumes* that the problem of building a general intelligence is solvable. Without significant *existing* intelligence an alleged "AI" will remain permanently stuck in the first epoch of holonic programming – it will remain nothing more than an optimizing compiler. It is true that so far attempts at computer-based intelligence have failed, and perhaps there is a barrier which states that while 750 megabytes of DNA can specify physical systems which learn, reason, and display general intelligence, no amount of human design can do the same.

But if no such barrier exists – if it is possible for an artificial system to match DNA and display human-equivalent general intelligence – then it seems very probable that seed AI is achievable as well. It would be the height of biological chauvinism to assert that, while it *is* possible for humans to build an AI and improve this AI to the point of roughly human-equivalent general intelligence, this same human-equivalent AI can never master the (humanly solved) programming problem of making improvements to the AI's source code.

Furthermore, the above statement misstates the likely interrelation of the epochs. An AI does not need to wait for full human-equivalence to begin improving on the programmer's work. An optimizing compiler can "improve" over human work by expending greater *relative* effort on the assembly-language level. That is, an optimizing compiler uses the programmatic advantages of *greater serial speed* and *immunity to boredom* to apply much greater design pressures to the assembly-language level than a human could exert *in equal time*. Even an optimizing compiler might fail to match a human at hand-massaging a small chunk of time-critical assembly language. But, at least in today's programming environments, humans no longer hand-massage most code – in part because the task is best left to optimizing compilers, and in part because it's extremely boring and wouldn't yield much benefit relative to making further high-level improvements. A sufficiently advanced AI that takes advantage of *massive serialism* and *freedom from evolutionary misoptimizations* may be able to apply massive design pressures to higher holonic levels of the system.

Even at our best, humans are not very good programmers; programming is not a task commonly encountered in the ancestral environment. A human programmer is metaphorically a blind painter – not just a blind painter, but a painter entirely lacking a visual cortex. We create our programs like an artist drawing one pixel at a time, and our programs are fragile as a consequence. If the AI's human programmers can master the essential design pattern of sensory modalities, they can gift the AI with a sensory modality for code-like structures. Such a modality might perceptually interpret: a simplified interpreted language used to tutor basic concepts; any internal procedural languages used by cognitive processes; the programming language in which the AI's code level is written; and finally the native machine code of the AI's hardware. An AI that takes advantage of a codic modality may not need to wait for human-equivalent *general* intelligence to beat a human in the *specific domain competency* of programming. Informally, an AI is native to the world of programming, and a human is not.

This leads inevitably to the question of how much programming ability would be exhibited by a seed AI with human-equivalent general intelligence *plus* a codic modality. Unfortunately, this leads into territory that is generally considered taboo within the field of AI. Some readers may have noted a visible incompleteness in the above list of seed AI epochs; for example, the last stage listed for human-driven and AI-driven improvement is "weak domination" of the improvement process by human programmers (the AI and the programmers make the same kind of improvements, but the programmers make more improvements than the AI). The obvious succeeding epoch is one in which AI-driven development roughly equals human development, and the epoch after that one in which AI-driven development exceeds human-driven development. Similarly, the discussion of epochs for recursive self-improvement stops at the point where AI-driven improvement sometimes opens up new portions of the opportunity landscape, but does not discuss the possibility of open-ended self-improvement: a point beyond which progress can continue in the absence of human programmers, so that by the time the AI uses up all the improvements visible at a given level, that improvement is enough to "climb the next step of the intelligence ladder" and make a new set of improvements visible. The epochs for overall intelligence define tool-level, prehuman, and infrahuman AI, but do not define human-equivalence or transhumanity.

3.3 Infrahumanity and Transhumanity: "Human-Equivalence" as Anthropocentrism

It is interesting to contrast the separate perspectives of modern-day Artificial Intelligence researchers and modern-day evolutionary psychologists with respect to the particular level of intelligence exhibited by *Homo sapiens sapiens.*

Modern-day AI researchers are strongly reluctant to discuss human equivalence, let alone what might lie beyond it, as a result of past claims for "human equivalence" that fell short. Even among those rare AI researchers who are still willing to discuss general cognition, the attitude appears to be: "First we'll achieve general cognition, then we'll talk human-equivalence. As for transhumanity, forget it."

In contrast, modern-day evolutionary theorists are strongly trained against Panglossian or anthropocentric views of evolution, i.e., those in which humanity occupies any special or best place in evolution. Here it is socially unacceptable to suggest that *Homo sapiens sapiens* represents cognition in an optimal or maximally developed form; in the field of evolutionary psychology, the overhanging past is one of Panglossian optimism. Rather than modeling the primate order and hominid family as evolving *toward* modern-day humanity, evolutionary psychologists try to model the hominid family as evolving somewhere, which then decided to call itself "humanity." (This view is beautifully explicated in Terrence Deacon's "The Symbolic Species" [18].) Looking back on the history of the hominid family and the human line, there is no reason to believe that evolution has hit a hard upper limit. *Homo sapiens* has existed for a short time by comparison with the immediately preceding species, *Homo erectus*. We look back on our evolutionary history from this vantage point, not because evolution stopped at this point, but because the subspecies *Homo sapiens sapiens* is the very first elaboration of primate cognition to cross over the minimum line that supports rapid cultural growth and the development of evolutionary psychologists. We observe human-level intelligence in our vicinity, not because human intelligence is optimal or because it represents a developmental limit, but because of the Anthropic Principle; we are the first intelligence smart enough to look around. Should basic design limits on intelligence exist, it would be an astonishing coincidence if they centered on the human level.

Strictly speaking, the attitudes of AI and evolutionary psychology are not irreconcilable. One could hold that achieving general cognition will be extremely hard and that this constitutes the immediate research challenge, while simultaneously holding that once AI is achieved, only ungrounded anthropocentrism would predict that AIs will develop to a human level and then stop. This hybrid position is the actual stance I have tried to maintain throughout this paper – for example, by decoupling discussion of developmental epochs and advantages of minds-in-general from the audacious question of whether AI can achieve a given epoch or advantage.

But it would be silly to pretend that the tremendous difficulty of achieving general cognition licenses us to sweep its enormous consequences under the rug. Despite AI's glacial slowness by comparison with more tractable research areas, Artificial Intelligence is still improving at an *enormously* faster rate than human intelligence. A human may contain millions or hundreds of millions of times as much processing power as a personal computer circa 2002, but computing power per dollar is (still) doubling every eighteen months, and human brainpower is not.

Many have speculated whether the development of human-equivalent AI, however and whenever it occurs, will be shortly followed by the development of transhuman AI [74, 102, 70, 56, 39, 30]. Once AI exists it can develop in a number of different ways; for an AI to develop to the point of human-equivalence and then remain at the point of human-equivalence for an extended period would require that all liberties be simultaneously blocked[37] at exactly the level which happens to be occupied by *Homo sapiens sapiens*. This is too much coincidence. Again, we observe *Homo sapiens sapiens* intelligence in our vicinity, not because *Homo sapiens sapiens* represents a basic limit, but because *Homo sapiens sapiens* is the very first hominid subspecies to cross the minimum line that permits the development of evolutionary psychologists.

Even if this were not the case – if, for example, we were now looking back on an unusually long period of stagnation for *Homo sapiens* – it would still be an unlicensed conclusion that the fundamental design bounds which hold for *evolution* acting on *neurons* would hold for *programmers* acting on *transistors*. Given the different design methods and different hardware, it would again be too much of a coincidence.

This holds doubly true for seed AI. The behavior of a strongly self-improving process (a mind with access to its own source code) is not the same as the behavior of a weakly self-improving process (evolution improving humans, humans improving knowledge). The ladder question for recursive self-improvement – whether climbing one rung yields a vantage point from which enough opportunities are visible that they suffice to reach the next rung – means that effects need not be proportional to causes. The question is not how much of an effect any *given* improvement has, but rather how much of an effect the improvement plus further triggered improvements and *their* triggered improvements have. It is literally a domino effect – the universal metaphor for small causes with disproportionate results. Our instincts for system behaviors may be enough to give us an intuitive feel for the results of any single improvement, but in this case we are asking not about the fall of a single domino, but rather about how the dominos are arranged. We are asking whether the tipping of one domino is likely to result in an isolated fall, two isolated falls, a small handful of toppled dominos, or whether it will knock over the entire chain.

If I may be permitted to adopt the antipolarity of "conservatism" – i.e., asking how *soon* things could conceivably happen, rather than how late – then I must observe that we have *no idea* where the point of open-ended self-improvement is located, and furthermore, *no idea* how fast progress will occur after this point is reached. Lest we overestimate the total amount of intelligence required, it should be noted that nondeliberate evolution did eventually

---

[37]This is a metaphor from the game Go, where you capture an opponent's group of stones by eliminating all adjoining clear spaces, which are known as "liberties."

stumble across general intelligence; it just took a very long time. We do not know how much improvement over evolution's incremental steps is required for a strongly self-improving system to knock over dominos of sufficient size that each one triggers the next domino. Currently, I believe the best strategy for AI development is to try for general cognition as a necessary prerequisite of achieving the domino effect. But in theory, general cognition might not be required. Evolution managed without it. (In a sense this is disturbing, since, while I can see how it would be theoretically possible to bootstrap from a nondeliberative core, I cannot think of a way to place such a nondeliberative system within the human moral frame of reference.)

It is conceptually possible that a basic bound rules out all improvement of effective intelligence past our current level, but we have no evidence supporting such a bound. I find it difficult to credit that a bound holding for minds in general on all physical substrates coincidentally limits intelligence to the exact level of the very first hominid subspecies to evolve to the point of developing computer scientists. I find it equally hard to credit bounds that limit strongly self-improving processes to the characteristic speed and behavior of weakly self-improving processes. "Human equivalence," commonly held up as the great unattainable challenge of AI, is a chimera – in the sense of being both a "mythical creature" and an "awkward hybrid." Infrahuman AI and transhuman AI are both plausible as self-consistent durable entities. Human-equivalent AI is not.

Given the tremendous architectural and substrate differences between humans and AIs, and the different expected cognitive advantages, there are no current grounds for depicting an AI that strikes an anthropomorphic balance of domain competencies. Given the difference between weakly recursive self-improvement and strongly recursive self-improvement; given the ladder effect and domino effect in self-enhancement; given the different limiting subjective rates of neurons and transistors; given the potential of minds-in-general to expand hardware; and given that evolutionary history provides no grounds for theorizing that the *Homo sapiens sapiens* intelligence range represents a special slow zone or limiting point with respect to the development of cognitive systems; therefore, there are no current grounds for expecting AI to spend an extended period in the *Homo sapiens sapiens* range of general intelligence. *Homo sapiens sapiens* is not the center of the cognitive universe; we are a noncentral special case.

Under standard folk psychology, whether a task is easy or hard or extremely hard does not change the default assumption that people undertaking a task do so because they expect positive consequences for success. AI researchers continue to try and move humanity closer to achieving AI. However near or distant that goal, AI's critics are licensed under folk psychology to conclude that these researchers believe AI to be desirable. AI's critics may legitimately ask for an immediate defense of this belief, whether AI is held to be five years away or fifty. Although the topic is not covered in this paper, I personally pursue general cognition as a means to seed AI, and seed AI as a means to transhuman AI, because I believe human civilization will benefit greatly from breaching the upper bounds on intelligence that have held for the last fifty thousand years, and furthermore, that we are rapidly heading toward the point where we *must* breach the current upper bounds on intelligence for human civilization to survive. I would not have written a paper on recursively self-improving minds if I believed that recursively self-improving minds were inherently a bad thing, whether I expected construction to take fifty years or fifty thousand.

4 Conclusions

> People are curious about how things began, and especially about the origins of things they deem important. Besides satisfying such curiosity, accounts of origin may acquire broader theoretical or practical interest when they go beyond narrating historical accident, to impart insight into more enduring forces, tendencies, or sources from which the phenomena of interest more generally proceed. Accounts of evolutionary adaptation do this when they explain how and why a complex adaptation first arose over time, or how and why it has been conserved since then, in terms of selection on heritable variation. [...] In such cases, evolutionary accounts of origin may provide much of what early Greek thinkers sought in an *arche,* or origin – a unified understanding of something's original formation, source of continuing existence, and underlying principle.

> Leonard D. Katz, ed., "Evolutionary Origins of Morality" [47]

On the cover of Douglas Hofstadter's *Gödel, Escher, Bach: An Eternal Golden Braid* are two trip-lets – wooden blocks carved so that three orthogonal spotlights shining through the 3D block cast three different 2D shadows – the letters "G", "E", "B". The trip-let is a metaphor for the way in which a deep underlying phenomenon can give rise to a number of different surface phenomena. It is a metaphor about intersecting constraints that give rise to a whole that is *deeper* than the sum of the requirements, the multiplicative and not additive sum. It is a metaphor for arriving at a solid core by asking what casts the shadows, and how the core can be stronger than the shadows by reason of its solidity. (In fact, the trip-let itself could stand as a metaphor for the different metaphors cast by the trip-let concept.)

In seeking the *arche* of intelligence, I have striven to neither overstate nor understate its elegance. The central shape of cognition is a messy 4D object that casts the thousand subfields of cognitive science as 3D shadows. Using the relative handful of fields with which I have some small acquaintance, I have tried to arrive at a central shape which is no more and no less coherent than we would expect of evolution as a designer.

I have used the levels of organization as structural support for the theory, but have tried to avoid turning the levels of organization into Aristotelian straitjackets – permitting discussion of "beliefs," cognitive content that combines the nature of concept structures and learned complexity; or discussion of "sequiturs," brainware adaptations whose function is best understood on the thought level. The levels of organization are visibly pregnant with evolvability and plead to be fit into specific accounts of human evolution – but this does not mean that our evolutionary history enacted a formal progress through Modalities, Concepts, and Thoughts, with each level finished and complete before moving on to the next. The levels of organization structure the functional decomposition of intelligence; they are not in themselves such a decomposition. Similarly, the levels of organization structure accounts of human evolution without being in themselves an account of evolution. We should not say that Thoughts evolved from Concepts; rather, we should consider a specific thought-level function and ask which specific concept-level functions are necessary and preadaptive for its evolution.

In building this theory, I have tried to avoid those psychological sources of error that I believe have given rise to past failures in AI; physics envy, Aristotelian straitjackets, magical analogies with human intelligence, and others too numerous to list. I have tried to give some explanation of past failures of AI, not just in terms of "*This* is the magic key we were missing all along (take two)," but in terms of "This is what the past researchers were looking at when they made the oversimplification, these are the psychological forces underlying the initial oversimplification and its subsequent social propagation, and this explains the functional consequences of the oversimplification in terms of the specific subsequent results as they appeared to a human observer." Or so I would *like* to say, but alas, I had no room in this chapter for such a complete account. Nonetheless I have tried, not only to give an account of some of AI's past failures, but also to give an account of how successive failures tried and failed to account for past failures. I have only discussed a few of the best-known and most-studied AI pathologies, such as the "symbol grounding problem" and "common-sense problem," but in doing so, I have tried to give accounts of their specific effects and specific origins.

Despite AI's repeated failures, and despite even AI's repeated failed attempts to dig itself out from under past failures, AI still has not dug itself in so deep that no possible new theory could dig itself out. If you show that a new theory does not contain a set of causes of failure in past theories – where the causes of failure include both surface scientific errors and underlying psychological errors, and these causes are together sufficient to account for observed pathologies – then this does not prove you have identified *all* the old causes of failure, or prove that the new theory will succeed, but it is sufficient to set the new approach aside from aversive reinforcement on past attempts. I can't promise that DGI will succeed – but I believe that even if DGI is slain, it won't be the AI dragon that slays it, but a new and different dragon. At the least I hope I have shown that, as a new approach, DGI-based seed AI is different enough to be worth trying.

As presented here, the theory of DGI has a great deal of potential for expansion. To put it less kindly, the present chapter is far too short. The chapter gives a descriptive rather than a constructive account of a functional decomposition of intelligence; the chapter tries to show evolvability, but does not give a specific account of hominid evolution; the chapter analyzes a few examples of past failures but does not fully reframe the history of AI. I particularly regret that the chapter fails to give the amount of background explanation that is usually considered standard for interdisciplinary explanations. In assembling the pieces of the puzzle, I have not been able to explain each of the pieces for those unfamiliar with it. I have been forced to the opposite extreme. On more than one occasion I have compressed someone else's entire lifework into one sentence and a bibliographic reference, treating it as a jigsaw piece to be snapped in without further explanation.

The only defense I can offer is that the central shape of intelligence is *enormous*. I was asked to write a chapter in a book, not a book in itself. Had I tried to describe interdisciplinary references in what is usually considered the minimum acceptable level of detail, this chapter would have turned into an encyclopedia. It is better to be accused of having failed to fully integrate a piece into the larger puzzle, than to leave that piece out entirely. If the chapter is unfinished then let it at least be *visibly* unfinished. This defies literary convention, but omitting facets of cognition is one of the chief sins of AI. In AI, it really is better to mention and not explain than to not mention and not explain – and at that, I have *still* been forced to leave things out. So to all those whose theories I have slighted by treating them in far less length than they deserve, my apologies. If it is any consolation, I have treated my own past work no differently than I have treated yours. The entire topic of Friendly AI has been omitted – except for one or two passing references to a "human moral frame of reference" – despite my feeling that discussion of the human moral frame of reference should not be severed from discussion of recursively self-improving generally intelligent minds.

I cannot promise that a book is on the way. At this point in the ritual progress of a general theory of cognition, there are two possible paths forward. One can embrace the test of fire in evolutionary psychology, cognitive psychology, and neuroscience, and try to show that the proposed new explanation is the most probable explanation for previously known evidence, and that it makes useful new predictions. Or, one can embrace the test of fire in Artificial Intelligence and try to build a mind. I intend to take the latter path as soon as my host organization finds funding, but this may not leave much time for writing future papers. Hopefully my efforts in this chapter will serve to argue that DGI is promising enough to be worth the significant funding needed for the acid test of building AI, although I acknowledge that my efforts in this chapter are not enough to put forth DGI as a strong hypothesis with respect to academia at large.

This chapter would not have been written without the support and assistance of a large number of people whose names I unfortunately failed to accumulate in a single location. At the least I would like to thank Peter Voss, Ben Goertzel, and Carl Feynman for discussing some of the ideas found in this chapter. Any minor blemishes remaining in this document are, of course, my fault. (Any major hideous errors or gaping logical flaws were probably smuggled in while I wasn't looking.) Without the Singularity Institute for Artificial Intelligence, this chapter would not exist. To all the donors, supporters, and volunteers of the Singularity Institute, my deepest thanks, but we're not finished with you yet. We still need to build an AI, and for that to happen, we need a lot more of you.

I apologize to the horde of authors whom I have inevitably slighted by failing to credit them for originating an idea or argument inadvertantly duplicated in this chapter; the body of literature in cognitive science is too large for any one person to be personally familiar with more than an infinitesimal fraction. As I was editing a draft of this chapter, I discovered the paper "Perceptual Symbol Systems" by Lawrence Barsalou [2]; as I submit this chapter I still have not read Barsalou's paper fully, but at minimum it describes a model in which concepts reify perceptual imagery and bind to perceptual imagery, and in which combinatorial concept structures create complex depictive mental imagery. Barsalou should receive full credit for first publication of this idea, which is one of the major theoretical foundations of DGI.

In today's world it is commonly acknowledged that we have a responsibility to discuss the moral and ethical questions raised by our work. I would take this a step farther and say that we not only have a responsibility to discuss those questions, but also to arrive at interim answers and guide our actions based on those answers – still expecting future improvements to the ethical model, but also willing to take action based on the best current answers. Artificial Intelligence is too profound a matter for us to have no better reply to such pointed questions as "Why?" than "Because we can!" or "I've got to make a living somehow." If *Homo sapiens sapiens* is a noncentral and nonoptimal special case of intelligence, then a world full of nothing but *Homo sapiens sapiens* is not necessarily the happiest world we could live in. For the last fifty thousand years, we've been trying to solve the problems of the world with *Homo sapiens sapiens* intelligence. We've made a lot of progress, but there are also problems that we've hit and bounced. Maybe it's time to use a bigger hammer.

References

1. Anderson T, Culler D, Patterson D, the NOW Team (1995) A Case for NOW (Networks of Workstations). *IEEE Micro*, 15(1):54–64.
2. Barsalou L (1999) Perceptual symbol systems. *Behavioral and Brain Sciences*, 22:577–609.

3. Becker D, Sterling T, Savarese D, Dorband J, Ranawak U, Packer C (1995) Beowulf: A Parallel Workstation for Scientific Computation. *Proceedings of the International Conference on Parallel Processing, 1995.*

4. Berlin B, Kay P (1969) *Basic Color Terms: Their Universality and Evolution.* University of California Press.

5. Bickerton D (1990) *Language and Species.* University of Chicago Press.

6. Bonmassar G, Schwartz E (1997) Space-variant Fourier Analysis: The Exponential Chirp Transform. *IEEE Pattern Analysis and Machine Vision,* 19:1080–1089.

7. Boynton R, Olson C (1987) Locating Basic Colors in the OSA Space, *Color Research and Application,* 12(2):94–105.

8. Brown R (1958) How Shall a Thing Be Called?. *Psychological Review,* 65:14–21.

9. Carey S (1992) Becoming a Face Expert. *Philosophical Transactions of the Royal Society of London,* 335:95–103.

10. Chalmers D, French R, Hofstadter D (1992) High Level Perception, Representation, and Analogy: A Critique of Artificial Intelligence Methodology. *Journal of Experimental and Theoretical AI,* 4(3):185–211.

11. Chalmers D (1995) Facing Up to the Problem of Consciousness. *Journal of Consciousness Studies,* 2(3):200–219.

12. Church RM, Meck WH (1984) *The Numerical Attribute of Stimuli.* In Roitblat H, Bever T, Terrace H (eds.), Animal cognition. Erlbaum.

13. Clearfield M, Mix K (1999) Number Versus Contour Length in Infants' Discrimination of Small Visual Sets. *Psychological Science,* 10(5):408–411.

14. Colton S, Bundy A, Walsh T (2000) On the Notion of Interestingness in Automated Mathematical Discovery. *International Journal of Human Computer Studies,* 53(3):351–375.

15. Cosmides L, Tooby J (1996) Are Humans Good Intuitive Statisticians After All? Rethinking Some Conclusions from the Literature on Judgement Under Uncertainty. *Cognition,* 58:1–73.

16. Dawkins R (1996) *Climbing Mount Improbable.* Norton.

17. Deacon T (1990) Rethinking Mammalian Brain Evolution. *American Zoologist,* 30:629–705.

18. Deacon T (1997) *The Symbolic Species.* Penguin.

19. Dedrick D (1998) *Naming the Rainbow: Colour Language, Colour Science, and Culture.* Kluwer Academic.

20. Dehaene S (1997) *The Number Sense: How the Mind Creates Mathematics,* Oxford University Press.

21. Dijkstra EW (1968) Go To Statement Considered Harmful. *Communications of the ACM,* 11(3):147–148.

22. Felleman D, van Essen D (1991) Distributed Hierarchical Processing in the Primate Cerebral Cortex. *Cerebral Cortex,* 1:1–47.

23. Finke RA, Schmidt MJ (1977) Orientation-specific Color After-effects Following Imagination. *Journal of Experimental Psychology: Human Perception and Performance,* 3:599–606.

24. Flack J, de Wall F (2000) *Any Animal Whatever: Darwinian Building Blocks of Morality in Monkeys and Apes.* In Katz L (ed.), Evolutionary Origins of Morality: Cross-Disciplinary Perspectives. Imprint Academic.

25. Garfinkel S (1997) Happy Birthday, Hal. *Wired, 5(1).*

26. Geist A, Beguelin A, Dongarra JJ, Jiang W, Manchek R, Sunderam VS (1994) PVM 3 User's Guide and Reference Manual. Technical Report ORNL/TM-12187, Oak Ridge National Laboratory.

27. Gould SJ, Lewontin R (1979) The spandrels of San Marco and the Panglossian Paradigm: A Critique of the Adaptationist Programme. *Proceedings of the Royal Society of London*, 205:281–288.

28. Gropp W, Lusk E, Skjellum A (1994) *Using MPI: Portable Parallel Programming with the Message-Passing Interface.* MIT Press.

29. Harnad S (1990) The Symbol Grounding Problem. *Physica D*, 42:335–346.

30. Hawking S (2001) Interview in *Focus Magazine.* Corrected English translation: http://www.kurzweilai.net/news/frame.html?main=news_single.html?id=495

31. Hayhoe M, Bensinger D, Ballard D (1998) Task Constraints in Visual Working Memory. *Vision Research*, 38(1):125–137.

32. Helmholtz H (1867) *Treatise on Physiological Optics.* Dover. First published in 1867.

33. Hexmoor H, Lammens J, Shapiro SC (1993) Embodiment in GLAIR: A Grounded Layered Architecture with Integrated Reasoning for Autonomous Agents. In *Proceedings of The Sixth Florida AI Research Symposium (FLAIRS 93)*.

34. Hochberg J (1957) Effects of the Gestalt Revolution: the Cornell Symposium on Perception. *Psychological Review*, 64(2):73–84.

35. Hofstadter D (1979) *Gödel, Escher, Bach: an Eternal Golden Braid.* Basic Books.

36. Hofstadter D (1985). Variations on a Theme as the Crux of Creativity. In Hofstadter D *Metamagical Themas: Questing for the Essence of Mind and Pattern*, Basic Books.

37. Hofstadter D, Mitchell M (1988) Conceptual Slippage and Analogy-making: A Report on the Copy-cat Project. In *Proceedings of the Tenth Annual Conference of the Cognitive Science Society.*

38. Hofstadter D, with the Fluid Analogies Research Group (1995) *Fluid Concepts and Creative Analogies.* Basic Books.

39. Hofstadter D (moderator) (2000) Stanford Symposium on Spiritual Robots Stanford University.

40. Holland P, Ingham P, Krauss S (1992) Development and Evolution: Mice and Flies Head to Head. *Nature*, 358:627–628.

41. Hopfield J, Tank D (1985) "Neural" Computation of Decisions in Optimization Problems. *Biological Cybernetics*, 52:141–152.

42. Hwang K, Xu Z (1998) *Scalable Parallel Computing.* McGraw-Hill.

43. Hurford J (1999) *The Evolution of Language and Languages.* In: Dunbar R, Knight C, Power C (eds.), The Evolution of Culture. Edinburgh University Press.

44. Hurvich L, Jameson D (1957) An Opponent-process Theory of Color Vision. *Psychological Review*, 64:384–390.

45. James W (1911) *The Meaning of Truth.* Longman Green and Co.

46. Judd D, MacAdam D, Wyszecki G (1964) Spectral Distribution of Typical Daylight as a Function of Correlated Color Temperature. *Journal of the Optical Society of America*, 54:1031–1040.

47. Katz L (2000) *Toward Good and Evil: Evolutionary Approaches to Aspects of Human Morality.* In Katz L (ed) Evolutionary Origins of Morality: Cross-Disciplinary Perspectives. Imprint Academic.

48. Koch C, Poggio T (1992) *Multiplying with Synapses and Neurons.* In McKenna T, Davis JL, Zornetzer SF (eds) Single Neuron Computation, Academic Press, Cambridge, MA.
49. Koch C (1999) *Biophysics of Computation.* Oxford University Press.
50. Koch C, Segev I (2000) The Role of Single Neurons in Information Processing. *Nature Neuroscience*, 3(Supp):1160–1211.
51. Koestler A (1967) *The Ghost in the Machine.* Hutchinson and Co.
52. Koffka K (1935) *Principles of Gestalt Psychology.* Harcourt Brace.
53. Kosslyn SM (1994) *Image and Brain: The Resolution of the Imagery Debate.* MIT Press.
54. Kosslyn A, Thompson, Maljkovic, Weise, Chabris, Hamilton, Rauch, Buonanno (1993) Visual Mental Imagery Activates Topographically Organized Visual Cortex: PET Investigations. *Journal of Cognitive Neuroscience*, 5:263–287.
55. Kumar V (1992) Algorithms for Constraint-satisfaction Problems: a Survey. *AI Magazine*, 13:32–44.
56. Kurzweil R (1999) The Age of Spiritual Machines: When Computers Exceed Human Intelligence. Viking Press.
57. Lakoff G (1987) *Women, Fire, and Dangerous Things: What Categories Reveal about the Mind.* University of Chicago Press.
58. Lakoff G, Johnson M (1999) *Philosophy In The Flesh: The Embodied Mind and Its Challenge to Western Thought.* Basic Books.
59. Lavie N, Driver J (1996) On the Spatial Extent of Attention in Object-based Visual Selection. *Perception and Psychophysics*, 58:1238–1251.
60. Lenat D (1983) EURISKO: A Program which Learns new Heuristics and Domain Concepts. *Artificial Intelligence*, 21.
61. Lenat D, Brown J (1984) Why AM and EURISKO Appear to Work. *Artificial Intelligence* 23(3):269–294.
62. Lenat D, Prakash M, Shepherd M (1986) CYC: Using Commonsense Knowledge to Overcome Brittleness and Knowledge Acquisition Bottlenecks. *AI Magazine*, 6:65–85.
63. Lowe D (1985) *Perceptual Organization and Visual Recognition.* Kluwer Academic.
64. Maloney LT (1986) Evaluation of Linear Models of Surface Spectral Reflectance with Small Numbers of Parameters. *Journal of the Optical Society of America* 3:1673–1683.
65. Marr D (1982) *Vision: A Computational Investigation into the Human Representation and Processing of Visual Information.* W.H. Freeman and Company.
66. McDermott D (1976) Artificial Intelligence Meets Natural Stupidity. *SIGART Newsletter*, 57.
67. Meck W, Church R (1983) A Mode Control Model of Counting and Timing Processes. *Journal of Experimental Psychology: Animal Behavior Processes*, 9:320–334.
68. Mervis C, Catlin J, Rosch E (1975) Development of the Structure of Color Categories. *Developmental Psychology*, 11(1):54–60.
69. Meyer D, Schvaneveldt R (1971) Facilitation in Recognizing Pairs of Words. *Journal of Experimental Psychology*, 90:227–234.
70. Minsky M (1994) Will Robots Inherit the Earth? *Scientific American*, 271(4):109–113.
71. Mitchell M (1993) *Analogy-Making as Perception.* MIT Press.

72. Moore G (1997) An Update on Moore's Law. Intel Developer Forum Keynote, San Francisco.
73. Moore C, Yantis S, Vaughan B (1998) Object-based Visual Selection: Evidence from Perceptual Completion. *Psychological Science*, 9:104–110.
74. Moravec H (1988) *Mind Children: The Future of Robot and Human Intelligence*. Harvard University Press.
75. Moravec H (1998) When will Computer Hardware Match the Human Brain? *Journal of Evolution and Technology*, 1.
76. Newell A (1980) Physical Symbol Systems. *Cognitive Science*, 4:135–183.
77. Newell A, Simon HA (1963) *GPS, a Program that Simulates Human Thought*, In Feigenbaum E, Feldman J (eds), Computers and Thought. MIT Press.
78. Palmer S, Rock I (1994) Rethinking Perceptual Organization: The Role of Uniform Connectedness. *Psychonomic Bulletin and Review*, 1:29–55.
79. Pearl J (1996) *Causation, Action, and Counterfactuals*. In Shoham Y (ed) Proceedings of the Sixth Conference on Theoretical Aspects of Rationality and Knowledge, Morgan Kaufmann.
80. Pearl J (2000) *Causality: Models, Reasoning, and Inference*. Cambridge University Press.
81. Pylyshyn ZW (1981) The Imagery Debate: Analogue Media Versus Tacit Knowledge. *Psychological Review*, 88:16–45.
82. Raymond ES (2003) Uninteresting. In *The On-line Hacker Jargon File*, version 4.4.7, 29 Dec 2003. http://www.catb.org/ esr/jargon/html/U/uninteresting.html
83. Raymond ES (2003) Wrong Thing. In *The On-line Hacker Jargon File*, version 4.4.7, 29 Dec 2003. http://www.catb.org/~esr/jargon/html/W/Wrong-Thing.html
84. Rehling J, Hofstadter D (1997) The Parallel Terraced Scan: An Optimization for an Agent-Oriented Architecture. *Proceedings of the IEEE International Conference on Intelligent Processing Systems*.
85. Ritchie G, Hanna F (1984) AM: A Case Study in AI Methodology. *Artificial Intelligence*, 23.
86. Rojer A, Schwartz E (1990) Design Considerations for a Space-variant Visual Sensor with Complex-logarithmic Geometry. In *Proc. 10th International Conference on Pattern Recognition*.
87. Rosch E, Mervis C, Gray W, Johnson D, Boyes-Braem P (1976) Basic Objects in Natural Categories. *Cognitive Psychology*, 8:382–439.
88. Rosch E (1978) *Principles of Categorization*. In Rosch E, Lloyd BB (eds), Cognition and Categorization. Erlbaum.
89. Rodman HR (1999) *Face Recognition*. In Wilson R, Keil F (eds) The MIT Encyclopedia of the Cognitive Sciences. MIT Press.
90. Sandberg A (1999) The Physics of Information Processing Superobjects. *Journal of Evolution and Technology*, 5.
91. Schwartz E (1977) Spatial Mapping in Primate Sensory Projection: Analytic Structure and Relevance to Perception. *Biological Cybernetics*, 25:181–194.
92. Schwartz E, Munsif A, Albright TD (1989) The Topographic Map of Macaque V1 Measured Via 3D Computer Reconstruction of Serial Sections, Numerical Flattening of Cortex, and Conformal Image Modeling. *Investigative Opthalmol.* Supplement, 298.
93. Shepard R (1992) *The Perceptual Organization of Colors*. In Barkow J, Cosmides L, Tooby J (eds), The Adapted Mind: Evolutionary Psychology and the Generation of Culture. Oxford University Press.

94. Sherman S, Koch C (1986) The Control of Retinogeniculate Transmission in the Mammalian Lateral Geniculate Nucleus. *Experimental Brain Research*, 63:1–20.
95. Shipley EF, Shepperson B (1990) Countable Entities: Developmental Changes. *Cognition*, 34:109–136.
96. Sober E (1984) *The Nature of Selection*. MIT Press.
97. Spelke ES (1990) Principles of Object Perception. *Cognitive Science*, 14(1):29–56.
98. Tooby J, Cosmides L (1992) *The Psychological Foundations of Culture*. In Barkow J, Cosmides L, Tooby J (eds), The Adapted Mind: Evolutionary Psychology and the Generation of Culture. Oxford University Press.
99. Tootell R, Silverman M, Switkes E, deValois R (1985) Deoxyglucose, Retinotopic Mapping and the Complex Log Model in Striate Cortex. *Science*, 227:1066.
100. Tversky A, Kahneman D (1974) Judgement Under Uncertainty: Heuristics and Biases. *Science*, 185:1124–1131.
101. Tversky A, Koehler D (1994) Support Theory: A Nonexistential Representation of Subjective Probability. *Psychological Review*, 101:547–567.
102. Vinge V (1993) Technological Singularity. VISION-21 Symposium `http://www.frc.ri.cmu.edu/~hpm/book98/com.ch1/vinge.singularity.html`
103. Voss P (2001) Presentation at the Fifth Convention of the Extropy Institute – Extro 5, San Jose, CA. Also this volume.
104. Wertheimer M (1923) Untersuchungen zur Lehre von der Gestalt, II. *Psychologische Forschung* 4:301–350. Condensed translation published as: Laws of Organization in Perceptual Forms, in Ellis WD (ed.) (1938) *A Sourcebook of Gestalt Psychology*. Harcourt Brace.
105. Winograd T (1972) *Understanding Natural Language*. Edinburgh University Press.
106. Wagner GP, Altenberg L (1996) Complex Adaptations and the Evolution of Evolvability. *Evolution*, 50:967–976.
107. Worden R (1995) A Speed Limit for Evolution. *Journal of Theoretical Biology*, 176:137–152.
108. Wulf W, McKee S (1995) Hitting the Memory Wall: Implications of the Obvious. *Computer Architecture News*, 23(1).
109. Yudkowsky E (2001) *Creating Friendly AI*. Publication of the Singularity Institute: `http://singinst.org/CFAI/`
110. Zemel R, Behrmann M, Mozer M, Bavelier D (2002) Experience-dependent Perceptual Grouping and Object-based Attention. *Journal of Experimental Psychology: Human Perception and Performance*, 28(1):202–217.

Index

Cognitive Technologies